Lecture Notes in Computer Science 8669

Commenced Publication in 1973
Founding and Former Series Editors:
Gerhard Goos, Juris Hartmanis, and Jan van Leeuwen

T0212731

Emilio Corchado   José A. Lozano
Héctor Quintián   Hujun Yin (Eds.)

Intelligent
Data Engineering and
Automated Learning –
IDEAL 2014

15th International Conference
Salamanca, Spain, September 10-12, 2014
Proceedings

 Springer

Volume Editors

Emilio Corchado
Héctor Quintián
University of Salamanca
Plaza de la Merced S/N
37008 Salamanca, Spain,
E-mail: {escorchado, hector.quintian}@usal.es

José A. Lozano
University of the Basque Country
Paseo Manuel de Lardizábal 1
20018 San Sebastián, Spain
E-mail: ja.lozano@ehu.es

Hujun Yin
University of Manchester
Sackville Street
Manchester, M13 9PL, UK
E-mail: hujun.yin@manchester.ac.uk

ISSN 0302-9743   e-ISSN 1611-3349
ISBN 978-3-319-10839-1   e-ISBN 978-3-319-10840-7
DOI 10.1007/978-3-319-10840-7
Springer Cham Heidelberg New York Dordrecht London

Library of Congress Control Number: 2014946729

LNCS Sublibrary: SL 3 – Information Systems and Application, incl. Internet/Web
and HCI

Typesetting: Camera-ready by author, data conversion by Scientific Publishing Services, Chennai, India

Printed on acid-free paper

Springer is part of Springer Science+Business Media (www.springer.com)

Preface

The IDEAL conference attracts international experts, researchers, leading academics, practitioners, and industrialists from communities of machine learning, computational intelligence, data mining, knowledge management, biology, neuroscience, bio-inspired systems and agents, and distributed systems. It has enjoyed a vibrant and successful history in the last 16 years, having been held in over 13 locations in seven different countries.

It continues to evolve to embrace emerging topics and exciting trends. This year IDEAL was held in the historical city of Salamanca (Spain), declared a UNESCO World Heritage Site in 1988 and European Capital of Culture in 2002.

This conference received about 120 submissions, which were rigorously peer-reviewed by the Program Committee members. Only the papers judged to be of highest quality were accepted and included in these proceedings.

This volume contains 60 papers accepted and presented at the 15th International Conference on Intelligent Data Engineering and Automated Learning (IDEAL 2014), held during September 10–12, 2014, in Salamanca, Spain. These papers provided a valuable collection of recent research outcomes in data engineering and automated learning, from methodologies, frameworks, and techniques to applications and case studies. The techniques include computational intelligence, big data analytics, social media techniques, multi-objective optimization, regression, classification, clustering, biological data processing, text processing, and image/video analysis.

The process of reviewing and selecting papers is extremely important to maintaining the high quality of the conference and therefore we would like to thank the Program Committee for their hard work in the rigorous reviewing process. The IDEAL conference would not exist without their help and professionalism.

IDEAL 2014 enjoyed outstanding keynote speeches by distinguished guest speakers: Prof. Francisco Herrera of University of Granada (Spain), Prof. Cesare Alippi of Politecnico di Milano (Italy), Prof. Juan Manuel Corchado of University of Salamanca (Spain), and Prof. Jun Wang of the Chinese University of Hong Kong (Hong Kong).

Our particular thanks also go to the conference main sponsors, the IEEE Spain Section, the IEEE Systems, Man and Cybernetics Society Spanish Chapter, AEPIA, Salamanca City Hall, Ciudad Rodrigo City Hall, University of Salamanca, and RACE project (USAL-USP), who jointly contributed in an active and constructive manner to the success of this conference. Finally, we would like to thank Alfred Hofmann and Anna Kramer of Springer for their continued collaboration on this publication.

September 2014

Emilio Corchado
José A. Lozano
Héctor Quintián
Hujun Yin

Organization

Honorary Chairs

Alfonso Fernández Mañueco   Mayor of Salamanca, Spain
Daniel Hernández Ruipérez   Rector of University of Salamanca, Spain
Marios Polycarpou   University of Cyprus, Cyprus

General Chair

Hujun Yin   University of Manchester, UK

Program Chair

Emilio Corchado   University of Salamanca, Spain

Program Co-chair

Jose A. Lozano   University of the Basque Country, Spain

Publicity Co-chairs

Jose A. Costa   Universidade Federal do Rio Grande do Norte, Brazil
Yang Gao   Nanjing University, China
Minho Lee   Kyungpook National University, Korea
Bin Li   University of Science and Technology of China, China

International Advisory Committee

Lei Xu   Chinese University of Hong Kong, Hong Kong, SAR China
Yaser Abu-Mostafa   CALTECH, USA
Shun-ichi Amari   RIKEN, Japan
Michael Dempster   University of Cambridge, UK
Nick Jennings   University of Southampton, UK
Soo-Young Lee   KAIST, South Korea
Erkki Oja   Helsinki University of Technology, Finland

Latit M. Patnaik   Indian Institute of Science, India
Burkhard Rost   Columbia University, USA
Xin Yao   University of Birmingham, UK

Steering Committee

Hujun Yin (Chair)   University of Manchester, UK
Emilio Corchado (Co-chair)   University of Salamanca, Spain
Laiwan Chan   Chinese University of Hong Kong, Hong Kong,
                                     SAR China
Guilherme Barreto   University of Brazil
Yiu-ming Cheung   Hong Kong Baptist University, Hong Kong,
                                     SAR China
Jose A. Costa   Federal University Natal, Brazil
Colin Fyfe   University of The West of Scotland, UK
Marc van Hulle   K.U. Leuven, Belgium
Samuel Kaski   Helsinki University of Technology, Finland
John Keane   University of Manchester, UK
Jimmy Lee   Chinese University of Hong Kong, Hong Kong,
                                     SAR China
Malik Magdon-Ismail   Rensselaer Polytechnic Institute, USA
Vic Rayward-Smith   University of East Anglia, UK
Peter Tino   University of Birmingham, UK
Zheng Rong Yang   University of Exeter, UK
Ning Zhong   Maebashi Institute of Technology, Japan

Program Committee

Aboul Ella Hassanien   Cairo University, Egypt
Adrião Duarte   Federal University, UFRN, Brazil
Agnaldo José da R. Reis   Federal University, UFOP, Brazil
Ajalmar Rêgo da Rocha Neto   Federal University, UFC, Brazil
Ajith Abraham   MirLabs, USA
Alberto Guillen   University of Granada, Spain
Alfredo Cuzzocrea   University of Calabria, Italy
Alfredo Vellido   University Politécnica de Cataluña, Spain
Alicia Troncoso   Universidad Pablo de Olavide, Spain
Álvaro Herrero   University of Burgos, Spain
Ana Belén Gil   University of Salamanca, Spain
Andre Carvalho   University of São Paulo, Brazil
André Coelho   University of Fortaleza, Brazil
Andreas König   University of Kaiserslautern, Germany
Andrzej Cichocki   Brain Science Institute, Japan
Anil Nerode   Cornell University, USA
Anne Canuto   Federal University, UFRN, Brazil
Anne Håkansson   Uppsala University, Sweden

Antônio de P. Braga	Federal University, UFMG, Brazil
Antonio Neme	Universidad Autonoma de la Ciudad de Mexico, Mexico
Ata Kaban	University of Birmingham, UK
Barbara Hammer	University of Bielefeld, Germany
Bernard de Baets	Ghent University, Belgium
Bernardete Ribeiro	University of Coimbra, Portugal
Bin Li	University of Science and Technology of China, China
Bogdan Gabrys	Bournemouth University, UK
Bruno Apolloni	University of Milan, Italy
Bruno Baruque	University of Burgos, Spain
Carla Möller-Levet	University of Manchester, UK
Carlos Pereira	ISEC, Portugal
Carmelo J.A. Bastos Filho	University of Pernambuco, POLI, Brazil
Chung-Ming Ou	Kainan University, Taiwan
Chunlin Chen	Nanjing University, China
Clodoaldo A.M. Lima	University of São Paulo, Brazil
Dan Dumitrescu	University of Babes-Bolyai, Romania
Daniel A. Keim	Universität Konstanz, Germany
Daniel Glez-Peña	University of Vigo, Spain
Dante I. Tapia	University of Salamanca, Spain
Darryl Charles	University of Ulster, UK
David Camacho	Universidad Autónoma de Madrid, Spain
David Hoyle	University of Manchester, UK
Davide Anguita	University of Genoa, Italy
Dongqing Wei	Shanghai Jiaotong University, China
Du Zhang	California State University, USA
Eiji Uchino	Yamaguchi University, Japan
Emilio M. Hernandez	University of São Paulo, Brazil
Ernesto Cuadros-Vargas	Universidad Católica San Pablo, Peru
Ernesto Damiani	University of Milan, Italy
Estevam Hruschka Junior	UFSCar - Federal University of Sao Carlos, Brazil
Eva Lorenzo	University of Vigo, Spain
Fabrice Rossi	National Institute of Research on Computer Science and Automatic, France
Felipe M.G. França	Federal University, UFRJ, Brazil
Fernando Buarque	University of Pernambuco, POLI, Brazil
Fernando Díaz	University of Valladolid, Spain
Fernando Gomide	Unicamp, Brazil
Florentino Fdez-Riverola	University of Vigo, Spain
Francesco Corona	Aalto University, Finland
Francisco Assis	Federal University, UFCG, Brazil
Francisco Ferrer	University of Seville, Spain

Francisco Herrera	University of Granada, Spain
Frank Klawonn	Ostfalia University of Applied Sciences, Germany
Gary Fogel	Natural Selection, USA
Gavin Brown	University of Manchester, UK
Gérard Dreyfus	École Supérieure de Physique et de Chimie Industrielles de Paris, France
Giancarlo Mauri	University of Milano-Bicocca, Italy
Héctor Quintián	University of Salamanca, Spain
Heloisa Camargo	Federal University, UFSCar, Brazil
Honghai Liu	University of Portsmouth, UK
Huiyu Zhou	Queen's University Belfast, UK
Hyoseop Shin	Konkuk University Seoul, Korea
Ignacio Rojas	University of Granada, Spain
Igor Farkas	Comenius University in Bratislava, Slovakia
Iñaki Inza	University of Pais Vasco, Spain
Ioannis Hatzilygeroudis	University of Patras, Greece
Ivan Silva	Federal University, USP, Brazil
J. Michael Herrmann	University of Edinburgh, UK
Jaakko Hollmén	Helsinki University of Technology, Finland
Jaime Cardoso	University of Porto, Portugal
James Hogan	Queensland University of Technology, Australia
Javier Bajo Pérez	Universidad Politécnica de Madrid, Spain
Javier Sedano	Instituto Tecnológico de Castilla y León, Spain
Jerzy Grzymala-Busse	University of Kansas, USA
Jesus Alcala-Fdez	University of Granada, Spain
Jing Liu	Xidian University, China
Joao E. Kogler Jr.	University of São Paulo, Brazil
Jochen Einbeck	Durham university, UK
John Gan	University of Essex, UK
John Qiang	University of Essex, UK
Jongan Park	Chosun University, Korea
Jorge Posada	VICOMTech, Spain
Jose A. Lozano	University of the Basque Country, UPV/EHU, Spain
Jose Alfredo F. Costa	Federal University, UFRN, Brazil
José C. Principe	University of Florida, USA
José C. Riquelme	University of Seville, Spain
Jose Dorronsoro	Autónoma de Madrid University, Spain
José Everardo B. Maia	State University of Ceará, Brazil
José F. Martínez	Instituto Nacional de Astrofisica Optica y Electronica, Mexico
José Luis Calvo Rolle	University of A Coruña, Spain
Jose M. Molina	Universidad Carlos III de Madrid, Spain
José Manuel Benítez	University of Granada, Spain

José Ramón Villar	University of Oviedo, Spain
José Riquelme	University of Seville, Spain
Jose Santos	University of A Coruña, Spain
Juan Botía	University of Murcia, Spain
Juan J. Flores	Universidad Michoacana de San Nicolas de Hidalgo, Mexico
Juan Manuel Górriz	University of Granada, Spain
Juán Pavón	Universidad Complutense de Madrid, Spain
Juha Karhunen	Aalto University School of Science, Finland
Ke Tang	University of Science and Technology of China, China
Keshav Dahal	University of Bradford, UK
Kunihiko Fukushima	Kansai University, Japan
Lakhmi Jain	University of South Australia, Australia
Lars Graening	Honda Research Institute Europe, Germany
Leandro Augusto da Silva	Mackenzie University, Brazil
Leandro Coelho	PUCPR/UFPR, Brazil
Lenka Lhotska	Czech Technical University, Czech Republic
Lipo Wang	Nanyang Technological University, Singapore
Lourdes Borrajo	University of Vigo, Spain
Lucía Isabel Passoni	Universidad Nacional de Mar del Plata, Argentina
Luis Alonso	University of Salamanca, Spain
Luiz Pereira Calôba	Federal University, UFRJ, Brazil
Luonan Chen	Shanghai University, China
Maciej Grzenda	Warsaw University of Technology, Poland
Manuel Graña	University of Pais Vasco, Spain
Marcelo A. Costa	Universidade Federal de Minas Gerais, Brazil
Marcin Gorawski	Silesian University of Technology, Poland
Márcio Leandro Gonçalves	PUC-MG, Brazil
Marcus Gallagher	The University of Queensland, Australia
Maria Jose Del Jesus	Universidad de Jaén, Spain
Mario Koeppen	Kyushu Institute of Technology, Japan
Marios M. Polycarpou	University of Cyprus, Cyprus
Mark Girolami	University of Glasgow, UK
Marley Vellasco	Pontifical Catholic University of Rio de Janeiro, Brazil
Matjaz Gams	Jozef Stefan Institute Ljubljana, Slovenia
Matthew Casey	University of Surrey, UK
Michael Herrmann	University of Edinburgh, UK
Michael Small	The University of Western Australia, Australia
Michal Wozniak	Wroclaw University of Technology, Poland
Ming Yang	Nanjing Normal University, China
Miroslav Karny	Academy of Sciences of Czech Republic, Czech Republic
Nicoletta Dessì	University of Cagliari, Italy

Olli Simula	Aalto University, Finland
Oscar Castillo	Tijuana Institute of Technology, Mexico
Pablo Estevez	University of Chile, Chile
Paulo Adeodato	Federal University of Pernambuco and NeuroTech Ltd., Brazil
Paulo Cortez	University of Minho, Portugal
Paulo Lisboa	Liverpool John Moores University, UK
Pei Ling Lai	Southern Taiwan University, Taiwan
Perfecto Reguera	University of Leon, Spain
Peter Tino	University of Birmingham, UK
Petro Gopych	Universal Power Systems USA-Ukraine LLC, Ukraine
Rafael Corchuelo	University of Seville, Spain
Ramon Rizo	Universidad de Alicante, Spain
Raúl Cruz-Barbosa	Tecnological University of the Mixteca, Mexico
Raúl Giráldez	Pablo de Olavide University, Spain
Regivan Santiago	UFRN, Brazil
Renato Tinós	USP, Brazil
Ricardo Del Olmo	Universidad de Burgos, Spain
Ricardo Linden	FSMA, Brazil
Ricardo Tanscheit	PUC-RJ, Brazil
Richard Chbeir	Bourgogne University, France
Richard Freeman	Capgemini, UK
Roberto Ruiz	Pablo de Olavide University, Spain
Rodolfo Zunino	University of Genoa, Italy
Romis Attux	Unicamp, Brazil
Ron Yang	University of Exeter, UK
Ronald Yager	Machine Intelligence Institute - Iona College, USA
Roque Marín	University of Murcia, Spain
Rudolf Kruse	Otto-von-Guericke-Universität Magdeburg, Germany
Salvador García	University of Jaén, Spain
Saman Halgamuge	The University of Melbourne, Australia
Sarajane M. Peres	University of São Paulo, Brazil
Seungjin Choi	POSTECH, Korea
Songcan Chen	Nanjing University of Aeronautics and Astronautics, China
Stefan Wermter	University of Sunderland, UK
Stelvio Cimato	University of Milan, Italy
Stephan Pareigis	Hamburg University of Applied Sciences, Germany
Sung-Bae Cho	Yonsei University, Korea
Sung-Ho Kim	KAIST, Korea
Takashi Yoneyama	ITA, Brazil
Tianshi Chen	Chinese Academy of Sciences, China

Tim Nattkemper   University of Bielefeld, Germany
Tzai-Der Wang   Cheng Shiu University, Taiwan
Urszula Markowska-Kaczmar   Wroclaw University of Technology, Poland
Vasant Honavar   Iowa State University, USA
Vasile Palade   Coventry University, UK
Vicente Botti   Polytechnic University of Valencia, Spain
Vicente Julian   Universidad Politécnica de Valencia, Spain
Wei-Chiang Samuelson Hong   Oriental Institute of Technology, Taiwan
Weishan Dong   IBM Research, China
Wenjia Wang   University of East Anglia, UK
Wenjian Luo   University of Science and Technology of China,
                                      China
Wu Ying   Northwestern University, USA
Yang Gao   Nanjing University, China
Yanira Del Rosario De
  Paz Santana   Universidad de Salamanca, Spain
Ying Tan   Peking University, China
Yusuke Nojima   Osaka Prefecture University, Japan

Local Organizing Committee

Emilio Corchado   University of Salamanca, Spain
Héctor Quintián   University of Salamanca, Spain
Álvaro Herrero   University of Burgos, Spain
Bruno Baruque   University of Burgos, Spain
José Luis Calvo   University of Coruña, Spain

Table of Contents

Erratum

MLeNN: A First Approach to Heuristic Multilabel Undersampling

Francisco Charte[1], Antonio J. Rivera[2],
María J. del Jesus[2], and Francisco Herrera[1]

[1] Dep. of Computer Science and Artificial Intelligence,
University of Granada, Granada, Spain
[2] Dep. of Computer Science, University of Jaén, Jaén, Spain
{fcharte,herrera}@ugr.es,{arivera,mjjesus}@ujaen.es
http://simidat.ujaen.es,http://sci2s.ugr.es

Abstract. Learning from imbalanced multilabel data is a challenging task that has attracted considerable attention lately. Some resampling algorithms used in traditional classification, such as random undersampling and random oversampling, have been already adapted in order to work with multilabel datasets.

In this paper MLeNN (*MultiLabel edited Nearest Neighbor*), a heuristic multilabel undersampling algorithm based on the well-known Wilson's Edited Nearest Neighbor Rule, is proposed. The samples to be removed are heuristically selected, instead of randomly picked. The ability of MLeNN to improve classification results is experimentally tested, and its performance against multilabel random undersampling is analyzed. As will be shown, MLeNN is a competitive multilabel undersampling alternative, able to enhance significantly classification results.

Keywords: Multilabel Classification, Imbalanced Learning, Resampling, ENN.

1 Introduction

Multilabel classification (MLC) [1] has many real-world applications, being a subject which has drawn significant research attention. That most multilabel datasets (MLDs) are imbalanced is something taken for granted. Many existent methods deal with this problem through MLC algorithms adaptations [2], aiming to perform some kind of adjustment in the training phase to take into account the imbalanced nature of MLDs. There are also some proposals relying on data resampling [3], generating new instances in which minority labels appear or deleting instances associated to the majority ones.

Until now, most of the published multilabel resampling algorithms are random based, including random undersampling (RUS). The aim of this paper is to propose a multilabel undersampling method which heuristically, rather than randomly, selects the instances for removing. The heuristic is founded on the Edited Nearest Neighbor (ENN) rule [4] and relies on two measures to assess

E. Corchado et al. (Eds.): IDEAL 2014, LNCS 8669, pp. 1–9, 2014.

the imbalance level in MLDs, as well as on a distance metric between the sets of labels (labelsets) appearing in each pair of instances.

The behavior of the proposed algorithm, called MLeNN, will be experimentally proved by applying it to six MLDs, and analyzing the results produced by three MLC methods. The significance of these results will be statistically evaluated, and the benefits produced by MLeNN will be demonstrated.

This paper is structured as follows. In Section 2 a brief introduction to multilabel classification and imbalanced learning is provided. Section 3 describes the proposed method, while all the experimentation details can be found in Section 4. Finally, Section 5 offers the final conclusions.

2   Preliminaries

The main characteristic of multilabeled data, when compared with data used in traditional classification, consists of associating a group of relevant labels to each instance, instead of only one class. This group is a subset of the whole set of labels present in the MLD. Therefore, the goal on any MLC algorithm is to predict the subset of labels which should be associated to the instances in an MLD. A general introduction to MLC can be found in [5]. Additionally, a recent review of MLC methods is offered in [1].

Imbalanced learning is a problem thoroughly studied in traditional classification, and many solutions, based on different approaches, have been proposed to face it. This problem emerges when there are many instances belonging to some classes (majority), but only a few representing others (minority). Usually, classifiers tend to be biased to the majority classes, in detriment of the minority ones. A comprehensive review of traditional imbalanced learning solutions is provided in [6].

Most studies assume that all MLDs are imbalanced. A triad of measures directed to assess the imbalance level in MLDs are proposed in [3], along with two random resampling algorithms, one for oversampling (LP-ROS) and anoher for undersampling (LP-RUS). Two of the measures will be detailed in the following section, since the heuristic used by MLeNN rely on them. Several methods aimed to face the learning from multilabel imbalanced data, based on ensembles of MLC classifiers [7] and non-parametric probabilistic models [2], are also available.

In general, the imbalance level in MLDs is noticeably higher than in traditional datasets. Moreover, algorithms aiming to cope with this problem have to take into account that each sample belongs to multiple labels. Thus, procedures such as the creation of new instances or deletion of existing ones will influence several labels, rather than only one class.

3   Heuristic Multilabel Undersampling with MLeNN

Undersampling algorithms usually perform worse than oversampling ones [8], since they cause a loss of information by removing instances. The information loss is even greater when undersampling is applied to MLDs, as each removed sample is representing not only one class but several labels. As a result, choosing the right instances to delete is of critical importance. Adapting ENN to work with MLDs needs to resolve two key points, how the candidates are selected and how the class differences among them and their neighbors are considered. MLeNN settles these points by firstly limiting the samples which can act as candidates to those in which none minority label appears, and secondly defining a metric distance to know what is the difference between any pair of labelsets.

Unlike LP-RUS [3], which has a random behavior, the MLeNN algorithm takes the samples to remove using a heuristic based on the two following bases:

- None of the minority labels can appear in the instance taken as reference to candidate for deletion.
- The labelset of the reference instance has to be different to that of its neighbors.

The second condition is based on the ENN rule [4], and adapted to the multilabel scenario as explained below. Algorithm 1 shows the MLeNN algorithm pseudo-code. The measures used and implementation details are discussed in the following subsections.

3.1 Candidate Selection

In order to choose which samples will act as candidates for removing, a method to know what labels are in minority is needed. Those instances in which any minority label appears will never be candidates, avoiding that some of the few samples representing a minority label are lost.

To complete this task, MLeNN relies on the measures proposed in [3]. Let D be an MLD, Y the full set of labels in it, and Y_i the labelset of the i-th instance. *IRLbl* (Equation 1) is a measure calculated individually to assess the imbalance level for each label. The higher is the *IRLbl* the larger would be the imbalance, allowing to know what labels are in minority or majority. *MeanIR* (Equation 2) is the average *IRLbl* for an MLD, useful to estimate the global imbalance level.

$$IRLbl(y) = \frac{\underset{y'=Y_1}{\overset{Y_{|Y|}}{\operatorname{argmax}}}(\sum_{i=1}^{|D|} h(y', Y_i))}{\sum_{i=1}^{|D|} h(y, Y_i)}, \quad h(y, Y_i) = \begin{cases} 1 & y \in Y_i \\ 0 & y \notin Y_i \end{cases}. \tag{1}$$

$$MeanIR = \frac{1}{|Y|} \sum_{y=Y_1}^{Y_{|Y|}} (IRLbl(y)). \tag{2}$$

MLeNN will iterate through all the MLD samples, taking as candidates those whose labelset does not contain any label with $IRLbl > MeanIR$. This way, all the instances containing a minority label will be preserved.

Algorithm 1. MLeNN algorithm pseudo-code.

Inputs: <Dataset> D, <Threshold> HT, <NumNeighbors> NN
Outputs: Preprocessed dataset

```
1: for each sample in D do
2:      for each label in getLabelset(D) do
3:          if IRLbl(label) > MeanIR then
4:              Jump to next sample              ▷ Preserve instance with minority labels
5:          end if
6:      end for
7:      numDifferences ← 0
8:      for each neighbor in nearestNeighbors(sample, NN) do
9:          if adjustedHammingDist(sample, neighbor) > HT then
10:             numDifferences ← numDifferences+1
11:         end if
12:     end for
13:     if numDifferences ≥ NN/2 then
14:         markForRemoving(sample)
15:     end if
16: end for
17: deleteAllMarkedSamples(D)
```

3.2 Labelset difference Evaluation

Wilson's ENN rule has been extensively used in traditional classification. The basic idea behind it is the following:

– Select a sample C as candidate.
– Look for $C's$ NN nearest neighbors. Usually $NN = 3$.
– If C class differs from the class of at least half of their neighbors (that is 2 when $NN = 3$), mark C for removing.

Since there is exclusively one class to compare with, the difference between the candidate class and that of any of its neighbors is either 0% (same class) or 100% (different classes). Therefore, the candidate will be removed when there is a 100% difference between its class and the class of half or more of its neighbors.

Table 1. Difference between the labelsets of two instances

label index	1	2	3	4	5	6	...	375	376
labelset1	0	1	1	0	0	1	0	0	0
labelset2	1	0	1	0	0	1	0	1	0
Dif. count	1	1	0	0	0	0	0	1	0

Multilabel instances have multiple labels associated, thus the difference between two samples labelsets could be 100%, but also any value below that and above 0%. Let us consider the two labelsets shown in Table 1 and how their differences could be evaluated. They belong to an MLD with 376 different labels, but each instance is associated only to 3 or 4 of them. This is the case of the corel5k dataset.

Using the Hamming distance, it could be concluded that a difference of 3 exists between the two labelsets. As the total length (number of labels) is 376, this would give a 0.798% difference. However, if only the active labels (those which are relevant) in either labelset are considered the result would be totally different, since there are only 7 active labels in total. The percentage of difference would be 42.857%, far higher than the previous one. There are many MLDs with hundreds of distinct labels, but a low number of active labels in each sample. Calculating differences using the usual Hamming distance will always produce extremely low values. Thus, considering only active labels makes sense when it comes to evaluate differences among labelsets.

MLeNN calculates an *adjusted* Hamming distance between the candidate and their neighbors labelsets, counting the number of differences and dividing it by the number of active labels. As a result, a value in the range of [0,1] is obtained. Applying a configurable threshold (HT in Algorithm 1), the algorithm determines which of its neighbors will be considered as distinct.

4   Experimentation and Analysis

This section describes the experimental framework used to test the performance of the proposed algorithm. Afterwards, obtained results and their analysis are provided.

4.1   Experimental Framework

The paper experimentation has been structured in two phases. The first goal is to determine if MLeNN is able to improve classification results. It compares the output of classifiers before and after preprocessing the same set of datasets. The second phase aims to compare MLeNN performance against that of multilabel random undersampling, preprocessing the original datasets with LP-RUS [3].

The six datasets whose characteristics are shown in Table 2 were used to experimentally assess MLeNN[1]. All of them are from the MULAN repository [9], and have been repeatedly used in the literature. They have been partitioned following a 2x5 strategy. As can be inferred from the *MeanIR* values, five of these datasets are truly imbalanced, with values ranging from 7.20 to 256.40. On the contrary, the emotions dataset could not be actually considered as imbalanced. It has been included in the experimentation to test the behavior of MLeNN when used with non-imbalanced MLDs. MLeNN was applied to all of them with $NN = 3$ (3 neighbors) and $HT = 0.75$ (75% labelset difference threshold).

---

[1] This paper has an associated website at http://simidat.ujaen.es/mlenn. Both dataset partitions and the MLeNN program can be downloaded from it. This website also offers full tables of results.

Regarding the MLC algorithms used, a basic Binary Relevance (BR [10]) transformation method and two more advanced classifiers, Calibrated Label Ranking (CLR [11]) and Random k-labelsets (RAkEL [12]), were selected. The well-known C4.5 tree-based classification algorithm was used as underlying classifier where needed.

Table 2. Characteristics of the datasets used in experimentation

	Dataset	Instances	Attributes	Labels	MaxIR	MeanIR
1	corel5k	5000	499	374	1120.00	189.57
2	corel16k	13766	500	161	126.80	34.16
3	emotions	593	72	6	1.78	1.48
4	enron	1702	753	53	913.00	73.95
5	mediamill	43907	120	101	1092.55	256.40
6	yeast	2417	198	14	53.41	7.20

4.2   Results and Analysis

The outputs produced by the MLC algorithms, learning from the base MLDs and those obtained after preprocessing with MLeNN and ML-RUS, have been evaluated with three measures: Accuracy, Macro-FMeasure and Micro-FMeasure. The former is a sample-based evaluation measure, thus all labels are given equal weight, whereas the other two are label-based. Macro-averaging measures tend to emphasize the results of rare labels, while micro-averaging does the opposite. How these measures assess prediction performance can be found in [5]. An intuitive visual representation of those results is offered in Figure 1.

First, it can be seen that the undersampling performed by MLeNN has improved base results in many cases. However, there are some exceptions. The most remarkable is that of the emotions dataset, whose results tend to be worse after MLeNN has been applied. This led to the conclusion that undersampling, whether random or heuristic, should not be applied to MLDs which are not truly imbalanced. Another fact that can be visually confirmed in Fig. 1 is that the performance of MLeNN is almost always better than that of ML-RUS.

Aiming to formally endorse these results, a Wilcoxon non-parametric statistical test was used to compare MLeNN with base results, firstly, and with ML-RUS, secondly. The results of these tests are shown in Table 3 and Table 4. A star at the right of a value denotes that it is the best ranking for a given measure. The symbol ⇆ indicates that there is no statistical difference between this ranking and the best one, whereas the symbol ⇊ states that a significant difference exists.

From the analysis of these results, that MLeNN is a competitive multilabel undersampling algorithm can be concluded, since it always achieves better performance than ML-RUS from an statistical point of view. Moreover, the undersampling conducted by MLeNN is able to improve classification results when compared with those obtained without preprocessing. MLeNN produced a statistically significant improvement in two of the three measures, despite the inclusion of an MLD such as emotions, which is not imbalanced, into the statistical study.

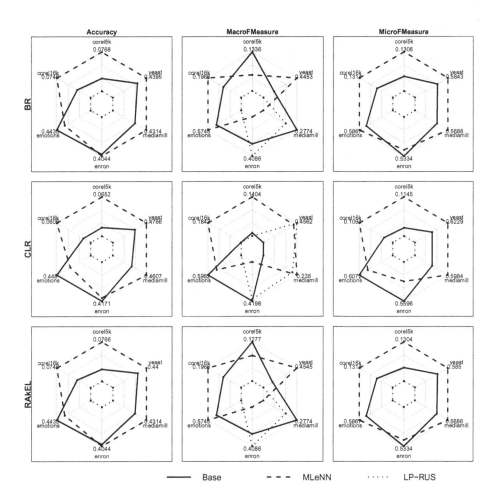

Fig. 1. Each plot shows classification results corresponding to a measure/algorithm combination

Table 3. First phase - Average Rankings

Algorithm	Accuracy		Macro-FM		Micro-FM	
Base	1.778	⇊	1.5556	⇆	1.778	⇊
MLeNN	1.222	⋆	1.4444	⋆	1.222	⋆

Table 4. Second phase - Average Rankings

Algorithm	Accuracy		Macro-FM		Micro-FM	
LP-RUS	2.000	⇊	1.6667	⇊	2.000	⇊
MLeNN	1.000	⋆	1.3333	⋆	1.000	⋆

5 Conclusions

The learning from imbalanced MLDs presents some serious difficulties. Several approaches have been proposed to overcome them, including random undersampling algorithms. This kind of methods does not usually work well, since they cause a significant loss of information potentially useful in the training process.

In this paper MLeNN, a novel multilabel heuristic undersampling algorithm, has been presented. The deleted instances are thoughtfully selected, instead of being randomly chosen. As the obtained results show, it is a technique able to improve classification results when applied to truly imbalanced MLDs. Moreover, it performs significant better than the random undersampling implemented by LP-RUS.

Acknowledgments. F. Charte is supported by the Spanish Ministry of Education under the FPU National Program (Ref. AP2010-0068). This work was partially supported by the Spanish Ministry of Science and Technology under projects TIN2011-28488 and TIN2012-33856 (FEDER Founds), and the Andalusian regional projects P10-TIC-06858 and P11-TIC-9704.

References

1. Zhang, M.-L., Zhou, Z.-H.: A review on multi-label learning algorithms. IEEE Trans. Knowl. Data Eng. (2013)
2. He, J., Gu, H., Liu, W.: Imbalanced multi-modal multi-label learning for subcellular localization prediction of human proteins with both single and multiple sites. PloS One 7(6), 7155 (2012)
3. Charte, F., Rivera, A., del Jesus, M.J., Herrera, F.: A first approach to deal with imbalance in multi-label datasets. In: Pan, J.-S., Polycarpou, M.M., Woźniak, M., de Carvalho, A.C.P.L.F., Quintián, H., Corchado, E. (eds.) HAIS 2013. LNCS, vol. 8073, pp. 150–160. Springer, Heidelberg (2013)
4. Wilson, D.L.: Asymptotic properties of nearest neighbor rules using edited data. IEEE Trans. on SMC-2(3), 408–421 (1972)
5. Tsoumakas, G., Katakis, I., Vlahavas, I.: Mining Multi-label Data. In: Maimon, O., Rokach, L. (eds.) Data Mining and Knowledge Discovery Handbook, ch. 34, pp. 667–685. Springer US, Boston (2010)
6. Haibo, H., Yunqian, M.: Imbalanced Learning: Foundations, Algorithms, and Applications. Wiley-IEEE Press (2013)
7. Tahir, M.A., Kittler, J., Bouridane, A.: Multilabel classification using heterogeneous ensemble of multi-label classifiers. Pattern Recognit. Lett. 33(5), 513–523 (2012)
8. García, V., Sánchez, J., Mollineda, R.: On the effectiveness of preprocessing methods when dealing with different levels of class imbalance. Knowl. Based Systems 25(1), 13–21 (2012)
9. Tsoumakas, G., Xioufis, E.S., Vilcek, J., Vlahavas, I.: MULAN: A Java Library for Multi-Label Learning. J. Mach. Learn. Res. 12, 2411–2414 (2011)

10. Godbole, S., Sarawagi, S.: Discriminative Methods for Multi-labeled Classification. In: Dai, H., Srikant, R., Zhang, C. (eds.) PAKDD 2004. LNCS (LNAI), vol. 3056, pp. 22–30. Springer, Heidelberg (2004)
11. Fürnkranz, J., Hüllermeier, E., Loza Mencía, E., Brinker, K.: Multilabel classification via calibrated label ranking. Mach. Learn. 73, 133–153 (2008)
12. Tsoumakas, G., Vlahavas, I.: Random k-labelsets: An ensemble method for multi-label classification. In: Kok, J.N., Koronacki, J., Lopez de Mantaras, R., Matwin, S., Mladenič, D., Skowron, A. (eds.) ECML 2007. LNCS (LNAI), vol. 4701, pp. 406–417. Springer, Heidelberg (2007)

Development of Eye-Blink Controlled Application for Physically Handicapped Children

Ippei Torii[*], Kaoruko Ohtani, Shunki Takami, and Naohiro Ishii

Dept. of Information Science, Aichi Institute of Technology, Aichi, Japan
{mac,ishii}@aitech.ac.jp
ruko2011@gmail.com
mail@takamin.net

Abstract. In this paper, we describe a new application operated with eye-blink for physically handicapped children who cannot speak to communicate with others. Process of detecting blinks is performed in the following steps. 1) To detect an eye area 2) To distinguish opening and closing of eyes 3) To add the method using saturation to detect blink 4) To decide by a conscious blink 5) To improve the accuracy of detection of a blink We reduce the error to detect a blink and pursue the high precision of the eye chasing program. The degree of disablement is varied in children. So we develop the system to be able to be customizes depends on the situation of users. And also, we develop the method into a communication application that has the accurate and high-precision blink determination system to detect letters and put them into sound.

Keywords: VOCA (Voice Output Communication Aid), Phsically handicapped children, OpenCv, Haar-like eye detection.

1   Introduction

Special support schools in Japan need some communication assistant tools especially for physically handicapped children. In this study, physically handicapped children are defined as children with permanent disablements of their trunks and limbs because of cerebral palsy, muscular dystrophy, spina bifida and so on. Their body movements are very limited and many of them also have mental disorders, so they cannot communicate with their families or caregivers. It prevents helpers from understanding what they really need or think.

Taking the situation into consideration, we develop a communication support tool operated by eye-blink for physically handicapped children. We use front cameras on tablets (iPad, iPad mini of Apple Inc. iOS5.0). After several verifications and examination, we have developed a contactless communication assistant application "Eye Talk". The application does not malfunction by surroundings, such as brightness or differences of eye shapes.

---

[*] Corresponding author.

E. Corchado et al. (Eds.): IDEAL 2014, LNCS 8669, pp. 10–17, 2014.

2   Purpose of the Study

The most common way to communicate with physically handicapped children is us-ing "Yes=○" or "No=×" cards. (Figure 1) For example, if a caregiver wants to ask a child whether he/she wants to drink water, the caregiver will ask him/her "Do you want to drink water?" and show him/her cards with ○ and × by turns. If the child takes a look at ○ card or put his/her eye on it longer than the other one, the caregiver will know he/she may want to drink water.

Fig. 1. Using "Yes=○" or "No=×" cards to communicate with a physically handicapped child

But caregivers have to predict what patients need or want to say by their expe-riences or circumstances in this method. So the questions made by caregivers can be totally different from what patients really want to say. And also, it is difficult to figure out the movement of eyes of patients. Sometimes caregivers have to just guess the answer.

Some communication support tools using movements of eyes for these physically handicapped people have been released already, such as TalkEye[1] or Let's Chat[2] . Most of them are relatively expensive because they require some special equipment. For example, TalkEye requires the executive head set to measure the movement of eyes.

3   Structure of the System

Process of detecting blinks is performed in the following steps,

1. To detect an eye area （By using Opencv Haar-like eye-detection）
2. To distinguish opening and closing of eyes （By using the complexity of binarized image）
3. To add the method using saturation to detect blink （Aiming more accurate detection）
4. To decide by a conscious blink （To define what is a "conscious blink"）
5. To improve the accuracy of detection of a blink

3.1   Detection of an Eye Area

There are many methods to detect an object. We choose OpenCv that is a library of programming functions for real time computer vision for image processing in this study. OpenCV's face detector uses a method that Viola, P. [3] developed first and Lienhart, R. [4] improved.

We use differences of brightness to figure out a face area and to remove other areas. Next, we detect an eye-area. Since an eye-area is inside a face area, we use data of a face area obtained by the method to obtain the average brightness of upper and lower part of eyes to define an eye-area.

3.2   Detection of Eye Opening and Closing

There are two methods to detect a blink of eyes. First one is calculating the size of the black area of eyes by using spiral labeling [5] and considering it as a blink when the size becomes smaller than the threshold. The second one is using the difference of the value of brightness of images to determine a blink. Figure 2 shows the image of spiral labeling. In spiral labeling, we search pixels from the starting pixel (1 in a square in Figure 5), calculate the medial level of the difference, and count the pixels within the threshold. In this method, we can reduce the time of processing if we can get the starting pixel because the area of processing is limited.

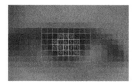

Fig. 2. Image of spiral labeling

But many physically handicapped children tend not to be able to open their eyes wide enough and their iris of the eye is relatively small, so it is difficult to detect the center of iris of the eye. So we use Value in HSV (Hue, Saturation, Value) in color space [6] to determine a blink. In this method, we cut an eye area based on the coordinate data obtained by Haar-like classifier and size it to reduce the load to a devise.

Then we obtain the average of brightness of the eye area. When someone closes his/her eyes, the average rises. We determine it as eye closing. Figure 3 shows the image of eye opening and closing.

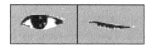

Fig. 3. Image of eye opening and closing

Usually, the classrooms of the most of special support schools are relatively dark to avoid giving extra impetuses to children, so we cannot get enough amount of light. So it is difficult to determine a blink, because there is no difference of darkness around the eye area and the average of brightness of white and black part of eyes that is close to the brightness of skin.

3.3   Detecting Method by Saturation in Color Space

Here we develop the method using saturation in color space. In this method, we calculate the average of saturation (0 to 255 in saturation of HSV) of area C (the center of the iris) and W (white part of the eye) in Figure 4. If the measured value is lower than the average of saturation, we determine it as eye closing. We collect many numbers of eye area data to calculate the average of saturation for eye opening.

Fig. 3. To Obtain the average of saturation

When a user starts the system, it calibrates the picture of eye opening. Based on this picture, the system determines it as eye closing when the saturation of eye area is lower than the threshold. We use the threshold (the average of saturation) between 5 flames to 14 flames before of the present flame. But on the other hand, it detects a small movement such as an unconscious blink or moving of the face. So we add complexity of image of the eye in last 3 flames including the present flame to exclude the error to determine a blink.

3.4   Detection of Eye Opening and Closing

We use the difference of the outlines between eye opening and closing. We determine it as eye closing when the pixel of the difference of amount of edge in the picture becomes flat. The threshold is based on the average value of the amount of edges of 10 flames (5 to 14 flames before of the present flame). We determine a blink with the difference between the threshold and the present flame. Figure5 shows the utilization to each frame of the process complexity and saturation of blink detection.

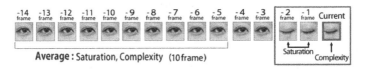

Fig. 5. Complexity of the image of the eye

The purpose of adding this method is to stop to detect an unconscious blink or a movement of the face.   First, we try to find the most optimum value of combination of the saturation and complexity to determine eye closing. The setting that has less error is LP=0.75, DP=0.86. We set this number as the threshold.

Figure 6 shows the image when the application starts. ABC shows the correlation values of complexity and DEF shows the correlation value of saturation. A is the value of complexity of the present flame. B is the average value of complexity of 5 to 14 flames before of the present flame. C is the value of complexity multiplying the threshold LP (0.75) to the value of the present flame. D is the value of saturation of the present flame. E is the average value of saturation of 5 to 14 flames before of the present flame. F is the value of saturation multiplying the threshold DP (0.86) to the value of the present flame. When B is larger than A, and E is larger than D, we determine it as eye closing.

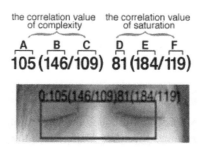

Fig. 6. Image and value of eye area

3.5   Developing Afterimage Method

After we developed the basic model of the system, we carried on a clinical experiment in the school for handicapped children. A subject suffering from spinal muscular atrophy has very weak muscle strength and cannot blink longer enough. So we need to improve the system to determine blinks even situations of users are different.

For users who can blink strongly enough, the method to determine blinks by the complexity and saturation is appropriate, but another detection method is required for users who do not have enough muscle strength. So we need to increase the sensitivity to detect blinks. Therefore, it is necessary to increase the processing power for detection of blinks. Since the tablet is fixed to the bed, we detect the position of eyes using OpenCV only when we start the system. It reduces the processing load and we can use 4 to 25 frames per second to determine blinks.

We use afterimages to determine weak and fast blinks instead of complexity and saturation.

By the method of using the afterimage, the number of the past frames to compare increases and we can capture changes with high accuracy. As a result, we can quantify the changes of series of movements "eye-opening → eye-closing → eye-opening". And it is possible to determine the situation of eyes as short blink, long blink, closing eyes continuously, except the malfunction due to fine movement of eyes. If a user opens and closes eyes within 50 flames, it is considered a blink. Figure 7 shows the determination of blinks by afterimage. The black area represents the current frame and the shaded area shows the afterimage. In the afterimage method, we determine the conscious blink and the unconscious blink using two thresholds. When we increase the sensitivity of the blink detector, the first threshold is defused to 1 frame maximum. The second threshold is set to 50 frames (approximately 2 seconds).

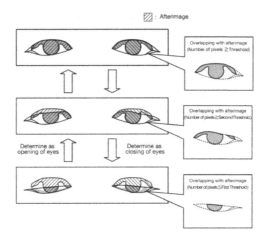

Fig. 7. Detection of unconscious blink by afterimages

In order to handle with face color and brightness of light, calibration is automatically activated when the black part in the image is doubled. Calibration means to detect the position of eyes and to set a threshold for binarizing. If the value of binarization threshold is too high or too low, it cannot detect blinks. We perform calibration to obtain the correct value for blink determination and set the threshold. As a result, it is possible to automatically capture the position of eyes in both a dark and a bright place or even if a user moves his/her face. It is possible to detect blinks using the value of the current frame, but if a user has long eyelashes, the outline becomes complicated because it acquires the contour of eyelashes. Therefore, we designed the method of using the afterimage to correspond to blinks of any user.

Table 1. List of errata of blinks

Acquisition of blinks from 30 healthy subjects			
Determination of eye-opening to eye-closing by complexity and intensity	Correct judgment 44.3%	Determination of eye-opening → eye-closing → eye-opening by afterimage sensitivity 80	Correct judgment 93%

Acquisition of blinks from 4 physically handicapped children (1 spina bifida, 1 cerebral palsy, 2 muscular dystrophy)													
	A	B	C	D		A	B	C	D		A	B	C	D
	○					○	○	○	×		○	○		○
				○		○	○	○	○		○		○	○
Determination of eye-opening to eye-closing by complexity and intensity					Determination of eye-opening → eye-closing → eye-opening by afterimage Sensitivity 80	○			○	Determination of eye-opening → eye-closing → eye-opening by afterimage sensitivity 93	○	○	○	
						○	○	○	×		○	○	○	○
						○			○		○	○	○	
						○		○	×		○		○	○
						○	○	○	○		○		○	○
							○		○		○	○	○	○
Correct judgment 5%				Correct judgment 72.5%					Correct judgment 87%				

We describe the comparison of the detection accuracy of two determination methods; by saturation and complexity and by the rate of change of characteristics with afterimage. For 30 healthy subjects, errata rate by saturation and complexity was around 40%, but error-free amount was more than 90% by afterimage.   When we carried out clinical study with 4 physically handicapped children (1 spina bifida, 1 cerebral palsy, 2 muscular dystrophy), their blinks were too weak to react. It is considered that their blinks were determined as unconscious blinks that we had excluded. When we used the determination method by afterimage, the rate of accurate judgments was 70% with the initial value and it increased nearly 90% by raising the sensitivity (Table 1).

4   Developing Communication Applications

We developed the blink determination system into new communication applications "Eye Talk" and "Eye Tell" for physically handicapped children.

4.1   Eye Talk, Eye Tell

In Eye Talk, a user can select a character in the least number of times. Figure 8 shows the image of the application. A user chooses a consonant in "column" first and a vowel in "row" next from the character table of Japanese of this application.

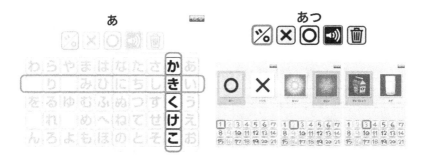

Fig. 8. Character table of Japanese of Eye Talk and Eye Tell

After choosing a letter, the frame cursor is moving along "Command" items (Fig. 8), which includes Voiced/Semivoiced sound symbols, Delite a letter, Select, Sound and Delite All, on the top of the screen.   If a user blinks when the frame cursor is on Select, the letter is chosen. By continuing this operation, a user can create a sentence as long as he/she needs. Then the voice sound comes out when a user choose Sound button.   Eye Tell is a modified communication application to indicate needs simply by "Yes = ○" and "No = ×" by blinks (Fig. 8). A user can make original symbols suitable for the level of handicap.   First, a helper makes 2 symbols that are suitable for the situation of a user. (Up to 21 sets.) Then the symbols on right and left sides of the screen turn on and off reciprocally. A user chooses the symbol when it turns on by a blink. The application judges the blink and put the sentence into voice sound.

Even a user who has weak muscle strength, the conscious blink is slower than the unconscious blink and speed of blink is different among individuals. In Eye Talk and Eye Tell, a user can adjust the sensitivity of determining blinks from 0 to 100.

A user can customize the following settings of the applications.

1) To show the picture of eyes (On or Off)
2) Sensitivity of detecting blinks (low 0 to high 100)
3) Speed of moving the frame cursor on the character table of Japanese (slow 0 to fast 100)
4) Speed of moving the frame cursor on Command items (slow 0 to fast 100)

The cursor moves 0 = 100 frames and 100 = 10 frames. About 24 frames of pictures are processed per a second, so the speed of movement of the cursor is 0 = about 4.16 seconds and 100 = about 0.41 seconds.

5 Conclusion

We are on the process of filing a patent for the high quality blink detection system and distributing it free of charge in Japan. For further development of this system, we research and develop the method of determination of gazing directions by analyzing the eye movement. If we can detect the gazing direction accurately on the tablet, burden on a user is reduced. It is applicable to any application by limiting the processing range to achieve higher speed and simplifying the process and high detection accuracy. To determine gazing directions, we use the afterimage method we have developed in this study. To determine gazing directions, we measure the rate of increase or decrease of the area of the white part of eyes by the afterimage method for processing an image of the eyes. In this way, we can determine gazing directions with high accuracy and less error. By combined the method to determine gazing directions with the blink detection system, a user will be able to get more unfettered communication.

The support applications are living necessities for handicapped people. They should be practical and easy to use. Those applications encourage physically handicapped people to communicate with others and lead the society to support each other regardless of handicap.

References

1. TalkEye, http://www.takei-si.co.jp/product/talkeye.html
2. Let's Chat, http://panasonic.co.jp/phc/products/home/communicationaids.html
3. Viola, P., Jones, M.J.: Rapid Object Detection using a Boosted Cascade of Simple Features. In: IEEE CVPR (2001)
4. Lienhart, R., Maydt, J.: An Extended Set of Haar-like Features for Rapid Object Detection. In: IEEE ICIP 2002, vol. 1, pp. 900–903 (September 2002)
5. Tsukada, A.: Automatic Detection of Eye Blinks using Spiral Labeling. In: Symposium on Sensing via Image Information (SSI 2003), vol. 9, pp. 501–506 (June 2003)
6. Joblore, G.H., Greenberg, D.: Color Spaces for Computer Graphics. Computer Graphics 12, 20–27

Generation of Reducts
Based on Nearest Neighbor Relation

Naohiro Ishii[1], Ippei Torii[1], Kazunori Iwata[2], and Toyoshiro Nakashima[3]

[1] Aichi Institute of Technology, Toyota, Japan
[2] Aichi University, Nagoya, Japan
[3] Sugiyama Jyogakuen University, Nagoya, Japan
{ishii,mac}@aitech.ac.jp, kazunori@vega.aichiu.ac.jp,
nakasima@sugiyama-u.ac.jp

Abstract. Dimension reduction of data is an important theme in the data processing and on the web to represent and manipulate higher dimensional data. Rough set is fundamental and useful to process higher dimensional data. Reduct in the rough set   is a minimal subset of features, which has the same discernible power as the entire features in the higher dimensional scheme. A nearest neighbor relation with minimal distance proposed here has a basic information for classification. In this paper, a new reduct generation method based on the nearest neighbor relation with minimal distance is proposed. To characterize the classification ability of reducts, we develop a graph mapping method of the nearest neighbor relation, which derives a higher classification accuracy.

Keywords: classification, reduct generation, nearest neighbor relation with minimal distance, mapping of nearest neighbor relation.

1   Introduction

Rough sets theory firstly introduced by Pawlak[1,2] provides us a new approach to perform data analysis, practically. Up to now, rough set has been applied successfully and widely in machine learning and data mining. The need to manipulate higher dimensional data in the web and to support or process them gives rise to the question of how to represent the data in a lower-dimensional space to allow more space and time efficient computation. Thus, dimension reduction of data still remains as an important problem. An important task in rough set based data analysis is computation of the attributes or feature reducts for the classification. By Pawlak[1,2]s rough set theory, a reduct is a minimal subset of features, which has the discernibility power as using the entire features. For the classification, nearest neighbor method[6] is simple and effective one.   We have problems for the application of the nearest neighbor method to reducts classification. Nearest neighbor relation with minimal distance between different classes is proposed here as a basic information for classification. We propose here a new reduct generation based on the nearest neighbor relation with minimal distance. Further, a graphical mapping of the nearest neighbor relation with minimal distance is developed for the weighted distance measure. Then, we propose the modified reducts , which are extended from reducts with redundant attributes It is shown that the modified reducts show a higher classification accuracy by mapping of the nearest neighbor relation.

E. Corchado et al. (Eds.): IDEAL 2014, LNCS 8669, pp. 18–26, 2014.

2   Nearest Neighbor Relation

In this paper, we propose a concept of a nearest neighbor relation with minimal distance. By the distance measure, nearest neighbors with minimal distance is discussed in the rough set classification. Then, the decision table [3,4] and the nearest neighbor relation are defined as follows.

Def.2.1 Let $T = \{U, A, C, D\}$ be a decision table, with $U = \{x_1, x_2, ...x_n\}$, which is a set of instances. By a discernibility matrix of T, denoted by $M(T)$, which is $n \times n$ matrix

$$m_{ij} = \{(a \in C : a(x_i) \neq a(x_j)) \\ \wedge (d \in D, d(x_i) \neq d(x_j))\} \, i, j = 1, 2, ... n \quad (1)$$

,where U is the universe of discourse, C is a set of features, A is a subset of C called condition, and D is a set of decision class features.

We can   define a new concept, a nearest neighbor relation with minimal distance,. δ.

Def. 2.2 A nearest   neighbor relation with minimal distance   is a set of pair instances

$$\{(x_i, x_j) : d(x_i) \neq d(x_j) \wedge (distance \ between \ x_i \ and \ x_j) \leq \delta\} \quad (2)$$

,where δ is the given positive value for the nearest neighbor distance.

In Table 2, for the instance x_1, the value of the attribute a, is a(x_1)=1. Similarly, that of the attribute b, is b(x_1)=0. Since a(x_1)=1 and a(x_5)=2, a(x_1) $\neq$ a(x_5) holds.

Table 1. Data example of decision table

Inst. & Attri.	a	b	c	d	class
x_1	1	0	2	1	1
x_2	1	0	2	0	1
x_3	2	2	0	0	2
x_4	1	2	2	1	2
x_5	2	1	0	1	2
x_6	2	1	1	0	1
x_7	2	1	2	1	2

Table 2. Dicernibility matrix of the decision table in Table 1

	x_1	x_2	x_3	x_4	x_5	x_6
x_1						
x_2	—	—				
x_3	b,c,d	b,c	a,b,c,d			
x_4	b	b,d	—			
x_5	a,b,c,d	a,b,c	—	—		
x_6	—	—	b,c	a,b,c,d	c,d	
x_7	a,b	a,b,d	—	—	—	c,d

In Table 1, to find the nearest neighbor relation with minimal distance δ, a lexicographical ordering for the instances is developed here as shown in Fig. 1. In Fig.1, the Euclidian distance between $x_6(2110)$ in class 1 and $x_7 = (2121)$ in class 2, becomes $\sqrt{2}$, where the first & second components are shown in ( ) and the third & fourth components in box . Similarly, the distance between $x_5(2101)$ and $x_6(2110)$ becomes $\sqrt{2}$, while the distance between $x_1(0021)$ and $x_7(2121)$ becomes $\sqrt{2}$.

$$(1,0)\frac{\boxed{2,1}x_1}{\boxed{2,0}x_2} \rightarrow (1,2)^{\boxed{2,1}x_4} \xrightarrow{\boxed{2,1}x_7} (2,1)\frac{\boxed{1,0}x_6}{\boxed{0,1}x_5} \rightarrow (2,2)_{\boxed{0,0}x_3} \tag{3}$$

Fig. 1. Lexicographical ordering for nearest neighbor relation

In Fig.1, (x_6, x_7) becomes an element of the nearest neighbor relation with a distance $\sqrt{2}$. Similarly, (x_5, x_6) and (x_1, x_7) are elements of the relation with a distance $\sqrt{2}$.

Thus, a nearest neighbor relation with minimal distance $\sqrt{2}$ becomes

$$\{(x_1, x_7), (x_5, x_6), (x_6, x_7)\} \tag{4}$$

3   Nearest Neighbor Relation for Reduct Generation

3.1   Generation of Reducts Based on Nearest Neighbor Relation with Minimal Distance

First, elements of nearest neighbor relation are picked up from the discernibility matrix. Assume that the set of elements of the nearest neighbor relation are $\{nn_{ij}\}$. Then , the following characteristics are shown.

Respective element of the set $\{nn_{ij}\}$ corresponds to the term of logical sum. As an example, the element {a,b,c}of discernibility matrix in the set $\{nn_{ij}\}$ corresponds to a logical sum (a+b+c). The following propositions hold in the relation.

Prop. 3.1. Respective Boolean term consisting of the set $\{nn_{ij}\}$ becomes a necessary condition to be reducts in the Boolean expression. This is trivial, since the logical product of respective Boolean term   becomes reducts in the Boolean expression.

Prop. 3.2. Logical product of respective terms corresponding to the set $\{nn_{ij}\}$ becomes a necessary condition to be reducts in the Boolean expression. This is also trivial by the reason of Prop.3.1. Thus, the relation between Prop.3.1 and Prop.3.2 is described as follows.

Prop. 3.3.   Reducts in the Boolean expression are included in the Boolean term in Prop.3.1and the logical product in Prop.3.2.

Fig. 2 shows that nearest neighbor relation with classification is a necessary condition for   reducts, but not sufficient condition.

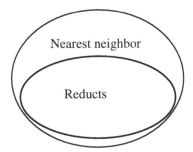

Fig. 2. Condition of nearest neighbor relation and reducts

Some reduction formula is classified to two classes (A) and (B) on the discernibility matrix.

(A) Attributes(variables) of the element in the discernibility matrix includes   those of any respective element in the set $\{nn_{ij}\}$.

(B) Attributes(variables) of the element in the discernibility matrix are different from   those of any respective element in the set $\{nn_{ij}\}$.

Then, the Boolean expression in the class (A) and (B) shows characteristic properties.

Prop. 3.4. The elements in the class (A) is absorbed in the   corresponding elements of the set $\{nn_{ij}\}$ in the Boolean expression.   When the element of the class (A) is expressed in Boolean expression, it is shown in disjunction of attributes(variables). The logical sum of attributes is absorbed in the logical sum element with the same fewer   attributes in the set $\{nn_{ij}\}$.

Prop. 3.5. · Within (B) class, the element with fewer attributes(variables)   plays a role of the absorption of the element with   larger attributes(variables). Then, the similar operation of   Prop.2.4   is carried out within (B) class.

Prop. 3.6.   After the absorption of   within (B) class, the remained   elements with fewer attributes(variables) in the (B) class product to the elements of the set $\{nn_{ij}\}$ in the Boolean expression. By the transformation of the Boolean expression   to logical sum of the product variables, reducts are obtained in the prime implicants.

Prop. 3.7. Nearest neighbor relation with minimum distance is a necessary condition for reducts in Boolean expression.   To solve the necessary and sufficient condition for reducts in Boolen expression, attributes(variables) in elements in the (B) class are applied as the   logical product of the nearest neighbor relation $\{nn_{ij}\}$ with minimal distance.

Let one attribute(variable) in the element   of the Class(B) be ε and the element of the $\{nn_{ij}\}$ be represented in the Boolean expression as

$$(\gamma_1 + \gamma_2 + ... + \gamma_m) \tag{5}$$

Then,   the logical   product $\varepsilon \cdot \gamma_1$ is carried out. If ε is absorbed in γ_1, $\varepsilon \cdot \gamma_1 = \gamma_1$. But, if $\varepsilon \cdot \gamma_1 \neq \gamma_1$, then a new candidate of reduct $\gamma_1^* = \varepsilon \cdot \gamma_1$ is made.

[Example 1] An example of reducts generated by the nearest neighbor relations is shown in the following. First, in Table 2, nearest neighbor relations with minimum distance are detected and shown in the shading cells $\{(x_1, x_7), (x_5, x_6), (x_6, x_7)\}$. The Boolean function of the nearest neighbor relation, $f_{NN}(E)$ of the logical   product of these three terms of (2) becomes

$$(a+b) \cdot (c+d) = a \cdot (c+d) + b \cdot (c+d) \tag{6}$$

These terms are   candidates of reducts.   These terms are simply expressed as

$$a \cdot c + a \cdot d + b \cdot c + b \cdot d \tag{7}$$

Second,   attributes of the class (B) is searched in the discernible matrix, Table 2. Thus, elements, $\{ b \}$ and $\{b,c\}$ and $\{b,d\}$ are found as in the class(B). Next, the logical product (b) and nearest relations (7) becomes

$$a \cdot b \cdot c + a \cdot b \cdot d + b \cdot c + b \cdot d$$

The first two two terms are absorbed to the third and the forth terms. Thus, the final reducts obtained is

$$b \cdot c + b \cdot d \tag{8}$$

The $\{b\}$ becomes the core of reducts.

[**Example 2**]. Let the attribute b of the instance x_3 be changed to 1 and the attribute c of the instance x_4 be changed to 1 in the values in Table 1.

The nearest neighbor relation with minimal distance is represented as

$$(a+bc)\cdot(d+bc) = bc + ad$$

In the Boolean expression. The element of the class (B) is only $\{b,c\}$ in the relation (x_3, x_2). Then, the following equations are derived.

$$b\cdot(bc+ad) = bc + abd \quad \text{and} \quad c\cdot(bc+ad) = bc + acd \qquad (9)$$

Thus, the derived reducts become $\{bc, abd, acd\}$.

4   Mapping of Nearest Neighbor Relation on Modified Reducts

To improve the classification accuracy using reducts followed by nearest neighbor, the modified reducts and weighting of the distance measure are proposed based on the mapping graph of the nearest neighbor. The weighted distance measure is introduced as shown in equation (10). The equation (10) shows a weighted Euclidean distance. To isolate neighbor points between different classes, the weight of the respective coordinate is introduced for the classification..

$$Weighted\ Distance = \sqrt{\sum_{i=1}^{n} \omega_i (x_i - y_i)^2} \qquad (10)$$

Here, we have problems whether the nearest neighbor relation is applicable to the characterization of reducts using the weighted distance[8].

Based on the reducts $\{b,c\}$ and $\{b,d\}$ based on the nearest neighbor relation, the following three extended reducts, called modified reducts here are derived on the discernibility matrixc,$\{b,c, a\}$,$\{b,c,d\}$ and $\{b,d,a\}$ as shown in Fig. 3.

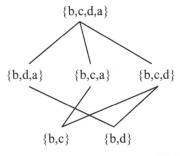

Fig. 3. Relation between reducts and modified reducts

4.1   Characterization of Reducts by Using Nearest Neighbor Relation

By using the nearest neighbor relation in Table 2., the weights of the modified reducts are computed by using   the variational method as follows. There exist three nearest neighbor relations, $(x_1, x_7), (x_5, x_6)$ and. (x_6, x_7). We consider their summation of the weighted distance in the equation (10) becomes

$$D = \sqrt{\omega_b(x_{12} - x_{72})^2 + \omega_c(x_{13} - x_{73})^2} + \sqrt{\omega_b(x_{52} - x_{62})^2 + \omega_c(x_{53} - x_{63})^2}$$
$$+ \sqrt{\omega_b(x_{62} - x_{72})^2 + \omega_c(x_{63} - x_{73})^2} \tag{11}$$

under   the condition   of weights

$$\omega_b^2 + \omega_c^2 = 1 \tag{12}$$

To obtain the optimum weights, the Lagrangian multiplier λ is introduced,

$$H = D + \lambda(\omega_b^2 + \omega_c^2 - 1) \tag{13}$$

Then, by the operations

$$\frac{\partial H}{\partial \omega_b} = 0, \ \frac{\partial H}{\partial \omega_c} = 0 \quad \text{and} \quad \frac{\partial H}{\partial \lambda} = 0 \tag{14}$$

the weights relation $\omega_b < \omega_c$ is obtained.

By this optimization process, it is sufficient to consider only different components between two inferences in the distance.

Prop. 4.1. Let the weights be $\omega_1, \omega_2, \omega_3, ... \omega_m$ and   the summation of difference of the distance in the respective component be $K_1, K_2, K_3, ... K_m$. If the components relation $K_1 \leq K_2 \leq K_3, ... \leq K_m$ holds,   the   weight   relation   becomes $\omega_1 \geq \omega_2 \geq \omega_3 ... \geq \omega_m$.

By the above Prop.4.1, $\omega_c > \omega_b > 0$ is derived for the   reduct {b,d}.

The instances are graphically mapped in the order of the nearest neighbor relation with Euclidean distance 1.   It is expected   to isolate the (0,1) in the class 1 from the (1,1) in the class 2, also the (1,0) from the (2,0).   Thus, $\omega_b > \omega_d > 0$ is derived.

Reducts based on the nearest neighbor classification   is developed. Based on the reducts {b, c} and {b, d}, the following three modified reducts are derived on the discernibility matrix, {b,c,a} , {b,c,d}   and   {b,d,a} [8].

5   Classification by Mapping of Nearest Neighbor Relation

By the mapping   of the nearest neighbor relation and using Prop.4.1, the following attribute weight relations are obtained.

$$\omega_c > \omega_b > \omega_a > \omega_d = 1 \qquad (15)$$

By three modified reducts, the extended modified reduct {a,b,c,d} is used. In this case, the accuracy becomes 1.0(100%).   The classification accuracy of reducts {b, d} , {b, c} and weighted modified reducts are compared in Fig. 4.

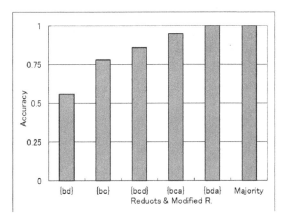

Fig. 4. Classification accuracy by reducts and weighted modified ones

6   Conclusion

Dimension reduction of data is an important problem as in web, data mining and image processing. Concept of rough set theory is a useful method to reduce the higher dimensional features. In this paper, we proposed a nearest neighbor relation with minimal distance. The nearest neighbor relation plays a   fundamental role of generation of reducts for between different classes First, a new generation method of reducts is proposed, which is based on the nearest neighbor relation with minimal distance. Second, characterization of   reducts was developed for the nearest neighbor classification by introducing the weighted distance measure. Then, this paper develops a graphical mapping method of the relation for the higher classification accuracy.

References

1. Pawlak, Z.: Rough Sets. International Journal of Computer and Information Science 11, 341–356 (1982)
2. Pawlak, Z., Slowinski, R.: Rough Set Approach to Multi-attribute Decision Analysis. European Journal of Operations Research 72, 443–459 (1994)

3. Skowron, A., Rauszer, C.: The Discernibility Matrices and Functions in Information Systems. In: Intelligent Decision Support-Handbook of Application and Advances of Rough Sets Theory, pp. 331–362. Kluwer Academic Publishers, Dordrecht (1992)
4. Skowron, A., Polkowski, L.: Decision Algorithms, A Survey of Rough Set Theoretic Methods. Fundamenta Informatica 30(3-4), 345–358 (1997)
5. Pawlak, Z., Skowron, A.: Rough sets and Boolean reasoning. Information Science 177, 41–73 (2007)
6. Cover, T.M., Hart, P.E.: Nearest Neighbor Pattern Classification. IEEE Transactions on Information Theory 13(1), 21–27 (1967)
7. Ishii, N., Morioka, Y., Bao, Y., Tanaka, H.: Control of Variables in Reducts - kNN Classification with Confidence. In: König, A., Dengel, A., Hinkelmann, K., Kise, K., Howlett, R.J., Jain, L.C. (eds.) KES 2011, Part IV. LNCS, vol. 6884, pp. 98–107. Springer, Heidelberg (2011)
8. Ishii, N., Torii, I., Bao, Y., Tanaka, H.: Modified Reduct Nearest Neighbor Classification. In: ACIS-ICIS, pp. 310–315. IEEE Comp. Soc. (2012)

Automatic Content Related Feedback
for MOOCs Based on Course Domain Ontology

Safwan Shatnawi[1], Mohamed Medhat Gaber[1], and Mihaela Cocea[2]

[1] School of Computing Science and Digital Media,
Robert Gordon University, Aberdeen, UK
{s.m.i.shatnawi,m.gaber1}@rgu.ac.uk
[2] School of Computing, University of Portsmouth, UK
mihaela.cocea@port.ac.uk

Abstract. MOOCs offer free access to educational materials, leading to large numbers of students registered in MOOCs courses. The MOOCs forums allow students to post comments and ask questions; due to the number of students, however, the course facilitators are not able to provide feedback in a timely manner. To address this problem, we identify content-knowledge related posts using a course domain ontology and provide students with timely informative automatic feedback. Moreover, we provide facilitators with feedback of students posts, such as frequent topics students ask about. Experimental results from one of the courses offered by *Coursera*[1] show the potential of our approach in creating a responsive learning environment.

Keywords: automatic feedback, topic detection, ontologies, clustering, MOOCs.

1 Introduction

Recently, Massive Open Online Courses (MOOCs) have become a hot topic in higher education [20]. MOOCs are free and open, i.e., no prerequisites are required to register. This led to the enrolment of a large number of students in MOOCs.

MOOCs forums allow collaborative discussions, which are a fertile environment for gaining insight into the cognitive process of the learners. The analysis of forums information enables us to obtain information about participants level of content knowledge, learning strategies, or social communications skills. A variety of participants exchanges exist in MOOCs forums, such as getting other participants' help, scaffolding others' understanding, or constructing knowledge between learners. Effective exchanges require communications and content knowledge utilisation and integration, leading to successful knowledge-building [11].

Current MOOCs settings do not provide participants (educators and learners) with any kind of timely analysis of forums contents. Consequently, the educators

---

[1] https://www.coursera.org/

E. Corchado et al. (Eds.): IDEAL 2014, LNCS 8669, pp. 27–35, 2014.
© Springer International Publishing Switzerland 2014

cannot reply to hundreds of thousands students sending questions or comments on the course materials in a timely manner. This, in turn, leads to delay in getting feedback to students, which could result in drop-out.

Feedback plays vital role in learning. Many studies researched the effects of feedback on students learning in both traditional and online settings. Online learning systems provide students with feedback related to close-ended questions, tests, or assignments. Types of feedback in online learning systems are either automatically generated or human generated feedback. However, provide students with content related feedback in online settings has not been researched. Content related feedback aims to build students content knowledge and to reduce the burden of obtaining information from multiple resources. Course facilitators in MOOCs settings can not provide timely informative content related feedback due to the massiveness feature of these courses.

In this research, we developed hybrid technique to provide students with timely content related feedback in MOOCs setting. Albeit we examined our technique on MOOCs, the proposed technique can be generalised to any knowledge acquisition settings. The system identifies a content-knowledge related posts using a course domain ontology. Then, it provides students with timely, informative content related feedback. Moreover, our system provides facilitators with feedback on students posts (e.g., frequently asked about topics) which results in clustering posts according to its topics in hierarchical clusters. The proposed system integrates domain ontology, machine learning, and natural language processing.

The paper is organised as follows: section 2 for related work, section 3 describes our proposed system, in section 4 we introduce the experimental work and results. Finally, in section 5, we summarise our work and the future of this research.

2   Related Work

In education, feedback is connected to the assessment process. Feedback is conceptualised information about student's performance or understanding. It aims to fill the gap between what is understood and what should be learnt [3]. In contrast, content feedback aims to improve learning by providing information for students to scaffold them toward the learning objectives [9]. A form of content feedback is providing students with hints and references to students questions, which is known as indirect feedback.

In online courses, peer feedback is adopted to promote learning. In MOOCs, this is introduced as a solution for the lack of facilitator feedback due to the massiveness feature of MOOC [16]. However, peer interactions only do not guarantee an optimum level of learning [8]. To facilitate learning, feedback should be provided to students timely and continuously [6]. Feedback systems provide students with feedback to structured or semi structured topics such as computer programming, spreadsheets, or mathematics [12,13,14,5]. The work presented in [12,13,14] guide the students to achieve the course objective based on preset

Table 1. Ontology Learning from Text

System	Process	Domain	Technique	Objective
Asium	semi -automated	Information extraction	linguistics and statistics	learn semantic knowledge from text
Text-To-Onto	semi -automated	Ontology management	linguistics and statistics	Ontology creation
TextStorm/Clouds	semi -automated	music and drawing	logic based and linguistics	build and refine domain ontology for musical pecies and drawings
Sndikate	fully automated	general ontology learning	linguistics based	build general domain ontology
OntoLearn	semi -automated	tourism	linguistics and statistics	develop interoperable infrastructure for tourism domain
CRCTOL	semi -automated	domain specific	linguistics and statistics	construct ontology from domain specific documents
Onto Gain	fully automated	general ontology learning	linguistics and statistics	build ontologies using unstructured text

scenarios. However, the work proposed by [5] analyses students work and provide students with dynamic feedback based on others solutions.

An ontology is an explicit formal specification of a shared conceptualisation of a domain of interest [18]. An ontology defines the intentional part of the underlying domain, while the extensional parts of the domain (knowledge itself or instances) are called the ontology population. Ontologies have been used in educational field to represent course content [4,22,1,2]. It can scaffold students learning due to its role in instructional design and curriculum content sequencing [3]. Also, ontologies have been used in intelligent tutoring systems [4], students assessments [10], and feedback [15]. An ontology-based feedback to support students in programming tasks was introduced by [15]. They suggested a framework for adaptive feedback to assist students overcoming syntax programming errors in program codes. In spite of describing their work as ontology based feedback, they did not describe the structure of their ontology nor the process of creating that ontology (manual/automated). Ontology building is a complex and time consuming task. It requires domain experts and knowledge engineers handcrafting knowledge sources and training data which is one of the major obstacles in ontology development. There are many attempts to automate or semi-automate the process of ontology building [21]. Table 1 summarises some of the tools developed to build domain ontologies from text. In our approach, we use the ontology to identify topics discussed by students in forums. Content analysis aims to describe the attribute of the message or post. The obtained attributes (clusters) should reflect the purpose of the research, be comprehensive, and be mutually exclusive [19]. An initial step in analysing forums content is to identify the topic and the role of the participants. An advantage of domain ontology based clustering is getting subjective clusters. One can get different clusters according to the desired perspective. In MOOCs setting, this will enable the system to acknowledge courses facilitators about topics students frequently ask about.

3   MOOC's Domain Ontology and Feedback

In this section we introduce formal definitions and specifications of course contents ontology. An ontology is formally defined as [7]:

Course Ontology 1. *A core ontology is a set of sign system* $\Theta := (T, P, C^*,$
$H, Root)$,
T: set of natural language terms of the Ontology
P: set of properties
C^*: *function that connects terms* $t \in T$ *to set of* $p \subset P$
H: hierarchy organisation connects term $t \in T$ *in a cyclic, transitive, directed*
relationships.
Root: is the top level root where all concepts $\in C^*$ *are mapped to it.*

3.1 Phase I: Building Domain Ontology

In this research, we represent course contents using domain ontology notations. The proposed system uses a course domain ontology to detect topics in students posts and topics' properties. As a result, automatic feedback is sent back to the student.

Building a domain ontology is an ontology engineering task and a time consuming process. We aim to allow domain experts (course facilitators) to build course domain ontologies in MOOCs setting. We started by identifying terms (concepts) related to the course knowledge. We use multiple knowledge sources to obtain the most frequent terms used in the knowledge sources. Next, we designed simple graphical user interface to build the terms (concepts map) hierarchy. For each term we add a set of properties (attributes). Some of these attributes connect two terms together (binary attributes) while others are unary attributes. For each property, we store a feedback that will be sent back to the student after processing his/her post. The aforementioned steps are called ontology population in ontology jargon. We used relational database to store all information about terms, properties and feedback. We also expanded terms and properties by storing its synonyms. Course facilitators can easily create the ontology.

Algorithm 1 builds course domain ontology. While Algorithm 2 converts the domain ontology into deterministic finite automata (DFAs) which will be used to process students posts in phase II.

3.2 Phase II: Processing Students Posts

In this process, we aim to discover all topics that appear in students' posts. Also, we discover topics' properties. In this phase, we rely on course domain ontology which was built earlier in Phase I as aforementioned. We generate a state table for all terms in the course domain ontology that represents terms deterministic finite automata. In an analogous manner we generate a state table for all properties.

Students posts are parsed and processed word by word (see Algorithm 3). We used a similar approach used in programming languages compilers to detect programming constructs. Instead, we are looking for terms and properties constructs. In case of multiple terms detection, we label the post to terms closest

---

Algorithm 1. Building Domain Ontology

$C \leftarrow$ Read Course's knowledge corpus.
$TDM \leftarrow$ Build terms document matrix.
$T \leftarrow$ Find most frequently terms.
$Root \leftarrow$ Ontology root node.
$Parent(Root) \leftarrow -1$
for all $t \in T$ **do**
  $parent \leftarrow$ parent(t)
end for
$P \leftarrow$ Set of All properties.
for all $t \in T$ **do**
  for all p$\in P$ and p $\subset$ t **do**
    Add(t,p)
  end for
end for

Algorithm 2. Concepts and Properties DFAs Generator

$C \leftarrow$ set of all concepts
$DT \leftarrow$ set of all distinct terms $\in C$
for all $c \in C$ **do**
  Parse c into set of individual words W
  $current\_state \leftarrow 0$
  $states \leftarrow 0$
  for int $i = 0$**to** $W.length()$ **do**
    if $state\_table[current\_state][W[i]] = 0$ **then**
      state_table[current_state][W[i]] $\leftarrow$ current_state
    else
      $states \leftarrow states + 1$
      $state\_table[current\_state][W[i]] = states$
    end if
  end for
  for int $j = 0$, $j < DT.length()$ **do**
    if $state\_table[current\_state][j] \neq 0$ **then**
      $state\_table[current\_state][j] = 999$ {999 means final state}
    end if
    $state\_table[current\_state][j] = identifier(c)$
  end for
end for

---

parent according to the domain ontology hierarchy. Later on, we cluster students posts according to its labels in hierarchical clusters. We use the same methodology to detect terms' properties. As a result, we have a set of terms and another set of properties. Both sets are used to send appropriate feedback to students. The following section envisages detailed description of feedback module.

3.3   Phase III: Feedback Generating

In this phase, we generate the feedback to be sent back to the student. We take
all terms and properties detected by phase II, and perform a simple search to
our domain ontology database. When we get a match, we send the feedback for
the user. Figure 1 shows the processes to generate the feedback.

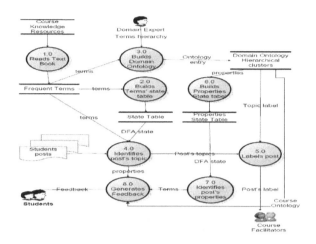

Fig. 1. Students posts Labelling and Feedback system

4   Experimental Setting and Results

in our experimental work, we used the 2013 version of "Introduction to Database
Management" course, offered by *Coursera*. We collected resources about this
topic using a text book, Wikipedia, and other Internet resources. Next, we extracted a list of topics (terms) appeared in these resources. After that, we play
the role of domain expert to create a concept hierarchy (Concept Map) using a
simple tree view user interface which allows us to re-organise these terms in tree
view structure. Then for every term we assigned a set of properties. Each property has associated feedback which will be sent to the students. The following
is an example that clarifies the course domain ontology that we created to test
our approach.

Algorithm 3. Process Students posts

Read (Post)
post_words ← parse(post)
identify(terms)
identify(properties)
Feedback ← Search ontology (terms, properties)
Generate_feedback(Feedback)

Course Ontology Example 1. $\Theta := (T, P, C^*, H, Root)$
T : "key", "primary key", "data", "information", "database management system", "foreign key", "relationship", "conceptual model",....
P: "definition", "type", "syntax", "use", "advantage",....
C*: "relationship is a conceptual model" , "schema consists of attributes" , "foreign key is part of relationship",...
H: Concept map hierarchy parent(DBMS,RDBMS), Parent(RDBMS,Table),...
Root: Database.

We populate the course ontology using the aforementioned content resources. As a result, every concept has many properties. Our proposed technique separates knowledge (Domain ontology) from implementation (driver function) which was described through Algorithms 1, 2, and 3. As a result, domain ontology can learn new knowledge and expand without any changes to the driver function.

We the prepared collection of questions which we use to test our system. Test collection was collected from database management textbooks and from database forums (learners questions). We used 438 posts from Coursera. We manually identified the post which is related to the course contents. For every post, we store its label and feedback. Then we run our system to assign a label and provide feedback for every post. We used precision, recall, and F-measure to validate our system. For posts labels, we used the binary classification method based on word to word similarity. On the other hand, we used semantic text similarity based on latent semantic analysis using SIMILAR [17] to evaluate the relevance of retrieved feedback to the stored feedback. The following are the equations used to validate the system.

$$Precision = \frac{A}{A + B} \tag{1}$$

$$Recall = \frac{A}{A + C} \tag{2}$$

$$F\text{-}measure = 2 \times \frac{Recall \times Precision}{Recall + Precision} \tag{3}$$

Where A is the number of correct labels obtained, B is the number of not retrieved labels and C is the number of incorrect labels retrieved. Table 2 shows the experimental results of the system. The results show the potential of the system in providing the students with timely feedback. The system achieved promising results in term of precision, recall, and *F-measure* as shown in Table 2. However, for some posts the system failed to label the post, consequently it failed to retrieve any feedback. A reason behind that is lack of domain knowledge where the posts were about technical issues related to the database system or about contents not related to the database management system. In some other cases, however, the system was able to successfully label the post, on the other hand, it failed in retrieving a relevant feedback. Some posts have multiple topics and properties; as a result the system retrieved extra feedback which is not relevant to the post. A possible solution for that is using part of speech tagging and divide the post into multiple statement.

Table 2. Experimental results

	LABELLING (%)	FEEDBACK (%)
Recall	82	72
Precision	91	84
F-measure	86	78

5   Summary and Future Work

Domain ontology and NLP can scaffold teaching and learning processes in MOOCs settings. Domain ontology is an effective representation of course content knowledge. We proposed a feedback system for MOOCs setting. Our system represents a MOOC's contents using domain ontology notations. We separated the knowledge part from the processing part. As a result, the system learns new knowledge without changing the processing part. We, also, generated deterministic finite automata using natural language expressions derived from domain ontology instances. We create simple tools to automate and mange domain ontology population. However, domain ontology creation still depends on domain experts to some extent. In the future, we will automate the process of creating a domain ontology. We will also explore the roles of MOOCs' participants in the forums.

References

1. Boyce, S., Pahl, C.: Developing domain ontologies for course content. Educational Technology & Society 10(3), 275–288 (2007)
2. Chi, Y.L.: Ontology-based curriculum content sequencing system with semantic rules. Expert Systems with Applications 36(4), 7838–7847 (2009)
3. Coll, C., Rochera, M.J., de Gispert, I.: Supporting online collaborative learning in small groups: teacher feedback on learning content, academic task and social participation. Computers & Education (2014)
4. Crowley, R.S., Medvedeva, O.: An intelligent tutoring system for visual classification problem solving. Artif. Intell. Med. 36(1), 85–117 (2006)
5. Dominguez, A.K., Yacef, K., Curran, J.: Data mining to generate individualised feedback. In: Aleven, V., Kay, J., Mostow, J. (eds.) ITS 2010, Part II. LNCS, vol. 6095, pp. 303–305. Springer, Heidelberg (2010)
6. Gibbs, G., Simpson, C.: Conditions under which assessment supports students' learning. Learning and Teaching in HIgher Education 1(1), 3–29 (2004)
7. Hotho, A., Maedche, A., Staab, S.: Ontology-based text document clustering. KI 16(4), 48–54 (2002)
8. Kanuka, H., Garrison, D.: Cognitive presence in online learning. Journal of Computing in Higher Education 15(2), 21–39 (2004)
9. Kulhavy, R., Stock, W.: Feedback in written instruction: The place of response certitude. Educational Psychology Review 1(4), 279–308 (1989)
10. Litherland, K., Carmichael, P., Martínez-García, A.: Ontology-based e-assessment for accounting: Outcomes of a pilot study and future prospects. Journal of Accounting Education 31(2), 162–176 (2013)

11. Mak, S., Williams, R., Mackness, J.: Blogs and forums as communication and learning tools in a MOOC. In: Dirckinck-Holmfeld, L., Hodgson, V., Jones, C., Laat, M.D., McConnell, D., Ryberg, T. (eds.) Proceedings of the 7th International Conference on Networked Learning 2010. University of Lancaster, Lancaster (2010)
12. Mathan, S., Anderson, J.R., Corbett, A.T., Lovett, C.: Recasting the feedback debate: Benefits of tutoring error detection and correction skills, pp. 13–20. Press (2003)
13. Mitrovic, A.: An intelligent sql tutor on the web. Int. J. Artif. Intell. Ed. 13(2-4), 173–197 (2003)
14. Narciss, S., Huth, K.: How to design informative tutoring feedback for multimedia learning. In: Instructional Design for Multimedia Learning, pp. 181–195 (2004)
15. Muñoz-Merino, P.J., Pardo, A., Scheffel, M., Niemann, K., Wolpers, M., Leony, D., Kloos, C.D.: An ontological framework for adaptive feedback to support students while programming. In: International Semantic Web Conference (2011)
16. Piech, C., Huang, J., Chen, Z., Do, C., Ng, A., Koller, D.: Tuned Models of Peer Assessment in MOOCs (July 2013)
17. Rus, V., Lintean, M., Banjade, R., Niraula, N., Stefanescu, D.: Semilar: The semantic similarity toolkit. In: Proceedings of the 51st Annual Meeting of the Association for Computational Linguistics: System Demonstrations, pp. 163–168. Association for Computational Linguistics, Sofia (2013)
18. Studer, R., Staab, S.: Handbook on Ontologies. In: International Handbooks on Information Systems. Springer (2009)
19. Sufian, A.J.M.: Methods and techniques of social research. University Press (2009)
20. Vardi, M.Y.: Will moocs destroy academia? Commun. ACM 55(11), 5 (2012)
21. Wong, W., Liu, W., Bennamoun, M.: Ontology learning from text: A look back and into the future. ACM Comput. Surv. 44(4), 20:1–20:36 (2012)
22. Zouaq, A., Nkambou, R., Frasson, C.: Building domain ontologies from text for educational purposes. In: Duval, E., Klamma, R., Wolpers, M. (eds.) EC-TEL 2007. LNCS, vol. 4753, pp. 393–407. Springer, Heidelberg (2007)

User Behavior Modeling in a Cellular Network Using Latent Dirichlet Allocation

Ritwik Giri[1], Heesook Choi[2], Kevin Soo Hoo[2], and Bhaskar D. Rao[1]

[1] Department of Electrical and Computer Engineering
University of California,
San Diego, California, USA
[2] Advanced Analytics Lab, Sprint
Burlingame, California, USA

Abstract. Insights into the behavior and preference of mobile device users from their web browsing/application activities are critical components of any successful dynamic content recommendation system, mobile advertisement platform, or web personalization initiative. In this paper we use an unsupervised topic model to understand the interests of the cellular users based upon their browsing profile. We posit that the length of time a user remains on a given website is positively correlated with the user's interest in the website's content. We propose an extended model to integrate this duration information efficiently by oversampling the URLs.

1 Introduction

In the competitive rush to provide better, more personalized web browsing experiences on mobile devices, cellular service providers have turned increasingly to user profiling to understand their subscribers better. User behavior modeling problem has been an interesting research topic in recent times [15] [4] [11] [3]. The simplest way to model the user behavior is using a keyword mining approach [7] [13]. In the keyword mining, the system looks for specific keywords in the URL and classifies it into a specific category. Though simplicity is its virtue, this technique has several shortcomings and disadvantages.

Recently researchers approached with a clustering technique [9] where the main objective was to group users who has similar interests. Keralapura et. al. [6] addressed the user profiling problem as a two-way clustering problem with an objective to cluster both users and browsing profiles simultaneously. In this work, an hour-glass model was used to cluster the similar type of websites beforehand manually. The hour-glass model, however, is not a feasible solution when dealing with a large dataset because of its manual clustering approach.

These issues gave us the motivation to look at this problem from a different perspective: Topic modeling, which was originally proposed for text modeling in large document collections. Several advanced generative models were developed for this task such as probabilistic Latent Semantic Analysis [2] and Latent Dirichlet Allocation (LDA) [1]. In this work, we use this topic modeling approach to analyze the browsing behavior of the users.

E. Corchado et al. (Eds.): IDEAL 2014, LNCS 8669, pp. 36–44, 2014.

Summarizing our contributions and advantages of our proposed model over the previous works,

* The similarity between a document classification problem and a user browsing behavior modeling problem has been identified, which has enabled to view our problem in hand, from a completely new perspective: Topic Modeling. After the inference procedure of our model, the extracted topics and also the inferred interests of users show that, this task has been performed efficiently.
* The amount of time a user is spending in a specific website has very important information about the interests of the user encoded in it. In [8] a deterministic approach has been followed to incorporate this information, whereas in our work we propose an oversampling based method, suited to our statistical model. Also, note that our work only involves the cellular users and their browsing behaviors.
* Using only few hours of data, i.e, browsing profile of the user for only few hours can only reveal the temporal interests of the user. To determine a more generic interest of a user we need to assess their browsing profiles over a significant amount of time. In this work we have used data over twelve days, which shows that scalability issue has been solved for our model to a great extent.

The rest of the paper is organized in the following way. Section 2 presents the user behavior modeling problem and introduces the Latent Dirichlet Allocation (LDA) model. In Section 3, we describe the dataset and details of the implementation. In Section 4, we present the novel extensions of LDA, required to incorporate the session duration, as well as to deal with the challenging problem of users interest modeling. In Section 5, we demonstrate the detailed experimental findings. Finally, Section 6 concludes the paper and talks about future directions of this work.

2 User Behavior Modeling

In this section, we study the nature of users' behavior profiling, along with the proposed model. We formulate the user behavior modeling problem as a topic modeling problem from a document classification perspective.

2.1 Analogy between Topic Modeling and User Behavior Modeling

We have the data consisting of a set of users and URLs of their visited web sites. For example, Figure 1 shows a history of a user's visited web sites. At first glance, though it looks like a set of random web sites, after taking a closer look, we could discover that the browsing history of a user is actually a mixture of multiple hidden topics such as ecommerce, online video streaming, etc.

In the user behavior modeling problem we need to identify the hidden interests of the users from their browsing profiles. Whereas, in the problem of topic

Sample Browsing Profile   Sample Document

distilleryimage8s3amazonawscom distilleryimage2instagramcom
distilleryimage7instagramcom distilleryimage3s3amazonawscom
distilleryimage3s3amazonawscom distilleryimage0instagramcom
distilleryimage8s3amazonawscom distilleryimage8s3amazonawscom
distilleryimage8s3amazonawscom distilleryimage8s3amazonawscom
distilleryimage8s3amazonawscom distilleryimage8s3amazonawscom
distilleryimage8s3amazonawscom distilleryimage3instagramcom
distilleryimage5instagramcom distilleryimage7instagramcom
distilleryimage0s3amazonawscom distilleryimage0s3amazonawscom
distilleryimage0s3amazonawscom distilleryimage6instagramcom
distilleryimage3instagramcom i2walmartimagescom
i2walmartimagescom iwalmartimagescom iwalmartcom
i2walmartimagescom i2walmartimagescom i2walmartimagescom
i2walmartimagescom iwalmartcom i2walmartimagescom
i2walmartimagescom ctschannelintelligencecom iwalmartimagescom
iwalmartcom iwalmartimagescom iwalmartimagescom iwalmartcom
iwalmartimagescom iwalmartimagescom iwalmartcom
iwalmartimagescom iwalmartimagescom iwalmartcom
iwalmartimagescom iwalmartimagescom iwalmartimagescom
iwalmartcom iwalmartcom omniturewalmartcom iwalmartcom
masndcdncom iiwalmartcom
itjs13201106151378646109iinitcedexisradarnet iwalmartcom
iwalmartcom iwalmartcom iwalmartcom i2walmartimagescom
i2walmartimagescom iwalmartimagescom iwalmartimagescom
i2walmartimagescom i2walmartimagescom tcv5cache8cyoutubecom
tcv5cache8cyoutubecom tcv5cache8cyoutubecom
tcv5cache8cyoutubecom tcv5cache8cyoutubecom
tcv5cache8cyoutubecom tcv5cache8cyoutubecom
tcv5cache8cyoutubecom

Comparisons of growth rates published in the scientific literature with
"predictions"
rom such relatively simple models for the same conditions of pH,
temperature and
water activity were often surprisingly close and encouraged further efforts.
Gradually, using models that had been validated by comparing outputs
with
independent data became recognised as just as reliable as accumulating
results from
the scientific literature or spending weeks generating more microbiological
data.
Occasionally it is important to have an "accurate" estimate of the
growth/survival, but
more often it is sufficient to have a "reasonable" estimate, but quickly. It is
necessary
to obtain quick and "good enough" estimations of the shelf-life of foods, in
which
pathogenic bacteria might grow, in new product development and in risk
assessment..

The above two-step approach to develop predictive models is still in use,
and not
only for death but also for growth curves. Commonly, the first step in the
developmental procedure is to establish the growth/death model in
constant
environment (primary model); the next step is to determine, how the
parameters of the
primary model are affected by environmental factors (secondary model –
see Fig.1).

Fig. 1. Sample Browsing profile vs sample document

modeling in a document classification task, main aim is to identify the hidden
topics of the corpus of documents using an intelligent unsupervised model. The
similarity of these two problems is evident from their respective definitions. User
browsing profile (URLs visited by the user) can be considered as a document
whereas each URL can be perceived as a unique word. So we will represent each
user's browsing profile as a random mixture of latent user interests which is analogous to the hidden topics. This problem formulation motivates us to use Latent
Dirichlet Allocation (LDA), an unsupervised modeling technique, to extract the
hidden user interests.

2.2 Overview of Latent Dirichlet Allocation

Latent Dirichlet Allocation (LDA) is a generative probabilistic model of a corpus
[14] [1]. A corpus is a set of all users browsing profiles. Each topic in the corpus
is characterized by a probabilistic distribution over URLs. It is assumed that the
browsing profile of each user is a mixture of interests (topics) and each URL in
the browsing profile is associated with one of those topics. LDA analyzes these
set of URLs to discover the hidden topics in the corpus and hidden interests
(topics) of users.

LDA assumes the following generative process to create a corpus of D user
browsing profiles with N_d URLs in user browsing profile d using K topics.

1 For each topic index $k \in 1, ...K$, draw topic distribution $\beta_k \sim Dirichlet(\eta_k)$
2 For each user profile $d \in 1, ...D$:
  (a) Draw the topic distribution of user profile $\theta_d \sim Dirichlet(\alpha)$
  (b) For each URL $n \in 1, ..., N_d$:
     i. Choose topic assignment $Z_{d,n} \sim Multinomial(\theta_d)$
     ii. Choose URL $w_{d,n} \sim Multinomial(\beta_{Z_{d,n}})$

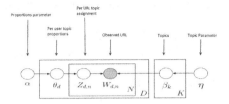

Fig. 2. Graphical Representation of LDA model

This generative process has also been shown in pictorial representation in Figure 2. In this work for the inference procedure we have used variational inference [5] [10] technique. It uses optimization to find a distribution over the latent variables that is close to the concerned posterior.

3 Overview of the Dataset and Implementation

In this section we describe our dataset used for our experiments and also discuss the implementation details to deal with the big dataset in a feasible manner.

3.1 Data Trace

An anonymized data set from a tier-1 cellular network provider in U.S has been collected from one switch in Texas. The data set includes the history of data usage for 12 days in April, 2012. It contains the aggregated information about subscriber and server IP addresses, anonymized identifier, URLs, session start time, session end time, user-agent, etc.

We used the identifier, session start time, session end time, and URL fields to analyze the users interests. Our dataset has around 280K (287728) users and 320K (323289) unique URLs and 126M(126154365) records.

3.2 Implementation: Mr.LDA

As discussed before, the one of the major shortcomings of the previous works for user behavior modeling problem is the scalability of those proposed models. Efficiency of the inference about the interest of the users is always proportional to the amount of data. Browsing profiles of the users over a longer period of time may help us to get a more general and clearer picture of the interests of the users instead of their temporal interests (For the case with only few hours of data). This motivation leads us to use the MapReduce Implementation of LDA (Mr. LDA) [16], which solves the issue of scalability to a great extent and enables us to use the data over twelve days, which includes more than 126 million records.

4 Discovering Hidden Interests of the Mobile Users

Here we present the extensions of the LDA model and how to extract the hidden/latent interests of the users based on their browsing history. At first, the browsing profiles along with the duration of each session are aggregated for each user from the original table of 126M records. Each user's browsing profile is now represented as a bag of URLs and the session duration information is encoded by oversampling the URLs. These bag of URLs have been used as input to train the Mr. LDA block.

4.1 Objectives

We are interested in the following two outputs of our Mr.LDA block:

1. $\gamma_{d,k}$: How much interest d^{th} user has in the kth topic.
2. $\beta_{v,k}$: How much URL v is associated with topic k

Thus from the first output we get the topic distribution of each user, i.e., how much interest each user has in different topics. The second output helps us to understand what the topics actually are, by checking the top most URLs of that topic.

4.2 Stop URLs: What Are They ?

During our first stage of the experimentation after sorting the output URL distribution for each topic we generate the top most URLs associated with each topic. But something is amiss as most of the topics are giving us a similar type mixed bag of URLs. After carefully assessing the browsing profiles of the users, we realize that most of the users are accessing a specific set of URLs which are appearing as the top most URLs of all the topics. The main reason for this shortcoming is the high frequency of these URLs, i.e., most of the users have accessed these URLs. This prevents our unsupervised model to extract the hidden user interests.

Similar problem has also been experienced in document modeling because of the prepositions, articles and other frequent common words. These words are known as **stop words** [12]. Similarly we refer to the high frequency URLs as **Stop URLs**. For our task we have separated 100 top most Stop URLs, that have been accessed by most number of unique users.

4.3 Encoding of Duration: Oversampling of URLs

We believe only considering the number of visits of URLs does not give a clear picture of the user's interests. The amount of time one user is spending in a specific website contributes significantly towards the interest of that user. Thus we also want to use this information efficiently to improve our model, which leads us to a key extension of our LDA model and we call it the Oversampling of URLs. Browsing profile of each user is represented by the bag of URLs. Bag of URLs representation means that the order of the URLs does not matter, we only consider that how many times a specific URL has been visited by the user. We propose to incorporate the session duration by oversampling the URLs at the time of constructing the bag of URLs, using the time that a user has spent in that website. We do not want to overemphasize on the duration information and choose to use 5 seconds (empirically chosen) as a single unit of sample. For example, suppose one user has visited a URL u and has spent 20 secs in that website. Then we will sample u $\frac{20}{5} = 4$ times in that bag of URLs. Hence we have encoded the duration information efficiently by this Oversampling of URLs.

4.4 Informed Prior

Despite all the similarities between a document and a user profile, there is a significant difference, which makes our problem of user browsing behavior modeling problem more challenging than a document classification task. While for a document it is easier to figure out the dominant topic, the interest of a user is more varied over several hidden interests. Hence to distinguish different topics, we need to provide the model some prior information before the inference procedure. This leads us to include an extension in the Mr.LDA block: **Informed Prior**.

The standard practice in topic modeling is to use a same symmetric prior. But for our task in hand we give few apriori information to our model, i.e, we start with a better initialization which will force the model to group some of the similar type of URLs in the same topic.

For example, suppose we want to specify frequently visited topmost weather related URLs to be grouped together in the same topic k. We define a list beforehand, $List_k = [gima.weather.com, imwx.com, ...m.weather.com]$. Now the prior is built as follows,

$$\eta_{v,k} = \begin{cases} 10, & \text{if } v \in List_k. \\ 0.01, & \text{otherwise.} \end{cases} \quad (1)$$

Where, $\eta_{v,k}$ is the informed prior for URL v of topic k. Note that this is just a better initialization technique. We are still learning from the available data at the time of training of our model, which is the key difference between our model and keyword mining approach.

5 Experimental Analysis

Extracted Topics: Now we use clean dataset, i.e., the one after removing the Stop URLs and also we have used the informed prior file. We need to choose the hyperparameter k, i.e, the number of topics carefully. For this task we have chosen $k = 9$ empirically after checking the loglikelihood for several other values of k. The extracted topics and 5 topmost URLs of each topic are shown in the Figure 3,

From the topmost URLs for each topic we have intuitively inferred about the topics and given the appropriate label. From the similarity of the nature of the URLs, we can see how well the user interests have been extracted. The only mixed bag topic is the eighth one, which gives us mostly URLs associated

1 : Weather	gimwx.com, mesonet.agron.iastate.edu, direct.weatherbug.com, map-tiles.**accuweather**.com, tiles.weatherbug.com
2. Pornography	img100.xvideos.com, m.**pornhub**.com, madrabbitsex.com, isanyoneup.com, video.xnxx.com
3. Social Networking	www.mocospace.com, cp.ifunny.mobi, **chat.groupme**.com, cdn-imageresize.mocospace.com, media.tumblr.com
4. Gaming/Sports	server.cardace.selfawaregames.com, static.storm8.com, api.kingdomsatwar.com, static.**mobile.espn**.go.com, jupiter.appads.com
5. Media	media-**cache**.pinterest.com, stream.iheart.com, i.mediatakeout.com, mobitv.com, radiojavan.com
6. E commerce	images.**craigslist**.org, images3.backpage.com, www.forever21.com, i.ebayimg.com, ecx.images-amazon.com
7. Chatting/Dating	i22.skout.com, images.**skout**.com, images.adam4adam.com, chat.imvu.com, msparp.com
8.Advertisements/Junk	t2.gstatic.com, jupiter.appads.com, f.chtah.com, iphone-sdk-31.**transpera**.com, m.yahoo.com
9. Images	distilleryimage11.**instagram**.com,distilleryimage31.instagram.com, twitpic.com, desmond.yfrog.com, api.plixi.com

Fig. 3. Extracted Topics/User interests

with several advertisements. We also have the topic distribution of each user, i.e, how much interest he or she has in each of these 9 topics. For visualization purpose we have randomly chosen 500 users from the whole set and we try to group them based on their top most interest. Figure 4(a) shows the plot of the

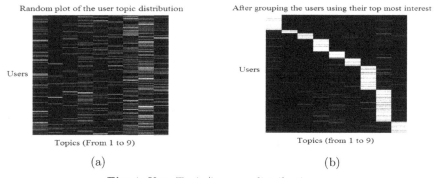

Random plot of the user topic distribution

After grouping the users using their top most interest

Users

Users

Topics (From 1 to 9)

Topics (from 1 to 9)

(a)

(b)

Fig. 4. User-Topic/interest distribution

topic/interest distribution of randomly chosen 500 users, and it can be seen some of the users have more than one interests. This is more clear after grouping the users based on their topmost interest, which is shown in Figure 4(b). As Figure 4 shows,the proposed topic modeling technique also groups users with similar interests. Besides, our topic model identifies more than one interest for a user, as well as we identify the intensity of users interest in each topic.

A sample browsing profile of a user is shown in the Figure 5(a), and the inferred user-topic distribution along with the topic names are in the Figure 5(b). We have omitted the Topic 8, as it is about advertisements and other junk, which we believe does not give us any important information regarding the interests of the user. In Figure 5(a) different URLs are shown in different colors depending on the topic they belong to, and in the plot of topic distribution of

(a)   (b)

Fig. 5. (a) Sample browsing profile and (b) Inferred User-Topic/interest distribution

that user, the probability bars are also shown using the corresponding colors. After minutely looking at the browsing profile of the user, we can see that he has visited mostly: youtube, walmart site, amazon and instagram. The topic distribution extracted by our model is also consistent as it assigns significant probabilities to three topics: Media, E commerce and Photos/Picture sharing.

6   Conclusions and Future Works

In this article we propose a novel way of modeling browsing behavior of cellular users and extracting their hidden interests, which can be very useful for mobile advertisements, recommendation systems, etc. We model the mobile user's browsing profile using an unsupervised topic model, and propose a clever extension to incorporate an important feature, session duration. The scalability issue has also been dealt with which permits us to use a large dataset, to extract more generic interests of the users.

References

1. Blei, D.M., Ng, A.Y., Jordan, M.I.: Latent dirichlet allocation. The Journal of Machine Learning Research 3, 993–1022 (2003)
2. Dumais, S.T.: Latent semantic analysis. Annual Review of Information Science and Technology 38(1), 188–230 (2004)
3. Gui, F., Adjouadi, M., Rishe, N.: A contextualized and personalized approach for mobile search. In: International Conference on Advanced Information Networking and Applications Workshops, WAINA 2009, pp. 966–971. IEEE (2009)
4. Han, X., Shen, Z., Miao, C., Luo, X.: Folksonomy-based ontological user interest profile modeling and its application in personalized search. In: An, A., Lingras, P., Petty, S., Huang, R. (eds.) AMT 2010. LNCS, vol. 6335, pp. 34–46. Springer, Heidelberg (2010)
5. Jordan, M.I., Ghahramani, Z., Jaakkola, T.S., Saul, L.K.: An introduction to variational methods for graphical models. Machine Learning 37(2), 183–233 (1999)
6. Keralapura, R., Nucci, A., Zhang, Z.L., Gao, L.: Profiling users in a 3g network using hourglass co-clustering. In: Proceedings of the Sixteenth Annual International Conference on Mobile Computing and Networking, pp. 341–352. ACM (2010)

7. Kosala, R., Blockeel, H.: Web mining research: A survey. SIGKDD Explor. Newsl. 2(1), 1–15 (2000), http://doi.acm.org/10.1145/360402.360406
8. Liu, H., Kešelj, V.: Combined mining of web server logs and web contents for classifying user navigation patterns and predicting users future requests. Data & Knowledge Engineering 61(2), 304–330 (2007)
9. Okazaki, S.: What do we know about mobile internet adopters? A cluster analysis. Information & Management 43(2), 127–141 (2006)
10. Pletscher, P.: Variational methods for graphical models (2006)
11. Sieg, A., Mobasher, B., Burke, R.: Web search personalization with ontological user profiles. In: Proceedings of the Sixteenth ACM Conference on Information and Knowledge Management, pp. 525–534. ACM (2007)
12. Steyvers, M., Griffiths, T.: Probabilistic topic models. In: Handbook of Latent Semantic Analysis, vol. 427(7), pp. 424–440 (2007)
13. Trestian, I., Ranjan, S., Kuzmanovic, A., Nucci, A.: Measuring serendipity: connecting people, locations and interests in a mobile 3g network. In: Proceedings of the 9th ACM SIGCOMM Conference on Internet Measurement Conference, pp. 267–279. ACM (2009)
14. Wei, X., Croft, W.B.: Lda-based document models for ad-hoc retrieval. In: Proceedings of the 29th Annual International ACM SIGIR Conference on Research and Development in Information Retrieval, pp. 178–185. ACM (2006)
15. Xie, Y., Yu, S.Z.: A large-scale hidden semi-markov model for anomaly detection on user browsing behaviors. IEEE/ACM Transactions on Networking 17(1), 54–65 (2009)
16. Zhai, K., Boyd-Graber, J., Asadi, N., Alkhouja, M.L.: Mr. lda: A flexible large scale topic modeling package using variational inference in mapreduce. In: Proceedings of the 21st International Conference on World Wide Web, pp. 879–888. ACM (2012)

Sample Size Issues in the Choice between the Best Classifier and Fusion by Trainable Combiners

Sarunas Raudys[1], Giorgio Fumera[2], Aistis Raudys[1], and Ignazio Pillai[2]

[1] Vilnius University, Faculty of Mathematics and Informatics
Naugarduko st. 24, LT-03225 Vilnius, Lithuania
{sarunas.raudys,aistis.raudys}@mif.vu.lt
[2] University of Cagliari, Dept. of Electrical and Electronic Eng.,
Piazza d'Armi, 09123 Cagliari, Italy
{fumera,pillai}@diee.unica.it
http://pralab.diee.unica.it

Abstract. We consider an open issue in the design of pattern classifiers, i.e., choosing between the best classifier among a given ensemble, and combining all the available ones using a trainable fusion rule. While the latter choice can in principle outperform the former, their actual effectiveness is affected by small sample size problems. This raises the need of investigating under which conditions one choice is better than the other one. We provide a first contribution, by deriving an analytical expressions of the expected error probability of best classifier selection, and by comparing it with the one of a well known linear fusion rule, implemented with the Fisher linear discriminant.

Keywords: data dimensionality, sample size, complexity, collective decision, expert fusion, accuracy, classification.

1 Introduction

Classifier ensembles have become a state-of-the-art approach for designing pattern classifiers (or "experts"), as an alternative to the traditional approach of using a single classification algorithm. One of the reasons is that identifying the best expert for a given task is often difficult, due to small sample size effects that arise when the data available for expert design are scarce. In this case, expert's performance cannot be reliably estimated, whereas combining all the available experts (e.g., by averaging their outputs, or by majority voting) can prevent the choice of the worst one. In principle, trainable fusion rules (e.g., weighted averaging or voting) can even outperform the best expert; in practice, since their parameters have to be estimated from a data set, also their performance is affected by small sample size. Sample size problems have been studied both theoretically (e.g., [10,5]), including the choice among fixed and trained fusion rules [1,6,7], and in applications like investment portfolio design [8]. Instead, the problem of choosing between best expert selection, and experts combination

E. Corchado et al. (Eds.): IDEAL 2014, LNCS 8669, pp. 45–52, 2014.

with trainable fusion rules, has been addressed so far only for regression tasks [3]. In this paper we address this problem in the context of classification tasks, and provide a first contribution toward the investigation of the conditions under which one choice is better than the other, in terms of the sample size, and of factors like the ensemble size, the performance of the individual experts, and their correlations. Our main contribution is the derivation of an analytical expression of the expected error probability resulting from the selection of the best expert of a given ensemble (Sect. 2). This allows us to compare it, in Sect. 4, with the error probability of a well known trainable fusion rule, the linear combination of experts' outputs by Fisher Linear Discriminant (summarized in Sect. 3), obtaining some preliminary insights and suggestions for future work.

2   Accuracy of Best Expert Selection

Problem formulation. We consider a common setting in classifier design, when m different experts $C_1, \ldots, C_m$ are available for a given classification task (obtained, e.g., using different classification algorithms), and the designer has to choose among using only the best (most accurate) individual expert, and combining all the available ones with a given trainable fusion rule. Each expert implements a decision function $C_i : \mathcal{X} \mapsto \mathcal{Y}$, where $\mathcal{X}$ is a given feature space and $\mathcal{Y}$ denotes the set of class labels. We assume that the experts have already been trained, and that a validation set V made up of n_V i.i.d. samples drawn from the (unknown) distribution $P(\mathbf{x}, y)$, $\mathbf{x} \in \mathcal{X}$, $y \in \mathcal{Y}$, is available for estimating their performance and for training the fusion rule. We assume that V is different from the training set, to avoid optimistically biased estimates. We denote as $e_{Gi} = P(C_i(\mathbf{x}) \neq y)$, $i = 1, \ldots, m$, the true ("genuine"), but unknown error probability of C_i, and with e_{Vi} the corresponding estimate computed as the error rate on V (i.e., the fraction of misclassified validation samples), which is a random variable (r.v.). The goal of this section is to derive an analytical expression of the expected error probability incurred by selecting the expert exhibiting the lowest error rate $\min_i e_{Vi}$ (i.e., the *apparent* best expert), assuming that ties are randomly broken. This is very difficult when the r.v. e_{Vi} are statistically dependent, as happens if they are computed on the same validation set. Therefore, we start by considering the simplest, albeit less realistic case in which a distinct and independent validation set of size n_V is used for each expert, which implies that also the e_{Vi}'s are independent. Then we will refine our results by developing a tractable model of their correlation.

Independent estimates of the error rate. Under this assumption, the joint probability of the e_{Vi}'s conditioned to the e_{Gi}'s is given by:

$$P(e_{V1}, \ldots, e_{Vm} | e_{G1}, \ldots, e_{Gm}) = \prod_{i=1}^{m} P(e_{Vi} | e_{Gi}) . \tag{1}$$

Let $r_i \in \{0, \ldots, n_V\}$ be a r.v. denoting the number of validation samples misclassified by C_i. Since each e_{Vi} is estimated on n_v i.i.d. samples as r_i/n_v, each of them follows a Binomial distribution:

$$P(r_i = s) = P\left(e_{Vi} = \frac{s}{n_V}\bigg|e_{Gi}\right) = \frac{n_V!}{(n_V - s)!s!}(e_{Gi})^s(1 - e_{Gi})^{n_V - s}. \tag{2}$$

To compute the error probability of the apparent best expert, we have to consider m disjoint events, denoted as $S_1, \ldots, S_m$, where S_k is the event that k different experts attain the same, smallest error rate, and thus, for $k > 1$, one of them is randomly selected as the best expert. Consider first the event S_1, and denote as C_i the expert that exhibits the smallest error rate. This means that C_i misclassifies $s \in \{0, \ldots, n_V - 1\}$ validation samples (with probability given by Eq. 2), and all the other experts C_j, $j \neq i$, misclassify more than s samples. The probability of the latter event, denoted as $P_i^{S_1}(s)$, is given by:

$$P_i^{S_1}(s) = \prod_{j=1, j\neq i}^{m} P(r_j > s), \tag{3}$$

where $P(r_j > s)$ can be computed as:

$$P(r_j > s) = P\left(e_{Vj} > \frac{s}{n_V}\bigg|e_{Gj}\right) = \sum_{s'=s+1}^{n_V} P(r_j = s'). \tag{4}$$

The probability that C_i exhibits the smallest error rate is thus given by:

$$P_i^{S_1} = \sum_{s=0}^{n_V - 1} P_i^{S_1}(s)P(r_i = s). \tag{5}$$

Consider now the event S_2, and denote as C_{i_1} and C_{i_2} the experts that exhibit the smallest error rate. Similarly to Eq. (3), the probability that all the other experts misclassify more than s samples is:

$$P_{i_1,i_2}^{S_2}(s) = \prod_{j=1, j\neq i_1, i_2}^{m} P(r_j > s), \tag{6}$$

and thus the probability that C_{i_1} and C_{i_2} exhibit the smallest error rate is:

$$P_{i_1,i_2}^{S_2} = \sum_{s=0}^{n_V - 1} P_{i_1,i_2}^{S_2}(s)P(r_{i_1} = s)P(r_{i_2} = s). \tag{7}$$

One can similarly derive the probabilities of events $S_3, \ldots, S_m$. Eventually, the expected error probability of best expert selection, denoted as P_{SEL}^{ind} ("ind" denotes the underlying independence assumption), is given by:

$$\begin{aligned} P_{SEL}^{ind} = \sum_{i=1}^{m} P_i^{S_1} e_{Gi} + \sum_{i_1=1}^{m} \sum_{i_2=1, i_2\neq i_1}^{m} P_{i_1,i_2}^{S_2} \frac{e_{Gi_1} + e_{Gi_2}}{2} + \\ \sum_{i_1=1}^{m} \sum_{i_2=1, i_2\neq i_1}^{m} \sum_{i_3=1, i_3\neq i_1, i_2}^{m} P_{i_1,i_2,i_3}^{S_3} \frac{e_{Gi_1} + e_{Gi_2} + e_{Gi_3}}{3} + \ldots \end{aligned} \tag{8}$$

When the e_{Gi}'s are assumed to be known for the purpose of a theoretical analysis, for moderate m values Eq. 8 can be computed exactly. Nevertheless, a good approximation can be obtained by considering only the first few terms of Eq. 8, depending on the sample size n_V, and on the values of $e_{G1}, \ldots, e_{Gm}$. As an example, we evaluated the accuracy of the approximation obtained using only the three terms explicitly shown in Eq. 8. To this aim we considered a two-class problem, an ensemble of $m = 7$ experts with $[e_{G1}, \ldots, e_{G7}] = [0.015, 0.02, 0.04, 0.07, 0.09, 0.11, 0.13]$, and different values of n_V, assuming that

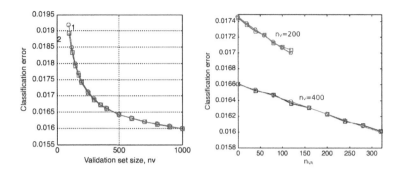

Fig. 1. Left: empirical error rate of best expert selection (squares), and theoretical value (circles) approximated with Eq. 8, as functions of n_V, for independent validation sets. Right: the same comparison for the case of dependent validation sets, for $n_V = 200$ and $n_V = 400$, as a function of n_{V_B}; theoretical values are comuputed using Eq. 9.

$n_V/2$ samples of each class are present in the validation sets. For each C_i we generated the number r_i of misclassified samples from a Binomial distribution with parameters e_{Gi} and n_V (see Eq. 2), and selected the apparent best expert. We then computed the average *true* error of the selected expert, over $100,000$ runs of the above procedure. We finally compared this value with the theoretical one of Eq. 8, considering only the first three terms. Fig. 1 (left) shows the empirical values of P_{SEL} and the approximated theoretical values, as functions of n_V. It can be seen that the approximation is very good, despite the approximation error tends to increase when the sample size n_V decreases.

Dependent estimates of the error rate. If the validation sets used for the m experts are not independent, the r.v. e_{Vi} are not independent either. In practice, the same validation set is typically used for all experts, but the analytical derivation of P_{SEL} in this case is infeasible, since all $m(m-1)/2$ pairwise correlations between the e_{Vi}'s must be taken into account. We therefore resort in this paper to a simplifying assumption which allows us to model the correlations in a tractable way, such that an analytical appoximation of P_{SEL} can be derived. The investigation of other correlation models that can lead to a better approximation is left as a future work. Let us denote as $I[C_i(\mathbf{x}) = y]$ the classification outcome (either a correct classification or a misclassification) of C_i on a given sample $(\mathbf{x}, y)$, where $I[A]$ denotes the identity function ($I[A] = 1$ if $A =$ True, and $I[A] = 0$ otherwise). We assume that the validation set of each expert is made up of two parts, V_A of size n_{V_A} and V_B of size $n_{V_B} = n_V - n_{V_A}$, such that the correlation ρ_C between the classification outcome of any pair of experts, $I[C_i(\mathbf{x}) = y]$ and $I[C_j(\mathbf{x}) = y]$, equals 1 for any $(\mathbf{x}, y) \in V_A$, and equals 0 for any $(\mathbf{x}, y) \in V_B$. In other words, any sample in V_A is either correctly classified or misclassified by all the experts, whereas the classification outcomes are independent on any sample in V_B (similarly to the case discussed above). Accordingly, the error rates of the experts on V_A are identical, and their ranking depends only on the samples in V_B. Let $e_{G_{\min}} = \min_i e_{Gi}$, and $P_{\text{SEL}}^{\text{ind}}$ denotes the error probability of the best expert selected on V_B (which can be computed as in Eq. 8, using n_{V_B} instead of n_V). The expected error probability of best expert selection under the above correlation model is:

$$P_{\text{SEL}}^{\text{corr}} = \frac{n_{V_A}}{n_V} e_{G_{\min}} + \frac{n_{V_B}}{n_V} P_{\text{SEL}}^{\text{ind}} . \tag{9}$$

To give an example of how accurately Eq. 9 approximates the error probability of best expert selection in a realistic scenario in which the same validation set is used for all experts, we consider again a two-class problem, an ensemble of $m = 7$ experts, and the same e_{Gi} values as in the example above. We first generated n_V artificial soft outputs for each expert, with identical $m(m - 1)/2$ pairwise correlations, computed the corresponding *true* error rates by thresholding the outputs at zero, and selected the best expert. The average, *true* error of the best expert selected was computed over $50,000$ runs of the above procedure, and was compared with the theoretical approximation of Eq. 9. Fig. 1 (right) shows this comparison for two validation set sizes, $n_V = 200$ and $n_V = 400$, as a function of n_{V_B}. In the considered case when the $m(m - 1)/2$ pairwise correlations are identical, the approximation turns out to be very good.

3   Accuracy of Linear Expert Fusion

To pursue our original goal, analytical expressions of the expected error probability of trainable fusion rules are needed, beside that of best expert selection. In the following we focus on the well known and widely used linear combination of soft outputs, whose expected error probability has already been analytically approximated in previous works, for the case when it is implemented with the Fisher Linear Discriminant (FLD) for two-class classification problems. Denoting as $y \in \{-1, +1\}$ the class label, and as $y_i(\mathbf{x}) \in \mathbb{R}$ the soft output of C_i, this fusion rule is defined as $f(\mathbf{x}) = \sum_{i=1}^{m} w_i y_i(\mathbf{x}) + w_0$, where $w_i \in \mathbb{R}$, $i = 0, \ldots, m$, and the decision function is $\text{sign}(f(\mathbf{x}))$. Let $\mathbf{w} = (w_1, \ldots, w_m)^\top$, the column vectors $\hat{\mu}_1$ and $\hat{\mu}_2$ denote the m-dimensional mean of the experts' soft outputs on class 1 and 2 estimated on validation samples, and $\hat{\boldsymbol{\Sigma}}_1$ and $\hat{\boldsymbol{\Sigma}}_2$ denote the estimates of the corresponding covariance matrices; moreover, let $\hat{\boldsymbol{\Sigma}} = \frac{1}{2}\left(\hat{\boldsymbol{\Sigma}}_1 + \hat{\boldsymbol{\Sigma}}_2\right)$, and let $\hat{\boldsymbol{\Sigma}}_F = (1 - \lambda)\hat{\boldsymbol{\Sigma}} + \lambda\mathbf{D}$ denote the estimate of the regularized pooled sample covariance matrix, where $\mathbf{D}$ is a diagonal matrix obtained by the diagonal of $\hat{\boldsymbol{\Sigma}}$, and $0 \leq \lambda \leq 1$ is a regularization parameter. Using the FLD, $\mathbf{w}$ and w_0 are computed as $\mathbf{w} = (\hat{\mu}_1 - \hat{\mu}_2)^\top \hat{\boldsymbol{\Sigma}}_F^{-1}$, and $w_0 = -\frac{1}{2}(\hat{\mu}_1 + \hat{\mu}_2)^\top \mathbf{w}$. The expected error probability is then [4]:

$$P_{\text{FLD}} = \Phi\left(-\frac{\delta}{2\sqrt{T_\mu T_\Sigma}}\right) , \tag{10}$$

where δ denotes the Mahalanobis distance of the two classes in the space of expert's outputs, and the scalars $T_\mu = 1 + \frac{2m}{n_V \delta^2}$ and $T_\Sigma = \frac{n_V}{n_V - m}$ account for the

Table 1. Numerical comparison between P_{FLD} (Eq. 10) and $P_{\text{SEL}}^{\text{corr}}$ (Eq. 9), for different classifier ensembles (see text for the details)

e_{G_1}	$e_{G_i}, i > 1$	m	n_V	ρ_S	ρ_C	P_{FLD}	$P_{\text{SEL}}^{\text{corr}}$
0.1	0.15	9	100	0.50	0.26	0.086	0.120
0.1	0.15	9	100	0.70	0.42	0.111	0.117
0.1	0.15	9	100	0.90	0.66	0.108	0.111
0.1	0.15	9	100	0.95	0.75	0.080	0.108
0.1	0.15	9	100	0.99	0.87	0.008	0.103
0.1	0.15	20	100	0.50	0.26	0.101	0.124
0.1	0.15	20	200	0.50	0.26	0.084	0.119
0.1	0.15	20	500	0.50	0.26	0.075	0.107
0.1	0.15	100	200	0.50	0.28	0.168	0.125
0.1	0.15	100	500	0.50	0.28	0.099	0.114
0.1	0.20	100	200	0.50	0.30	0.201	0.117
0.1	0.20	100	500	0.50	0.30	0.127	0.101

inexact, sample-based estimates of the mean vectors μ_1 and μ_2, and covariance matrix Σ, respectively. Expression 10 turns out to be the best approximation known so far of the expected error probability of the FLD combiner [11,9].

4   Analytical and Empirical Comparison: An Example

In the following we show an example of how, exploiting the above results, one can compare the performance attained by the two considered design choices (selecting the bext expert, and combining the available ones with the LDF fusion rule), aimed at understanding the conditions under which one is preferable than the other. We remind the reader that the eventual, practical outcome of such an investigation is the derivation of guidelines for the design of pattern classifiers, analogous, e.g., to the guidelines derived in [2] for the choice between simple and weighted average fusion rules in ensemble design. We first carry out a numerical comparison of the analytical expressions of $P_{\text{SEL}}^{\text{corr}}$ and P_{FLD} (respectively Eqs. 9 and 10). We then carry out an empirical comparison using artificial data, that allows one to assess the validity of the conclusions that can be drawn from an analytical comparison, also when the underlying assumptions are not satisfied.

Numerical comparison. In this example we consider different ensembles made up of $m = 9$, 20, and 100 experts, in which the *true* best expert (say, C_1) has an error probability $e_{G_1} = 0.1$, and the remaining ones have identical error probabilities $e_{G_2} = \ldots = e_{G_m} = 0.15$, and 0.20. We also consider validation set sizes $n_V = 200$, 400, 1000, different pairwise correlations ρ_S between experts' classification outputs, and different pairwise correlations ρ_S between the corresponding soft outputs, as shown in the left-most columns of Table 1. The right-most columns show the values of P_{FLD} (Eq. 10), and of $P_{\text{SEL}}^{\text{corr}}$ (Eq. 9).

The most evident fact from the example in Table 1 is that the FLD trainable combiner tends to outperform the best expert selection for smaller ensemble

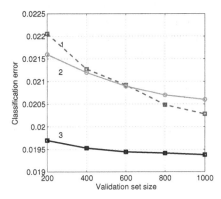

Fig. 2. Expected error probability of best expert selection (1: empirical, 2: theoretical approximation with Eq. 9), and of the FLD combiner (3: empirical), as functions of validation set size n_V

sizes m, while the opposite happens for larger m. As one can expect, increasing the validation set size n_V is beneficial for both solutions; instead, for a fixed n_V, increasing the ensemble size is detrimental, especially for the FLD trainable combiner. Instead, no clear pattern related to the effect of the correlations emerges from this simple example: a wider investigation is required to this aim.

Experiments on artificial data. In this example, we consider an ensemble of $m = 9$ experts for a two-class problem. We artificially generated their soft outputs from a Gaussian distribution with identical covariance matrices $\Sigma_1 = \Sigma_2$, such that $[e_{G_1}, \ldots, e_{G_9}] = [0.139, 0.083, 0.021, 0.019, 0.021, 0.025, 0.026, 0.027, 0.038]$, and $\rho_C = 0.3$. We therefore set $n_{V_A} = 0.3 n_V$. For the FLD combiner, the regularization parameter was set to $\lambda = 0.2$. In Fig. 2 we show the error probability of best expert selection and of the FLD combiner, as a function of validation set size, assuming that the number of validation samples of both classes is the same. The empirical values were computed as averages over 1,000 independent runs of the above procedure. The theoretical values were approximated using Eqs. 9 and 10. In this example the FLD combiner outperforms the best expert selection notably (note that in this case $n_V \gg m$). The theoretical values of P_{FLD} (Eq. 10), not reported here, turned out to be an almost exact approximation of the empirical values (curve 3 in Fig. 2). It can also be seen that the approximation of $P_{\text{SEL}}^{\text{corr}}$ by Eq. 9 exhibits a slightly lower, but still good accuracy, than in the example of Fig. 1 (right).

5   Concluding Remarks

In this paper we started the investigation of a relevant open issue related to pattern classifier design, and to multiple classifier systems in particular. It consists of investigating the conditions under which the selection of the apparent best expert, out of a given ensemble, is a better solution than combining all the available experts using a trainable fusion rule. The main contribution of this paper is the derivation of the first analytical expression known so far of the expected classification probability of the best expert selection strategy, capable of taking into account small sample size effects due to the use of a finite set of hold-out samples for performance estimation. We also developed a simple model that accounts for the pairwise correlations between experts' misclassifications, that in practical settings are unlikely to be statistically independent. Our results can be exploited in future works for a thorough analytical comparison with the expected error probability of trained combiners. In this paper we made an example of such a comparison, involving the well known and widely used FLD linear combiner. Our example pointed out the role of validation set size, ensemble size, and correlations between the experts, in determining which of the considered design choices is most effective in finite sample size situations. The development of more accurate models of the error probability of best expert selection, taking into account experts' correlations, as well as the analytical derivation of the expected error probability of other trained combiners, are relevant issues for future work.

Acknowledgment. This research was funded by grant MIP 057/2013 from Research Council of Lithuania.

References

1. Duin, R.P.W.: The Combining Classifier: To Train Or Not To Train. In: Proc. 16th Int. Con. Pattern Recognition, vol. II, pp. 765–770 (2002)
2. Fumera, G., Roli, F.: A Theoretical and Experimental Analysis of Linear Combiners for Multiple Classifier Systems. IEEE Trans. Pattern Analysis and Machine Intelligence 27(6), 942–956 (2005)
3. Rao, N.S.V.: On fusers that perform better than best sensor. IEEE Trans. Pattern Analysis and Machine Intelligence 23(8), 904–909 (2001)
4. Raudys, S.: On the amount of a priori information in designing the classification algorithm. Engineering Cybernetics N4, 168–174 (1972) (in Russian)
5. Raudys, S.: Statistical and Neural Classifiers. Springer, London (2001)
6. Raudys, S.: Experts' Boasting in Trainable Fusion Rules. IEEE Trans. Pattern Analysis and Machine Intelligence 25(9), 1178–1182 (2001)
7. Raudys, S.: Trainable fusion rules. I. Large sample size case. Neural Networks 19, 1506–1516 (2006); Trainable fusion rules. II. Small sample-size effects. Neural Networks 19, 1517–1527 (2006)
8. Raudys, S.: Portfolio of automated trading systems: Complexity and learning set size issues. IEEE Trans. Neural Networks Learning Systems 24(3), 448–459 (2013)
9. Takeshita, T., Toriwaki, J.: Experimental study of performance of pattern classifiers and the size of design samples. Patt. Rec. Lett. 16, 307–312 (1995)
10. Vapnik, V.N.: The Nature of Statistical Learning Theory. Springer, Berlin (1995)
11. Wyman, F., Young, D., Turner, D.: A comparison of asymptotic error rate expansions for the sample linear discriminant function. Patt. Rec. 23, 775–783 (1990)

On Interlinking Linked Data Sources
by Using Ontology Matching Techniques
and the Map-Reduce Framework

Ana I. Torre-Bastida[1], Esther Villar-Rodriguez[1], Javier Del Ser[1],
David Camacho[2], and Marta Gonzalez-Rodriguez[1]

[1] TECNALIA, OPTIMA Unit, E-48160 Derio, Spain
{isabel.torre,esther.villar,javier.delser,marta.gonzalez}@tecnalia.com
[2] Universidad Autonoma de Madrid, 28049 Cantoblanco, Madrid, Spain
david.camacho@uam.es

Abstract. Interlinking different data sources has become a crucial task
due to the explosion of diverse, heterogeneous information repositories
in the so-called Web of Data. In this paper an approach to extract relationships between entities existing in huge Linked Data sources is presented. Our approach hinges on the Map-Reduce processing framework
and context-based ontology matching techniques so as to discover the
maximum number of possible relationships between entities within different data sources in an computationally efficient fashion. To this end
the processing flow is composed by three Map-Reduce jobs in charge for
1) the collection of linksets between datasets; 2) context generation; and
3) construction of entity pairs and similarity computation. In order to
assess the performance of the proposed scheme an exemplifying prototype is implemented between DBpedia and LinkedMDB datasets. The
obtained results are promising and pave the way towards benchmarking the proposed interlinking procedure with other ontology matching
systems.

Keywords: Semantic Web, Linked Data Sources, Web of Data, ontology
matching, data interlinking.

1 Introduction

The Semantic Web in general and, in particular, renowned initiatives such as
Linked Open Data (LOD) are responsible for motivating and encouraging data
publication, interlinking and sharing on the Web. Nowadays the amount and size
of published RDF datasets are steadily growing in the benefit of organizations,
companies and the academia [1]. These increased data scales yields the global
conception of the Web as a huge information repository consisting of data from
different domains and their interconnections: in other words, the so-called Web
of Data. In this context, a crucial task to ensure interoperability and coherence
between different datasets is to interlink their constituent entities considering
semantically enriched relationship models and similarity measures.

E. Corchado et al. (Eds.): IDEAL 2014, LNCS 8669, pp. 53–60, 2014.

Notwithstanding the noted relevance for dataset interlinking in the Web of Data, in practice the large amount of published data makes it of utmost necessity to derive and develop new mechanisms capable of discovering interlinks in a non-supervised and computationally efficient manner. At this point two problems arise [2]: on the one hand, the heterogeneous datasets are described under different ontologies, which require to be reconciled and aligned (ontology matching and alignment). On the other hand, the interlinking procedure should identify and connect different instances that refer to the same concept in real life (also referred to as data interlinking).

More concisely, there exist nowadays a large number of links (643,753,224) between different datasets (2,122) of the LOD cloud, according to statistics collected within the LOD2 project (`http://stats.lod2.eu`). Taking these numbers into account, this paper proposes to capitalize on these linksets and existing ontology matching techniques towards improving the heterogeneous datasets integration process. However, such datasets and ontologies are currently very large, thus ontology matching techniques applied to such datasets undergo severe performance problems. Bearing this rationale in mind, this paper delves into an approach implemented onto the Map-Reduce framework to integrate heterogeneous datasets based on existing linksets and using ontology matching techniques based on context-based tools for LOD and semantic similarity between entities.

Both ontology matching and data interlinking are research paradigms that have so far grasped the attention of the research community as evinced by the related literature: for instance, Scarlet [3] is a context-based ontology matcher, whereas Silk [4] is a link discovery framework. However, to the knowledge of the authors very few attention has been paid to the derivation of systems jointly implementing both processes. This is indeed the research gap covered in this manuscript: ontology matching and interlinking operate collaboratively so as to produce a more cohesive integration between datasets, as well as to take advantage of interlinking tasks previously performed in the LOD cloud.

The rest of this paper is structured as follows: first the main concepts related to the proposed approach – LOD, ontology matching, data interlinking and the Map-Reduce framework – are reviewed in Section 2 and supported by a literature survey of the closest contributions to this research. Next, Section 3 describes the proposed scheme and its principal features. Section 4 discusses a study use case to evaluate the performance of the scheme and finally, Section 5 ends the paper by drawing some concluding remarks and outlining future research lines.

2   Background and Related Work

This section elaborates on the fundamentals of LOD and existing techniques and processes aimed at relating formal entities, such as data interlinking and ontology matching. Next the Map-Reduce framework is introduced as the selected programming and parallel processing model to accommodate the computational complexity required for matching and interlinking datasets of huge dimensionality. Finally, the state of the art on discovering relationships among datasets is surveyed and put in context of the present contribution.

2.1 Linked Data: Interlinking and Ontology Matching

Quoting Heath and Bizer in [5], the term Linked Data refers to "a set of best practices for publishing and interlinking structured data on the Web [...]". By this general definition they defined the Linked Data paradigm and provided a mechanism for building the Web of Data, which is based on semantic web technologies and may be considered as a simplified version of the so-called Semantic Web. The data model for representing interlinked data is RDF [2], where data is represented as node-and-edge-labeled directed graphs. Each *node-edge-node* relationship is called a *triple* and its components are referred to as *subject, predicate* and *object*, respectively. Subjects, predicates and objects are *resources* and every resource is labeled with a URI. Due to the heterogeneity of datasets, the processes devoted to relating entities from different data sources and establishing links between them lay at the functional core of the Web of Data. Two main processes can be unfolded from the former task, as presented in [6]:

- Ontology matching: the Web of Data covers a wide range of domains and their datasets are defined by different ontologies associated to these domains, which in general may overlap with each other. These ontologies are composed by different classes, properties and instances, denoted as entities. Therefore, when operating with several different datasets, ontology matching techniques identify which entities defined in different ontologies refer to the same object.
- Data interlinking: this process stands for the identification of the same entity across different datasets and the publication of a link between them.

As mentioned in the introduction, the scheme proposed in this paper leverages the existing linksets in the LOD datasets as a guidance for the ontology matching and data interlinking processes.

2.2 Map-Reduce Framework

Nowadays there is a trend in different communities of Computer Science towards the development of new technologies that allow analyzing efficiently very large amounts of data. In this context, new models of parallelization are emerging, as the case of the Map-Reduce framework implemented in tools such as Hadoop. Specifically, Map-Reduce [7] is a programming model and an associated implementation for parallel data processing with the objective of analyzing large volumes of data. Map-Reduce programs are executed in parallel over computing cluster: a Map-Reduce program is denoted as job, and is composed of map and reduce tasks. Summarizing, a job takes as an input a set of key/value pairs and produces a set of key/value pairs as output. As denoted by its denomination, a Map-Reduce program implements the computation as two sequential processes:

1. *Map* performs filtering and sorting (such as ordering triples by entities that appear in these into queues, one queue for each entity). In this task the input from the sources is mapped to key/value pairs.
2. *Reduce* executes a summary operation, e.g. counting the number of triples in each queue. The input key/value pairs are sorted and clustered by the key.

Map-Reduce is a powerful parallelization framework with great potential in improving the computation performance of many different problems and research fields centered in processing large amounts of raw data, such as crawled documents, web request logs and open datasets. In our case, the ontology matching is a computationally intensive process, since iterative tasks like crawling and entity matching for each possible pair of entities are complex specially when dealing with large sets of data. For instance, a similarity based matcher on an n-sized set of entities requires repeating the process $n(n-1)/2$ times [8], which unveils the complexity level of a simple similarity evaluation task and the imperative necessity of massively parallel computational frameworks like Map-Reduce.

2.3   Related Work

This section analyzes relevant literature in the field of ontology matching and data interlinking, and those in which both disciplines collaborate together. To begin with, in regard to ontology matching two different flavors can be distinguished: *content* based and context based matching. In content based matchers the ontology entities are compared by its *external* content (i.e. annotations, structure and semantics, among others). In context based, however, the contextual information of the ontologies is used to implement the matching (i.e. annotated resources, dictionaries like Wordnet, external datasets and other features alike). Our approach falls within the second family of ontology matchers, since the dataset contexts are used to discover relational paths between entities. This group we can find alternatives such as the aforementioned Scarlet [3] or Blooms [9]. As for data interlinking, the most popular tool to discover and specify links among datasets is Silk [4]. When processing large ontologies as assumed in this manuscript, the work by Zhang et al in [8] can be regarded as the closest to our technical approach due to the use of Map-Reduce and a similarity measure. However, a major drawback of their approach is that they ignore instances. Furthermore in our case the selection of entities for comparison is driven by those links previously collected from the datasets, which improves the time performance by limiting the number of pairs of entities to match.

3   Proposed Interlinking Approach

As anticipated in the introduction of this paper, the proposed interlinking scheme builds upon a parallel matching approach based on the Map-Reduce framework and exploiting existing linksets between datasets to improve the ontology matching and data interlinking processes. Figure 1 depicts the architecture of our approach, which consists of three interrelated components that permit to discover relationships between entities or *alignments*. Before this three-fold process is started, a pre-processing stage is first implemented so as to translate the two datasets – labeled with *source* and *target* – in files formatted with N-triples.

Such formatted files are first processed through a linkset collection stage where linksets are found by using a Map-Reduce job that explores both datasets

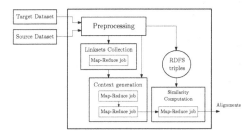

Fig. 1. Architecture overview of the proposed system

(source/target) and collects triples with `owl:sameAs` predicates. Once we have the links for each dataset, another Map-Reduce job filters triples related to each link, i.e. those in which one of the linked entities appears as a subject, predicate or object of the triple. Then a second Map-Reduce job is executed to gather related triples from the source dataset and related triples from the target dataset, grouped by the link by which they are related to each other. Once this is done, a similarity computation procedure operates on the list of related triples for each link, based on which a set of entities pairs and their relative similarity is produced. In what follows a more detailed description of each compounding processing stage is given.

3.1 Ontology Matching through Datasets Linksets

The extraction and collection of linksets and the context generation over which pairs of entities are inferred is exemplified in Figure 2. On the one hand, the input to the collection of linksets is the whole set of datasets triples (source and target), and is implemented in the form of a Map-Reduce job. In the Map task, each record corresponds to a triple statement, and the function is in charge of determining 1) whether the predicate is of the type `owl:sameAs`; and 2) if the statement links one entity of the source dataset with one of the target dataset. As an output the Map task emits a key/value pair where the subject and object concatenation is the key and the statement is the value. During the subsequent Reduce stage links are grouped by key and assigned a unique identifier.

In the second process (context generation) two Map-Reduce jobs are responsible for obtaining the list of related triples for each link. The first job is executed once for each dataset having this as input. The output of this job is a set of key/value pairs where the key is the link identifier with the linked entity of the corresponding dataset, and the list of triples related to this entity. The second is a join Map-Reduce job, where the two output files of the previous jobs are combined in an output by the first part of the key (link identifier).

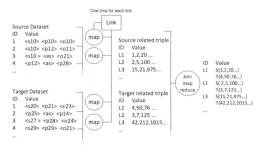

Fig. 2. Example of a construction data flow of related triples list grouped by link

3.2 Matching Entities by Similarity Measure

This section delves into the third process where similarity metrics are computed and alignments are subsequently produced. As shown in Figure 3, the input to this process is the list of related triples by link. For each of these records, all entity-pairs possibilities between source and target triples are combined in the Map task, achieving as an output an ID consisting of the concatenation of the entities/concepts to be compared (subjects [s10_ s29], predicates [p10_p29] or objects [O10_O21]) and as VALUE, the triples from which entities have been extracted. Next, the Reduce task computes its similarity metric $\vartheta(e_s, e_t)$ and outputs a list of alignments defined by their entity pairs and their associated similarity metric (e.g. s10, s29, $\vartheta(s10, s29)$). The similarity function is formulated as

$$\vartheta(e_s, e_t) = \alpha\lambda_{sin}(e_s, e_t) + (1 - \alpha)\lambda_{ont}(e_s, e_t); \tag{1}$$

where $\lambda_{sin}(\cdot, \cdot)$ denotes the syntactic similarity defined by the so-called Levenshtein string measure; $\lambda_{ont}(\cdot, \cdot)$ stands for the ontological similarity calculated based on the ontologies followed by the datasets; and α is an arbitrary parameter to balance between both composite similarities, by default set to $\alpha = 0.5$. The TF/IDF technique is used [10] for the computation of the ontological similarity, being TF the frequency of a word in the descriptor of the entity and IDF the frequency of the word in the set of source ontology and target ontology. To obtain the set of TF/IDF values, the RDF schema triples are extracted in the pre-processing phase, from which a file is created and used in the Reduce task by virtue of the Map-Reduce distributed cache feature. Based on this file the descriptor is created for each entity with the words extracted from the value of the RDFS properties (e.g. label, comment, `subclassOf`) associated to the entity. The cosine distance is selected to quantify the similarity $\vartheta(\cdot, \cdot)$ between related entities. Based on a threshold ϑ_{th} final alignments are filtered out in the form $(alg_n, [e_1, e_2, \vartheta(e_1, e_2)])$, being alg_n the unique identifier of the alignment.

4   Experiments

To assess the performance of the proposed interlinking approach a prototype has been implemented using Java 1.6 and the 2.1.2 version of MapR. Map-Reduce jobs are run over a 4-node MapR cluster, each featuring 2 processing cores running at 2 GHz with 8 GB RAM. The considered prototype deals with the integration of datasets belonging to DBpedia, a large multi-domain ontology derived from Wikipedia with approximately 232 million English RDF triples that relate more than 3.5 million entities. The objective of the experiment to be next discussed is to find alignments between this dataset and the so-called Linked Movie Database (LinkedMDB), an open semantic web data source dedicated to movie-related information with more than 6 million triples. The number of already defined links between both datasets is 13800 (DBpedia $\mapsto$ LinkedMDB) and 30354 (LinkedMDB $\mapsto$ DBpedia), which can be regarded as a baseline indicator to which to benchmark the performance of the proposed interlinking scheme.

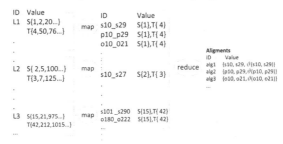

Fig. 3. Example data flow of the semantic similarity computation

Having said this, Table 1 shows the number of related entity pairs and alignments discovered as a function of different threshold values ϑ_{th} for the similarity metric. The execution time in seconds for this test has been 1928 seconds. The discussion starts by noting that for low values of ϑ_{th} the number of produced alignments results to be significantly higher than for high values of the same variable; however, it is important to stress on the fact that those alignments featuring a low similarity metric may correspond to loosely related entities. Based on this rationale, a practical approach to extending interlinks via our proposed scheme should balance between the quality of the alignments (high value of ϑ_{th}) and the coverage of the produced interlinks, which stands for the number of entities effectively reflected in the new alignment set.

Table 1. Performance results

ϑ_{th}	0.1	0.2	0.3	0.4	0.5	0.6	0.7	0.8	0.9
Alignments	405879	157361	186303	59128	2326	435	179	113	21

Finally, the alignment set obtained by the proposed approach when calculating alignments can be exemplified by the meaningful relationship found between an specific DBpedia entity and its LinkedMDB counterpart at a similarity metric equal to $\vartheta_{th} = 0.9$. This relationship is established from the following triples:

```
(linkedmdb:9736, owl:sameAs, dbpedia:Orson_Welles)
(dbpedia:The_Other_Side_of_the_Wind, dbpedia:producer, dbpedia:Orson_Welles)
(linkedmdb:46921, linkedmdb:producer, linkedmdb:9736)
(alig_15,(dbpedia:producer, linkedmdb:producer, 0.95))
(alig_16,(dbpedia:The_Other_Side_of_the_Wind, linkedmdb:46921, 0.98))
```

which, as of version 3.6, is not contained in the DBpedia linkset and hence, buttresses the utility of the proposed tool with respect to existing procedures for interlink discovery (e.g. Silk).

5   Conclusions and Future Research Lines

This paper has elaborated on a novel context-based matching and data inter-linking approach, implemented on the Map-Reduce framework. The aim of this approach is to discover relationships between entities belonging to different se-mantic data sources based on matching their respective ontologies and comput-ing similarity metrics in a computationally efficient manner. The preliminary results presented in this manuscript are promising and unveil the potentiality of the proposed scheme to ease and improve currently used ontology matching and data interlinking processes in forthcoming evolutions of the Web of Data.

Future work will be devoted to arranging further tests over datasets of in-creased size with respect to those used in this manuscript. Furthermore, a per-formance comparison of the proposed scheme will be done with state-of-the-art ontology matching (Scarlet, Blooms) and interlinking (Silk) techniques imple-mented as serial processing stages isolated from each other, which will shed light on the benefits of the semantic definition of alignments utilized in our scheme.

References

1. Bauer, F., Kaltenböck, M.: Linked Open Data: The Essentials, edition mono/monochrom, Vienna, Austria (2011)
2. Shvaiko, P., Euzenat, J.: Ontology Matching: State of the Art and Future Challenges. IEEE Transactions on Knowledge and Data Engineering 25(1), 158–176 (2013)
3. Sabou, M., d'Aquin, M., Motta, E.: SCARLET: SemantiC RelAtion DiscoveRy by Harvesting OnLinE OnTologies. In: Bechhofer, S., Hauswirth, M., Hoffmann, J., Koubarakis, M. (eds.) ESWC 2008. LNCS, vol. 5021, pp. 854–858. Springer, Heidelberg (2008)
4. Volz, J., Bizer, C., Gaedke, M., Kobilarov, G.: Silk – A Link Discovery Framework for the Web of Data. In: WWW 2009 Workshop on Linked Data on the Web (2009)
5. Heath, T., Bizer, C.: Linked Data: Evolving the Web into a Global Data Space. Synthesis Lectures on the Semantic Web: Theory and Technology, pp. 1–136 (2011)
6. Scharffe, F., Euzenat, J.: Linked Data meets Ontology Matching: Enhancing Data Linking through Ontology Alignments. In: International Conference on Knowledge Engineering and Ontology Development (KEOD), pp. 279–284 (2011)
7. Dean, J., Ghemawat, S.: MapReduce: Simplified Data Processing on Large Clus-ters. Communications of the ACM 51(1), 107–113 (2008)
8. Zhang, H., Hu, W., Qu, Y.: Constructing Virtual Documents for Ontology Match-ing Using MapReduce. In: Pan, J.Z., Chen, H., Kim, H.-G., Li, J., Wu, Z., Horrocks, I., Mizoguchi, R., Wu, Z. (eds.) JIST 2011. LNCS, vol. 7185, pp. 48–63. Springer, Heidelberg (2012)
9. Jain, P., Hitzler, P., Sheth, A.P., Verma, K., Yeh, P.Z.: Ontology Alignment for Linked Open Data. In: Patel-Schneider, P.F., Pan, Y., Hitzler, P., Mika, P., Zhang, L., Pan, J.Z., Horrocks, I., Glimm, B. (eds.) ISWC 2010, Part I. LNCS, vol. 6496, pp. 402–417. Springer, Heidelberg (2010)
10. Aizawa, A.: An Information-Theoretic Perspective of TFIDF Measures. Informa-tion Processing and Management 39(1), 45–65 (2003)

Managing Borderline and Noisy Examples in Imbalanced Classification by Combining SMOTE with Ensemble Filtering

José A. Sáez[1], Julián Luengo[2], Jerzy Stefanowski[3], and Francisco Herrera[1]

[1] Department of Computer Science and Artificial Intelligence, University of Granada,
CITIC-UGR, 18071, Granada, Spain
{smja,herrera}@decsai.ugr.es
[2] Department of Civil Engineering, LSI, University of Burgos,
09006, Burgos, Spain
jluengo@ubu.es
[3] Institute of Computing Science, Poznań,
University of Technology, ul. Piotrowo 2, 60-965 Poznań, Poland
Jerzy.Stefanowski@cs.put.poznan.pl

Abstract. Imbalance data constitutes a great difficulty for most algorithms learning classifiers. However, as recent works claim, class imbalance is not a problem in itself and performance degradation is also associated with other factors related to the distribution of the data as the presence of noisy and borderline examples in the areas surrounding class boundaries.

This contribution proposes to extend SMOTE with a noise filter called Iterative-Partitioning Filter (IPF), which can overcome these problems. The properties of this proposal are discussed in a controlled experimental study against SMOTE and its most well-known generalizations. The results show that the new proposal performs better than exiting SMOTE generalizations for all these different scenarios.

Keywords: Classification, imbalanced data, SMOTE, class noise, noise filters.

1 Introduction

Real-world classification problems from many fields present a highly imbalanced distribution of examples among the classes. In imbalance data one class is represented by a much smaller number of examples than the other classes. The minority class is usually the most interesting one [2] and thus class imbalance becomes a source of difficulty for most learning algorithms which assume an approximately balanced class distribution. As a result, minority class examples usually tend to be misclassified. Re-sampling methods modify the balance between classes by taking into account local properties of examples. Among these methods, the *Synthetic Minority Over-sampling Technique* (SMOTE) [5] is one of the most well-known.

E. Corchado et al. (Eds.): IDEAL 2014, LNCS 8669, pp. 61–68, 2014.

Even though SMOTE achieves a better distribution of the number of examples in each class it presents several drawbacks related to its *blind* oversampling. These drawbacks may aggravate even more the difficulties produced for noisy and borderline examples as some researchers have shown that the class imbalance ratio (IR) is not a problem itself. For instance in [14] the influence of noisy and borderline examples on classification performance in imbalanced datasets is experimentally studied. Borderline examples are those located either very close to the decision boundary between minority and majority classes or located in the area surrounding class boundaries where classes overlap. In [14] noisy examples are referred as those located deep inside the region of the other class.

The main aim of this contribution is to examine a new extension of SMOTE, where the IPF noise filter is combined with it resulting in SMOTE-IPF, compared to other re-sampling methods also based on generalizations of SMOTE. Its suitability for handling noisy and borderline examples in imbalanced data will be particularly evaluated as they are one of the main sources of difficulties for learning algorithms. In order to control the noise scenario, the experimental study will be carried out with special synthetic datasets containing different shapes of the minority class example boundaries and levels of borderline examples as considered in related studies [14]. After preprocessing these datasets, the performances of the classifiers built with C4.5 will be evaluated and they will be also contrasted using the proper statistical tests.

The rest of this contribution is organized as follows. Section 2 presents the imbalanced dataset problem and the motivations of our extension of SMOTE. Section 3 describes the experimental framework and includes the analysis of the experimental results. Finally Section 4 presents some concluding remarks.

2   Borderline and Noisy Examples in Imbalanced Datasets

In this section, first the problem of imbalanced datasets related to borderline and noisy examples is described in Section 2.1. Next the details of the proposed extension of SMOTE to solve these issues are given in Section 2.2.

2.1   Imbalanced Classification with Borderline and Noisy Examples

The main difficulty of imbalanced datasets is that standard classifiers tend to misclassify examples belonging to the minority class. This is because *accuracy* does not distinguish between the number of correct labels of different classes, and thus measures without these drawbacks have been proposed in the literature [10]. This paper considers the usage of the *Area Under the ROC Curve* (AUC) measure, which provides a single-number summary for the performance of learning algorithms and it is recommended in many other works on imbalanced data.

Imbalance ratio is not the only source of difficulty for classifiers. Recent works have indicated other relevant issues related to the degradation of performance. Closely related to the overlapping between classes, in [14] another interesting problem in imbalanced domains is pointed out: the higher or lower presence of examples located in the area surrounding class boundaries, which are called borderline examples. Researchers have found that misclassification often occurs near class boundaries where overlapping usually occurs as well and it is hard to find a feasible solution for it [9].

The authors in [14] showed that classifier performance degradation was strongly affected by the quantity of borderline examples and that the presence of other noisy examples located farther outside the overlapping region was also very difficult for re-sampling methods. To clarify terminology, one must distinguish (inspired by [14,13]) between safe, borderline and noisy examples:

- *Safe examples* are placed in relatively homogeneous areas with respect to the class label.
- *Borderline examples* are located in the area surrounding class boundaries, where either the minority and majority classes overlap or these examples are very close to the difficult shape of the boundary - in this case, these examples are also difficult as a small amount of the attribute noise can move them to the wrong side of the decision boundary [13].
- *Noisy examples* are individuals from one class occurring in the safe areas of the other class. According to [13] they could be treated as examples affected by class label noise.

This contribution focuses on studying the influence of noisy and borderline examples on generalizations of SMOTE considering the synthetic datasets used in [14] where safe, borderline and noisy examples are distinguished. The examples belonging to the two last groups often do not contribute to correct class prediction [11]. Therefore removing them partially or completely should improve classification performance and we propose to use noise filters to achieve this goal. We are particularly interested in ensemble filters as they are the most careful while deciding whether an example should be viewed as noise and removed.

2.2   Combining SMOTE and IPF

SMOTE is one of the most well-known and used re-sampling techniques. It generates new synthetic examples of the minority class by along the linear space between every minority example and some of its randomly selected k-nearest neighbors. One of the main shortcomings of SMOTE is overgeneralization as SMOTE blindly generalizes regions of the minority class without checking positions of the nearest examples from the majority classes. These problems may be aggravated with some distributions of data when imbalanced datasets are suffering from noisy and borderline examples. As result SMOTE is usually combined with an additional cleaning to remove noisy and borderline examples [11].

Combining SMOTE with an additional step of under-sampling aims to remove mislabeled data from the training data after the usage of SMOTE. However, they do not perform this task as well as they should in all cases. Specific and more powerful methods designed to eliminate mislabeled examples are thus required to successfully deal with noise data in imbalanced domains. Noise filters are preprocessing mechanisms designed to detect and eliminate noisy examples in the training set [3,12,15]. The result of noise elimination in preprocessing is a reduced training set which is then used as an input to a machine learning algorithm.

In addition, there are many other noise filters based on the usage of ensembles [7]. Similar techniques have been widely developed considering the building of several classifiers with the same learning algorithm [8,17]. Instead of using multiple classifiers learned from the same training set, in [8] a *Classification Filter* approach is suggested, in which the training set is partitioned into n subsets, then a set of classifiers is trained from the union of any $n - 1$ subsets; those classifiers are used to classify the examples in the excluded subset, eliminating the examples that are incorrectly classified. This paper proposes to extend SMOTE with one of this ensemble filters that has proven to work specially well: the IPF filter [12].

IPF removes noisy examples in multiple iterations until a stopping criterion is reached. The iterative process stops when, for a number of consecutive iterations k, the number of identified noisy examples in each of these iterations is less than a percentage p of the size of the original training dataset. Initially, the method starts with a set of noisy examples $A = \emptyset$. The basic steps of each iteration are the following:

1. Split the current training dataset E into n equal sized subsets.
2. Build a classifier with the C4.5 algorithm over each of these n subsets and use them to evaluate the whole current training dataset E.
3. Add to A the noisy examples identified in E using a voting scheme (consensus or majority).
4. Remove the noisy examples: $E \leftarrow E \setminus A$.

Two voting schemes can be used to identify noisy examples: consensus and majority. The former removes an example if it is misclassified by all the classifiers, whereas the latter removes an example if it is misclassified by more than half of the classifiers.

The parameter setup for the implementation of IPF used in this work has been determined experimentally in order to better fit it to the characteristics of imbalanced datasets with noisy and borderline examples once they have been preprocessed with SMOTE. More precisely, the majority scheme is used to identify the noisy examples, $n = 9$ partitions with random examples in each one are created and $k = 3$ iterations for the stop criterion and $p = 1\%$ for the percentage of removed examples are considered.

In short, the SMOTE algorithm balances the class distribution and it helps to fill in the interior of sub-parts of the minority class whereas IPF removes the noisy examples originally present in the dataset and also those created by SMOTE cleaning up the boundaries of the classes making them more regular.

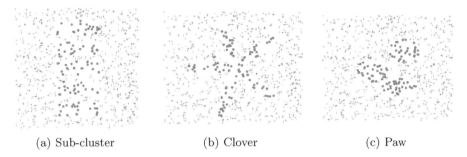

(a) Sub-cluster   (b) Clover   (c) Paw

Fig. 1. Shapes of the minority class

3   Experimental Analysis

In this section we present the details of the experimental study developed. Section 3.1 shows how the synthetic imbalanced datasets with borderline examples were built. Then Section 3.2 presents the results and their analysis.

3.1   Datasets and Re-sampling Techniques for Comparison

This paper uses the family of synthetic datasets earlier used in research on the role of borderline examples [14] and generated by special software and evaluated [16] on the role of borderline examples for different basic classifiers, such as C4.5, and re-sampling methods.

The synthetic datasets consist of two classes (the minority versus the majority class) with examples randomly and uniformly distributed in the two-dimensional real-value space. Datasets with 600 examples and $IR = 5$ and datasets with 800 examples and $IR = 7$ are considered.

Three different shapes of the minority class examples are used. *Sub-cluster* (Figure 1a) where the examples from the minority class are located inside rectangles following related works on small disjuncts. *Clover* (Figure 1b) represents a more difficult, non-linear setting, where the minority class resembles a flower with elliptic petals. In *Paw* (Figure 1c) the minority class is decomposed into 3 elliptic subregions of varying cardinalities, where two subregions are located close to each other, and the smaller sub-region is separated.

By increasing the ratio of borderline examples, i.e., the disturbance ratio, from the minority class subregions we consider 5 levels of DR: 0%, 30%, 50%, 60% and 70%. The width of the borderline overlapping areas is comparable to the width of the safe parts of sub-regions.

The proposal SMOTE-IPF is compared using these datasets against well-known re-sampling techniques to adjust the class distribution: SMOTE-ENN [1] is *filtering-based* approach based on extending SMOTE with an additional filtering, whereas SL-SMOTE [4] is based on directing the creation of the positive examples (*change-direction* methods). The aforementioned preprocessing techniques will be analyzed comparing the AUC results obtained by C4.5 for our

Table 1. AUC results obtained by C4.5 on synthetic datasets

Dataset	SMOTE	SMOTE-ENN	SL-SMOTE	SMOTE-IPF
sub-cluster $IR=5,\ DR=0$	**93.00**	90.00	91.50	92.70
sub-cluster $IR=5,\ DR=30$	81.90	81.80	82.20	**83.40**
sub-cluster $IR=5,\ DR=50$	**80.80**	77.80	80.40	77.80
sub-cluster $IR=5,\ DR=60$	78.60	77.00	78.10	**79.70**
sub-cluster $IR=5,\ DR=70$	77.50	77.00	**82.30**	80.10
sub-cluster $IR=7,\ DR=0$	94.79	93.07	90.79	**95.21**
sub-cluster $IR=7,\ DR=30$	81.57	79.64	81.71	**83.00**
sub-cluster $IR=7,\ DR=50$	78.29	75.29	**81.29**	78.57
sub-cluster $IR=7,\ DR=60$	80.21	75.71	**81.36**	79.79
sub-cluster $IR=7,\ DR=70$	**82.50**	78.21	81.21	80.93
clover $IR=5,\ DR=0$	83.50	**86.80**	85.70	85.50
clover $IR=5,\ DR=30$	83.80	83.40	81.10	**85.30**
clover $IR=5,\ DR=50$	**83.40**	81.00	83.00	82.40
clover $IR=5,\ DR=60$	80.80	80.50	78.70	**82.20**
clover $IR=5,\ DR=70$	77.20	76.10	77.10	**79.00**
clover $IR=7,\ DR=0$	87.93	87.79	88.21	**91.64**
clover $IR=7,\ DR=30$	83.64	83.14	84.36	**85.71**
clover $IR=7,\ DR=50$	81.21	**82.86**	78.71	81.93
clover $IR=7,\ DR=60$	78.79	77.71	77.21	**82.14**
clover $IR=7,\ DR=70$	78.29	76.07	78.29	**80.21**
paw $IR=5,\ DR=0$	**96.30**	94.90	92.80	94.50
paw $IR=5,\ DR=30$	84.40	**86.40**	84.50	84.90
paw $IR=5,\ DR=50$	84.60	83.30	**85.00**	84.90
paw $IR=5,\ DR=60$	81.00	81.30	82.30	**83.30**
paw $IR=5,\ DR=70$	82.80	**83.50**	82.70	83.40
paw $IR=7,\ DR=0$	93.43	93.93	92.93	**94.29**
paw $IR=7,\ DR=30$	84.29	84.93	83.50	**85.29**
paw $IR=7,\ DR=50$	85.00	84.79	84.29	**85.79**
paw $IR=7,\ DR=60$	**83.36**	80.71	83.00	82.14
paw $IR=7,\ DR=70$	82.57	80.57	84.14	**85.71**

approach against applying SMOTE alone, SMOTE-ENN and SL-SMOTE. Additionally, statistical comparisons in each of these cases will be also performed using Wilcoxon's signed ranks statistical test [6].

3.2 Results on Synthetic Datasets

Table 1 presents the AUC results obtained by C4.5 on each synthetic dataset when preprocessing with each re-sampling approach considered in this paper. The best case for each dataset is remarked in bold. From these results, we can observe that increasing DR, fixing a shape of the minority class and an IR, strongly deteriorates the performance of C4.5 in all cases. SMOTE-IPF obtains better results than the rest of the re-sampling methods in 16 of the 30 datasets considered and obtains results close to the best performances in the rest of the cases. Highest improvements of SMOTE-IPF are obtained in non-linear datasets, since 12 of the 16 overall best performance results are obtained in them.

Table 2 collects the results of applying Wilcoxon's signed ranks statistical test between SMOTE-IPF versus the re-sampling techniques. As the p-values and the sums of ranks show, SMOTE-IPF produces a statistically significant improvement in the results obtained. From these results we can conclude that SMOTE-IPF performs better than other SMOTE versions when dealing with

Table 2. Wilcoxon's test results for the comparison of SMOTE-IPF (R^+) versus SMOTE variants (R^-) from AUC results obtained by C4.5

Methods	R^+	R^-	$p_{Wilcoxon}$
SMOTE-IPF vs. SMOTE	366.0	99.0	0.0050
SMOTE-IPF vs. SMOTE-ENN	408.0	27.0	< 0.0001
SMOTE-IPF vs. SL-SMOTE	362.5	102.5	0.0070

the synthetic imbalanced datasets built with borderline examples, particularly in those with non-linear shapes of the minority class.

4   Concluding Remarks

This work proposes to extend SMOTE by a new element, the IPF noise filter, to control the presence of noisy and borderline examples and the noise introduced by the balancing between classes produced by SMOTE and to make the class boundaries more regular. Synthetic imbalanced datasets with different shapes of the minority class, imbalance ratios and levels of borderline examples have been used to analyze the suitability of this approach. All these datasets have been preprocessed with SMOTE-IPF and several well-known re-sampling techniques.

AUC values using C4.5 over the preprocessed datasets and the supporting statistical test have shown that our proposal has a notably better performance when dealing with imbalanced datasets with noisy and borderline examples, especially in the non-linear synthetic datasets. This work opens future efforts on analyzing the proposal on real datasets with class and attribute noise, as well as using more and different classifiers to check the suitability of the proposed method.

Acknowledgment. Supported by the National Project TIN2011-28488 and Regional Projects P10-TIC-06858, P11-TIC-9704, P12-TIC-2958 and NCN-2013 /11/B/ST6/00963. José A. Sáez holds an FPU scholarship.

References

1. Batista, G., Prati, R., Monard, M.: A study of the behavior of several methods for balancing machine learning training data. ACM SIGKDD Explorations Newsletter 6(1), 20–29 (2004)
2. Bhowan, U., Johnston, M., Zhang, M.: Developing new fitness functions in genetic programming for classification with unbalanced data. IEEE T. Syst. Man Cy. B 42(2), 406–421 (2012)
3. Brodley, C.E., Friedl, M.A.: Identifying Mislabeled Training Data. Journal of Artificial Intelligence Research 11, 131–167 (1999)
4. Bunkhumpornpat, C., Sinapiromsaran, K., Lursinsap, C.: Safe-Level-SMOTE: Safe-Level-Synthetic Minority Over-Sampling TEchnique for Handling the Class Imbalanced Problem. In: Theeramunkong, T., Kijsirikul, B., Cercone, N., Ho, T.-B. (eds.) PAKDD 2009. LNCS, vol. 5476, pp. 475–482. Springer, Heidelberg (2009)

5. Chawla, N.V., Bowyer, K.W., Hall, L.O., Kegelmeyer, W.P.: SMOTE: synthetic minority over-sampling technique. J. Artif. Intell. Res. 16, 321–357 (2002)
6. Demšar, J.: Statistical Comparisons of Classifiers over Multiple Data Sets. J. Mach. Learn. Res. 7, 1–30 (2006)
7. Gamberger, D., Lavrac, N., Dzeroski, S.: Noise Detection and Elimination in Data Preprocessing: experiments in medical domains. Appl. Artif. Intell. 14, 205–223 (2000)
8. Gamberger, D., Boskovic, R., Lavrac, N., Groselj, C.: Experiments With Noise Filtering in a Medical Domain. In: Proceedings of the Sixteenth International Conference on Machine Learning, pp. 143–151. Morgan Kaufmann Publishers (1999)
9. García, V., Alejo, R., Sánchez, J.S., Sotoca, J.M., Mollineda, R.A.: Combined effects of class imbalance and class overlap on instance-based classification. In: Corchado, E., Yin, H., Botti, V., Fyfe, C. (eds.) IDEAL 2006. LNCS, vol. 4224, pp. 371–378. Springer, Heidelberg (2006)
10. He, H., Garcia, E.: Learning from imbalanced data. IEEE T. Knowl. Data En. 21(9), 1263–1284 (2009)
11. Kermanidis, K.L.: The effect of borderline examples on language learning. J. Exp. Theor. Artif. In. 21, 19–42 (2009)
12. Khoshgoftaar, T.M., Rebours, P.: Improving software quality prediction by noise filtering techniques. J. Comput. Sci. Technol. 22, 387–396 (2007)
13. Kubat, M., Matwin, S.: Addresing the curse of imbalanced training sets: one-side selection. In: Proc. of the 14th Int. Conf. on Machine Learning, pp. 179–186 (1997)
14. Napierała, K., Stefanowski, J., Wilk, S.: Learning from imbalanced data in presence of noisy and borderline examples. In: Szczuka, M., Kryszkiewicz, M., Ramanna, S., Jensen, R., Hu, Q. (eds.) RSCTC 2010. LNCS, vol. 6086, pp. 158–167. Springer, Heidelberg (2010)
15. Sáez, J.A., Luengo, J., Herrera, F.: Predicting noise filtering efficacy with data complexity measures for nearest neighbor classification. Pattern Recogn. 46(1), 355–364 (2013)
16. Stefanowski, J.: Overlapping, rare examples and class decomposition in learning classifiers from imbalanced data. In: Ramanna, S., Howlett, R.J. (eds.) Emerging Paradigms in ML and Applications. SIST, vol. 13, pp. 277–306. Springer, Heidelberg (2013)
17. Verbaeten, S., Van Assche, A.: Ensemble methods for noise elimination in classification problems. In: Windeatt, T., Roli, F. (eds.) MCS 2003. LNCS, vol. 2709, pp. 317–325. Springer, Heidelberg (2003)

TweetSemMiner: A Meta-Topic Identification Model for Twitter Using Semantic Analysis

Héctor D. Menéndez*, Carlos Delgado-Calle, and David Camacho

Departamento de Ingeniería Informática. Escuela Politécnica Superior,
Universidad Autónoma de Madrid,
C/Francisco Tomás y Valiente 11, 28049 Madrid, Spain
{hector.menendez,david.camacho}@uam.es,carlos.delgadoc@estudiante.uam.es
http://aida.ii.uam.es

Abstract. The information contained in Social Networks has become increasingly important over the last few years. Inside this field, Twitter is one of the main current information sources, produced by the comments and contents that their users interchange. This information is usually noisy, however, there are some hidden patterns that can be extracted such as trends, opinions, sentiments, etc. These patterns are useful to generate users communities, which can be focused, for example, on marketing campaigns. Nevertheless, the identification process is usually blind, difficulting this information extaction. Based on this idea, this work pretends to extract relevant data from Twitter. In order to achieve this goal, we have desgined a system, called TweetSemMiner, to classify user comments (or *tweets*) using general topics (or *meta-topics*). There are several works devoted to analize social networks, however, only *Topic Detection* techniques have been applied in this context. This paper provides a new approach to the problem of classification using semantic analysis. The system has been developed focused on the detection of a single *meta-topic* and uses techniques such as Latent Semantic Analysis (LSA) combined with semantic queries in DBpedia, in order to obtain some results which can be used to analyze the effectiveness of the model. We have tested the model using real users, whose comments were subsequently evaluated to check the effectiveness of this approach.

Keywords: Twitter, tweets, meta-topic, Topic Detection, DBpedia, LSA, semantic analysis.

1 Introduction

Twitter is a microblogging service, i.e., a service that enables its users to send and publish short messages, sharing information freely. This type of service allows expressing short reviews and comments or share a higher content using URLs,

---

* This work has been partly supported by: Spanish Ministry of Science and Education under project TIN2010-19872 and Savier an Airbus Defense & Space project (FUAM-076914 y FUAM-076915).

E. Corchado et al. (Eds.): IDEAL 2014, LNCS 8669, pp. 69–76, 2014.

which makes it a source of information on countless topics, as is used by a variety of users of different ages, gender and social status. Currently, there are a large number of Social Networks available (many come and go every day), but Twitter remains since its inception and launch in 2006 (by Jack Dorsey, Evan Williams, Biz Stone and Noah Glass) as one of the most popular. For this reason, it has been chosen as a source of information for the development of this work.

Owing to each user usually has many interests (such as sports, video games, movies and music, among others) and users often follow different social roles (classmates/work, professional athletes and record companies, amongst others), the extraction of useful specific information about user preferences becomes very complex [3]. This work seeks to obtain publications from the tweets of the users followed by other user, organize them according to a general topic covering several themes (or *meta-topic*) -as could be sports, video games, movies, etc.

Thus, this work proposes a system, called TweetSemMiner, to perform a semantic analysis of the publications (or tweets) drawn from these users, for further analysis, and it will be focused on finding relationships between the *tweets* and the *meta-topic* selected (preset for this model). TweetSemMiner will be used to generate a ranking of tweets by proximity to the *meta-topic*.

This information is useful for grouping users which share common interests and investigate trends prevailing in each moment, apart from the communities that are formed from these trends, etc. TweetSemMiner can be applied to carry out more effective marketing campaigns, better organization at the enterprise level, better products targeted to their recipients, etc.

To address these issues, text mining techniques (or Data Mining) [1,9] are often used, as well as TDT (Topic Detection and Tracking) [1]. This allows information extraction from texts. Some tools such as **Relfinder** [5] and **Wikipedia Miner** [8] are based on semantic analysis. The former is a system that shows visually (using graphs) relations obtained through a dataset (often obtained from DBpedia) between different terms. The latter is formed by a set of tools developed around Wikipedia, to extract relevant information from it (search terms, semantic matching between them, disambiguation of topics, etc.).

The main goal of this work is to apply semantic techniques in order to find relationships using dictionaries -such as Wordnet [7] or DBpedia [4] combined with Latent Semantic Analysis (LSA) [2]. The system is composed by an architecture which extracts the desired Twitter tweets and semantically analyzes them to return a tweet ranking with respect to the *meta-topic* preselected by these techniques.

The problems that arise are many and varied. On the one hand, semantic analysis application has never been simple and the results are far from being completely successful (as may occur in other branches of Artificial Intelligence). On the other hand, dealing with semi-structured information imposes a preliminary meta-analysis of each topic to be treated, which, added to the above has led to the selection of a single meta-topic when developing the first model. Previous problems make extremely difficult to define a general model to represent any possible *meta-topic* in Twitter, hence it was decided to perform an initial model that uses a unique meta-topic, so that specific data that can be analyzed are obtained for later, to develop a more general model. Note that the architecture of the project is flexible so that it is only need to modify one module to suit other *meta-topics*.

The rest of the paper is structured as follows: Section 2 introduces the system architecture, Section 3 describes the experimental results, and, finally, the last section presents the conclusions, as well as the future work.

2   TweetSemMiner: The Twitter *Meta-topic* Analysis Architecture

This section describes the model developed. The main goal is to design a semantic analyzer that allows automatically to identify series of texts that could be included in a given context (*meta-topic*). Given a user name, with a public profile, *Twitter* extracts some *tweets* of each of the users which the selected user follows. After, the model will perform some operations using SPARQL queries and using LSA [2], obtaining a ranking of *tweets* ordered by their similarity to the *meta-topic* desired.

The designed model has been applied to a particular *meta-topic*: **video games**, leaving for future work extending the list of possible *meta-topics*.

2.1   TweetSemMiner Architecture

The system architecture is divided into six different modules (see Fig. 1 left). The modules of the architecture are:

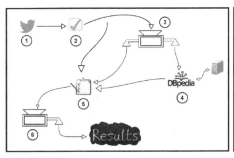

Fig. 1. System architecture (left) and Semantic tree for video-games (right) based on DBpedia categories hierarchy

- **_Tweets_ Extraction (1)**: Extract the user *tweets*. To perform this operation, the Twitter official REST API [6] has been used.
- **Text Preprocessing (2)** Take the *tweets* extracted and preprocess them by eliminating non-ASCII characters, stopwords, etc.

- **LSA analysis (3)**: LSA [2] is applied to the preprocessed data, obtaining the *keywords* and term - term matrix, which provides information about the similarity between terms. This will provide semantic relationships between those terms that will not be found in DBpedia.
- **DBpedia Queries (4)**: Collects the *keywords* obtained from the LSA module and uses them to establish relations with the *meta-topic* by querying DBpedia using SPARQL queries. These queries have been designed using a semi-structured tree for the chosen meta-topic (see Fig. 1 right). It allows to find the similarity level between the concept and the *meta-topic*. Fig. 1 (right) shows the structure associated to the "video-games" *meta-topic*.
- **Final Keywords Rating (5)**: This module is related to Algorithm 1. It complements the values obtained from DBpedia with the terms that have been semantically connected according to LSA. Thus, every keyword has a value which is taken from DBpedia or their connection with other concepts.
- **Ranking Application (6)**: Taking the evaluation obtained from the previous module, a value for each *tweet* is generated, obtaining a ranking which depends on the adequacy of the *tweet* to *meta-topic*.

Algorithm 1 Merge Stats Algorithm

---

1. Let $key\_value\_table = sparql\_key\_value$ and set $i = \emptyset$
2. **for** $row \in matrix$ **do**
3. $key = keys[i]$
4. $j = \emptyset$
5.   **for** $col \in row$ **do**
6. $sub\_key = keys[j]$
7. $aux = eval(col) * sparql\_key\_value[key]$
8.   **if** $aux > key\_value\_table[key]$ **then**
9. $key\_value\_table[key] = aux$
10. $j++$
11. $i++$

All modules developed in this architecture are generic and could be used for any *topic* except the **DBpedia Queries** module, which requires special treatment according to the *meta-topic* semi-structure (i.e., for a new *meta-topic*, a new semi-structure tree needs to be designed).

2.2   TweetSemMiner Execution

The system works as follows (see Fig. 1 left): First, the user name is introduced through the client **(1)**. This sends the information to the server, which sends the requests to *Twitter* to extract the *timeline* of users that are followed by this user. The Twitter server returns the *tweets* related to such *timelines* **(2)**. The server obtains the keywords from the *tweets* **(3)** and sends search queries to DBpedia Prefix Searcher. DBpedia returns an URL to the server to find matches between the terms and the DB. With these URLs the server sends SPARQL queries to DBpedia **(4)**, which returns the results to the server. With this information, TweetSemMiner makes an assessment on the relationships of each term to the *meta-topic* **(5)**. This information is supplemented by the term - term matrix obtained applying LSA, which evaluates those terms that are not in DBpedia. Once the rating is completed, a ranking is generated and the tweets are ordered by its relevance or connection with the *meta-topic* **(6)**. Finally, the client displays this ranking to the user.

3   Experimental Results

The experiments have been applied to 24 public Twitter profiles. All tests were conducted using "video-games" as *meta-topic*. From the final tweets ranking, 100 tweets were selected per user to perform a hand made assessment (2400 tweets). The criteria used to evaluate the tweets takes values from **0** (the tweet is not related to the *meta-topic*) to **2** (the tweet is clearly related to the *meta-topic*).

Ranking values have been truncated to be compared against the hand made assessment as follows: **Less than 0.5** (the tweet in no related to the *meta-topic*), **Between 0.5 and 0.7** (It is not clear if the tweet is related or not to the *meta-topic*) and **Greater than 0.7** (the tweet is related to the *meta-topic*).

To evaluate the behaviour of our approach, four different metrics have been used: Accuracy, Precision, Recall and F-measure metrics. Accuracy measures the total number of coincidences between the expected classification and the system classification, Precision measures when any instance is wrongly classified into a class (false positive), Recall measures when any instance is properly classified according to its class (true positive) and F-measure serves as a trade-off metric between Precision and Recall.

Values of Precision, Recall and F-measure corresponding to the evaluation of '1' class will not be deeply analyzed, since these are fuzzy values and represent uncertainty. The user name of our test subjects has been removed and replaced by the 'userXX' tag.

Table 1, shows both *tweets* tagged in each evaluation class (0, 1 and 2), and the accuracy value (Acc) for each user. This evaluation emphasizes the Accuracy in order to discover when the system obtains the best results and its potencial problems.

Users with **accuracy greater than 0.9** (user04, user07, user11, user12, user13, user19 and user20) do not speak of "video-games" either have a few *tweets* dealing with them (less than 5, see Table 1). Precision, Recall and F-measure values (Tables 2 and 3) show that F-measure values are very high (greater than 0.95, see Table 2). This means that for users who are not concerned with the *meta-topic* chosen (in this case "Video games"), our model makes very few false positives mistakes.

For users with an **Accuracy between 0.66 and 0.9** there are two relevant cases. On the one hand, for users who just talk about video games (user09 and user10) also have a high F-measure value for the '0' class (0.94 and 0.82,

Table 1. This table shows the three different evaluation classes and the Accuracy (Acc) for each user

Name	0	1	2	Acc	Name	0	1	2	Acc
user01	17	16	67	0.66	user13	94	3	3	0.919192
user02	26	3	71	0.7326733	user14	39	9	52	0.55
user03	16	14	70	0.7227723	user15	46	12	42	0.4554455
user04	96	1	3	0.98	user16	34	6	60	0.6336634
user05	77	6	17	0.6262626	user17	66	12	22	0.7373737
user06	97	1	2	0.8383838	user18	32	26	42	0.4141414
user07	97	1	2	0.989899	user19	97	1	2	0.919192
user08	11	11	78	0.8019802	user20	92	5	3	0.939394
user09	96	2	2	0.8888889	user21	78	8	14	0.7979798
user10	91	2	7	0.71	user22	31	6	63	0.6633663
user11	100	0	0	1	user23	24	10	66	0.6930693
user12	100	0	0	1	user24	64	3	33	0.5353535

Table 2. This table shows Precision(P), Recall(R) and F-measure(F) values for '0'class and user

Name	P0	R0	F0	Name	P0	R0	F0
user01	0.05882353	1	0.1111111	user13	0.9462366	1	0.9723757
user02	0.03846154	1	0.07407407	user14	0.02702703	1	0.05263158
user03	0.0625	1	0.1176471	user15	0.02222222	1	0.04347826
user04	0.9895833	0.9895833	0.9895833	user16	0.03030303	1	0.05882353
user05	0.6363636	0.9607843	0.765625	user17	0.7230769	1	0.8392857
user06	0.8645833	0.9764706	0.917127	user18	0.09375	0.75	0.1666667
user07	1	0.9896907	0.9948187	user19	0.9270833	0.9888889	0.9569892
user08	0.0909091	1	0.1666667	user20	0.989011	0.9782609	0.9836066
user09	0.9157895	0.9666667	0.9405405	user21	0.8181818	0.9545455	0.8811189
user10	0.7362637	0.9305556	0.8220859	user22	0.03333333	1	0.06451613
user11	1	1	1	user23	0.04347826	1	0.08333333
user12	1	1	1	user24	0.4285714	0.8181818	0.5625

respectively). However, it has some false positive for class '2' (see Table 3). This is because the system has not obtained enough relevant information from these *tweets* through DBpedia and, therefore, semantic relationships found by LSA are poor. This makes sense, due to when the data set is very small, it is very difficult to obtain semantic relations that are useful. On the other hand, for users that often speak of videogames (class '2' greater than 65) (user01, user02, user03, user08 and user23) several conclusions have been extracted: the first is that the F-measure value is quite high (greater than 0.8, see Table 3), so is shown that the model is properly managing the true positives. Moreover, deeply research (analyzing the profiles of those users) shows that all of them represent a company blog or video game or one of its employees. From these data we can deduce that for a user to obtain such high values on a *meta-topic*, it is normal to have a professional involvement in this issue. If we analyze the value of F-measure for the *tweets* evaluated as '0', it has

Table 3. This table shows Precision(P), Recall(R) and F-measure(F) values for '2'class and user

Name	P2	R2	F2	Name	P2	R2	F2
user01	0.9253731	0.7126437	0.8051948	user13	1	0.3333333	0.5
user02	1	0.7272727	0.8421053	user14	0.9074074	0.5697674	0.7
user03	1	0.7171717	0.8352941	user15	1	0.4444444	0.6153846
user04	0.6666667	0.6666667	0.6666667	user16	1	0.6262626	0.7701863
user05	0.8125	0.5416667	0.65	user17	0.9090909	0.5405405	0.6779661
user06	0	0	0	user18	0.725	0.3972603	0.5132743
user07	0.5	1	0.6666667	user19	1	0.3333333	0.5
user08	1	0.7979798	0.8876404	user20	0.6666667	0.3333333	0.4444444
user09	0.5	0.25	0.3333333	user21	0.7142857	0.4347826	0.5405405
user10	0.4285714	0.1111111	0.1764706	user22	1	0.6565657	0.792683
user11	-	-	-	user23	1	0.6868687	0.8143713
user12	-	-	-	user24	0.7272727	0.6	0.6575342

problems to correctly detect the true negatives (less than 0.17, see Table 2). This is the same problem that was presented to detect the F-measure value of class '2', in the first case. The problem is similar, but, in this case, the excess of information obtained from DBpedia (which causes them to properly assess the *tweet* speaking of video games) can cause others incorrectly evaluation, by creating strong semantic relationships between terms which not always necessarily have to refer to video games.

The users having a more balanced assessments obtain the **worst Accuracy values (less than 0.66)** (user14, user15, user16, user18, user22 and user24) present values of F-measure more varied. The range extends from reasonable values for detecting both positive and negative (as in the case of user24 with '0' class value equal to 0.56 and '2' class value equal to 0.65), through high detection values of positive and low detection of negatives (user14, user15, user16 and user22, see Tables 2 and 3) until very fuzzy values in the case of user22. After analyzing the profiles of those users looking for a logical explanation for such a variety of values, it has come to the following conclusion: the greater the variety of topics that address users followed by the user, the more complex *meta-topic* discriminating, due to, when the amount of data is small for each topic the results of DBpedia are poor to draw reasonable semantic relations.

From this analysis it can extract several conclusions: first, connections obtained through DBpedia are favored by the amount of relevant information found within the data; in addition, LSA is also influenced by the quality and quantity of data. Finally, when the user messages are clearer (i.e., they are clearly talking about the *meta-topic*, or not), it will be easy to classify the tweets and the results generated by the model will be better.

4   Conclusions and Future Work

This work has introduced a semantic-based system for automatic *meta-topic* identification, named TweetSemMiner, that allows to analyze data previously extracted from the Twitter social network and related to a specific *meta-topic*. To carry out the analysis, the extracted data are processed and relevant terms obtained are sent to DBpedia queries obtaining relations between the terms and the *meta-topic*, for those terms not found by DBpedia, LSA is applied. It obtains a term-term matrix, which represents the semantic relationships between terms. With the information extracted from these techniques, the tweets are ranked respect to their relevance to the *meta-topic* selected ("video-games").

Even if the model shows some limilations about the amount of data which can be extrated an processed from Twitter, the system shows that the approach is able to detect those conversations which are not delaing with the *meta-topic*, and those ones that are totally connected, having some troubles with the noisy cases.

There are also some issues that could be studied in the future; the SPARQL semi-structured tree might be improved, more *meta-topics* should be added to the system, and different clustering techniques migth be applied instead of LSA to obtain different results.

References

1. Berry, M.: Survey of Text Mining: Clustering, Classification, and Retrieval. Springer (September 2003)
2. Dumais, S.T.: Latent semantic analysis. Annual Review of Information Science and Technology 38(1), 188–230 (2004)
3. Jung, J.J.: Contextual synchronization for efficient social collaborations in enterprise computing: A case study on tweetpulse. Concurrent Engineering: R&A 21(3), 209–216 (2013)
4. Lehmann, J., Isele, R., Jakob, M., Jentzsch, A., Kontokostas, D., Mendes, P.N., Hellmann, S., Morsey, M., van Kleef, P., Auer, S., Bizer, C.: DBpedia - a large-scale, multilingual knowledge base extracted from wikipedia. Semantic Web Journal 1, 1–29 (2012)
5. Lohmann, S., Heim, P., Stegemann, T., Ziegler, J.: The relfinder user interface: Interactive exploration of relationships between objects of interest. In: Proceedings of the 14th International Conference on Intelligent User Interfaces (IUI 2010), pp. 421–422. ACM, New York (2010)
6. Makice, K.: Twitter API: Up and Running: Learn How to Build Applications with the Twitter API, 1st edn. O'Reilly Media, Inc. (April 2009)
7. Miller, G.A.: WordNet: A Lexical Database for English. Commun. ACM 38(11), 39–41 (1995)
8. Milne, D., Witten, I.H.: An open-source toolkit for mining wikipedia. Artificial Intelligence 194, 222–239 (2013)
9. Bello, G., Menéndez, H., Okazaki, S., Camacho, D.: Extracting collective trends from twitter using social-based data mining. In: Bădică, C., Nguyen, N.T., Brezovan, M. (eds.) ICCCI 2013. LNCS, vol. 8083, pp. 622–630. Springer, Heidelberg (2013)

Use of Empirical Mode Decomposition
for Classification of MRCP Based Task Parameters

Ali Hassan[1], Hassan Akhtar[1], Muhammad Junaid Khan[1], Farhan Riaz[1],
Faiza Hassan[2], Imran Niazi[3], Mads Jochumsen[3], and Kim Dremstrup[3]

[1] Department of Computer Engineering, College of Electrical and Mechanical Engineering,
National University of Sciences and Technology, Pakistan
[2] Margalla Institute of Health Sciences, University of Health Sciences, Pakistan
[3] Department of Health Science and Technology (HST), Aalborg University, Denmark

Abstract. Accurate detection and classification of force and speed intention in
Movement Related Cortical Potentials (MRCPs) over a single trial offer a great
potential for brain computer interface (BCI) based rehabilitation protocols. The
MRCP is a non-stationary and dynamic signal comprising a mixture of frequencies with high noise susceptibility. The aim of this study was to develop efficient preprocessing methods for denoising and classification of MRCPs for
variable speed and force. A proprietary dataset was cleaned using a novel application of Empirical Mode Decomposition (EMD). A combination of temporal,
frequency and time-frequency techniques was applied on data for feature extraction and classification. Feature set was analyzed for dimensionality reduction using Principal Component Analysis (PCA). Classification was performed
using simple logistic regression. A best overall classification accuracy of 77.2%
was achieved using this approach. Results provide evidence that BCI can be potentially used in tandem with bionics for neuro-rehabilitation.

Keywords: Empirical mode decomposition, principal component analysis,
movement related cortical potential, brain computer interface.

1   Introduction

With the advent of powerful signal processing techniques, usage of Electroencephalo-
graphy (EEG) offers a powerful tool for Brain Computer Interface (BCI). MRCP
signals are essentially EEG signals acquired from the surface of the skull via invasive
or non-invasive electrodes. The correlation of electrode positioning on the skull surface with corresponding limb movement has been established experimentally [1].

The MRCP is a negative shift signal having low frequency contents of up to 10 Hz
[2]. It occurs approximately 1 s before the onset of executed and imagined movement
[2] and has great potential in computer assisted motor skill learning [1,2]. The negative shift is typically followed by a rapidly rising potential. The initial negative phase
of MRCP is influenced by a multitude of factors [3,4] with significant contribution of
intended force and speed of the movement. This behavior can be processed for
determination of a forthcoming movement intention using a BCI system [5,6]. The

E. Corchado et al. (Eds.): IDEAL 2014, LNCS 8669, pp. 77–84, 2014.
© Springer International Publishing Switzerland 2014

classification of MRCP signals for decoding of intended force and speed has been performed previously [7,8,9]. These studies have utilized various frequency, time and time-frequency techniques. In the time domain, the employment of higher order statistics has been demonstrated for feature extraction, achieving better classification in non-stationary signals [10]. Also, various frequency techniques such as wavelets [11] have been used for feature extraction. Time-frequency methods have been studied for enhanced classification accuracies in [12].

Our work focuses on use of Empirical Mode Decomposition (EMD) for preprocessing of non-stationary EEG for denoising; powerful features extraction in time, frequency and time-frequency domains, and subsequent application of PCA to feature set before feeding it to a linear regression classifier.

The outline of this paper is as follows: a description of dataset is provided in Section 2. Feature extraction and classification is performed as per [9] in Section 3, which also depicts steps for employment of our novel method of denoising and feature extraction for classification improvement. Classification results are presented in Section 4 and Section 5 provides a discussion of the achieved results.

2   Dataset

2.1   Subjects

The dataset comprises 12 healthy subjects (8 males and 4 females, 27±6 years old), with each subject performing 50 trials of 4 different types of movements resulting in 12x50x4=2400 total samples. All subjects had given their informed consent and the data acquisition was duly approved by Denmark ethical committee (N-20100067).

2.2   Signal Acquisition Details

Electrodes were attached to scalp of subjects as per international 10-20 system at the standardized locations i.e. FP1, F3, F4, Fz, C3, C4, Cz, P3, P4 and Pz. The channels were sampled at 500 Hz and converted to digital format using 32 bits accuracy. Signals were band-pass filtered from 0.05 to 10 Hz through a Butterworth filter and subsequently a large Laplacian filter was applied to form a surrogate channel. All further processing was performed on surrogate channel. For further details regarding signal acquisition, readers are referred to [9].

2.3   Dataset Detail

Signals are acquired using 10 electrodes whose response was processed for movement intention detection. All subjects provided in dataset were seated in an electromagnetic interference shielded setting and had a force transducer pedal attached to right foot. Subjects were directed to undertake isometric dorsiflexions of right ankle. Maximum Voluntary Contraction (MVC) was performed in beginning of the experimental session for later use. Each subject performed 50 repetitions of cued movement for following four categories: (i) 0.5 seconds to achieve 20% MVC (referred to as Fast20 or F20) (ii) 0.5 seconds to achieve 60% MVC (referred to as Fast60 or F60) (ii) 3 seconds to achieve 20% MVC (referred to as Slow20 or S20) (iii) 3 seconds to achieve 60% MVC (referred to as Slow60 or S60). Dataset captured both force and speed of movement of all 12 subjects in 4. Moreover, all subjects were instructed to concentrate on a point on a screen located 2 meters from the seating chair in order to minimize unwanted eye movement artifacts. The recorded samples corrupted by the eye movement were rejected to make the total data samples to be equal to 2145.

3   Methodology

3.1   Initial Feature Extraction and Results

For starting reference, 6 features listed in [9] were extracted from dataset (i) point of maximum negativity, (ii) mean amplitude, (iii) and (iv) slope and intersection of a linear regression from −2 s to the point of detection or −0.1 s, (v) and (vi) slope and intersection of a linear regression from −0.5 s to point of detection or −0.1 s. Classification was performed using SVMs with Gaussian kernels.

Implementation results following the similar methodology mentioned in [9] are presented in Table 1. The results are not matched to those presented by Jochumsen et al [9], because of different methods for preprocessing and classification in Jochumsen et al [9] study, optimization was done for both feature space and Gaussian SVM kernel for classification. However, the main emphasis of current work was to explore the effect of proposed preprocessing techniques (denoising, dimension reduction and novel combination of features) on classifier performance.

Table 1. Comparison of 2-class six combinations of our implementaition versus Jochumsen et al. [9]. Our results also show sensitivity (Sen) and specifity (Spec).

Method	Jochumsen et al [9] Accuracy	Our Implementation Accuracy	Sensitivity	Specificity
F20 vs F60	0.79	0.58	0.47	0.70
F20 vs S20	0.85	0.66	0.67	0.65
F20 vs S60	0.85	0.66	0.64	0.69
F60 vs S20	0.89	0.66	0.50	0.83
F60 vs S60	0.74	0.64	0.46	0.82
S20 vs S60	0.73	0.55	0.42	0.66

3.2   Pre-processing Requirement

As per characteristics of MRCP discussed in [2], its typical waveform rapidly dips by an order of -8 to -12 μ volts and then rapidly rises to its steady state. It is a challenging task to separate this signal at single trial level from background noise. Data of 12 subjects was grouped class wise as per four classes described earlier. A mean of the data was plotted to discern the extent of average noise and shape of waveform (see figure 1). It is clear from figure that single trial MRCP signals are noisy. Moreover, Slow20 and Slow60 signals are very similar over their averages. Hence their preprocessing was required for efficient feature extraction and classification.

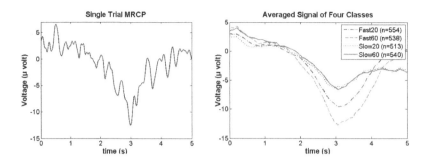

Fig. 1. Comparison of single trial MRCP signal verses average over all subjects of each task. Total number of trials added across all 12 subjects are shown as 'n' against each class.

3.3   Empirical Mode Decomposition (EMD)

EMD is an adaptive filter which requires low processing resources and separates a signal in frequency bands. The main advantage of EMD is that it performs well for both stationary and non-stationary data which makes it ideal for non-stationary MRCP signals. It is essentially a transform which decomposes a signal iteratively into its sub-bands in order of decreasing frequency contents, named as Intrinsic Mode Functions (IMF). The pseudo code for extracting IMFs from a signal is given in Algorithm 2.1.

Algorithm 2.1 Pseudo Code for Empirical Mode Decomposition (EMD)

Input: $x(t)$ % Input Signal
Output: $r(t)$ % Output IMFs
1: $r_1(t) \leftarrow x(t)$
2: $i \leftarrow 1$
3: **while** $r(t)$ is monotonic **do**
4:     Determine the max and min of $r_i(t)$
5:     Generate upper and lower envelopes $e_m(t)$ and $e_l(t)$ using cubic splines
6:     Calculate local mean $m(t) = \frac{e_m(t) + e_l(t)}{2}$
7:     Extract the detail $d_i(t) = r_i(t) - m(t)$
8:     **if** $d_i(t)$ is IMF **then**
9:         $r_{i+1}(t) = r_i(t) - d_i(t)$
10:    **else**
11:        Repeat Steps 4-7
12:    **end if**
13: **end while**

3.4   Tailored Preprocessing Method Employed

As described in data acquisition details i.e. Section 2.2; all samples of surrogate channel were bandpass filtered from 0.05 till 10 Hz. However, as shown in Figure-1, the waveform of each sample is still very noisy. In order to improve signal to noise ratio we perform denoising through employment of EMD on each sample.

As explained in Section 3.3, EMD application results in a number of IMFs whose number depends upon the frequency content present in the signal, with the last IMF containing a monotonic function. Hence IMFs are essentially adaptive filters with progressively decreasing frequencies. In order to improve the signal, a number of combinations were tried which yield the best classification accuracies with different classifiers. It was empirically determined that optimal values were achieved after first three IMFs were subtracted from the original signal. Frequency analysis of the signals revealed that average frequency content of the signals after subtraction of first three IMFs is 0.5 Hz, which is a very low frequency signal.

The resultant signal after subtraction of first three IMFs is shown in Figure-2.

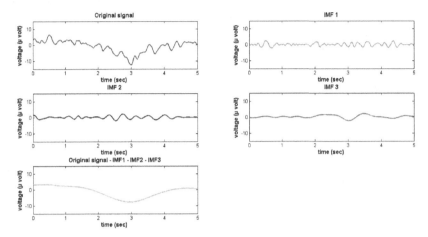

Fig. 2. An original sample of data, its first three IMFs and cleaned signal through subtraction of first three IMFs from original signal

Mean of the four class data after subtraction of first three IMFs was again plotted to determine the extent of noise reduction in data (shown in figure 3). A comparison amongst figures 1 and 3 reveals that the data is smoothened and denoised.

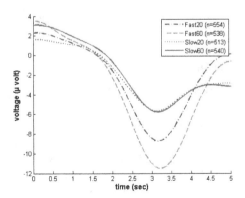

Fig. 3. Mean of four classes after denoising through subtraction of first three IMFs from original data. These waveforms contain identical number of samples 'n'.

3.5   Enhanced Feature Extraction

A feature set comprising 280 features was extracted. In the time domain, three features of skewness, kurtosis, variance and six features listed in [9] were extracted. Seven levels of wavelet decomposition were performed using code available at [14] to extract 255 frequency domain features. For time-frequency features, 16 power spectral density features were extracted as per [11], utilizing Reduced Interference (RI) distribution where the utilized kernel is $\int_{-\infty}^{+\infty} h(t)\, e^{-j2\pi v\tau}\, dt$. Classification was performed using SVM in Weka data mining toolkit [14] since same classifier was used by [9] in their work. In order to reduce data dimensionality, principal component analysis (PCA) was applied with 95% variance retention.

4   Results

4.1   Results Using Data Cleaned through IMF Subtraction

Two-class classification for six combinations of four available classes was performed using cleaned data (through first three IMF subtraction). However it is apparent that classification accuracy increased nominally.

Table 2. Classification accuracies utilizing 280 feature set with and without PCA, using SVM. Classification accuracies versus six combinations of 2-class problem along with their respective specifity (Spec) and sensitivity (Sen) figures are listed below. Last column shows percentage change between baseline and our results using clean data with PCA.

Method	Without PCA for cleaned data			With PCA for cleaned data			% change
	Accuracy	Sen	Spec	Accuracy	Sen	Spec	
F20 vs F60	0.56	0.35	0.78	0.56	0.50	0.62	-3.9
F20 vs S20	0.64	0.76	0.51	0.64	0.64	0.59	-1.9
F20 vs S60	0.64	0.59	0.65	0.66	0.71	0.59	-1.4
F60 vs S20	0.67	0.76	0.58	0.68	0.71	0.63	2.6
F60 vs S60	0.65	0.60	0.70	0.65	0.64	0.66	1.7
S20 vs S60	0.50	0.40	0.60	0.55	0.40	0.60	0.9

4.2   Further Analysis of Dataset

Dataset was further analyzed through visual plotting of all samples and it was found out that more than 30% of samples were out of conformance with a typical waveform associated with MRCP signals. These samples were attributed towards either incorrect detection of the start of MRCP signal during data recording or problematic electrodes. The entire dataset was visually inspected and almost 30% samples in total nonconformance with a typical MRCP signal were rejected. Feature extraction was again performed to formulate a 280 feature vector and various classifiers were applied on data. It was found out that simple logistic regression (SLR) provided best classification accuracy when coupled with PCA, whose results are shown in Table 3.

Table 3. Classification accuracies utilizing 280 feature set with and without PCA for cleaned data. Classification accuracies versus six combinations of 2-class problem along with their respective specifity (Spec) and sensitivity (Sen) figures. Last column shows percentage change between baseline [9] and our results using clean data with PCA and SLR.

Method	Without PCA using SLR			With PCA using SLR			% change wrt Jochumsen et al [9]
	Accuracy	Sen	Spec	Accuracy	Sen	Spec	
F20 vs F60	0.61	0.62	0.60	0.61	0.67	0.54	3.8
F20 vs S20	0.75	0.74	0.76	0.77	0.76	0.78	16.4
F20 vs S60	0.72	0.75	0.68	0.70	0.79	0.57	5.3
F60 vs S20	0.71	0.80	0.62	0.74	0.81	0.66	10.8
F60 vs S60	0.74	0.81	0.63	0.76	0.83	0.66	19.2
S20 vs S60	0.59	0.89	0.17	0.59	0.99	0.01	9.0

As per figure 3, Slow20 and Slow60 signals average out to be almost similar, therefore this corroborates the results shown in table 3 that their classification accuracy versus each other is poorest. However, Fast20 and Fast60 classes can be clearly discerned from both Slow20 and Slow60. Therefore, we propose that Fast20 and Fast60 be classified against combined slow movement with results in table 4.

Table 4. Percent Classification accuracies utilizing 280 feature set with and without PCA for dataset with 30% dropped samples. Classification accuracies versus combinations of 2-class problem annotating utilized classifier are listed along with their respective specifity (Spec) and sensitivity (Sen) figures.

Method	Without PCA for SLR			With PCA for SLR		
	Accuracy	Sen	Spec	Accuracy	Sen	Spec
F20 vs Slow	0.74	0.61	0.82	0.74	0.60	0.83
F60 vs Slow	0.76	0.58	0.88	0.77	0.61	0.88
Fast vs Slow	0.73	0.83	0.60	0.73	0.84	0.59

5   Discussion

Results of table 3 and 4 directly corroborate the visual evidence regarding four classes. Slow20 and Slow 60 give worst classification accuracy of 60.5% whereas greatest difference exists between Fast60 and Slow 60, giving a classification accuracy of 76.2%. However, overall best classification accuracy of 77.2% achieved through Fast60 versus Slow exists because combined total number of training samples for Slow20 and Slow60 train the classifier better.

6   Future Work

Same methodology can be applied for denoising of dataset with larger number of samples which will improve efficacy of feature set for better classification. Moreover, entire scheme can also be improved and implemented in hardware for providing classification in a single trial for stroke patients' rehabilitation with help of bionics.

References

1. Nascimento, O., Farina, D.: Movement-Related Cortical Potentials Allow Discrimination of Rate of Torque Development in Imaginary Isometric Plantar Flexion. IEEE Transactions on Biomedical Engineering 55(11) (2008)
2. Niazi, I., Jang, N., Tiberghein, O., Nielsen, F., Dremstrup, K., Farina, D.: Detection of Movement Intention from Single-Trial Movement-Related Cortical Potentials. Journal of Neural Engineering 8, 066009 (2011)
3. Shibasaki, H., Hallett, M.: What is the Bereitschaftspotential? Clinical Neurophysiology 117, 2341–2356 (2006)
4. Jahanshahi, M., Hallett, M.: The Bereitschaftspotential: Movement-related Cortical Potentials, 1st edn. Springer, Berlin (2003)
5. Wolpaw, J., Birbaumer, N., McFarland, D., Pfurtscheller, G., Vaughan, T.: Brain-computer interfaces for communication and control. Clinical Neurophysiology 113, 767–791 (2002)
6. Daly, J., Wolpaw, J.: Brain-computer interfaces in neurological rehabilitation. Lancet Neurology 7, 1032–1043 (2008)
7. Farina, D., Nascimento, O., Lucas, M., Doncarli, C.: Optimization of wavelets for classification of movement-related cortical potentials generated by variation of force-related parameters. Journal of Neuroscience Methods 162(1-2), 357–363 (2007)
8. Gu, Y., Nascimento, O., Lucas, M., Farina, D.: Identification of task parameters from movement-related cortical potentials. Medical & Biological Engineering & Computing 47(12), 1257–1264 (2009)
9. Jochumsen, M., Niazi, I., Mrachacz-Kersting, N., Farina, D., Dremstrup, K.: Detection and classification of movement-related cortical potentials associated with task force and speed. Journal of Neural Engineering 10(5), 056015 (2013)
10. Shafiul Alam, S., Bhuiyan, M.: Detection of Seizure and Epilepsy Using Higher Order Statistics in the EMD Domain. IEEE Journal of Biomedical and Health Informatics 17(2) (2013)
11. Robinson, N., Vinod, A., KengAng, K., Tee, K., Guan, C.: EEG-Based Classification of Fast and Slow Hand Movements Using Wavelet-CSP Algorithm. IEEE Transactions on Biomedical Engineering 60(8) (2013)
12. Tzallas, A., Tsipouras, M., Fotiadis, D.: Epileptic Seizure Detection in EEGs Using Time–Frequency Analysis. IEEE Transactions on Information Technology in Biomedicine 13(5) (2009)
13. http://www.mathworks.com/matlabcentral/fileexchange/33146-feature-extraction-using-multisignal-wavelet-packet-decomposition/content/getmswpfeatV00.m
14. Hall, M., Frank, E., Holmes, G., Pfahringer, B., Reutemann, P., Witten, I.: The WEKA Data Mining Software: An Update. SIGKDD Explorations 11(1) (2009)

Diversified Random Forests
Using Random Subspaces

Khaled Fawagreh, Mohamed Medhat Gaber, and Eyad Elyan

IDEAS, School of Computing Science and Digital Medial, Robert Gordon University,
Garthdee Road, Aberdeen, AB10 7GJ, UK

Abstract. Random Forest is an ensemble learning method used for classification and regression. In such an ensemble, multiple classifiers are used where each classifier casts one vote for its predicted class label. Majority voting is then used to determine the class label for unlabelled instances. Since it has been proven empirically that ensembles tend to yield better results when there is a significant diversity among the constituent models, many extensions were developed during the past decade that aim at inducing some diversity in the constituent models in order to improve the performance of Random Forests in terms of both speed and accuracy. In this paper, we propose a method to promote Random Forest diversity by using randomly selected subspaces, giving a weight to each subspace according to its predictive power, and using this weight in majority voting. Experimental study on 15 real datasets showed favourable results, demonstrating the potential of the proposed method.

1 Introduction

Random Forest (RF) is an ensemble learning technique used for classification and regression. Ensemble learning is a supervised machine learning paradigm where multiple models are used to solve the same problem [22]. Since single classifier systems have limited predictive performance [27] [22] [18] [24], ensemble classification was developed to overcome this limitation [22] [18] [24], and thus boosting the accuracy of classification. In such an ensemble, multiple classifiers are used. In its basic mechanism, majority voting is then used to determine the class label for unlabelled instances where each classifier in the ensemble is asked to predict the class label of the instance being considered. Once all the classifiers have been queried, the class that receives the greatest number of votes is returned as the final decision of the ensemble.

Three widely used ensemble approaches could be identified, namely, boosting, bagging, and stacking. Boosting is an incremental process of building a sequence of classifiers, where each classifier works on the incorrectly classified instances of the previous one in the sequence. AdaBoost [12] is the representative of this class of techniques. However, AdaBoost is prone to overfitting. The other class of ensemble approaches is the bootstrap aggregating (bagging) [7]. Bagging involves building each classifier in the ensemble using a randomly drawn sample of the data, having each classifier giving an equal vote when labelling unlabelled

E. Corchado et al. (Eds.): IDEAL 2014, LNCS 8669, pp. 85–92, 2014.

instances. Bagging is known to be more robust than boosting against model over-fitting. Random Forest (RF) is the main representative of bagging [8]. Stacking (sometimes called stacked generalisation) extends the cross-validation technique that partitions the dataset into a held-in data set and a held-out data set; training the models on the held-in data; and then choosing whichever of those trained models performs best on the held-out data. Instead of choosing among the models, stacking combines them, thereby typically getting performance better than any single one of the trained models [26].

The ensemble method that is relevant to our work in this paper is RF. RF has been proved to be the state-of-the-art ensemble classification technique. Since it has been proven empirically that ensembles tend to yield better results when there is a significant diversity among the models [16] [9] [1] [25], this paper investigates how to inject more diversity by using random subspaces to construct an RF that is divided into a number of sub-forests, where each of which is based on a random subspace.

This paper is organised as follows. First, an overview of RFs is presented in Section 2. This is followed by Section 3 where our method is presented. Experimental results demonstrating the superiority of the proposed extension over the standard RF is detailed in Section 4. In Section 5, we describe related work. The paper is then concluded with a summary and pointers to future directions in Section 6.

2   Random Forests: An Overview

Random Forest is an ensemble learning method used for classification and regression. Developed by Breiman [8], the method combines Breiman's bagging approach [7], and the random selection of features, introduced independently by Ho [14] [15] and Amit and Geman [2], in order to construct a collection of decision trees with controlled variation. Using bagging, each decision tree in the ensemble is constructed using a sample with replacement from the training data. Statistically, the sample is likely to have about 64% of instances appearing at least once in the sample. Instances in the sample are referred to as in-bag instances, and the remaining instances (about 36%), are referred to as out-of-bag instances. Each tree in the ensemble acts as a base classifier to determine the class label of an unlabeled instance. During the construction of the individual trees in the RF, randomisation is also applied when selecting the best node to split on. Typically, this is equal to $\sqrt{F}$ where F is the number of features in the dataset.

Breiman [8] introduced additional randomness during the construction of decision trees using the classification and regression trees (CART) technique. Using this technique, the subset of features selected in each interior node is evaluated with the Gini index heuristics. The feature with the highest Gini index is chosen as the split feature in that node. Gini index has been introduced by Breiman et al. [19]. However, it has been first introduced by the Italian statistician Corrado Gini in 1912. The index is a function that is used to measure the impurity of data, i.e., how uncertain we are if an event will occur. In classification, this event would be the determination of the class label [4].

In the basic RF technique [8], it was shown that the RF error rate depends on *correlation* and *strength*. Increasing the correlation between any two trees in the RF increases the forest error rate. A tree with a low error rate is a strong classifier. Increasing the strength of the individual trees decreases the RF error rate. Such findings seem to be consistent with a study made by Bernard et al. [5] which showed that the error rate statistically decreases by jointly maximising the strength and minimising the correlation.

Key advantages of RF over its AdaBoost counterpart are robustness to noise and overfitting [8] [17] [23] [6]. In the original paper about RF, [8] outlined four other advantages of an RF. First, its accuracy is as good as Adaboost and sometimes better. Second, it is faster than bagging or boosting. Third, it gives useful internal estimates of error, strength, correlation and variable importance. Last, it is simple and easily parallelised.

3   Diversified Random Forests

To some extent, the standard RF already applies some diversity to the classifiers being built during the construction of the RF. In a nutshell, there are two levels of diversity being applied. The first level is when each decision tree is constructed using sampling with replacement from the training data. The samples are likely to have some diversity among each other as they were drawn at random. The second level is achieved by randomisation which is applied when selecting the best node to split on.

The ultimate objective of this paper is to inject more diversity in an RF. From the training set, we create a number of subspaces. The number of subspaces is determined as follows.

$$Subspaces = \alpha \times Z \qquad (1)$$

where α denotes the subspace factor such that $0 < \alpha \le 1$, and Z is the size of the diversified RF to be created. Each subspace will contain a fixed randomised subset of the total number of features and will correspond to a sub-forest. A projected training dataset will be created for each subspace and will be used to create the trees in the corresponding sub-forest. The number of trees in each sub-forest is given by the equation

$$Trees = \frac{Z}{Subspaces} \qquad (2)$$

We will refer to the resulting forest as the Diversified Random Forest (DRF) as shown in Figure 1.

A weight is then assigned to each projected training dataset using the *Absolute Predictive Power (APP)* given by Cuzzocrea et al. [11]. Given a dataset S, the APP is defined by the following equation

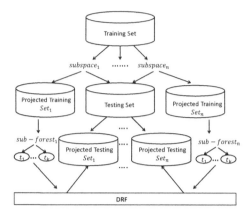

Fig. 1. Diversified Random Forest

$$APP(S) = \frac{1}{|Att(S)|} \times \sum_{A \in Att(S)} \frac{I(S, A)}{E(S)} \tag{3}$$

where $E(S)$ is the entropy of a given dataset S having K instances and $I(S, A)$ is the information gain of a given attribute A in a dataset S. $E(S)$ is a measure of the uncertainty in a random variable and is given by the following equation

$$E(S) = \sum_{i=1}^{K} -p_i(x_i) \log_2 p_i(x_i) \tag{4}$$

where x_i refers to a generic instance of S and p_{x_i} denotes the probability that the instance x_i occurs in S. $I(S, A)$ is given by

$$I(S, A) = E(S) - \sum_{v \in Val(A)} \left(\frac{|S_v|}{|S|} \right) E(S_v) \tag{5}$$

where $E(S)$ denotes the entropy of S, $Val(A)$ denotes the set of possible values for A, S_v refers to the subset of S for which A has the value v, and $E(S_v)$ denotes the entropy of S_v.

This weight will be inherited by the corresponding sub-forest and will be used in the voting process. This means that the standard voting technique currently used in the standard RF is going to be replaced by a weighted voting technique. Algorithm 1 summarises the main steps involved in the construction of an DRF.

To measure diversity, we can use the entropy in equation 4 to find the average entropy over a validation set V having K instances as given in the following equation:

$$Diversity(V) = \frac{E(V)}{K} \tag{6}$$

Algorithm 1. Diversified Random Forest Algorithm

{User Settings}
input Z: the desired size of the DRF to be created
input S: the number of features in the sub-forest
input α: the subspace factor
{Process}
Create an empty vector $\overrightarrow{DRF}$
Create an empty vector $\overrightarrow{Weights}$
Using Equation 1, create N subspaces each containing 70% of the features chosen at random from all features F
For each subspace i in the previous step, create a projected training set TR_i
Repeat the previous step to create a projected testing set TS_i for each subspace (this is required for the testing phase)
Using Equation 3, assign a weight to each projected training set and add this weight to $\overrightarrow{Weights}$
Using Equation 2, determine the number of trees in each subspace:
$$treesPerSubspace \Leftarrow Z/N$$
for $i = 1 \rightarrow N$ **do**
  for $j = 1 \rightarrow treesPerSubspace$ **do**
    Create an empty tree T_j
    repeat
      Sample S out of all features in the corresponding projected training set TR_i using Bootstrap sampling
      Create a vector of the S features $\overrightarrow{F_S}$
      Find Best Split Feature $B(\overrightarrow{F_S})$
      Create a New Node using $B(\overrightarrow{F_S})$ in T_j
    until No More Instances To Split On
    Add T_j to the $\overrightarrow{DRF}$
  end for
end for
{Output}
A vector of trees $\overrightarrow{DRF}$
A vector of weights $\overrightarrow{Weights}$ (to replace standard voting by weighted voting during the testing phase)

Two examples will be given to clarify this equation. Assume we have n validation instances and an DRF of 500 trees where each tree predicts the class label. In the first example, assume we have instance x. Since we have 500 trees, we will have $(c_1, c_2, .., c_{500})$ class labels for this instance. Assuming that all trees agree on one class, the entropy in equation 4 yields -1 log 1 = 0. If we use equation 6 to find the average for the n testing instances, we can measure the diversity. In the second example, let us consider another instance y, where half of the trees chose one class label and the other half chose another one. For such an instance, the entropy in equation 4 yields -0.5 log 0.5 + -0.5 log 0.5 = 1. We can see that when the trees diversified, the entropy increased. To summarize what was demonstrated in the previous examples, the entropy increases if the trees disagree, therefore, the higher the disagreement, the higher the entropy.

4 Experimental Study

RF and DRF were tested on 15 real datasets from the UCI repository [3] and both had a size of 500 trees. For our DRF, we used a subsapce factor of 2%, by equation 1, this produced 10 subspaces and hence 10 sub-forests. We used a random 70% of the features for each subspace. By equation 2, each sub-forest contained 50 trees. Results of the experiment, based on the average of 10 runs, are depicted in Table 1 below (we have highlighted in boldface all the datasets where DRF outperformed RF).

Table 1 shows that DRF has outperformed RF on the majority of the datasets. It is worth noting that our technique has shown exceptionally better performance on the medical datasets, namely, *diabetes* and *breast-cancer*, both with 9 and 10 features respectively. It is worth further investigating this interesting outcome.

As shown in Table 1, in the few cases that RF outperformed DRF, the difference was by a very small fraction ranging from 0.01% to 2.22%. On the other hand, DRF has performed better with a difference that ranged from 0.18% to 5.63%. Applying the paired t-test on the 8 cases that DRF outperformed RF has shown statistical significance of the results with *p-value* = 0.01772.

Table 1. Performance Comparison of RF & DRF

Dataset	Number of Features	RF	DRF
soybean	36	77.543106%	75.43103%
eucalyptus	20	20.0%	**22.76%**
car	7	62.108845%	59.93197%
credit	21	75.97059%	**76.147064%**
sonar	61	0.7042253%	**0.9859154%**
white-clover	32	63.333332%	**63.809525%**
diabetes	9	73.71648%	**79.34865%**
glass	10	12.328766%	**17.123287%**
vehicle	19	73.81944%	73.40277%
vote	17	97.97298%	97.36487%
audit	13	96.30882%	96.29411%
breast-cancer	10	71.855675%	**75.46391%**
pasture	23	41.666668%	40.833336%
squash-stored	25	55.555553%	53.333344%
squash-unstored	24	60.000004%	**61.11111%**

5 Related Work

The random subspace method was initially introduced by Ho [15]. However, and unlike the approach proposed in this paper, the subspaces were not weighted according to their predictive power. The method became so popular that many researchers adopted it to enhance ensemble techniques. Breiman [8], for example, combined his bagging approach [7], and the random subspace method, in order to construct an RF containing a collection of decision trees with controlled variation. Cai et al. [10] proposed a weighted subspace approach to improve performance but their approach was tailored for bagging ensembles [7]. García-Pedrajas and Ortiz-Boyer [13] presented a novel approach to improve boosting ensembles that combined the methods of random subspace and AdaBoost [12].

Opitz [20] developed an ensemble feature selection approach that is based on genetic algorithms. The approach not only finds features relevant to the learning task and learning algorithm, but also finds a set of feature subsets that will promote disagreement among the ensemble's classifiers. The proposed approach demonstrated improved performance over the popular and powerful ensemble approaches of AdaBoost and bagging.

Panov and Džeroski [21] presented a new method for constructing ensembles of classifiers that combines both bagging and the random subspace method with the added advantage of being applicable to any base-level algorithm without the need to randomise the latter.

6   Conclusion and Future Work

In this paper, we have proposed an extension to Random Forest called *Diversified Random Forest*. The new proposed forest was constructed using sub-forests where each of which was built from a randomly selected subspace whose corresponding projected training dataset was later assigned a weight. We have used the *Absolute Predictive Power (APP)* to weigh each projected dataset. As demonstrated in Section 4, the extended RF outperformed the standard RF on the majority of the datasets.

In our experiments, we have used a subspace factor of 2%, a size of 500 trees for the forest to be created, and 70% of the features in each subspace. In the future, we will attempt different values for these parameters. Generally, we expect our extended forest to work well with higher dimensional datasets since the more features there are, the more likely the subspaces are going to be diverse.

References

1. Adeva, J.J.G., Beresi, U., Calvo, R.: Accuracy and diversity in ensembles of text categorisers. CLEI Electronic Journal 9(1) (2005)
2. Amit, Y., Geman, D.: Shape quantization and recognition with randomized trees. Neural Computation 9(7), 1545–1588 (1997)
3. Bache, K., Lichman, M.: UCI machine learning repository (2013)
4. Bader-El-Den, M., Gaber, M.: Garf: towards self-optimised random forests. In: Huang, T., Zeng, Z., Li, C., Leung, C.S. (eds.) ICONIP 2012, Part II. LNCS, vol. 7664, pp. 506–515. Springer, Heidelberg (2012)
5. Bernard, S., Heutte, L., Adam, S.: A study of strength and correlation in random forests. In: Huang, D.-S., McGinnity, M., Heutte, L., Zhang, X.-P. (eds.) ICIC 2010. CCIS, vol. 93, pp. 186–191. Springer, Heidelberg (2010)
6. Boinee, P., De Angelis, A., Foresti, G.L.: Meta random forests. International Journal of Computationnal Intelligence 2(3), 138–147 (2005)

7. Breiman, L.: Bagging predictors. Machine Learning 24(2), 123–140 (1996)
8. Breiman, L.: Random forests. Machine Learning 45(1), 5–32 (2001)
9. Brown, G., Wyatt, J., Harris, R., Yao, X.: Diversity creation methods: a survey and categorisation. Information Fusion 6(1), 5–20 (2005)
10. Cai, Q.-T., Peng, C.-Y., Zhang, C.-S.: A weighted subspace approach for improving bagging performance. In: IEEE International Conference on Acoustics, Speech and Signal Processing, ICASSP 2008, pp. 3341–3344. IEEE (2008)
11. Cuzzocrea, A., Francis, S.L., Gaber, M.M.: An information-theoretic approach for setting the optimal number of decision trees in random forests. In: 2013 IEEE International Conference on Systems, Man, and Cybernetics (SMC), pp. 1013–1019. IEEE (2013)
12. Freund, Y., Schapire, R.E.: A decision-theoretic generalization of on-line learning and an application to boosting. Journal of Computer and System Sciences 55(1), 119–139 (1997)
13. García-Pedrajas, N., Ortiz-Boyer, D.: Boosting random subspace method. Neural Networks 21(9), 1344–1362 (2008)
14. Ho, T.K.: Random decision forests. In: Proceedings of the Third International Conference on Document Analysis and Recognition, vol. 1, pp. 278–282. IEEE (1995)
15. Ho, T.K.: The random subspace method for constructing decision forests. IEEE Transactions on Pattern Analysis and Machine Intelligence 20(8), 832–844 (1998)
16. Kuncheva, L.I., Whitaker, C.J.: Measures of diversity in classifier ensembles and their relationship with the ensemble accuracy. Machine Learning 51(2), 181–207 (2003)
17. Liaw, A., Wiener, M.: Classification and regression by randomforest. R News 2(3), 18–22 (2002)
18. Maclin, R., Opitz, D.: Popular ensemble methods: An empirical study. arXiv preprint arXiv:1106.0257 (2011)
19. Breiman, L., Friedman, J.H., Olshen, R.A., Stone, C.J.: Classification and regression trees. Wadsworth International Group (1984)
20. Opitz, D.W.: Feature selection for ensembles. In: AAAI/IAAI, pp. 379–384 (1999)
21. Panov, P., Džeroski, S.: Combining bagging and random subspaces to create better ensembles. Springer (2007)
22. Polikar, R.: Ensemble based systems in decision making. IEEE Circuits and Systems Magazine 6(3), 21–45 (2006)
23. Robnik-Šikonja, M.: Improving random forests. In: Boulicaut, J.-F., Esposito, F., Giannotti, F., Pedreschi, D. (eds.) ECML 2004. LNCS (LNAI), vol. 3201, pp. 359–370. Springer, Heidelberg (2004)
24. Rokach, L.: Ensemble-based classifiers. Artificial Intelligence Review 33(1-2), 1–39 (2010)
25. Tang, K., Suganthan, P.N., Yao, X.: An analysis of diversity measures. Machine Learning 65(1), 247–271 (2006)
26. Wolpert, D.H.: Stacked generalization. Neural Networks 5(2), 241–259 (1992)
27. Yan, W., Goebel, K.F.: Designing classifier ensembles with constrained performance requirements. In: Defense and Security, pp. 59–68. International Society for Optics and Photonics (2004)

Fast Frequent Pattern Detection Using Prime Numbers

Konstantinos F. Xylogiannopoulos[1], Omar Addam[1],
Panagiotis Karampelas[2,4], and Reda Alhajj[1,3]

[1] Dept. of Computer Science, University of Calgary, Calgary, Alberta, Canada
[2] Dept. of Information Technology, Hellenic American University, Manchester, NH, USA
[3] Dept. of Computer Science, Global University, Beirut, Lebanon
[4] Dept. of Informatics & Computers, Hellenic Air Force Academy, Attica, Greece

Abstract. Finding all frequent itemsets (patterns) in a given database is a challenging process that in general consumes time and space. Time is measured in terms of the number of database scans required to produce all frequent itemsets. Space is consumed by the number of potential frequent itemsets which will end up classified as not frequent. To overcome both limitations, namely space and time, we propose a novel approach for generating all possible frequent itemsets by introducing a new representation of items into groups of four items and within each group, items are assigned one of four prime numbers, namely 2, 3, 5, and 7. The reported results demonstrate the applicability and effectiveness of the proposed approach. Our approach satisfies scalability in terms of number of transactions and number of items.

Keywords: frequent pattern mining, association rules, data mining, prime representation.

1 Introduction

Data mining is the process of discovering and predicting hidden and unknown knowledge by analyzing known databases. It is different from querying in the sense that querying is a retrieval process, while mining is a discovery process. Data mining has received considerable attention over the past decades and a number of effective mining techniques already exist. They are well investigated and documented in the literature. However, existing and emerging applications of data mining motivated the development of new techniques and the extension of existing ones to adapt to the change. Data mining has several applications, including market analysis, pattern recognition, gene expression data analysis, spatial data analysis, among others.

Despite the availability of a large number of algorithms described in the literature, we argue there is still room for improvement in the process employed to determine frequent itemsets in a database of transactions. Consequently, in this paper we propose a new approach for producing frequent itemsets by using a different representation of the input data. We mainly divide items into groups of four items each; then we employ the prime numbers 2, 3, 5, and 7 to represent items in each group. This allows us to represent and store each four items as a product of prime numbers. The single value is used to produce back the items by finding its prime factors which are the primes that correspond to the items coded in the single value. We compared the time and space requirements of the proposed approach with Apriori and FP-Growth as implemented in WEKA; we will

E. Corchado et al. (Eds.): IDEAL 2014, LNCS 8669, pp. 93–101, 2014.
© Springer International Publishing Switzerland 2014

consider other methods as future work. The reported results demonstrate the applicability and effectiveness of the proposed approach as an attractive technique.

The rest of this paper is organized as follows. Section 2 is related work. Section 3 covers the problem definition. Section 4 presents the proposed method. Section 5 describes the experiments and reports the results. Section 6 is conclusions.

2   Related Work

The problem of finding associations among itemsets was first introduced by Agrawal et al. [2] in 1993; other algorithms include Tree Projection [1], Trie-based approach by Amir et al. [3], FP-Growth [6], and Viper [11].

Parallelism was expected to remove the bottleneck of the sequential Pattern mining techniques leading to better and faster processing of massive datasets. The literature contains many methods where parallelism is incorporated. Hadoop [4] and MapReduce distributed framework are used to handle large, complex and unstructured datasets in scalable manner. The work in [5] uses a distributed Apriori algorithm where the data is preprocessed then replicated on different nodes. Zaki [12] discusses the problems facing the parallel and distributed associate rule and frequent pattern mining. Recently, Leung and Hayduk [8] realized the increasing interest in big data analysis and proposed a tree-based algorithm that uses MapReduce to mine frequent patterns from big uncertain data. Boyd and Crawford [5] highlight critical questions for big data handling. They argue that it is necessary to critically interrogate the assumptions and biases related to big data. Kang et al. [7] discuss the need for mining big graphs. Lin and Ryaboy [9] discuss the case of twitter for scaling big data mining infrastructure. Marz and Warren [10] cover principles and best practices of scalable real-time data systems.

3   Problem Definition

In the current paper, we try to address two problems related with the first phase of the association rules mining which is the frequent itemsets detection. The first problem concerns the reduction of the required space needed for the database to store all itemsets produced by transactions. This is very important because smaller database size implies also faster data analysis since DBMS can optimize the loading of the database in memory. The second problem we try to address is to perform faster frequent itemsets detection. Similarly, this problem is connected with the database size and how many times we need to scan the database to detect all frequent itemsets. The main drawback of very well-known algorithms for association rules mining, such as Apriori and FP-Growth, is the need to load the database in memory $n + 1$ times (where n is the length of the largest frequent itemset) in Apriori, and at least 2 times in FP-Growth.

4   The Proposed Approach

Our method named Prime Numbers Itemsets Detection (PriNID) is divided into three phases: (a) the creation of a dictionary with items organized in groups of four items assigned with a specific prime number, (b) the creation of the products of prime numbers of each group and subsequently the detection of all repeated products (and eventually all frequent products), and (c) the conversion of the products of the prime numbers back to the original itemsets. The most important phase with respect to time and size is the second step because we need to load the database from the disk and perform the analysis to detect all repeated itemsets. The first and third phases are significantly lesser in terms of size and time, as we will see below because they can be directly executed in memory. The first phase is actually common for every algorithm (such as Apriori and FP-Growth) since the items need somehow to be coded before processing. The same stands for the third phase since after the analysis the coded itemsets should be converted back to the original items from which they were composed. We have named our method "itemsets detection" and not "frequent itemsets detection" because, as we will see later in the paper, our method detects all repeated itemesets with just one scan without the need to predefine any kind of parameters. This is very useful because there is no need to repeat the analysis for different initial parameters and, therefore, reload the database file from the disk, which saves time, and furthermore even with this type of execution is faster than other algorithms like Apriori and FP-Growth by returning all repeated itemsets and not just the frequenter. Having the full set of itemsets at the end of the analysis helps us to better determine problem solving situations and even discover patterns in itemsets which otherwise, with limited results, it could be difficult if not impossible.

In the first phase we have to create the dictionary by creating groups of 4 items and then we assign to each item one of the first four prime numbers 2, 3, 5 and 7, respectively, and always in the specific order for all groups (i.e., the first item is prime number 2, the second 3, the third 5 and the fourth 7). We categorize the full list of items into groups of four for two reasons. First, the product of the first four prime numbers 2, 3, 5 and 7 is 210. This means that we need only 1 byte to store the largest product we can get which is 210. Instead of needing 4 bytes to store in the database a quartet of 1s and 0s, which represent if the item exist or not in the itemset, we need only 1 byte to store a number between 0 and 210. This is enough because, as we will see later in the paper, the number of the product hides the specific information inside it and it is very easy to extract it. Therefore, we can reduce the total size of the database that we need to hold all transactions for all items up to four times. This allows the DBMS to perform faster analysis since it has a smaller table to load in memory. The second reason is that if we use more than four items in a group we can store less information since the product of the prime numbers rapidly grows very large and, therefore, we need more bytes per column in the database. For example, if we use groups of eight items then the product of the first eight prime numbers (2, 3, 5, 7, 11, 13, 17, 19) is 9,699,690 which requires three bytes per group or actually four bytes for a 32 bit integer. Doing this we can use one more prime (number 23) and have actually 9 items per group using the maximum capacity of a 32bit integer. In case we want to use up to 64bit integers then we can store in eight bytes the first 15 primes. While in such cases we need four bytes per nine items or eight bytes per 15 items, with our method we need three bytes (one per group of four items) to accommodate 12 items (three more than the first of the previous examples) or 4 bytes to accommodate 16 items (one more then the second of the previous example). As we can observe the optimal case is the use of only the first four primes in any scenario.

Creating the product of the four first primes and storing it to the database is the key characteristic of our method. The reason is that every combination of products that can be created out of the four primes ($\sum_{i=1}^{4} \binom{4}{i} = \binom{4}{1} + \binom{4}{2} + \binom{4}{3} + \binom{4}{4} = 4 + 6 + 4 + 1 = 15$) can be created in a unique way and, therefore, it is a one-to-one assignment to each itemset that can be created from the four different items of the specific group.

This way we have only to analyze the product back to prime numbers and find the corresponding itemset. For example, in the case of an itemset with product 42 the number can be analyzed as a product of primes 2, 3 and 7 only. Other approaches suggested in the literature, like [51], which use products of prime numbers for large itemsets or use products based on more occurrences of same primes for weighted analysis, are impossible. A very simple example can be illustrated as follows. If we want to use larger itemsets, e.g., more than 15 items then the product of the first 16 primes up to the prime number 53 will be almost two times larger than 2^{64} which is impossible to be stored in any computer system using 64bit technology. Furthermore, if we want to use weighted method and therefore represent the products as powers of primes then even for very small itemsets without significant weight per item we will receive again products that no 64bit system can store in memory or database. For example, if we have 8 items and each item has been used in the itemset 3 times then based on [13] we have to calculate the following product $2^3 \times 3^3 \times 5^3 \times 7^3 \times 11^3 \times 13^3 \times 17^3 \times 19^3$, this is 50 times larger than 2^{64}.

The total number of groups we can create is $\frac{n}{4}$ if the result of the operation $(n \bmod 4)$ equals to 0 (the total number of items is an exact product of 4) or $\left\lfloor \frac{n}{4} \right\rfloor + 1$ (floor of the division plus 1) if the result of the operation $(n \bmod 4)$ is different than 0. After creating the groups we assign each item to a specific group and a prime number in each item of the group. For example, for the 26 letters of the English alphabet we need $\left\lfloor \frac{26}{4} \right\rfloor + 1 = 6 + 1 = 7$ groups. We fill the first group with the first four letters of the alphabet and assign the prime numbers as A=2, B=3, C=5 and D=7. For the second group we will have the next four letters and the same prime numbers are assigned to them as E=2, F=3, G=5 and H=7. We can continue the process of creating groups and assign prime numbers to them until we assign all items into groups of four items each. We can do the same process for every kind of items by creating a dictionary to hold as information the item, the group that the item belongs to, and the prime number assigned to the specific item. The dictionary can be dynamic and grows as more items are added in the process.

When an itemset is added to the database we have first to transform it into the appropriate product before it is saved. We use the predefined dictionary to detect to which group each item in the itemset belongs, and in case we have more than one item per group we have to calculate the product. For example, if an itemset consists of items A, B and G then in the first group we will store the value 6, which is the product of the two prime numbers 2 and 3 assigned to A and B, respectively, and in the second group the value 5. The process' time of creating the products, before saving the itemsets in the database, is a very easy to performed multiplication of small integers and, therefore, does not affect the whole storing process time. Furthermore, the detection of each product in the dictionary, in order to create the product, is also a very easy and fast operation to be performed by using binary search with $O(\log n)$ time complexity, given the fact also that the dictionary is a very small list of items (therefore n is similarly very small).

In the second phase, we need to query the database in order to return all products. We can perform the specific action in several steps by executing the appropriate query or with just one query, which means that we need only one database scan. In the first case, we have to query the database for each group and count all occurrences of each number in the specific group. By performing a single query in the second case, we can have as results all the frequent products without the need to repeat the process. The specific method does not need any previous information and it does not produce the results incrementally as the Apriori algorithm. The query we have to execute on the first group will give us all the products that have modulus 0 with all the four prime numbers. So, in the pre-discussed example of the alphabet letters instead of searching directly, e.g., for string AB or itemset 110000000000000000000000000, which are represented by the product 6, we just query the first group of the itemset (1100) for $mod(2) = 0$ or $mod(3) = 0$. With $mod(2) = 0$ we will take all products that can be divided by 2 and with the second all products that can be divided by 3. In both cases, we will get the product 6 which is the product of the first two prime numbers 2 and 3 (Algorithm 1).

The last phase of the method is to analyze the results and extract the actual itemsets based on the alphabet (Algorithm 2). This phase is very important since in the case of AB for instance, we need to factorize 6 and get the primes 2 and 3, with which we can directly find in the dictionary the corresponding items A and B. Further, the specific method can return overlapping results which have to be further analyzed to extract the accurate number of occurrences of each itemset. The fact that our method returns not just the itemsets but also detects and reports overlapping itemsets can be possibly used for different kind of analysis to extract further characteristics of the itemsets or any kind of other useful information. However, if we want we can pally filters in this phase and get back results of specific frequency. The filtering process is extremely fat and can be redone many times without time consumption and of course without having the need to reanalyze the data as other algorithms do when initial parameters have to be changed.

The proposed method's advantage is that it has to read the database only once and can analyze million or even billion of itemsets. There is no need to read each group individually since it is represented in the database as a single field. Moreover, the specific method is not depended on initial values as, e.g., the frequency of the itemsets we want discover. Then it is much easier to detect itemsets based on a variable frequency until we discover the desired frequency value. This can be done without the need for any more analysis. However, in cases where the number of items is large there is a drawback in the calculations since we have to calculate each combination of groups. However, as we will see from the experiments, the method can perform an analysis 10 or more times faster than any known method without been affected by the calculations even when the number of items is large. These calculations have to be performed on memory which is a much faster process than re-reading a large database file from disk like other algorithms require. The problem of large number of calculations can be easily addressed can be easily addressed if the grouping will be performed in a way that the most frequent products are together or when products of the same kind are very difficult to occur in the same itemset. For example, different brands of beers are very rare to be bought together in a supermarket since consumers prefer one of the many different brands. Doing this many fields in the database will have no products and the value will be 0, something that will accelerate the arithmetic calculations.

```
Algorithm 1. Find All Frequent Products (FAFP)
Input: table of transactions
Output: an array of all frequent products and their number of occurrences
1        FAFP (table Transactions)
2.1      for each group and combination of groups
2.2          find all products which satisfy mod(2)==0 OR
             mod(3)==0 OR mod(5)==0 OR mod(7)==0
2.3      end for
3        end FAFP
```

Example 1. Assume we have 4 items A, B, C and D. Since we have exactly 4 items, we can form only one group and assign the prime numbers as A=2, B=3, C=5 and D=7. If we assume that the itemset ACD is added to the database and has to be stored for future frequent itemset detection, then we have to calculate the product of the three items which will be 2*5*7=70. In this case we will save in the database the number 70. In the frequent itemset detection phase we will get back the number 70 with just one occurrence. Then we have to analyze the number 70 and we get back the product 2*5*7, which represents the itemset ACD. The advantage of the specific method is that while we store an itemset as product it can be analyzed back with one way since the factors are all prime numbers.

Example 2. Assume that after the frequent itemset detection phase we get the result reported in Table 1. Based on the results we analyze the numbers and create an array as reported in Table 2. In Table 2 we have to create a column for each one of the four prime numbers and we have to assign in each row 1 if it exists in the product and 0 if not. We have also added at the beginning a column that shows the actual itemset to help us with the description of the method.

Algorithm 2. Products Factorization (PF)
Input: array of all frequent products and their number of occurrences
Output: an array of all frequent itemsets and their number of occurrences

```
1       PF (string X, LERP)
2.1     for each row in array
2.2         find all occurrences of products in row
2.3         add all occurrences
2.4     end for
3       end PF
```

Table 1. Products Results

Itemset	Product	Occurrences
A	2	12
C	5	9
D	7	17
AC	10	3
AD	14	3
BC	15	2
BD	21	4
CD	35	2
ACD	70	4
BCD	105	4

Table 2. Products Analysis Array

Itemset	2=A	3=B	5=C	7=D	Occurrences
A	1	0	0	0	12
C	0	0	1	0	9
D	0	0	0	1	17
A	1	0	1	0	3
A	1	0	0	1	3
B	0	1	1	0	2
B	0	1	0	1	4
C	0	0	1	1	2
A	1	0	1	1	4
B	0	1	1	1	4

Starting from the first row we have the number 2 which occurs 12 times. However, we can observe that number 2 exists in two more rows which represent the itemsets AC and AD. Therefore, the total number of occurrences of item A is 12+3+3=18. We continue the process with the number in the second row, which is 5 and exists also in five more rows. Therefore, the total occurrences of item C are 9+3+2+2+4+4=24. We can repeat the same process for the third row of the array and number 7 which represent the letter D. In this case, we have 17+3+4+2+4+4=34 total occurrences. For the fourth record, we have the number 10 which is the itemset AC and occurs in one more row. The total number of occurrences is 3+4=7. Next we have 14 which represents AD and exists in two rows. The total number of occurrences is 3+4=7. Then we continue with 15 (BC) and the total number of occurrences is 2+4=6, 21 (BD) with total number of occurrences 4+4=8, 35 (CD) with total number of occurrences 2+4+4=10, 70 (ACD) with total number of occurrences 4, and finally 105 (BCD) with total number of occurrences 4. When we reach the last row of the array we are done with the itemset detection. The advantage of this method is that with a single database scan and a query over the resulting array we can have all itemsets and then filtered them as we need for detect the frequenter based on different frequency parameter, without the need to have any previous knowledge or redo the process in an incremental way as it is done in the Apriori algorithm.

5   Experimental Results

In order to examine the performance of our approach we have conducted extensive experiments on five different databases. The first database has one million transactions, the second two million transactions the third three million transactions, the fourth four million transactions and the last with eight million transactions. All experiments have been conducted with 8 items in order to make possible the comparison of our method with Apriori and FP-Growth. In order to create the transactions we used the first eight letters of the English alphabet and randomly created strings of variable size, which represent transactions. For every experiment we created five files. The first is a standard text file with 1 in each column to indicate that the specific item of the transaction exists and "?" if it does not exists. The reason we used the question mark symbol is because we load the specific files to the very well-known open source application WEKA in order to perform frequent itemsets detection with the Apriori and FP-Growth algorithms. We have also created five tables in a DBMS for each one of the experiments to be used in our method. The reason we used only eight items for the experiments is that for some reason WEKA couldn't manage to scale up the experiments with more items and perform analysis in feasible time with the hardware we have. Even more, as we will observe from the experimental results, we faced the same problem for the two last experiments with four and eight million itemsets, which again WEKA couldn't scale up and produce results in reasonable amount of time. For the experiments with WEKA we have used a computer with an i7, quad core, multithreading CPU with 8 logical cores at 3.4 GHz, 32 GB of RAM, 64bit operating system and the 64bit version of WEKA 3.6. For the experiments with our method we have used an i5, quad core CPU with 4 logical cores at 3.2 GHz, 8 GB RAM, 64bit operating system and a 64bit DBMS.

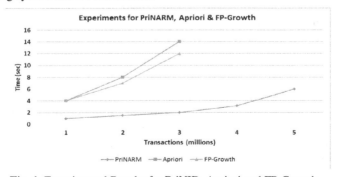

Fig. 1. Experimental Results for PriNID, Apriori and FP-Growth

As we can see in "Fig.1" although we run our method on a less powerful computer than the one we used to run WEKA the results were produced much faster. Furthermore, even for larger datasets our method managed to perform very fast in linear time complexity and return results, while WEKA didn't manage to scale up and return results in feasible time for the specific hardware. More specifically, for the one million transactions we needed less than one second while with WEKA we needed 4 seconds for both Apriori and FP-Growth algorithms. For the two million transactions we needed again one second for our method while WEKA needed 10 and 8 seconds for Apriori and FP-Growth, respectively. For the three million transactions we needed less than 2 seconds while with WEKA we needed 14 and 12 seconds for Apriori and FP-Growth, respectively. For the two last experiments of four and eight million itemsets we didn't manage to have actual time measurement for Apriori and FP-Growth. However, our method performed the analysis in approximately three and six seconds respectively

All experiments have been run 5 times for each different size of datasets and we have taken the average time. We have to mention that WEKA first loads the whole dataset in memory and then performs all the calculations, something that it couldn't be time-measured and, therefore, the times of the experiments represent the actual analysis process. For our method we execute the queries and the total time encapsulates the time the DBMS needs to read the data from the disk. Furthermore, in order to avoid caching from the database, after each execution of the experiment we restarted the database engine service in order to clear memory form previously loaded data.

6   Conclusion and Future Work

This paper introduced a method for the fast detection of repeated itemsets in a database of transactions with the use of prime numbers. The detection of the repeated patterns can help us to perform frequent itemset detection like many other very well-known algorithms. The results compared to other commonly used algorithms such as Apriori and FP-Growth have proved that the proposed method is significantly faster and scale up much better. Furthermore, by using prime numbers and storing their products instead of the classical representation of transactions with 1 and 0 we can reduce the size of the database needed. This allows less space on disk but also smaller size of data to be loaded in memory to perform the analysis. Moreover, we have shown that our method can be performed with only one database scan, something which can significantly reduce the overall time of the analysis since the database loading in memory is the most time consuming part of all frequent itemsets detection algorithms. Currently, we are trying to further expand the proposed method by using more properties of the prime numbers.

References

1. Agarwal, R., Aggarwal, C., Prasad, V.: A Tree Projection Algorithm for Generation of Frequent Item Sets. Journal of Parallel and Distributed Computing 61(3), 350–371 (2001)
2. Agrawal, R., Imieliski, T., Swami, A.: Mining association rules between sets of items in large databases. In: Proc. of ACM SIGMOD, pp. 207–216 (1993)
3. Amir, A., Feldman, R., Kashi, R.: A new and versatile method for association generation. Information Systems 22(6-7), 333–347 (1997)

4. Bialecki, A., Cafarella, M., Cutting, D., Malley, O.: Hadoop: a framework for running applications on large clusters built of commodity hardware, `http://hadoop.apache.org/`

5. Boyd, D., Crawford, K.: Critical Questions for Big Data. Information, Communication and Society 15(5), 662–679 (2012)

6. Han, J., Pei, J., Yin, Y.: Mining frequent patterns without candidate generation. In: Proc. of ACM SIGMOD, pp. 1–12 (2000)

7. Kang, U., Chau, D.H., Faloutsos, C.: Big graph mining: Algorithms and discoveries. SIGKDD Explorations 14(2) (2012)

8. Leung, C.K.-S., Hayduk, Y.: Mining Frequent Patterns from Uncertain Data with MapReduce for Big Data Analytics. In: Meng, W., Feng, L., Bressan, S., Winiwarter, W., Song, W. (eds.) DASFAA 2013, Part I. LNCS, vol. 7825, pp. 440–455. Springer, Heidelberg (2013)

9. Lin, J., Ryaboy, D.: Scaling big data mining infrastructure: The twitter experience. SIGKDD Explorations 14(2) (2012)

10. Marz, N., Warren, J.: Big Data: Principles and best practices of scalable realtime data systems. Manning Publications (2013)

11. Shenoy, P., Haritsa, J.R., Sudarshan, S., Bhalotia, G., Bawa, M., Shah, D.: Turbo-charging vertical mining of large databases. In: ACM SIGMOD (2000)

12. Zaki, M.J.: Parallel and distributed association mining: a survey. IEEE Concurrency 7(4) (2002)

13. Tohidi, H., Hamidah, I.: Using Unique-Prime-Factorization Theorem to Mine Frequent Patterns without Generating Tree. American Journal of Economics and Business Administration 3(1), 58–65 (2011)

Multi-step Forecast Based on Modified Neural Gas Mixture Autoregressive Model

Yicun Ouyang and Hujun Yin

School of Electrical and Electronic Engineering, The University of Manchester,
Manchester, M13 9PL, UK
yicun.ouyang@postgrad.manchester.ac.uk, hujun.yin@manchester.ac.uk

Abstract. Neural networks have been increasingly applied to financial time series forecasting. For the challenging multi-step forecasting there are usually two strategies: iterative and independent. The iterative method repeatedly performs single step predictions and thus accumulates the prediction errors. The independent method considers each forecast horizon as an independent task and thus excludes the relationship between various future points. This paper deals with these problems by using a modified neural gas mixture autoregressive (NGMAR) trained with input vectors segments of different forecast horizons. Some neurons will become specialised for some specific-step-ahead prediction. Additional parameters are used to monitor each neuron's updating patterns and its suitability for various step-ahead predictions. Experimental results on several financial time series and benchmark data demonstrate the effectiveness of proposed method and markedly improvement performances over many existing neural networks.

Keywords: financial time series, multi-step forecast, neural gas mixture autoregressive.

1 Introduction

As it is known, the long-term or multi-step forecast of time series is still an open problem for researchers in the field of time series analysis [1]. In general, there are two kinds of method, iterative and independent. The iterative method (e.g.[2]) is to repeat calculating predicted outputs by the same model,

$$\hat{x}_{t+1} = f(x_{t-l+1}, \cdots, x_{t-1}, x_t) \tag{1}$$

$$\hat{x}_{t+2} = f(x_{t-l+2}, \cdots, x_{t-1}, x_t, \hat{x}_{t+1}) \tag{2}$$

$$\cdots \qquad \cdots$$

$$\hat{x}_{t+h} = f(x_{t-l+h}, \cdots, x_t, \hat{x}_{t+1}, \cdots, \hat{x}_{t+h-1}) \tag{3}$$

where l is the size of time window and h is the prediction horizon. This method is intuitive and natural, but it is not widely used for long-term time series prediction because the estimation errors produced in each step are accumulated along

E. Corchado et al. (Eds.): IDEAL 2014, LNCS 8669, pp. 102–109, 2014.

with time. That is, the past errors generated in the earlier periods still have effects in later periods, thus degenerating potential of the method for long-term time series forecast.

The other method, independent method (e.g.[3]) builds individual models for each step forecast instead of using the same prediction model iteratively. That means, for each specific prediction horizon, one unique model is to be found,

$$\hat{x}_{t+1} = f_1(x_{t-l+1}, \cdots, x_{t-1}, x_t) \tag{4}$$

$$\hat{x}_{t+2} = f_2(x_{t-l+1}, \cdots, x_{t-1}, x_t) \tag{5}$$

$$\cdots \qquad \cdots$$

$$\hat{x}_{t+h} = f_h(x_{t-l+1}, \cdots, x_{t-1}, x_t) \tag{6}$$

Since each prediction is calculated based on actual values $x_{t-l+1}, \cdots, x_{t-1}, x_t$, estimated errors are not accumulated along with time. Therefore, this method has the potential to outperform the iterative method as the forecast horizon increases. However, it has significantly increased number of models thus making the computation much more intensive. Moreover, the complex dependencies between neighbour points $\hat{x}_{t+j-1}$ and $\hat{x}_{t+j}$ are excluded because they are independently predicted by using different models with the past values $x_{t-l+1}, \cdots, x_{t-1}, x_t$. This may reduce prediction accuracy.

Besides the above methods, there are other methods being developed, such as the joint method [4], multiple-output method [1] as well as parametric function method [5]. Neural networks have attracted a great deal of attention for time series modelling, such as vector SOM (VSOM) [6], multilayer perceptron (MLP) [7], support vector regression (SVR) [8], neural gas (NG) [9], self-organizing mixture autoregressive (SOMAR) [6] and neural gas mixture autoregressive (NGMAR) [10]. Most of them were proposed for one-step prediction. However, they can be framed into the iterative or independent method for multi-steps, though this may give poor performance because of the drawbacks mentioned before. In this paper, we propose a new method based on modified neural gas mixture autoregressive(NGMAR) for multi-step forecast tasks, solving the problems such as accumulated errors and lack of dependencies between points.

2 Methodology

2.1 Training Vectors

To perform one-step-ahead prediction by standard neural networks, the training time series is divided into segments of the same length l, which is also called embedding dimension,

$$\mathbf{x}(t) = [x_{t-l+1}, \cdots, x_{t-1}, x_t], t = 1, 2, \cdots, T \tag{7}$$

The last point x_t is used as the desired output and the first $l - 1$ points $x_{t-l+1}, \cdots, x_{t-1}$ are used as the input vector. For multi-step forecast, the iterative method accumulates errors along with time. To overcome this drawback, we directly build the relationship between past values and future values as the independent way to discover the patterns inside. The newly constructed input vectors for training can be described as

$$\mathbf{x}^1(t) = [x_{t-l+1}, \cdots, x_{t-1}, x_t] \tag{8}$$

$$\mathbf{x}^2(t) = [x_{t-l+1}, \cdots, x_{t-1}, x_{t+1}] \tag{9}$$

$$\cdots \qquad \cdots$$

$$\mathbf{x}^h(t) = [x_{t-l+1}, \cdots, x_{t-1}, x_{t+h-1}] \tag{10}$$

The input vectors $[\mathbf{x}^1(t), \mathbf{x}^2(t), \cdots, \mathbf{x}^h(t)]_{t=1}^T$ are used for training the model.

2.2   Structure of Neurons

Each neuron of NGMAR has a reference vector of length l, with the last component of the vector used as the prediction output in the predicting stage,

$$\mathbf{w}_i(t) = [w_{i,1}, \cdots, w_{i,l-1}, w_{i,l}] \tag{11}$$

In our model, we add a new vector $\mathbf{g}_i(t)$ into neuron i to judge whether this neuron should be used for predicton tasks. The new vector, termed status vector, is constructed as

$$\mathbf{g}_i(t) = [g_{i,1}, g_{i,2}, \cdots, g_{i,h}] \tag{12}$$

where $g_{i,j}$ is the percentage of neuron i has been updated by input vectors $\mathbf{x}^j(t)$.

$$g_{i,j} = \frac{T_{i,x^j}}{T} \tag{13}$$

where T_{i,x^j} is the times of neuron i updated by $\mathbf{x}^j(t)$. For h-step-ahead prediction, given a threshold thd_h, if $g_{i,h} \geq thd_h$, it means this neuron has been trained by input vector $\mathbf{x}^h(t)$ sufficiently and should be used for h-step-ahead prediction.

2.3   Training Procedure

In the training stage, the sum of autocorrelation of coefficients (SAC) between an input and the reference vectors of all neurons are calculated and the neuron with the minimum distance is chosen as the best matching unit (BMU). The reference vectors of the BMU and its neighbours are updated following the updating rule,

$$\mathbf{w}_{i_k}(t+1) = \mathbf{w}_{i_k}(t) + \gamma(t)exp(-k/\lambda(t))e_{i_k}(t)\mathbf{x}(t) \tag{14}$$

where $\gamma(t)$ is the learning rate, $\lambda(t)$ is the neighbourhood range, k is the ranking of neuron i_k and $e_{i_k}(t)$ is the modelling error at node i_k.

In our model, the updating rule is similar as that of the NGMAR. As mentioned previously, there is another vector $\mathbf{g}_i(t)$, associated with the percentage of updating times. Since only some neurons in the network are updated, we adopt a parameter α to represent the percentage of these neurons. Assuming $\alpha = 10\%$, only the nearest $M' = \alpha M = 0.1M$ neurons are updated with each input vector. In the meantime, these neurons need to update their status vectors as

$$g_{i,j}(t) = \frac{g_{i,j}(t-1)T + 1}{T} \tag{15}$$

2.4 Predicting Procedure

In the predicting stage, the input vectors are constructed by dividing the test series into segments similar to the training stage but with the last value left out. The threshold is introduced to evaluate whether the neuron has been trained sufficiently for multi-step forecast. For one-step-ahead forecast, the threshold thd_1 means the lowest percentage being updated by $\mathbf{x}^1(t)$ any neuron should exceed if it can be used for this prediction task. Only the neurons satisfying the condition $g_1 \geq thd_1$, should be used to select the BMU. These neurons make up a set named the candidate set. Similarly, the j-step-ahead forecast can be calculated with the threshold set $\{thd_1, thd_2, \cdots, thd_h\}$ in the same way

$$C_j = \{\mathbf{w}_i | g_{i,j} \geq thd_j\}, j = 1, 2, \cdots, h \tag{16}$$

For the h-step-ahead forecast, the candidate neurons set is $C_h = \{\mathbf{w}_i | g_{i,h} \geq thd_h\}$. Among these neurons, the BMU i^* is selected and its last, l-th component is used as the predicted value

$$\hat{x}_{t+h-1} = w_{i^*,l} \tag{17}$$

3 Experimental Results

We used several financial time series and benchmark data to evaluate the performance of the new model. For each time series, the experiment was run for 10 times over the test set and average results were obtained for comparison.

3.1 Foreign Exchange Rates

The financial time sequences used in experiments are foreign exchange (FX) rates downloaded from PACIFIC exchange rate service. They were used to evaluate and compare the performances of different models. They consist of the daily closing prices of British Pound against US Dollar, Euro and JP Yen over 12 years from 2002 to 2013. Each sequence contains 3000 points, of which 2700 points were used for training and the rest for testing. The neuron number was 100 and the size of sliding window was selected as 10 according to BIC.

As reviewed in the literature [11,12], there are several methods to measure the prediction accuracy. Compared to mean absolute error (MAE), normalized root mean squared error (NRMSE) gives more weights rather than same ones

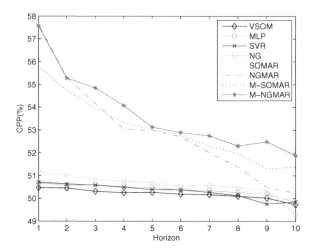

Fig. 1. CPP(%) results for different horizons on GBP/USD

to large errors as well as measuring values on a unified scale. Therefore, in the experiments, the prediction performance was measured by NRMSE

$$NRMSE = \sqrt{\frac{(1/N_{test}) \sum_{t=1}^{N_{test}} (x(t) - \hat{x}(t))^2}{(1/N_{test}) \sum_{t=1}^{N_{test}} (x(t) - \bar{x})^2}} \tag{18}$$

where $\hat{x}_t$ and x_t denote the prediction output and actual value, $\bar{x}$ is the sample mean, and N_{test} is the number of test examples.

Besides NRMSE, another method correct prediction percentage (CPP) is also used for measuring prediction accuracy, esp for trend prediction

$$CPP = \frac{Number\ of\ Correct\ Direction\ Predictions}{Total\ Number\ of\ Predictions} \tag{19}$$

Fig. 1 plots the CPP results for different horizons on GBP/USD. The prediction results are presented in Tables 1-3. All the neural networks except the M-versions were based on their independent models for multi-step forecast, since the testing errors are much lower than that of the iterative method. M-version is the proposed method applied to different neural networks either SOMAR or NG-MAR. The proposed methods are better than all other competitors in terms of NRMSE and CPP. The results have been demonstrated statistically significant with corresponding p-values of t-test.

The improved performance of the new method can be briefly explained by the direct relationship built between the past values and the future value for each forecast horizon, as well as dependencies among sequenced time series points. As described in the previous sections, the input vectors were used for training the model over all forecast horizons instead of some specific forecast horizon.

Table 1. Multi-step Forecast Performance Comparison of Neural Networks Based on Independent Method and Modified SOMAR/NGMAR on GBP/USD (h=10)

Multi-step method	CPP(%)	p-value	NRMSE	p-value
VSOM	50.23	**2.95e-4**	0.2403	**1.77e-6**
MLP	50.41	**2.21e-4**	0.2380	**8.38e-5**
SVR	50.46	**1.77e-4**	0.2252	**7.71e-5**
NG	50.68	**1.37e-4**	0.1866	**5.33e-5**
SOMAR	52.07	**0.0064**	0.1537	**4.52e-4**
NGMAR	53.22	**0.0119**	0.1369	**1.13e-4**
M-SOMAR	53.83	**0.0175**	0.1435	**0.0167**
M-NGMAR	**54.38**	N/A	**0.1197**	N/A

Table 2. Multi-step Forecast Performance Comparison of Neural Networks Based on Independent Method and Modified SOMAR/NGMAR on GBP/EUR (h=10)

Multi-step method	CPP(%)	p-value	NRMSE	p-value
VSOM	50.12	**3.31e-5**	0.1387	**5.57e-6**
MLP	50.17	**1.07e-5**	0.1372	**5.34e-6**
SVR	50.40	**7.11e-4**	0.1295	**1.74e-6**
NG	50.64	**5.25e-4**	0.1073	**7.20e-5**
SOMAR	51.41	**0.0062**	0.0801	**3.31e-5**
NGMAR	52.42	**0.0154**	0.0739	**4.93e-5**
M-SOMAR	52.35	**0.0134**	0.0683	**4.02e-4**
M-NGMAR	**53.10**	N/A	**0.0577**	N/A

Table 3. Multi-step Forecast Performance Comparison of Neural Networks Based on Independent Method and Modified SOMAR/NGMAR on GBP/JPY (h=10)

Multi-step method	CPP(%)	p-value	NRMSE	p-value
VSOM	50.10	**3.66e-5**	0.1531	**5.62e-5**
MLP	50.11	**7.87e-4**	0.1515	**7.87e-4**
SVR	50.39	**1.07e-5**	0.1364	**3.69e-4**
NG	50.44	**4.41e-4**	0.1305	**4.81e-4**
SOMAR	51.25	**1.29e-4**	0.1028	**0.0024**
NGMAR	51.80	**3.67e-4**	0.0924	**0.0035**
M-SOMAR	51.98	**0.0018**	0.0835	**0.0191**
M-NGMAR	**52.47**	N/A	**0.0674**	N/A

3.2   Benchmark Data

The prediction performance of the proposed method was also evaluated and compared to other methods on the two benchmark data sets, Mackey-Glass and Lorenz. A sample Mackey-Glass sequence consisting of 2000 points was generated from the following equation with parameter $\alpha = 0.2$, $\beta = -0.1$ and $\delta = 17$

$$\frac{dx(t)}{dt} = \beta x(t) + \frac{\alpha x(t - \delta)}{1 + x(t - \delta)^{10}} \tag{20}$$

Table 4. Multi-step Forecast Performance Comparison of Neural Networks Based on Independent Method and Modified SOMAR/NGMAR on Mackey-Glass (h=10)

Multi-step method	CPP(%)	p-value	NRMSE	p-value
VSOM	50.26	**4.77e-5**	0.1089	**8.86e-5**
MLP	50.37	**2.42e-5**	0.0936	**6.61e-5**
SVR	50.43	**3.73e-5**	0.0835	**4.60e-5**
NG	50.41	**2.29e-5**	0.0787	**5.43e-4**
SOMAR	51.73	**6.33e-4**	0.0718	**0.0071**
NGMAR	52.61	**5.27e-4**	0.0679	**0.0095**
M-SOMAR	52.85	**3.04e-4**	0.0645	**0.0136**
M-NGMAR	**54.27**	N/A	**0.0581**	N/A

Table 5. Multi-step Forecast Performance Comparison of Neural Networks Based on Independent Method and Modified SOMAR/NGMAR on Lorenz (h=10)

Multi-step method	CPP(%)	p-value	NRMSE	p-value
VSOM	50.29	**1.76e-7**	0.0951	**2.74e-7**
MLP	50.35	**2.55e-6**	0.0892	**3.12e-7**
SVR	50.44	**1.77e-6**	0.0803	**5.60e-6**
NG	50.48	**5.53e-5**	0.0787	**5.73e-5**
SOMAR	51.53	**4.14e-4**	0.0715	**8.77e-4**
NGMAR	52.51	**3.82e-4**	0.0644	**9.74e-4**
M-SOMAR	52.74	**0.0071**	0.0627	**0.0049**
M-NGMAR	**54.07**	N/A	**0.0540**	N/A

where $x(t)$ is the sample time series at time t. The length of training sequence is 1800 (validation set is 180), and remaining 200 points were used as the test set. The parameters of the new model were chosen by 10-fold cross-validation.

Another benchmark data used in the experiment is Lorenz time series. A sample sequence was generated with parameters chosen as $\sigma = 10$, $\rho = 28$ and $\beta = 8/3$ from the following system

$$\frac{dx(t)}{dt} = \sigma(y(t) - x(t)) \tag{21}$$

$$\frac{dy(t)}{dt} = x(t)(\rho - z(t)) - y(t) \tag{22}$$

$$\frac{dz(t)}{dt} = x(t)y(t) - \beta z(t) \tag{23}$$

where $x(t)$, $y(t)$ and $z(t)$ are the values of time series at time t.

The size of sliding window was set as 12 according to BIC. The results are shown in Tables 4-5. The proposed method outperforms other multi-step forecast methods significantly with the lowest NRMSE and highest CPP. The new methods were compared with each other, and the results in Tables 4 and 5 demonstrate that M-NGMAR is slightly better than M-SOMAR.

4   Conclusion and Future Work

Multi-step forecasting is a challenging task. This paper presents a new model based on NGMAR to improve their performance in multi-step forecast by introducing new training and predicting strategies. The good performance of the new models on FX rates demonstrat the effectiveness of these strategies. The following are the main conclusions drawn. However, computationally, the new method consumes much more time because of its way of validating parameters and training model.

Recently, the echo state network (ESN) [13] has gained much interests from researchers. It performs better prediction than other neural networks over various time series. Therefore, our future work is to apply the new strategies to ESN.

References

1. Ben Taieb, S., Sorjamaa, A., Bontempi, G.: Multiple-output modeling for multi-step-ahead time series forecasting. Neurocomputing 73, 1950–1957 (2010)
2. Hill, T., O'Connor, M., Remus, W.: Neural network models for time series forecasts. Management Science 42, 1082–1092 (1996)
3. Sorjamaa, A., Hao, J., Reyhani, N., Ji, Y., Lendasse, A.: Methodology for long-term prediction of time series. Neurocomputing 70, 2861–2869 (2007)
4. Zhang, L., Zhou, W.-D., Chang, P.-C., Yang, J.-W., Li, F.-Z.: Iterated time series prediction with multiple support vector regression models. Neurocomputing 99, 411–422 (2013)
5. Cheng, H., Tan, P.-N., Gao, J., Scripps, J.: Multistep-Ahead Time Series Prediction. In: Ng, W.-K., Kitsuregawa, M., Li, J., Chang, K. (eds.) PAKDD 2006. LNCS (LNAI), vol. 3918, pp. 765–774. Springer, Heidelberg (2006)
6. Ni, H., Yin, H.: A self-organising mixture autoregressive network for FX time series modelling and prediction. Neurocomputing 72(16-18), 3529–3537 (2009)
7. Hornik, K., Stinchcombe, M., White, H.: Multilayer feedforward networks are universal approximators. Neural Networks 2(5), 359–366 (1989)
8. Cortes, C., Vapnik, V.: Support-vector networks. Machine Learning 20(3), 273–297 (1995)
9. Martinetz, T.M., Berkovich, S.G., Schulten, K.J.: "Neural-gas" network for vector quantization and its application to time-series prediction. IEEE Transactions on Neural Networks 4(4), 558–569 (1993)
10. Ouyang, Y., Yin, H.: A neural gas mixture autoregressive network for modelling and forecasting FX time series. Neurocomputing 135, 171–179 (2014)
11. De Gooijer, J.G., Hyndman, R.J.: 25 years of time series forecasting. International Journal of Forecasting 22, 443–473 (2006)
12. Hyndman, R.J., Koehler, A.B.: Another look at measures of forecast accuracy. International Journal of Forecasting 22, 679–688 (2006)
13. Li, D., Han, M., Wang, J.: Chaotic time series prediction based on a novel robust echo state network. IEEE Transactions on Neural Networks and Learning Systems 23(5), 787–799 (2012)

LBP and Machine Learning
for Diabetic Retinopathy Detection

Jorge de la Calleja, Lourdes Tecuapetla, Ma. Auxilio Medina,
Everardo Bárcenas, and Argelia B. Urbina Nájera

Universidad Politécnica de Puebla, Puebla, México, C.P. 72640
{jorge.delacalleja,lourdes.tecuapetla,
maria.medina,ismael.barcenas,argelia.urbina}@uppuebla.edu.mx

Abstract. Diabetic retinopathy is a chronic progressive eye disease associated to a group of eye problems as a complication of diabetes. This disease may cause severe vision loss or even blindness. Specialists analyze fundus images in order to diagnostic it and to give specific treatments. Fundus images are photographs taken of the retina using a retinal camera, this is a noninvasive medical procedure that provides a way to analyze the retina in patients with diabetes. The correct classification of these images depends on the ability and experience of specialists, and also the quality of the images. In this paper we present a method for diabetic retinopathy detection. This method is divided into two stages: in the first one, we have used local binary patterns (LBP) to extract local features, while in the second stage, we have applied artificial neural networks, random forest and support vector machines for the detection task. Preliminary results show that random forest was the best classifier with 97.46% of accuracy, using a data set of 71 images.

Keywords: machine learning, local binary patterns, medical image analysis.

1 Introduction

Currently there are approximately 382 million people worldwide with diabetes mellitus or simply diabetes. It is expected within 11 to 25 years, that this disease will increase to 592 million people according to the World Health Organization (WHO) and the International Diabetes Federation (IDF). Also, diabetes is the leading cause of blindness in people of working age between 40 and 60 years old, generating a health care spending of about $548 billion [7], [15].

The main complications of a person diagnosed with diabetes is that glucose levels are high, affecting several organs including kidney, nerves, brain and eyes. Particularly, diabetes can cause damage to the retina, known as diabetic retinopathy. At least the 80% of diabetic persons with 10-20 years of being diagnosed present any symptoms related to diabetic retinopathy [15].

Several works using machine learning and image processing methods have been proposed in order to classify some types of eye diseases, including diabetic

E. Corchado et al. (Eds.): IDEAL 2014, LNCS 8669, pp. 110–117, 2014.

retinopathy. In 2000 Ege B.M. et al [4] introduced a tool to provide automatic analysis of digital images of diabetic retinopathy. They tested several statistical classifiers, such as Bayesian, Mahalanobis and k-nn. Their best results were obtained with the Mahalanobis classifier. Osareh A. et al [13] proposed an automatic method to detect exudate regions. They introduced comprising image color normalization, enhancing the contrast between the objects and background, segmenting the color retinal images into homogenous regions using fuzzy c-means clustering, and classifying the regions into exudates and non exudates patches using artificial neural networks. They reported 92% of sensitivity and 82% of specificity. Niemeijer M. et al [10] proposed an automated system able to detect exudates and cotton-wool spots in color images. In their work they used k-nearest neighbors, reporting a operating characteristic curve (ROC) of 0.95 and sensitivity/specificity pairs of 0.95/0.88 for the detection of bright lesions of any type; 0.95/0.86, 0.70/0.93 and 0.77/0.88 for the detection of exudates, cotton-wool spots and drusen, respectively. In 2010, Silberman N. et al [14] proposed an automated system to detect diabetic retinopathy from retinal images. This approach used support vector machines to recognize exudates obtaining 98.4% of accuracy. In 2011, Gowda A. et al [5] attempted to detect exudates using backpropagation neural networks. The significant features were identified from preprocessed images by using two methods: decision trees and genetic algorithms. Their approach showed a classification accuracy of 98.45%. Kavitha S. and Duraiswamy K. [8] focused on automatic detection of diabetic retinopathy exudates in color fundus retinal images. Experiments on classification of hard and soft exudates were performed using image processing techniques. Exudates were detected with the aid of thresholding color histogram. The overall sensitivity, specificity and accuracy were 89.78%, 99.12% and 99.07%, respectively.

In contrast to previous approaches, we present a method to detect only diabetic retinopathy based on fundus images. Also, with the purpose of knowing how well local binary patterns performs on this domain, we have applied it to obtain local features from the images. The well-known machine learning methods of artificial neural networks, random forest and support vector machines were applied for the detection task. The paper is organized as follows: Section 2 describes a brief introduction of diabetic retinopathy. In Section 3 we describe the methods. In Section 4 we show preliminary experimental results and finally in Section 5 conclusions and future work are presented.

2   Diabetic Retinopathy

Diabetes mellitus is a metabolic disorder which results from a defect in insulin synthesis and secretion or from a resistance of the receptors on target tissues for this hormone. The most significant complication of diabetes mellitus involving the eye and which develops in 85% of all diabetics eventually is the retinopathy [6].

Diabetic retinopathy is an abnormality involving the small blood vessels that targets the central region, for example the macula. In fact, the diabetic retinopathy is a progressive disease and this is the main factor that causes blindness.

Classification of diabetic retinopathy according to the Early Treatment Diabetic Retinopathy Study (ETDRS) [1] is divided as: Diabetic Retinopathy Non Proliferative (DRNP), which is subdivided into mild, moderate, severe and very severe; and Diabetic Retinopathy Proliferative (DRP), which is subdivided into early, high-risk and advanced.

3   The Method

In order to classify fundus images as healthy or sick, we propose an automated method that is divided into two stages: 1) Extraction of features and 2) Classification of the images (Figure 1). In the first stage the characterization of the images is performed using local binary patterns considering uniform patterns, rotation invariant and rotation invariant uniform patterns. Then, a set of features is provided to machine learning algorithms to perform the classification task. In next subsections we explain the algorithms.

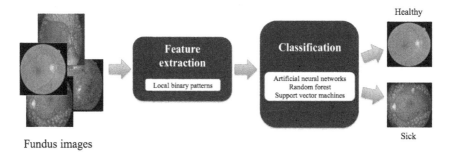

Fig. 1. Stages of the proposed method

3.1   Feature Extraction

The general idea for image classification is to extract relevant information (features) in a image, encode it as efficiently as possible, and compare one image encoding with a data base of similarly encoded images. Traditional approaches such as principal component analysis or linear discriminant analysis, look at the whole image as a high-dimensional vector in order to extract features. High dimensionality is not good, thus a lower-dimensional subspace is identified where useful information is preserved. Other alternative is extracting local features from images, i.e. instead of process the whole image as a high-dimensional vector, the idea is to describe only local features of an object. A method for performing this local feature extraction is local binary patterns explained below.

Local Binary Patterns. LBP was introduced as a texture description operator by Ojala et al. in 1996 [11]. This technique labels the pixels of an image by thresholding a 3×3 neighborhood of each pixel with the center value to yield a binary number (Figure 2). These binary numbers, called local binary patters

(LBP), codify local primitives including several types of curved edges and flat areas. The basic LBP operator only consider a small neighborhood, therefore it can not capture dominant features with large scale structures. Hence this operator was extended by using circular neighborhoods and interpolating the pixel values in order to allow any radius and number of pixels in the neighborhood [12].

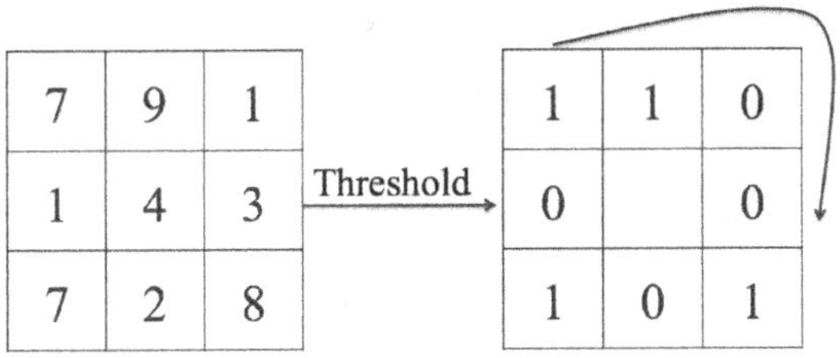

7	9	1
1	4	3
7	2	8

Threshold →

1	1	0
0		0
1	0	1

Fig. 2. The basic LBP Operator. The binary pattern 11001010 is equivalent to decimal 202.

The LBP operator produces 2^P different binary patterns for the P neighbors pixels, some of them contain more information than others, thus it is possible to use only a subset of fundamental patterns called uniform patterns to describe the texture of images. A LBP is uniform if it contains at most two bitwise transitions from 0 to 1 or vice versa when the binary string is considered circular. For example, 00000000, 01111110 and 11100111 are uniform patterns [12]. Accumulating the patterns which have more than 2 transitions into a single bin yields an LBP operator, with less than 2^P bins. For example, the number of labels for a neighborhood of 8 pixels is 256 for the standard LBP. After labeling a image with the LBP operator, a histogram of the labeled image $f_l(x, y)$ can be defined as

$$H_i = \sum_{x,y} I(f_l(x, y) = i), i = 0, ..., n - 1. \tag{1}$$

where n is the number of different labels produced by the LBP operator and

$$I(F) = \begin{cases} 1 & F \text{ is true} \\ 0 & F \text{ is false} \end{cases} \tag{2}$$

This LBP histogram contains information about the distribution of the local micro-patterns, thus they can be used to describe image characteristics.

Fundus images can be seen as a composition of micro-patterns which can be described by the LBP histograms. Therefore, we have used LBP features to represent fundus images, which were equally divided into small regions $R_0, R_1, ...R_m$ to extract LBP histograms. The LBP features extracted from each sub-region are concatenated into a single histogram (Figure 3),

$$H_{i,j} = \sum_{x,y} I(f_l(x, y) = i)I(x, y) \in R_j, i = 0, ..., n - 1; j = 0, ..., m - 1. \tag{3}$$

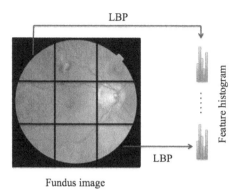

Fundus image

Fig. 3. A fundus image is divided into small regions. The LBP histograms are extracted and concatenated into a single one.

3.2 Machine Learning Algorithms

For the classification task we have used supervised learning, i.e. we have used a set of labeled images classified as healthy or sick. In next subsections we briefly describe artificial neural networks, random forest and support vector machines; we recommend the reader to review the references for details of these algorithms.

Artificial Neural Networks (ANN). This method has been one of the most used for learning tasks. ANN are based on the observation of biological neural systems which are formed by sets of units called neurons or nodes that are interconnected. Generally the architecture of an artificial neural network is divided into three parts called layers. These layers are the input layer, hidden layers, and the output layer. The way in which an artificial neural network works is as follows: each node (except for the input nodes) receives the output of all nodes in the previous layer and produces its own output, which then feeds the nodes in the next layer [9].

Random Forest (RF). This algorithm is an ensemble of unpruned classification trees, induced from bootstrap samples of the training data, using random feature selection in the tree induction process [2]. Prediction is made by aggregating the predictions of the ensemble. Random forest generally yields better performance than single tree classifiers such as C4.5.

Support Vector Machines (SVM). This method is based on the structural risk minimization principle from computational learning theory [3]. This principle provides a formal mechanism to select a hypothesis from a hypothesis space for learning from finite training data sets. The aim of SVMs is to compute the hyperplane that best separates a set of training examples. Two cases are analyzed: the linear separable case and the non-linear separable case. In the first case we are looking for the optimal hyperplane in the set of hyper-planes separating the given training examples. The optimal hyperplane maximizes the sum of the distances to the closest positive and negative training examples (considering only two classes). The second case is solved by mapping training examples to a high-dimensional feature space using kernel functions. In this space the decision boundary is linear and we can apply the first case. There are several kernels such as polynomial, radial basis functions, neural networks, Fourier series, and splines, among others; that are chosen depending on the application.

4   Experimental Results

We tested our method using images from the Messidor database[1]. These images were taken using a color video 3CCD camera on a Topcon TRC NW6 non-mydriatic retinograph with a 45 degree field of view. From the original data set of 1200 images, we selected a subset of 100 images at random. However, as we commented previously, specialists classify the images based on their ability and experience, but also considering the quality of them. Thus, we had to selected only 71 images due to their quality and visual information. Therefore we experimented with two data sets: 100 and 71 images.

In order to experiment with different binary patterns we used uniform patterns (59 bins), rotation invariant patterns (36 bins) and rotation invariant uniform patterns (9 bins) [12] considering 8 pixels in a 3×3 neighborhood.

We used artificial neural networks, random forest and support vector machines that are implemented in Weka[2] using default parameters and uniquely modifying the seeds in each iteration.

The metrics used to evaluate the performance of machine learning algorithms were accuracy and f-measure defined as $accuracy = (TP+TN)/(TP+FP+TN+FN)$, $f-measure = 2 \times (recall \times precision)/(recall + precision)$, $precision = TP/(TP+FP)$, $recall = TP/(FN+TP)$; where TP (True Positive) is the number of correct predictions of a positive example, FP (False Positive) is the number of incorrect predictions of a positive example, TN (True Negative) is the number of correct predictions of a negative instance and FN (False Negative) is the number of correct predictions of a positive instance.

Tables 1 and 2 show the results for each machine learning algorithm and local binary approach. These results were obtained by averaging ten runs of 10-fold cross-validation for each algorithm; that is, we randomly divided the original data set into ten equally sized subsets and performed 10 experiments, using in each experiment one of the subsets for testing and the other nine for training. The columns UP, RIP and RIUP denote uniform patterns, rotation invariant patterns and rotation invariant uniform patterns, respectively.

---

[1] MESSIDOR is a project within the scope of diabetic retinopathy.
  `http://messidor.crihan.fr/download-en.php`
[2] Weka is a collection of machine learning algorithms for data mining tasks.
  `http://www.cs.waikato.ac.nz/ml/weka/`

Table 1. Accuracy and F-measure for the algorithms using a data set of 100 images

	UP		RIP		RIUP	
	Acc.(%)	F − m	Acc.(%)	F − m	Acc.(%)	F − m
ANN	74.2	0.7408	77.6	0.7806	76.6	0.7668
RF	73.2	0.7274	74.2	0.7374	73.2	0.7306
SVM	77.2	0.7786	77.6	0.7822	78.8	0.7917

As we can see in Table 1, when using a data set of 100 images, the best classifier was support vector machines with 78.8% of accuracy and 0.7917 for $f - measure$ using rotation invariant uniform patterns. However, analyzing the results of the Table 2, we can observe that random forest obtained the best results with 97.4% of accuracy using uniform patterns and rotation invariant uniform patterns; and 0.974 for $f - measure$. We want to highlight that for this data set, all algorithms significantly improve their performance with over 93% for accuracy and 0.93 for $f - measure$. We can also comment that using a set of images with better quality for training, helps to obtain better results, i.e. the improvement of correct classification increases almost 18%.

Table 2. Accuracy and F-measure for the algorithms using a data set of 71 images

	UP		RIP		RIUP	
	Acc.(%)	F − m	Acc.(%)	F − m	Acc.(%)	F − m
ANN	93.2	0.9328	94.6	0.9468	94.9	0.9496
RF	97.4	0.9740	96.6	0.9660	97.4	0.9740
SVM	95.7	0.9580	95.7	0.9580	95.7	0.9580

5   Conclusions

We have presented a method for diabetic retinopathy detection using fundus images. Our preliminary experimental results show that the best machine learning algorithm was random forest with above 97% of accuracy. In addition, local binary patterns was useful in reducing data, in fact 9 bins using rotation invariant uniform pattern produced the best classification results. Future work includes: repeating the experiments with a larger data set, classifying several types of diabetic retinopathy, comparing other machine learning algorithms and feature extraction techniques, and finally to develop a software tool for specialists.

References

1. Bonafarte, S., et al.: Retinopatía diabética. Elsevier (1997)
2. Breiman, L.: Random Forests. Machine Learning 45(1), 5–32 (2001)

3. Burges, C.: A tutorial on support vector machines for pattern recognition. Proceedings of Data Mining and Knowledge Discovery 2, 121–167 (1998)
4. Ege, B.M., Hejlesen, O.K., Larsen, O.V., Moller, K., Jennings, B., Kerr, D., Cavan, D.A.: Screening for diabetic retinopathy using computer based image analysis and statistical classification. Computer Methods and Programs in Biomedicine 62(3), 165–175 (2000)
5. Gowda, A., et al.: Exudates detection in retinal images using back propagation neural networks. International Journal of Computer Applications 25 (2011)
6. Hem, K., et al.: Fluorescein angiosgraphy: A user's manual (2008)
7. International Diabetes Federation. IDF Diabetes Atlas153, 7–68 (2013)
8. Kavitha, D., Duraiswamy, K.: Automatic detection of hard and soft exudates in fundus images using color histogram thresholding. European Journal of Science Research 48, 493–504 (2011)
9. Mitchell, T.: Machine learning. McGrawHill (1997)
10. Niemeijer, M., et al.: Automatic detection and differentiation of drusen, exudates and cotton-wool spots in digital color fundus photographs for diabetic retinopathy diagnosis. Investigative Opthalmology and Visual Science 48 (2007)
11. Ojala, T., Pietikainen, M., Harwood, D.: A comparative study of texture measures with classification based on featured distribution. Pattern Recognition 29(1), 51–59 (1996)
12. Ojala, T., Pietikainen, M., Maenpaa, T.: Multiresolution gray-scale and rotation invariant texture classification with local binary patterns. IEEE Transactions on Pattern Analysis and Machine Intelligence 24(7), 971–987 (2002)
13. Osareh, A., et al.: Automatic recognition of exudative maculopathy using fuzzy c-means clustering and neural networks. Medical Image Understanding and Analysis 3, 49–52 (2001)
14. Silberman, N., et al. Case for automated detection of diabetic retinopathy. In AAAI Spring Symposium on AI for Development, (2010)
15. World Health Organization. Diabetes mellitus. Media Centre (2014)

Automatic Validation of Flowmeter Data in Transport Water Networks: Application to the ATLLc Water Network

Diego Garcia[1], Joseba Quevedo[1], Vicenç Puig[1], Jordi Saludes[1],
Santiago Espin[2], Jaume Roquet[2], and Fernando Valero[2]

[1] Advanced Control Systems (SAC), Universitat Politècnica de Catalunya (UPC),
Campus de Terrassa, Rambla Sant Nebridi, 10,
08222 Terrassa, Barcelona, Spain
{diego.garcia,joseba.quevedo,vicenc.puig,jordi.saludes}@upc.edu
[2] ATLL Concessionària de la Generalitat de Catalunya S.A.
Sant Martí de l'Erm, 30,
08970 Sant Joan Despí, Barcelona, Spain

Abstract. In this paper, a methodology for data validation and reconstruction of flow meter sensor data in water networks is presented. The raw data validation is inspired on the Spanish norm (AENOR-UNE norm 500540). The methodology consists in assigning a quality level to data. These quality levels are assigned according to the number of tests that data have passed. The methodology takes into account not only spatial models but also temporal models relating the different sensors. The methodology is applied to real-data acquired from the ATLLc Water Network. The results demonstrate the performance of the proposed methodology in detecting errors in measurements and in reconstructing them.

1 Introduction

In any water network, a telecontrol system must acquire, store and validate data obtained in real time from sensors periodically (e.g. every few minutes) to achieve an accurate monitoring of the whole network.

The sensor measures a physical quantity and converts it into a signal that can be read by an instrument. The measuring system then converts the sensor signals to values aiming to represent certain "real" physical quantities. These values, known as "raw data", need to be validated before they can be used in a reliable way for several network water management tasks, namely: planning, investment plans, operations, maintenance and billing/consumer services and operational control (Quevedo et al., 2010a).

Frequent operation problems in the communication system between the set of the sensors and the data loggers, or in the telecontrol itself, generate missing data during some periods of time. Therefore, missing data should be replaced by a set of estimated data obtained from other spatially related sensors.

E. Corchado et al. (Eds.): IDEAL 2014, LNCS 8669, pp. 118–125, 2014.

A second common problem is the lack of reliability of the water system meters (e.g. due to offset, drift and breakdowns) producing false flow data readings. These false data must also be detected and replaced by estimated data.

According to the nature of the available knowledge, different types of data validation can be implemented, with varying degrees of sophistication. In general, one may distinguish between elementary signal-based ("*low-level*") methods and model-based ("*higher level*") methods (see, e.g. Denoeux et al., 1997; Mourad & Bertrand-Krajeswski, 2002). Elementary signal based methods use simple heuristics and limited statistical information of a given sensor (Burnell, 2003; Jorgensen et al., 1998; Maul-Kotter & Einfalt, 1998).

Typically, these methods are based on validating either signal values or signal variations. In the signal value-based approach, data are assessed as valid or invalid according to two thresholds (a high one and a low one); outside these thresholds data are assumed invalid. On the other hand, methods based on signal variations look for strong variations (peaks in the curve) as well as lacks of variation (flat curve).

Model-based methods rely on the use of models to check the consistency of sensor data. This consistency check is based on computing the difference between the predicted value from the model and the real value measured by the sensors. Then, this difference, known as residual, will be compared with a threshold value (zero in the ideal case). When the residual is bigger than the threshold, it is determined that there is a problem in the sensor or in the system. Otherwise, it is considered that everything is working properly.

The result of data validation may be either a binary variable indicating whether the data are considered valid or not, or a continuous validity index interpreted as a degree of confidence in the data. Moreover, a sub-product of using model-based approaches for sensor data validation is that the prediction provided by the model can be used to reconstruct the faulty sensor.

2   Proposed Methodology

This section presents a methodology for data validation/correction of sensor data taking into account not only spatial models but also temporal models (time-series of each flow meter) and internal models of the several components in the local units (pumps, valves, flows, levels, etc.). This proposal allows for robust isolation of wrong sensor data which should be replaced by adequate estimated data. The methodology is applied to flow and level meters, since it exploits their temporal redundancy of data.

2.1   Data Validation Methodology

Raw data validation is inspired on the Spanish norm (AENOR-UNE norm 500540). The methodology is based on assigning a quality level to the considered sensor dataset. Quality levels are assigned according to the number of tests that have been passed, as represented in Figure 1. An explanation of each level is as follows:

- *Level 0*: The **communications** level simply monitors whether the data are recorded at the fixed sampling time taking used for the supervisory system to collect data (e.g. this could not be the case due to problems in the communication system).

- *Level 1*: The **bounds** level checks whether the data are inside their physical range. For example, the maximum values expected by the flow meters are obtained by pipes' maximum flow parameters.

- *Level 2:* The **trend** level monitors the data rate. For example, level sensor data cannot change more than several centimetres per minute in a real tank.

- *Level 3:* The models level uses three parallel models:

  - **Local station related variables model**: the local station model supervises the possible correlation existing between the different variables in the same local station (i.e. flow and the command in the same valve).

  - **Time series model**: This model takes into account a data time series for each variable (Blanch et al., 2009). For example, analysing historical flow data in a pipe, a time series model can be derived and the output of this time series model is used to compare and validate the recorded data.

  - **Spatial model***: The up-downstream model checks the correlation models between historical data of sensors located in different but near local stations in the same pipe (Quevedo et al., 2010b, 2012). For example, data of flow meters located at different points of the same pipe of the water network allows checking the sensor set reliability.

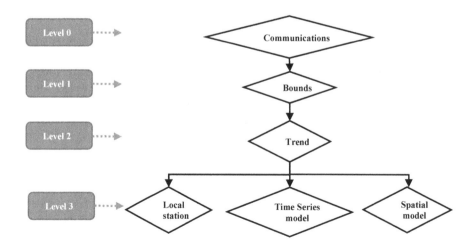

Fig. 1. Raw flowmeter data validation tests

2.2   Data Reconstruction Methodology

The levels 0, 1, 2, 3a, 3b and 3c in Figure 1 are used to validate the raw data from the sensors. If any of these levels does not validate the raw data, reconstructed data is provided by the best of the three models considered in level 3 (see Section 3). The best of these three models considered is used to reconstruct the non-validated data at time k, according to their Mean Square Error (MSE)

$$MSE = \frac{1}{L} \sum_{i=k-L}^{k-1} (y(i) - \hat{y}(i))^2 \tag{1}$$

where y is the non-validated data, $\hat{y}$ is the reconstructed data and L is the number of previous data samples used to compute the MSE.

3   Models for Data Validation and Reconstruction

In this section, the different models used for data validation and reconstruction will be described.

3.1   Spatial Model

The water network model constitutive elements and their basic relationships are introduced in this section. The mass balance expression for the i-th tank is stated as a discrete-time difference equation

$$y_i(k+1) = y_i(k) + \frac{\Delta t}{A_i} \left(q_{in_i}(k) - q_{out_i}(k) \right) \tag{2}$$

where $y_i(k)$ is the tank level, A_i is the tank section, $q_{in_i}(k)$ is the manipulated inflow and $q_{out_i}(k)$ is the outflow, which may include manipulated tank outflow and consumer demands, both given in m^3/s.

Moreover, in a water network system nodes are represented as intersections of mains, which mass balance may be expressed as the static equation

$$\sum_i q_{in_i}(k) = \sum_i q_{out_i}(k) \tag{3}$$

where, similarly to Equation (2), $q_{in_i}(k)$ and $q_{out_i}(k)$ correspond to the inflow and outflow of the i-th subnet node, also given in m^3/s.

3.2   Time-Series Model

Usually the flow in the pipes have a daily repetitive behaviour that can modelled using a Time Series (TS) model. TS models take advantage of the temporal redundancy of the measured variables. Thus, for each sensor with periodic behaviour, a TS model can be derived:

$$\hat{y}_{ts}(k) = g(y_m(k-1), ..., y_m(k-L))$$ (4)

where g is the TS model, for data exhibiting a periodicity of L samples.

The aggregate hourly flow model may be built on the basis of a time series modelling approach using ARIMA modelling (Box &Jenkins, 1970) or using Holt-Winters Time Series Model. A TS analysis is carried out on several daily aggregate series, which consistently showed a daily seasonality, as well as the presence of deterministic periodic components. A general expression for the hourly time series model can be derived using three main components (Quevedo, 2010a):

- One-day-period oscillating signal with zero average value to cater for cyclic deterministic behaviour, implemented using a second-order (two-parameter) model with two oscillating modes, in s-plane $s_{1-2}=+/-2\pi/24\ j$ or equivalently, in z-plane: $z_{1-2} = cos(2\ \pi/24)+/- sin(2\ \pi/24)j$. The oscillating polynomial is

$$y(k)= 2cos(2\pi/24)\ y(k-1) - y(k-2)$$ (5)

- An integrator that taking into account possible trends and non-zero mean values of the flow data is described by

$$y(k) = y(k-1)$$ (6)

- An autoregressive component of order 21 to consider the influence of previous values within the series is considered

$$y(k) = -a_1y(k-1) - a_2y(k-2) - a_3y(k-3)- ... - a_{21}y(k-21)$$ (7)

Component (6) plus the orders of the two components presented in (4) and (5) leads to a final order of 24 (i.e. number of samples within a day for sampling period of 1 h) for the obtained model with the following structure

$$y_p(k) = -b_1y(k-1) - b_2y(k-2) - b_3y(k-3) - b_4y(k-4) - b_5y(k-5)$$
$$- b_6y(k-6) - ...- b_{24}y(k-24)$$ (8)

Thus, this TS model of order 24 is consistent with the daily pattern (see Figure 3).

4   Application to the ATLLc Water Network

The methodology presented in previous section has been applied to ATTLc Water Network. The methodology presented in previous section exploits the "*spatial redundancy*" existing in the networks by means of spatial models relating upstream and downstream flow meters. The methodology is applied through the following steps to search outliers and reconstruct data when they are found. First, in case of two flow meters in the same pipe, a linear model given by

$$\sum_{j=1}^{n_{in}} q_{in_j}(t) = K \sum_{l=1}^{n_{out}} q_{out_l}(t) + M \tag{9}$$

is found, where $\sum_{j=1}^{n_{in}} q_{in_j}(t)$ and $\sum_{l=1}^{n_{out}} q_{out_l}(t)$ are the hourly flows measured by the input and output sensors, respectively (see Fig. 2). If there is a tank between the input and output sensors, data from the sensor level is included in the input sensor data.

Fig. 2. Two flowmeters in the same pipe

Parameters K and M are estimated by using real data and using the least-squares method. In the ideal case, those parameter should be $K = 1$ and $M = 0$, respectively. Then, with the residuals obtained by this model and using a threshold of 3σ (three times the standard deviation), outliers can be found and removed.

Additionally, a 24 hours ARIMA time series models are found for both input and output sensors. They are used to determine if the outlier values belongs to the input or to the output sensor. Finally, the invalidated data are been reconstructed by the model that provides better prediction according to MSE in (1).

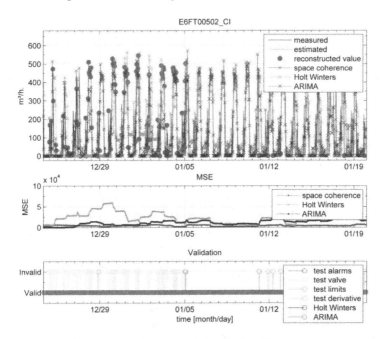

Fig. 3. Validation and reconstruction results of the flowmeter E6FT00502

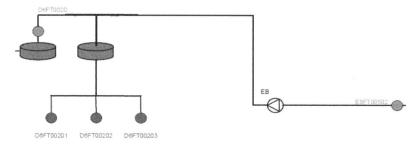

Fig. 4. Spatial relationship between 5 flowmeters and 2 level sensors of a tank

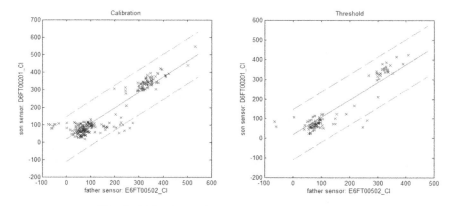

Fig. 5. Calibration and threshold of the spatial model

Figure 3 shows the results of a flow meter with a spatial model including two level sensors in case of a tank with two bodies and five flow meters (Figure 4) and two time series models for the reconstruction phase. Most of the data invalidated by limits test and valves flow meter incoherence test have been reconstructed by the spatial model presented in Figure 5.

5   Conclusions

In this paper, a methodology for automatic data validation and reconstruction of sensor data of the water network has been developed taking into account not only spatial models but also temporal models (time series of each flowmeter) and internal models of the several components in the local units (pumps, valves, flows, levels, etc.). The methodology consists in assigning a quality level to data and quality levels are assigned according to the number of tests that have been passed.

Acknowledgements. This work is partially supported by CICYT SHERECS DPI-2011-26243 of the Spanish Ministry of Education, by EFFINET grant FP7-ICT-2012-318556 of the European Commission and by AGAUR Doctorat Industrial 2013-DI-041.

References

1. Blanch, J., Puig, V., Saludes, J., Quevedo, J.: ARIMA Models for Data Consistency of Flowmeters in Water Distribution Networks. In: 7th IFAC Symposium on Fault Detection, Supervision and Safety of Technical Processes, pp. 480–485 (2009)
2. Box, G.E.P., Jenkins, G.M.: Time series analysis forecasting and control. Holden-Day (1970)
3. Burnell, D.: Auto-validation of district meter data. In: Maksimovic, Butler, Memon (eds.) Advances in Water Supply Management, Swets & Zeitlinger Publish. (2003)
4. Denoeux, T., Boudaoud, N., Canu, S., Dang, V.M., Govaert, G., Masson, M., Petitrenaud, S., Soltani, S.: High level data fusion methods. Technical Report CNRS/EM2S/330/11-97v1.0, Université de Technologie de Compiègne (1997)
5. Jörgensen, H.K., Rosenörn, S., Madsen, H., Mikkelsen, P.: Quality control of rain data used for urban run-off systems. Water Science and Technology 37, 113–120 (1998)
6. Maul-Kötter, B., Einfalt, T.: Correction and preparation of continuously measured rain gauge data: a standard method in North Rhine-Westphalia. Water Science and Technology 37(11), 155–162 (1998)
7. Mourad, M., Bertrand-Krajeswski, J.L.: A method for automatic validation of long time series of data in urban hydrology. Water Science and Technology 45(4-5), 263–270 (2002)
8. Quevedo, J., Pascual, J., Puig, V., Saludes, J., Espin, S., Roquet, J.: Data validation and reconstruction of flowmeters to provide the annual efficiency of ATLL transport water network in Catalonia. New Developments in IT & Water (2012)
9. Quevedo, J., Puig, V., Cembrano, G., Blanch, J.: Validation and reconstruction of flowmeter data in the Barcelona water distribution network. Control Engineering Practice Journal 18(6), 640–651 (2010a)
10. Quevedo, J., Blanch, J., Puig, V., Saludes, J., Espin, S., Roquet, J.: Methodology of a data validation and reconstruction tool to improve the reliability of the water network supervision. In: Water Loss Conference 2010, Sao Paulo, Brazil (2010b)
11. UNE: Redes de estaciones meteorológicas automáticas: directrices para la vali-dación de registros meteorológicos procedentes de redes de estaciones automáticas: validación en tiempo real. AENOR UNE 500540 (2004)

Data Analysis for Detecting
a Temporary Breath Inability Episode

María Luz Alonso[1], Silvia González[2], José Ramón Villar[3], Javier Sedano[2],
Joaquín Terán[1], Estrella Ordax[1], and María Jesús Coma[1]

[1] Hospital Universitario de Burgos, 09006 Burgos,
Unidad de Sueño y Unidad de Investigación, Spain
joaquinteransantos@yahoo.es
[2] Instituto Tecnológico de Castilla y León. C/ López Bravo 70, Pol. Ind.
Villalonquejar, 09001 Burgos, Spain
javier.sedano@itcl.es
[3] Computer Science Department, University of Oviedo, ETSIMO, c/Independencia
13, 33004 Oviedo, Spain
villarjose@uniovi.es

Abstract. This research is focused on a real world problem: the identification of a specific type of apnea disorder. Recently, a technique for
evaluating and diagnosing of a certain type of apnea has been published
carried out. In a brief, this technique proposes that a subject is given
with two belts, to be placed on the thorax and on the abdomen, respectively; each belt includes a 3D accelerometer. In a sleep laboratory, the
subject is monitored while sleeping and the apnea episodes are manually
discovered and registered. Besides, during the test, the data from the
sensors is gathered and segmented. The hypothesis of this study is that
the diagnose of the apnea episodes can be accomplished using the data
from a single 3D acceleration sensor. If successful, this technique for the
diagnose of the apnea might reduce the costs of the tests as well as allow
evaluating challenging cases, as those related with children or with elder
people.
This study focus on the time series (TS) analysis to extract the most
relevant patterns corresponding with the apneas episodes.Focusing on
the analysis of the TS, this study will apply a well-known TS technique
to extract the most relevant patterns. The main contributions of this
study are i) to determine if a previous step for estimating the posture is
needed, which is a very important decision in the design of the embedded
solution, and ii) the evaluation of the hypothesis of diagnosing by means
of a single 3D accelerometer.

Keywords: HOT-SAX, Apnea diagnosis, Wearable Sensors, Ambient
Assisted Living.

1 Introduction

The used of 3D accelerometers in the diagnosis of relevant events in the subject everyday life is getting the focus within the scientific community [12,15,16],

E. Corchado et al. (Eds.): IDEAL 2014, LNCS 8669, pp. 126–133, 2014.

specially after the development of bracelets that are available in the market at reduced fees, [4] among others.

The general process is to place one or more sensors in different parts of the body of the subject, and then start to register the acceleration raw data (RD). This RD is a three component acceleration that includes the acceleration due to the Earth gravity plus the acceleration of the part of the body where the sensor is located. Consequently, the placement of the sensor plays an important role in the selection of which are the most interesting RD transformations [11]. Besides, the selection of a location and the most suitable transformations are problem dependent and, up to our knowledge, no general solution can be envisage.

According to the problem faced and the sensor locations several design decisions must be taken, for instance, the pre-processing and transformations that best suite the problem, or the inclusion of modules for estimating the posture, among others. The posture of a part of the body can be easily measured with a specific inertial sensor, but that means that two sensors should be used: the 3D accelerometer and the inertial sensor [15,16]. However, extracting the gravity from the RD and simple vector multiplications would lead to estimate the current posture, though the estimated posture would lack in an uncertainty level [8].

From now on, this study will focus one specific problem: the detection of patterns due to a certain type of apnea sleeping disorder. Currently, a novel method for the apnea diagnosis has been proposed based on the measurements gathered from two 3D accelerometers [1,2]. This study is a consequence of the one presented in [13], where some features were chosen for detecting the breath recovery after an apnea study. The aim of this study is twofold: on the one hand, it is desired to determine if the posture detection is needed or not. On the other hand, the hypothesis of setting up an apnea episode diagnose with a single 3D accelerometer is evaluated.

The structure of this study is as follows. The next section presents the most representative methods for apnea diagnosis. Afterwards, the analysis to be performed are detailed in Sect. 3. Moreover, an algorithm for assisting in the diagnosis of the apnea is depicted. Sect. 4 deals with the experiments and results. Finally, the Sect. 5 draw the conclusions and future work of this study.

2   A Brief Description of the Methods for the Apnea Diagnosis

The Spanish Society of Pneumology and Thoracic Surgery defines Obstructive Apnea Hypopnea Syndrome (OAHS) as "a picture of excessive sleepiness, cognitive-behavioral disorders, respiratory, cardiac, metabolic or inflammatory secondary to repeated episodes of obstruction of the upper airway during sleep". These episodes are measured with the Respiratory Disturbance Index (RDI): an apnea diagnose is positive if the RDI value, computed for those symptoms related with the apnea that have not been explained by other causes, is higher than 5. However, this definition is rather controversial.

The most referenced method for diagnosing sleep disorders, the OSAHS among them, is the Conventional Polysomnography (PSG), where technicians monitor subjects while sleeping in a specific sleep labs. The PSG quantifies both the amount and quality of sleep and breathing disorders as well as the physiological implications. In order to identify the sleep stages several variables are recorded with the PSG: the electroencephalogram, the electrooculogram, the chin electromyagram, the pulse oximetry, the naso-bucal airflow (by nasal cannula and thermistor), the thoracic and abdominal snoring movements and breathing disorders and the electrocardiogram. The PSG should be performed at night or during the subject's usual period of sleeping, recording at least 6.5 hours provided more than 3 sleeping hours are included. The PSG is a relatively expensive, highly laborious and technically complex technique that is not universally available. Nowadays, the demand for centers exploration and the high prevalence of the OSAS make impossible to fulfill with all the service appliances.

Besides, the apnea is defined as an absence or reduction higher than 90 % of the respiratory signal detected by thermistors, nasal cannula or pneumotachograph and a duration longer than 10 seconds in the presence of respiratory effort detected by thoraco-abdominal bands. The respiratory effort includes both the abrupt inhalation and exhalation. During the inspiration stage, the outer intercostal muscles contract, raising both the chest cavity and the ribs. Conversely, when exhaling, the inner intercostal muscles contract bringing the diaphragm and the ribs back to their original position.

All these movements, as well as the subjects posture, can be registered using 3D accelerometers. Recently, a study devoted to tackle these problems using a novel method called Respiratory Polygraphy (RP) has been published [1,2]. This method proposes the use of a set of two portable sensors for recording the respiratory variables. Two 3D accelerometers are placed, one on the thorax and another on the abdomen, in order to detect episodes of apnea. According to the American Academy of Sleep Medicine [3,7,10] , the RP is a Type 3 study, where breathing and pulse oximetry thoracoabdominal effort for a total between 4 to 7 channels are recorded.

Using the RP it is possible to register the apnea episodes when experts are running the experiment in the sleep laboratory. The challenge is to register the apnea cases using only the information from a single 3D accelerometer, which may drastically reduce the costs of the experiments and make it portable. Provided this challenge is achieved, the RP could be specially useful when special subjects should be studied, like children and elder people performing apneas episodes. If the RP perform successfully, then the apnea studies would be easier and costless as the subjects might have the controls in their own bedroom. It is worth noticing that current portable sleep monitoring devices includes plenty of sensors and are really uncomfortable. On the other hand, the most similar study in the literature proposes the use of two accelerometers instead of a single one in order to detect respiratory rate abnormalities [8].

3   Decisions on Posture and Remarkable Patterns

As mentioned before, the aim of this study includes not only the extraction of the relevant patterns for the apnea's breath recovery episodes but also to determine if the posture estimation is needed. Briefly, the RD is filtered so the gravity (G) and the body acceleration (BA) are extracted [12]. Provided that both the sensor's axis alignment and the sensor placement are known, it is possible to use the G components to determine the body posture using simple scalar products between vectors. However, the expected results lack in imprecision because i) filtering introduces errors in the G and BA, and ii) no perfect alignment between the axis and the acceleration vectors will be obtained: the posture becomes imprecise as long as several possible postures can be possible for a certain acceleration value.

Managing uncertainty in embedded devices is possible and plenty approaches can be found in the literature. Nevertheless, the computer requirements would increase in case of embedding the posture estimation, as well as the power consumption. Consequently, it is not an option to determine if the block of code for estimating the posture is needed.

In order to determine if this posture estimation is needed or not, the following reasoning has been designed. Firstly, the RP experiment including the 3D accelerometers must be carried out for a subject suffering apneas episodes. This data must be segmented considering the postures and the apnea episodes should me clearly marked.

Secondly, an interval of the RP experiment for which the subject had suffered several apnea episodes in the same posture is chosen. Only the data gathered from the 3D accelerometer placed on the thorax is considered. For this period of time, two main series of data are to be analyzed: i) the data for the most representative RD's component within the current posture, and ii) the data from the modulus of the acceleration. The former is manually chosen by means of visualizing the data and then selecting the axis that better resembles the apnea episodes. The latter TS is simply the square root of the sum of the squares of the three acceleration components. Therefore, for both TS the segmentation holds and the apneas episodes can be perfectly located.

Thirdly, an algorithm for extracting discords must be applied to both TS and the patterns for anomalous behavior should be found. In this study, discords are defined as patterns within TS that are not usually performed. Several algorithms can be found in the literature [5,6]; in this study, the anomalous behavior is the breath recovery after an apnea episode and the HOT-SAX [6] approach has been chosen due to its simplicity. In case that the apnea episodes can not be detected with the modulus but with the acceleration component the body posture will be required.

Finally, either with the acceleration component or with the modulus, the extracted patterns should be analyzed for their generality. In other words, these patterns for anomalous behavior -the breath recovery periods- can be general to all the patients. If this premise is true, then an algorithm for detecting apnea episodes can be easily developed. This algorithm is outline in the next subsection. Consequently, if general patterns are found then the cost of the apnea evaluation test would be highly reduced as the RP with the two accelerometers would be valid enough.

3.1 Deployment of the Relevant Patterns

A very simple algorithm is just needed to determine the existence of an apnea episode: it is based on the K-Nearest Neighbour algorithm. Any TS is represented using SAX. Each type of apnea abnormality induces a set of patterns extracted by HOT-SAX. Then, the certainty of an apnea episode is aggregated with the similarity of the current TS with the episode's set of registered patterns. Therefore, a set of SAX patterns is collected for each analyzed abnormal behavior -in this study, only the breath recovery episodes are considered-. In order to assert that a abnormal behavior has been detected, each of its stored SAX patterns are compared with the current activity and the certainty of the class is incremented accordingly. An specific certainty model need to be specified: from the simple case of a match infers the alarm to the more elaborate certainty aggregation models, fuzzy models for instance. Again this issue needs further analysis.

Algorithm 1. Breath recovery recognition algorithm

Require: X, a normalized TS of size n
Ensure: Computes E, the possible episodes and their certainties
 Determine $\widehat{X} = SAX(X)$
 for each $A \in$ Available abnormality to Detect **do**
  for each $Y_{pattern} \in$ Set of Abnormal Patterns for A **do**
   Compute similarity between $\widehat{X}$ and $Y_{pattern} \rightarrow simXY$
   Update the certainty of $\widehat{X}$ of class A according to $simXY$
  end for
 end for
 E={}, the set of episodes found
 for each $A \in$ Available abnormality to Detect **do**
  if certainty of A is higher that 0 **then**
   insert A and its certainty in E
  end if
 end for

---

4   Experimentation and Results

For this study a pair of belts have been developed. Each belts includes a data logger device that integrates a triaxial accelerometer [14], which is sampled at a frequency of 16 Hz, and wireless communications with a laptop. Each sample is a tuple with the three components of the RD $\{a_{i,j}\}$, $\forall i \in \{x, y, z\}$ and $j = 1, 2, \ldots, t, \ldots$ the time stamp.

One middle-age volunteer has been enrolled in the study, which was carried out in the sleep laboratory. For the essay, the belts with the accelerometers were placed on the thorax and on the abdomen as stated by RP. The data gathered from the subject were monitored, segmented according the posture and stored. We then focused on detecting the apneas in the decubito supine posture: in this posture, the most representative axis for apnea episode registering is the accelerometer $z-$axis. Only the data from the 3D accelerometer located on the thorax was considered.

The HOT-SAX algorithm was executed 15 consecutive times for the detection of as many discordant patterns as possible. Once a pattern was detected by this algorithm, it was marked to avoid obtaining repeated results from the following goes of the algorithm. The HOT-SAX parameters -number of symbols and number of intervals- were set to 3 as stated in the original contribution; the window size was set to 120 samples.

Results are depicted in Fig. 1 for both the $z-$axis -the upper figure- and for the module -the lower figure-. In both of them, the dashed line marks when the apnea episodes took place. The dotted boxes represent the discords extracted from the 15 goes of the HOT-SAX algorithm; these boxes have been enlarged for the sake of the visualization. The patterns at the beginning and at the end of the TS correspond with changes in the posture of the subject while sleeping and are the limits of the interval to consider. Besides, using the $z-$axis, the first apnea episode was not identified, while episodes 2, 3 and 4 were extracted in the 6th, 11th and 2nd goes, respectively. On the other hand, when using the modulus of the acceleration, all the apneas episodes were extracted at run 13th, 4th, 8th and 5th, correspondingly.

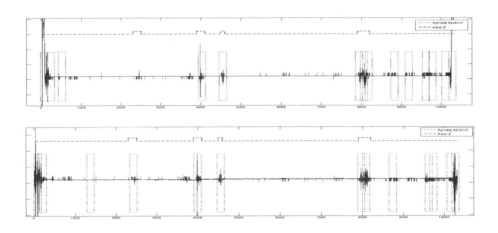

Fig. 1. Discord patterns detected after the 15 goes of the HOT-SAX algorithm for the $z-$axis -upper part- and the module -lower part-

The first remarkable result is that the $z-$axis is not able to detect the whole set of apnea episodes. However, all the apnea episodes are identified and the corresponding patterns are extracted when HOT-SAX is run on the modulus of the RD. This is a very interesting result as far as it seems that using the posture estimation might not be needed. However, the results found so far need to be tested with several subjects and during the whole RP test, considering the total apnea episodes in any possible posture. Furthermore, it was found that several patterns describe the same apnea episode. This means that i) the length of an apnea episode is relatively higher than the window size, and ii) the algorithm presented in subsection 3.1 must consider evaluating the certainty each pattern induces on each abnormal behavior and the aggregation of the certainties when determining the abnormal behaviors the subject suffered.

5   Conclusions and Future Research Lines

This study deals with the extraction of patterns that describe apnea episodes. The well-known HOT-SAX algorithm is used for the discord discovery, and an algorithm for classifying the TS windows is briefly outline. Moreover, the study focus on the need of a posture estimation module for the apnea discovery. Results showed that posture estimation seems useless as long as using the acceleration modulus has outperformed the results obtained for the most representative acceleration component. In addition, the occurrence of several distorts describing the same apnea episode may induce the certainty management in the classification algorithm: each matched pattern might introduce a certainty for several abnormal behaviors; these certainties must be aggregated for determining the most promising classes, if any.

Future work includes studying the discord discovery from several subjects and during the whole RP test in order to confirm the results obtained from this study. In addition, a more specific algorithm should be designed in order to detect apnea episodes, including the certainty each pattern suggests. Once this algorithm is finished, then the validation of the method (RP plus intelligent analysis) can be carried out.

Acknowledgments. This research has been funded by the Spanish Ministry of Science and Innovation, under project TIN2011-24302, and has been supported through the SACYL 2013 GRS/822/A/13.

References

1. Alonso, M.L., Canet, T., Cubel, M., Estivill, E., Fernandez, E., Gozal, D., Jurado, M.J., Lluch, A., Martínez, F., Merino-Andreu, M., Pin-Arboledas, G., Roure, N., Sanmartí, F., Sans, O., Segarra, F., Tomás, M., Terán, J.: Documento de consenso del síndrome de apneas-hipopneas durante el sueño en niños. Archivos de Bronconeumología 47, 1–18 (2011), http://www.sciencedirect.com/science/article/pii/S0300289611700254
2. Alonso, M.L., Terán, J., Cordero, J., González, M., Rodríguez, L., Viejo, J.L., Marañón, A.: Fiabilidad de la poligrafía respiratoria domiciliaria para el diagnóstico del síndrome de apneas/hipopneas durante el sueño. Análisis de costes Arch. Bronconeumol. 44(1), 22–28 (2008)
3. Berry, R., Brook, S.R., Gamaldo, C.E., Harding, S.M., Marcus, C.L., Vaughn, V.B., Tangredi, M.M.: The AASM Manual for the Scoring of Sleep and Associated Events: Rules, Terminology and Technical Specification. American Academy of Sleep Medicine (2012), http://www.aasmnet.org/Store/ProductDetails.aspx?pid=176

4. FITBIT co: Fitib website (2014), http://www.tbit.com
5. Hills, J., Lines, J., Baranauskas, E., Mapp, J., Bagnall, A.: Classification of time series by shapelet transformation. Data Mining and Knowledge Discovery 28(4), 851–881 (2014)
6. Keogh, E., Lin, J., Fu, A.: Hot sax: Efficiently finding the most unusual time series subsequence. In: Proceedings of the Fifth IEEE International Conference on Data Mining (ICDM 2005). IEEE Press (2005)
7. Kushida, C.A., Littner, M.R., Morgenthaler, T., Alessi, C.A., Bailey, D., Coleman, J.J., Friedman, L., Hirshkowitz, M., Kapen, S., Kramer, M., Lee-Chiong, T., Loube, D.L., Owens, J., Pancer, J.P., Wise, M.: Practice parameters for the indications for polysomnography and related procedures: an update for 2005. Sleep 28, 499–521 (2005)
8. Lapi, S., Lavorini, F., Borgioli, G., Calzolai, M., Masotti, L., Pistolesi, M., Fontana, G.A.: Respiratory rate assessments using a dual-accelerometer device. Respiratory Physiology and Neurobiology 191, 60–66 (2014)
9. Li, J., Keogh, E., Leonardi, S., Chiu, B.: A symbolic representation of time series, with implications for streaming algorithms. In: Proceedings of the 8th ACM SIGMOD Workshop on Research Issues in Data Mining and Knowledge Discovery (2003)
10. Lloberes, P., Durán-Cantolla, J., Martínez-García, M.A., Marín, J.M., Ferrer, A., Corralf, J., Masa, J.F., Parra, O., Alonso-Álvarez, M.L., Terán, J.: Normativa sobre diagnóstico y tratamiento del síndrome de apneas-hipopneas del sueño. Arhivos de Bronconeumología 47 (2011)
11. Villar, J.R., Chira, C., González, S., Sedano, J., Trejo, J.M.: Stroke onset detection by means of hybrid artificial intelligent systems. Submitted to Integrated Computer Aided Engineering (2014)
12. Villar, J.R., González, S., Sedano, J., Chira, C., Trejo, J.M.: Human activity recognition and feature selection for stroke early diagnosis. In: Pan, J.-S., Polycarpou, M.M., Woźniak, M., de Carvalho, A.C.P.L.F., Quintián, H., Corchado, E. (eds.) HAIS 2013. LNCS (LNAI), vol. 8073, pp. 659–668. Springer, Heidelberg (2013)
13. Villar, J.R., González, S., Sedano, J., Chira, C., Trejo, J.M.: Hybrid systems for analyzing the movements during a temporary breath inability episode. In: Polycarpou, M., de Carvalho, A.C.P.L.F., Pan, J.-S., Woźniak, M., Quintian, H., Corchado, E. (eds.) HAIS 2014. LNCS (LNAI), vol. 8480, pp. 549–560. Springer, Heidelberg (2014)
14. Xie, H., Fedder, G.K., Sulouff, R.E.: 2.05 - accelerometers. In: Comprehensive Microsystems, pp. 135–180. Elsevier (2008), http://www.sciencedirect.com/science/article/pii/B9780444521903000537
15. Zhou, H., Hu, H.: Upper limb motion estimation from inertial measurements. International Journal of Information Technology 13(1) (2007)
16. Zhou, H., Hu, H.: Reducing drifts in the inertial measurements of wrist and elbow positions. IEEE Transactions on Instrumentation and Measurement 59(3), 575–585 (2010)

CPSO Applied in the Optimization of a Speech Recognition System

Amanda Abelardo, Washington Silva, and Ginalber Serra

Federal Institute of Education, Science and Technology
Department of Electroelectronics,
Laboratory of Computational Intelligence Applied to Technology
Av. Getlio Vargas, 04, Monte Castelo, CEP: 65030-005, São Luis, Maranhão, Brazil
http://www.ifma.edu.br

Abstract. This paper proposes an optimization of a fuzzy inference system for the automatic recognition of numerical commands of voice using Chaotic Particle Optimization (CPSO). In addition preprocessing the speech signal with mel-frequency cepstral coefficients, we use the discrete cosine transform (DCT) to generate a two-dimensional temporal matrix used as input to a system of fuzzy implication to generate the pattern of the words to be recognized.

Keywords: Speech Recognition, Fuzzy Systems,Swarm Particle Optimization.

1 Introduction

Recently bio inspired algorithms based in meta-heuristics populations has been used to solve problems of search and optimization in many problem domains, for which robust solutions are hard or impossible to find using tradicionals methods. These algorithms are inspired in biological mechanisms of evolution. A algorithm based in this principle, which is going to discussed in this paper is a Chaotic Particle Swarm (CPSO) that is a variation of the original PSO, this method of heuristic search relative recent inspired by the cooperative behavior of group of biological populations [3]. The PSO was began for James Kennedy and Russel Elberhart since 1995 for solve problems in the continuous domain. A social cognitive theory is behind of the PSO, each individual of a population has its own experience and is able of the estimate the quality of her. How the individuals are social, they also have knowledge about its neighbors behave. These two types of information correspond the individual learning (cognitive) and the cultural transmission (social), respectively. Then, the probability of decision a certain will be a function of its past performance and the best performance of its neighbors [1]. From this principle is define a search space, where the particles have its individual velocity and position, that each iteration is refreshed according to several individual and collective variables. In this speech recognition system nonlinear techniques of optimization is tried for decrease the intrinsic confusion of the generated models for represent the voice. To find a separability metric among the patterns for optimize the recognition is the function of the CPSO.

E. Corchado et al. (Eds.): IDEAL 2014, LNCS 8669, pp. 134–141, 2014.

2   Methodology

In this proposal, a speech signal is encoded and parametrized in a two-dimensional time matrix with four parameters of the speech signal. After coding, the mean and variance of each pattern are used to generate the rule base of Mamdani fuzzy inference system. The mean and variance are optimized using particle swarm in order to have the best performance of the recognition system. This paper consider as patterns the Brazilian locutions (digits): **0, 1, 2, 3, 4, 5, 6, 7, 8, 9**. The Discrete Cosine Transform (DCT) [11],[12] is used to encoding the speech patterns. The use of DCT in data compression and pattern classification has been increased in recent years, mainly due to the fact its performance is much closer to the results obtained by the Karhunen-Loève transform which is considered optimal for a variety of criteria such as mean square error of truncation and entropy [13],[14]. This paper demonstrates the potential of DCT, CPSO and fuzzy inference system in speech recognition and the viability of them[15].

2.1   Pre-processing Speech Signal

Initially, the speech signal is digitized, so it is divided in segments they which are windowed and encoded in a set of parameters defined by the order of mel-cepstral coefficients (MFCC). The DCT coefficients are computed and the two-dimensional time DCT matrix is generated, based on each speech signal to be recognized. When a window is applied to a given signal, it selects a small portion of this signal, named frame, to be analyzed. The duration of the frame T is defined as the total of time over which a set of parameters is considered valid. The duration of the frame is used to determine the total of time from successive calculations of parameters [8]. This paper uses the hamming window with duration time (frames) of 10ms with 50% overlap between frames, thus, only a fraction of the signal is changed for each new frame. Experiments on human perception have shown that complex sound frequencies within a certain bandwidth of a nominal frequency should not be individually identified. When one of the components of this sound is out of bandwidth, this component can not be distinguished. Normally, it is considered a critical bandwidth for speech from 10% to 20% of the center frequency of the sound. One of the most popular way to map the frequency of a given sound signal for perceptual frequencies values, i.e., to be capable of exciting the human hearing range is the Mel-Scale [9].

2.2   Two-Dimensional Time Matrix DCT Coding

The two-dimensional time matrix, as the result of DCT in a sequence of T mel-cepstral coefficients observation vectors on the time axis, is given by:

$$C_k(n, T) = \frac{1}{N} \sum_{t=1}^{T} mfcc_k(t) cos\frac{(2t-1)n\pi}{2T} \tag{1}$$

where $mfcc$ are the mel-cepstral coefficients, and $k, 1 \le k \le K$, is the k-th (line) component of t-th frame of the matrix and $n, 1 \le n \le N$ (column) is the order of DCT.

Thus, the two-dimensional time matrix, where the interesting low-order coefficients k and n that encode the long-term variations of the spectral envelope of the speech signal is obtained [10]. Thus, there is a two-dimensional time matrix $C_k(n, T)$ for each input speech signal. The elements of the matrix are obtained as follows: For a given spoken word P (digit), ten examples of utterances of P are gotten. This way it has itself $P_0^0, P_1^0, ..., P_9^0, P_0^1, P_1^1 ..., P_9^1, P_0^2, P_1^2, ..., P_9^2, ..., P_m^j$, where $j = 0, 1, 2, ..., 9$ and $m = 0, 1, 2, ..., 9$. Each frame of a given example of the word P generates a total of K mel-cepstral coefficients and the significant features are taken for each frame along time. The N-th order DCT is computed for each mel-cepstral coefficient of same order within the frames distributed along the time axis, i.e., c_1 of the frame t_1, c_1 of the frame t_2, ...,, c_1 of the frame t_T, c_2 of the frame t_1, c_2 of the frame t_2, ...,, c_2 of the frame t_T, and so on, generating elements $\{c_{11}, c_{12}, c_{13}, ..., c_{1N}\}$, $\{c_{21}, c_{22}, c_{23}, ..., c_{2N}\}$, $\{c_{K1}, c_{K2}, c_{K3}, ..., c_{KN}\}$, and the matrix given in equation (1). Therefore, a two-dimensional time matrix DCT is generated for each example of the word P, represented by C_{kn}^{jm}. Finally, the matrices of mean CM_{kn}^j (2) and variances CV_{kn}^j (3) are generated. The parameters of CM_{kn}^j and CV_{kn}^j are used to produce gaussians matrices C_{kn}^j which will be used as fundamental information for implementation of the fuzzy recognition system. The parameters of this matrix will be optimized by CPSO.

$$CM_{kn}^j = \frac{1}{M} \sum_{m=0}^{M-1} C_{kn}^{jm} \tag{2}$$

$$CV_{kn}^j(var) = \frac{1}{M-1} \sum_{m=0}^{M-1} \left[C_{kn}^{jm} - \left(\frac{1}{M} \sum_{m=0}^{M-1} C_{kn}^{jm} \right) \right]^2 \tag{3}$$

where M=10.

2.3   Rule Base Used for Speech Recognition

Given the fuzzy input set A, the fuzzy output set B, should be obtained by the relational max-t composition. This relationship is given by.

$$B = A \circ Ru \tag{4}$$

Where Ru is a fuzzy relational rules base.

The fuzzy rule base of practical systems usually consists of more than one rule. There are two ways to infer a set of rules: Inference based on composition and inference based on individual rules [17], [19]. In this paper the compositional inference is used. Generally, a fuzzy rule base is given by:

$$Ru^l : IF \ x_1 \ is \ A_1^l \ and...and \ x_n \ is \ A_n^l \ THEN \ y \ is \ B^l \tag{5}$$

where A_i^l and B^l are fuzzy set in $U_i \subset \Re$ and $V \subset \Re$, and $x = (x_1, x_2, ..., x_n)^T \in U$ and $y \in V$ are input and output variables of fuzzy system, respectively. Let M be the number of rules in the fuzzy rule base; that is, $l = 1, 2, ...M$.

From the coefficients of the matrices $\mathbf{C}_{kn}^j$ with $j = 0, 1, 2, ..., 9$, $k = 1, 2$ and $n = 1, 2$ generated during the training process, representing the mean and variance of each pattern j a rule base with $M = 40$ individual rules is obtained and given by:

$$Ru^j : IF \ C_{kn}^j \ THEN \ y^j \tag{6}$$

In this paper, the training process is based on the fuzzy relation Ru^j using the Mamdani implication. The rule base Ru^j should be considered a relation $R(X \times Y) \to [0, 1]$, computed by:

$$\mu_{Ru}(x, y) = I(\mu_A(x), \mu_B(y)) \tag{7}$$

where the operator I should be any t-norm [18]. Given the fuzzy set A' input, the fuzzy set B' output might be obtained by **max-min** composition, [17]. For a minimum t-norm and max-min composition it yields:

$$\mu_{(B')} = max_x min_{x,y}(\mu_{A'}(x), \mu_{(Ru)}(x, y)) \tag{8}$$

2.4   Generation of Fuzzy Patterns

The elements of the matrix C_{kn}^j were used to generate gaussians membership functions in the process of fuzzification. For each trained model j the gaussians memberships functions $\mu_{c_{kn}^j}$ are generated, corresponding to the elements c_{kn}^j of the two-dimensional time matrix $\mathbf{C}_{kn}^j$ with $j = 0, 1, 2, 3, 4, 5, 6, 7, 8, 9$, where j is the model used in training. The training system for generation of fuzzy patterns is based on the encoding of the speech signal $s(t)$, generating the parameters of the matrix C_{kn}^j. Then, these parameters are fuzzified, and they are related to properly fuzzified output y^j by the relational implications, generating a relational surface $\mu_{(Ru)}$, given by:

$$\mu_{Ru} = \mu_{c_{kn}^j} \circ \mu_{y^j} \tag{9}$$

This relational surface is the fuzzy system rule base for recognition optimized by particle swarm to maximize the speech recognition. The decision phase is performed by a fuzzy inference system based on the set of rules obtained from the mean and variance matrices of two dimensions time of each spoken digit. In this paper, a matrix with minimum number of parameters (2×2) in order to allow a satisfactory performance compared to pattern recognizers available in the literature. The elements of the matrices $\mathbf{C}_{kn}^j$ are used by the fuzzy inference system to generate four gaussian membership functions corresponding to each element $c_{kn}^j \big|^{k=1,2;n=1,2}$ of the matrix. The set of rules of the fuzzy relation is given by:

$$IF \ c_{kn}^j \big|^{k=1,2;n=1,2} \ THEN \ y^j \tag{10}$$

$$IF \quad c'^{j}_{kn} \big|^{k=1,2;n=1,2} \quad THEN \quad y'^{j} \tag{11}$$

From the set of rules of the fuzzy relation between antecedent and consequent, a data matrix for the given implication is obtained. After the training process, the relational surfaces is generated based on the rule base and implication method. The speech signal is encoded to be recognized and their parameters are evaluated in relation to the functions of each patterns on the surfaces and the degree of membership is obtained. The final decision for the pattern is taken according to the $max - min$ composition between the input parameters and the data contained in the relational surfaces. The process of defuzzification for the pattern recognition is based on the **mean of maxima (mom)** method given by:

$$\mu_{y'^{j}} = \mu_{c'^{j}_{kn}} \circ \mu_{(Ru)} \tag{12}$$

$$y' = mom(\mu_{y'^{j}}) = mean\{y|\mu_{y'^{j}} = max_{y \in Y}(\mu_{y'^{j}})\} \tag{13}$$

3   Optimization of Relational Surface with Particle Swarm

The algorithm Particle Swarm (PSO) is a meta algorithm of local optimization because it tries optimize a set of values, but frequently it is found a local maximum rather than a global maximum [2]. In this paper, the 40 variables of mean and the 40 variables of variance codificated and fuzzicated were optimized for the CPSO, being 4 for each of the digits $(0 - 9)$ to which the recognition is done. The CPSO is motivated for the social behavior through of the competition and the cooperation among individuals, so as in nature can bring many benefits, finding good solutions efficiently and maintaining a certain simplicity in the optimization process [5]. The basic elements of the technique are:Position of the particle j $(X_j(t))$, Population $Pop(t)$, Velocity of the particle $j(V_j(t))$, Cost function $f(X(t))$, The better position of the particle j until time $t(X*(t))$, Cognitive parameter c_1, Social parameter c_2, Number of iterations and Weight of inertia. The motion of the particle is affected for two factors: the better solution found until that moment by the particle *(pbest)* and the better solution found by the swarm *(gbest)*. This factors, designated cognitive and social components influence in the locomotion of the particles by the space of search are creating attraction forces. As a result the particles interacts with all its neighbors and save in its memory the location of optimal solutions. After each iteration *pbest* and *gbest* are updated if found by the particle or by the swarm,respectively. This process repeats until the optimum result is obtained or number of iterations is reached. [4]

The basic algorithm to implement the PSO suggests that the particles approach is the optimal point of this space with the inertia term w (factor that influence the movimentation of the particle in the space) monotonically decreasing. With this term only suffering decay (only energy loss), case there is a lot of particles in a local minimum wont have how exit this point, because the particle doesn't accelerate [6], presenting problems to approach of the optimum solution. To solve this problem was added the control of inertial not monotonic. In the traditional method the inertial weight ,w, decrease linearly from 0.4 to 0.9, with the inertial control its value oscillates according the cosine function between 0.4 and 1.2 and is given by the following equation, where i is the current iteration and I is the total number of iterations.

$$ w = \left[\cos(\frac{\Pi i}{2I})m \right] + s \tag{14} $$

where $m = \frac{(w_{max}+w_{min})}{2}$ and $s = m + w_{min}$. With the value of the inertia weigh assuming a behavior not monotonic, the optimization would happen more fast and provides a global search in a moment that would less likely find another best solution.

With that, the inertial w can act on diversification and intensification of the PSO. The number of oscillations of the function that control the inertial along of the PSO is the number of attempts of leaks of local minimums, since each time w overcome 1, the swarm can scattering [7]. For accelerate the convergence process was held also the exchange between random numbers and chaotic numbers.This numbers are generated by deterministic formulas of iterations, how the logistic map, one of the chaotic maps more simple generates chaotic numbers using the follows formula: $Cr(t + 1) = k * Cr(t) * (1 - Cr(t))$. The chaos theory can be applied how technique as efficient in practical applications since can be described how a non-linear limited system with behavior dynamic deterministic that have ergodic properties and stochastic. Mathematically, a chaotic variable is aleatory and unforeseeable, but also has a element of regularity. A analysis statistical that justified the use of the logistic map in the PSO showed that the logistic map with $k = 4$, generates more aleatory numbers next of the two extremes of the interval $[0, 1]$. So the use of this characteristic allows that the particles can give bigger jumps for escape of local optimum or leave a situation of stagnation more easily, how also can give smaller jumps enabling further refinement of research[16]. The use of chaotic numbers causes the PSO receives the designation of CPSO (Chaotic Particle Optimization).

4   Results

To prove the effects of the advances in the PSO were made tests in many stages of the algorithm. The characteristics of the tests and the better cost are shown in the table 1.

How is show in the table in some tests was taken the automatic withdrawal of silence and the temporal dynamic alignment of the bank of data. How isn't a focus of this paper this process will not be discussed in detail. The best individual in the process of training by PSO and CPSO , related to base of rules of *fuzzy* inference system for recognition of the voice command, is showed in the Figure 4. The total performance using the CPSO was 89 digits identified correctly in the training process.

Table 1. Results

Chaotic Numbers	c_1	c_2	Weight control	Turb.	No silence	DTW	GBEST
NOT	3	1	NOT	NOT	NOT	NOT	29%
NOT	2	1	NOT	NOT	NOT	NOT	58%
NOT	2	1.5	NOT	NOT	YES	YES	74%
NOT	2	1.5	NOT	NOT	YES	YES	75%
NOT	2	1.5	YES	NOT	YES	YES	76%
NOT	2	1.5	YES	YES	YES	YES	80%
NOT	2	1.5	YES	YES	YES	YES	83%
YES	2	1.5	YES	YES	YES	YES	89%

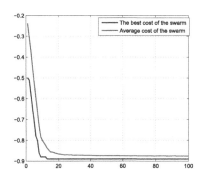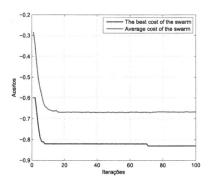

Fig. 1. The best result for the PSO and CPSO

5 Conclusion

Analyzing the tests realized with the CPSO with some alterations how was proposed in this paper, observed also that the best result achieved was in the last test, where all studied advances were apply : Automatic Withdrawal of Silence, Temporal Alignment using DTW, control of inertial weight and Turbulence, with the use of this advances noted also that the algorithm converged faster than the traditional CPSO. How future proposal has been the include of the others advances and modify of the current advances, for the algorithm achieve better values for the local maximum.

Acknowledgment. The authors would like to thank FAPEMA (Foundation for Research Support and Scientific Development of Maranhão), the financial support to this research and the Group of Applied Electronics Instrumentation Technology, Federal Institute of Education, Science and Technology of Maranhão by the infrastructure used in development of this research.

References

1. Serapião, A.: Fundamentos de Otimização por Inteligência de Enxames: Uma visão geral. Revista Controle & Automação 20(03) (2009)
2. Coppin, B.: Inteligência Artificial. In: Tradução e Revisão Técnica de Jorge Duarte Pires Valério, LTC. Rio de Janeiro, Brasil (2010)
3. da Luz, E.: Estimação de Fonte de Poluição Atmosférica Usando Otimização por Enxame de Partículas. Instituto Nacional de Pesquisas Espaciais, Tese de Mestrado, Brasil, São José dos Campos (2008)
4. Chiachia, G., Penteado, B., Marana, A.: Fusão de Métodos de Reconhecimento Facial Através da Otimização por Enxame de Partículas. In: UNESP - Faculdade de Ciências. Bauru - São Paulo, Brasil
5. Silveira, T., de Oliveira, H., da Silva, L.: Controle de Inércia para Fuga de Mínimos Locais de Funções Não-Lineares na Otimização por Enxame de Partículas. Universidade Federal de Alfenas, Brasil, Minas Gerais
6. Silveira, T., Miranda, V., de Oliveira, H.: Controle de Inércia Não-Monotônico para Fuga de Mínimos Locais na Otimização por Enxame de Partículas. Universidade Federal de Alfenas, Brasil, Minas Gerais
7. Silveira, T., de Oliveira, H., da Silva, L., Salgado, R.: Controle de Inércia Não Monotônico na Otimização por Enxame de Partículas. In: Interdisciplinary Studies in Computer Science, pp. 69–82 (July/December 2009)
8. Picone, J.W.: Signal modeling techiniques in speech recognition. IEEE Transactions on Computer 79(4), 1214–1247 (1991)
9. Rabiner, L., Biing-Hwang, J.: Fundamentals of Speech Recognition. Prentice Hall, New Jersey (1993)
10. Fissore, L., Laface, P., Rivera, E.: Using word temporal structure in HMM Speech recongnition. In: ICASSP 1997, Munich-Germany, vol. 2, pp. 975–978 (April 1997)
11. Ahmed, N., Natarajan, T., Rao, K.: Discrete Cosine Trasnform. IEEE Transaction on Computers c-24 edn. (January 1974)
12. Zhou, J., Chen, P..: Generalized Discrete Cosine Transform. In: 2009-Pacific-Asia Conference on Circuits, Communications and System, Chengdu, China (May 2009)
13. Hua, Y., Liu, W.: Generalized Karhunen–Loeve Transform. IEEE Signal Processing Letters 5(6) (June 1998)
14. Effros, M., Feng, H., Zeger, K.: Suboptimality of the Karhunen Lové Transform for Transform Coding. IEEE Transactions on Information Theory 50(8) (August 2004)
15. Zeng, J., Liu, Z.-Q.: Type-2 Fuzzy Hidden Markov Models and Their Application to Speech Recognition. IEEE Transactions on Fuzzy Systems 14(3) (June 2006)
16. Filho, O.P.C., Martinhon, C.A., Cabral, L.A.F.: Uma nova abordagem híbrida do algoritmo de Otimização por Enxame de Partículas com Busca Local Iterada para o problema de Clusterização de Dados. In: X Congresso Brasileiro de Inteligência Computacional, Porto de Galinhas - Ipojuca, Pernambuco, Brasil (2013)
17. Wang, L.-X.: A course in Fuzzy Systems and Control. Prentice Hall (1994)
18. Babuska, R.: Fuzzy Modeling for Control. Kluwer Academic Publishers (1998)
19. Chen, G.: Discussion of Approximation Properties of Minimum Inference Fuzzy System. In: Proceedings of the 29th Chinese Control Conference, Beijing, China, July 29-31 (2010)

Object-Neighbourhood Clustering Ensemble Method

Tahani Alqurashi and Wenjia Wang

University of East Anglia, Norwich Research Park, Norfolk, UK
T.Alqurashi@uea.ac.uk, wjw@uea.ac.uk

Abstract. Clustering is an unsupervised learning and clustering results are often inconsistent and unreliable when different clustering algorithms are used. In this paper we have proposed a clustering ensemble framework, named Object-Neighbourhood Clustering Ensemble (ONCE), to improve the consistency, reliability and quality of the clustering result. The core of the ONCE is a new consensus function that addresses the uncertain agreements between members by taking the neighbourhood relationship between object pairs into account in the similarity matrix. The experiments are carried out on 11 benchmark datasets. The results show that our ensemble method outperforms the co-association method, when the Average linkage is used. Furthermore, the results show that our ensemble method is more accurate than the baseline algorithm, and this indicates that the clustering ensemble method is more consistent and reliable than a single clustering algorithm.

Keywords: Clustering Ensemble, Consensus Function, Neighbourhood similarity.

1 Introduction

Clustering is the process of identifying natural groups within data by assigning each object to a specific group or cluster, so that objects in the same cluster are more similar to each other than objects from any other group. However, due to the fact that there is normally no prior knowledge about the underlying structure of clusters or about any particular properties that we want to find and what clusters we consider to be good about the data [5], different algorithms may produce quite inconsistent clustering results and there is not a single clustering algorithm that always performs well for all kinds of data. These issues make that selecting an appropriate clustering algorithm for a given non-trivial problem can be a challenge, and the enumeration of all possible solutions is computationally expensive, and, in practice, infeasible for large datasets [10]. Consequently, a promising technique - clustering ensemble, has emerged in the past decade.

In general, a clustering ensemble combines multiple partitions $P_i, i = 1...M$, of the same data into a single final partition π^* without accessing the original features. The main challenge in clustering ensemble is to define a consensus function CF that combines a set of members into a better partition π^* in terms

E. Corchado et al. (Eds.): IDEAL 2014, LNCS 8669, pp. 142–149, 2014.

of consistency, reliability and accuracy, compared with the individual members in the ensemble.

In this paper, a new consensus function is presented based on a novel similarity measure, which considers the average neighbourhood similarity when calculating the similarity matrix between pairs of objects. The derived algorithm is hence called Object Neighbourhood Clustering ensemble, ONCE, in short.

The rest of the paper is organized as follows. Section 2 briefs the related works. Section 3 describes the proposed method. Section 4 presents the experiment design and results. Conclusions are provided in Section 6.

2 Related Work

Some studies have focused on combining members by mapping them onto a new representation in which the pairwise similarity between objects is calculated, which is called the Co-association matrix (Co) [2]. It seems ideal for collecting all the information available in the clustering ensemble but in fact the original matrix [2] takes into consideration just the pairwise relationship between objects in the ensemble members. However, there is a situation where roughly half of the members place some objects pairs in the same cluster but the other half place them in a different cluster. In this case, we call them uncertain pairs of objects, which cause problems in generating reliable consensus clustering results.

Recently, researchers have realised that more information within the generated members can be obtained to create this matrix. Wang et.al, [11] extended Co by considering the cluster size and the dimensions of the objects within the data when calculating the similarity matrix; they call it probability accumulation (PA).

In PA, a more informative similarity matrix is obtained, which means that the chance of obtaining several pairs of objects with the same similarity score is less than when using Evidence Accumulation(EA). Vega-Pons et.al, [10] proposed a weighted-association matrix that takes three different factors into consideration; these are: the number of elements in the cluster to which a pair of objects belongs, the number of clusters in the ensemble member analysed, and the similarity value between the objects that were obtained by this member. They call this method Weighted Evidence Accumulation (WEA).

Most recently, Yi et.al, [12] highlighted an issue that is often overlooked by other methods: how to handle the uncertain data pairs when calculating the similarity matrix. They addressed this issue by proposing a new clustering ensemble, based on the matrix completion theory, where they filtered the similarity value of the uncertain pairs in the EA matrix, and then they estimated their value to complete the matrix by applying a matrix completion algorithm namely Augmented Lagrangian as proposed by [7]. In fact, they did not deal with this issue, they avoided it through a matrix completion process, which may cause some information loss.

3   Object-Neighbourhood Clustering Ensemble

As mentioned earlier, a key task in clustering ensemble is to design a consensus function for combining multiple partitions to generate a single and hopefully better partition. We defined a consensus function based on our novel similarity measure that addresses the limitation of co-association matrix by considering the neighbourhood similarity.

The key idea of our similarity definition is derived from Jarvis and Patrick [6], who defined the similarity between object pairs as the number of nearest neighbours that the pair shares as long as the objects themselves belong to their common neighbourhood. They call it Shared Nearest Neighbour (SNN).

Generally, the neighbourhood is the region in the data space covering the pair of objects in question. Therefore, objects in the same cluster are all considered to be in the same neighbourhood region, and objects in different clusters are not considered to be in the same neighbourhood region.

Definition 1. *The common neighbours to a pair of objects are the other objects in the same cluster as the pair itself.*

Thus, the similarity measurement between any two objects is defined as:

Definition 2. *The more common neighbours that two objects have, the more similar they are.*

The difference between our definition and Jarvis and Patrick definition [6] is that the latter is based on the number of shared nearest neighbours, determined by any similarity/distance measure, whereas we take the similarity score of all the shared/common neighbours into consideration when we calculate the similarity between pairs of objects.

Assume that x_a is a common neighbour to x_i and x_j, and Z is the set of all common neighbours between x_i and x_j, for each pair of objects, the average similarity, $B(x_i, x_j)$, of their common neighbours is defined as follows:

$$B(x_i, x_j) = \frac{\sum_{x_a \in Z} A(x_a, x_i) + A(x_a, x_j)}{2 \mid Z(x_i, x_j) \mid} \tag{1}$$

Where, $A(x_a, x_i)$ and $A(x_a, x_j)$ are the neighbourhood association of x_i and x_j to their common neighbour x_a respectively and can be calculated from the co-association matrix defined as:

$$A(x_i, x_j) = \frac{1}{M} \sum_{m=1}^{M} \delta(P_m(x_i), P_m(x_j)) \tag{2}$$

Where $\delta(P_m(x_i), P_m(x_j))$ is 1 if x_i and x_j are placed in the same cluster in partition P_m, and 0 otherwise.

$B(x_i, x_j) \in [0, 1]$, when $B(x_i, x_j) = 0$, it means that there are no common neighbours between x_i and x_j, and when $B(x_i, x_j) = 1$, it means that x_i, x_j and their common neighbours are placed in the same by all ensemble members.

When $0 < B(x_i, x_j) < 1$, it means that x_i and x_j have some common neighbours placed in same cluster by some members in the ensemble.

Then, the overall neighbourhood similarity W if defined by the following equation:

$$W(x_i, x_j) = B(x_i, x_j) + A(x_i, x_j) \tag{3}$$

W should be scaled to $[0,1]$.

A corresponding ensemble framework is derived with a hierarchical clustering algorithm as its consensus function based on the neighbourhood similarity. The framework is named Object-Neighbourhood Clustering Ensemble (ONCE).

4 Experimental Design

4.1 Experiment Procedure

For a given dataset, an ensemble is built with seven members generated by using 7 different algorithms respectively, which are k-means, agglomerative hierarchical clustering using single and average linkage, k-medoids, c-means, Kernel k-means and the Normalised cut algorithm. As for the consensus function of the hierarchical clustering algorithm, just for the purpose of comparison, Single (SI), Complete (Cm) and Average (Av) Linkage) are used to obtain the final clustering result. Following a common practice in clustering research field, the number of cluster is set to the number of the classes for all datasets.

We also calculate the co-association matrix and for a fair comparison we used the same three different hierarchical clustering algorithms as used in the ONCE to obtain the final clustering result.

Since the labels of the data are known, we use the Normalised Mutual Information (NMI) [8] to evaluate the clustering results. Then, we run each experiment ten times with different initial conditions, and compute the average accuracy and the standard deviation to verify the consistency and the reliability of the ensembles. Moreover, we report the average performance for each method across all datasets as well as the standard deviation. We also run the experiment using k-means as a baseline algorithm. The statistical significance of the results is evaluated by employing some non-parametric tests described in [4] and [1].

4.2 Datasets

The experiment is conducted using eleven datasets. The D31 and R15 datasets were generated by [9], while the aggregation dataset was generated by [3]. The real datasets are from the UCI repository[1]. The characteristics of these datasets are given in Table 1.

---

[1] http://archive.ics.uci.edu/ml/datasets.html

Table 1. Details of the datasets used in the experiments

Dataset	# Objects	# Features	# Cluster	Dataset type	Modified
D31	3100	2	31	Artificial	No
R15	600	2	15	Artificial	No
Aggregation	788	2	7	Artificial	No
Iris	150	4	3	Real	No
Wine	178	13	3	Real	No
Thyroid	215	5	3	Real	No
User modelling (Um)	399	5	4	Real	Yes
Multiple Features (Mfeatures)	2000	2	10	Real	No
Glass	214	19	6	Real	No
Breast Cancer Wisconsin (Bcw)	683	9	2	Real	Yes
Contraceptive Method Choice (Cmc)	1473	9	3	Real	No

5   Results and Analysis

Table 2 presents the results of the NMI; each entry in the table represents the average accuracy of ten runs, followed by the standard deviation. The results of our method are compared with the Co method and in order to make a fair comparison we compare the result of two like-to-like algorithms; in other words, the two algorithms use the same linkage method. This is done in order to determine whether the uncertain pair affects the final clustering result of the Co ensemble, and if so to what extent our method is able to improve the quality of the final result.

The bold value in each row shows the best result comparing like-to-like algorithms, and the underlining value represents the highest accuracy for each dataset. The last two rows show the average accuracy for each algorithm over all the datasets, and the Wins (W)/Ties (T)/Losses (L) row counts the W/T/L (in terms of accuracy) comparing the two like-to-like algorithms. We also counted W/T/L in terms of the highest accuracy in order to compare ONCE with Co, and to compare the ensemble method with the baseline algorithm as well as with the member average.

As we can see, the accuracy of ONCE-SI in most of the datasets was improved, relatively to Co-SI; in particular, eight out of the eleven datasets in total were improved in terms of accuracy, whereas for the remaining datasets (Bcw, R15 and Thyroid) the accuracy was decreased.

In the case of the Aggregation dataset, the accuracy of ONCE-SI, ONCE-Cm and ONCE-Av were increased, relative to Co-SI, Co-Cm and Co-Av, respectively. On the other hand, for the Um dataset, the accuracy of ONCE-Cm and ONCE-Av were decreased, and the accuracy of ONCE-SI was slightly improved. However, in general this dataset also achieved low accuracy using k-means as well as the member average. We noticed that this is also the case in the Cmc dataset, where we obtained low accuracy with most of the ensemble methods, and we noticed that the accuracy for the member average is also very low as well as for k-means, which indicate that these datasets are not suitable for clustering analysis (or they may need a special distance/similarity measurement).

Table 2. The average accuracy of ten runs for each dataset measured by NMI on 11 datasets. Including the average performance of each method across 11 datasets and the W/T/L for each ensemble method comparing the two like-to-like algorithms.

Dataset	ONCE-Sl	Co-Sl	ONCE-Cm	Co-Cm	ONCE-Av	Co-Av	Ave-mem	k-means
D31	**0.912 ± 0.013**	0.911 ± 0.018	**0.961 ± 0.004**	**0.961 ± 0.005**	0.961 ± 0.005	**0.965 ± 0.002**	0.774 ± 0.328	0.916
R15	0.989 ± 0.009	**0.991 ± 0.007**	**0.994 ± 0.000**	0.989 ± 0.018	**0.994 ± 0.000**	**0.994 ± 0.000**	0.850 ± 0.272	0.918
Aggregation	**0.950 ± 0.002**	0.935 ± 0.022	**0.974 ± 0.010**	0.941 ± 0.038	**0.984 ± 0.006**	0.967 ± 0.029	0.767 ± 0.341	0.864
Bcw	0.026 ± 0.008	**0.047 ± 0.044**	0.457 ± 0.250	**0.702 ± 0.154**	**0.741 ± 0.003**	0.736 ± **0.002**	0.455 ± 0.341	**0.748**
Cmc	**0.028 ± 0.005**	0.012 ± 0.007	**0.032 ± 0.001**	**0.032 ± 0.001**	**0.032 ± 0.001**	**0.032 ± 0.001**	0.025 ± 0.013	0.032
Iris	**0.768 ± 0.027**	0.733 ± 0.034	0.766 ± 0.017	**0.774 ± 0.021**	**0.771 ± 0.021**	0.763 ± 0.022	0.630 ± 0.282	0.742
Glass	**0.394 ± 0.040**	0.374 ± 0.037	**0.395 ± 0.029**	0.382 ± 0.021	**0.394 ± 0.008**	0.383 ± 0.021	0.366 ± 0.133	0.368
Um	**0.040 ± 0.003**	0.039 ± 0.003	0.241 ± 0.080	**0.245 ± 0.090**	0.290 ± 0.133	**0.359 ± 0.102**	0.176 ± 0.150	0.342
Wine	**0.435 ± 0.000**	0.407 ± 0.109	0.422 ± 0.009	**0.424 ± 0.003**	**0.434 ± 0.011**	0.429 ± 0.003	0.321 ± 0.187	0.426
Mfeatures	**0.319 ± 0.087**	0.142 ± 0.094	**0.472 ± 0.034**	0.454 ± 0.031	**0.479 ± 0.001**	**0.479 ± 0.002**	0.374 ± 0.230	0.478
Thyroid	0.127 ± **0.074**	**0.195 ± 0.108**	**0.446 ± 0.075**	0.368 ± 0.080	**0.403 ± 0.096**	0.358 ± 0.080	0.228 ± 0.149	0.443
Average	**0.454 ± 0.024**	0.435 ± 0.044	0.560 ± 0.046	**0.570 ± 0.042**	**0.589 ± 0.026**	0.588 ± 0.024	0.451 ± 0.221	0.571
W/T/L	8/0/3	3/0/8	5/2/4	4/2/5	6/3/2	2/3/6	0/0/11	1/1/9

In Bcw, we believe that improving the uncertain pairs of objects makes the Single and Complete linkage not appropriate for this dataset; in general, the Single linkage with the Co and ONCE matrices achieved very low accuracy. However, the greatest improvement of our method was in the Glass dataset, which gave the highest NMI score using the Single, Complete and Average linkage methods, comparing them (like-to-like) with Co. This indicates that the uncertain pair of object affects the Co methods.

As expected, the ensemble method performs better than a single clustering algorithm in this experiment; it performs better than the k-means except in the case of the Bcw dataset. In total, this confirms the perception that the performance of the Ensemble method is more accurate than a single algorithm. In comparing the consistency, we found the ensemble method to be more reliable than the single algorithm, where the latter achieved a high standard definition in all our tested datasets. Comparing the three linkage clustering methods used, it can be observed that the Average linkage performed better than the other two linkage methods, which gave the highest average accuracy using the ONCE and Co matrices. Looking to Wins/Ties/Losses, we observe that in total our method wins more often than the Co method in respect of comparing the like-to-like methods. Comparing the highest accuracies, our method wins four times and ties three times (two of which were highest accuracies) and loses four times across the total of eleven datasets. Co wins three times and loses five times, and k-means wins once. Finally, we observe that the highest accuracy is achieved by ONCE in six datasets; these are R15, Aggregation, Wine, Thyroid, Mfeatures and Glass.

We applied the Iman-Davenport test [4] to assess all three versions of ONCE against all three variations of the Co method under the null hypothesis that the mean ranks are equal for all methods. We also applied it on the like-to-like algorithms in order to compare their performance. As suggested by Demšar [1], if there are statistically significant differences, we will proceed with the Nemenyi test as a post hoc test for a pairwise comparison between them in order to discover where the differences lie.

In the first test, we can reject the null hypothesis of the mean rank being equal for all methods (the result of Iman-Davenport test was equal to 5.72, which gives a negligible p-value). On the other hand, in the other three tests, we cannot reject the null hypothesis that the two like-to-like algorithms are equal. We believe that the reason behind why we cannot reject the null hypothesis is that the difference in accuracy is not large enough between the two like-to-like algorithms, so it is harder to detect significant differences. Figure 1 shows the

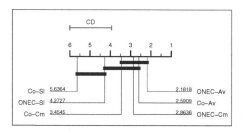

Fig. 1. The Critical difference diagram of the critical level of 0.1 in which it shows the Comparison of all six ensemble methods using 11 datasets

result of post-hoc Nemenyi test in the critical differences diagram of the critical level of 0.1, which was introduced by [1]. This diagram shows the mean rank order of each method on a linear scale. The solid bars in these diagrams show a group of algorithms in cliques, indicating that there are no significant differences in rank from one to another.

Nevertheless, we can identify three groups of algorithms; the first group (top clique) consists of ONCE-Av, Co-Av, ONCE-Cm and Co-Cm. The second group excludes the ONCE-Av and include ONCE-SI which means that at 10% there is a significant differences between the algorithms in the first group and Co-SI but not between them and ONCE-SI. Furthermore, we can observe that using the Single linkage with the ONCE and Co matrices is significantly worse than using the Average linkage with the ONCE. We observe that ONCE-Av achieves the highest rank under this experiment setup and that there is a significant difference between ONCE-Av and CO-SI, and there are not a significant differences with other algorithms.

6   Conclusion

We have introduced a novel a similarity measure based on the neighbourhood association and presented a corresponding ensemble approach called Object-Neighbourhood Clustering Ensemble (ONCE), in order to address the problem of uncertain pairs of objects, produced by the uncertain agreement between ensemble members about an object that is hard to cluster. The experimental results indicate that on average the ONCE-Av is better than the Co-Av and the other methods. In addition, the results show that our ensemble method is more accurate than the baseline algorithm and more consistent and reliable than the single clustering algorithm. From the significance test, we can conclude that there is a significant difference between ONCE-Av and CO-SI, ONCE-SI, but not to the other algorithms under this experimental setup.

In general, it is interesting to note that the effect of the uncertain pairs of objects varied with the different datasets. This is due to the fact that not every dataset is affected by uncertain pairs of objects, even though these are in fact hard objects to cluster. This is where some of the datasets are improved by our method, such as Iris, which has overlapping clusters, and Aggregation, which has uneven-sized clusters with difficult boundaries. Therefore, the further work may investigate these two situations using our method with more of this kind of datasets, and also the idea of filtering out bad members in terms of quality from a collection of ensemble members in order to improve the quality of the clustering ensemble.

References

1. Demšar, J.: Statistical comparisons of classifiers over multiple data sets. The Journal of Machine Learning Research 7, 1–30 (2006)
2. Fred, A.L., Jain, A.K.: Combining multiple clusterings using evidence accumulation. IEEE Transactions on Pattern Analysis and Machine Intelligence 27(6), 835–850 (2005)
3. Gionis, A., Mannila, H., Tsaparas, P.: Clustering aggregation. ACM Transactions on Knowledge Discovery from Data (TKDD) 1(1), 4 (2007)
4. Iman, R.L., Davenport, J.M.: Approximations of the critical region of the fbietkan statistic. Communications in Statistics - Theory and Methods 9(6), 571–595 (1980)
5. Jain, A.K., Murty, M.N., Flynn, P.J.: Data clustering: a review. ACM Computing Surveys (CSUR) 31(3), 264–323 (1999)
6. Jarvis, R.A., Patrick, E.A.: Clustering using a similarity measure based on shared near neighbors. IEEE Transactions on Computers 100(11), 1025–1034 (1973)
7. Lin, Z., Chen, M., Ma, Y.: The augmented lagrange multiplier method for exact recovery of corrupted low-rank matrices. arXiv preprint arXiv:1009.5055 (2010)
8. Strehl, A., Ghosh, J.: Cluster ensembles—a knowledge reuse framework for multiple partitions. The Journal of Machine Learning Research 3, 583–617 (2003)
9. Veenman, C.J., Reinders, M.J.T., Backer, E.: A maximum variance cluster algorithm. IEEE Transactions on Pattern Analysis and Machine Intelligence 24(9), 1273–1280 (2002)
10. Vega-Pons, S., Ruiz-Shulcloper, J., Guerra-Gandón, A.: Weighted association based methods for the combination of heterogeneous partitions. Pattern Recognition Letters 32(16), 2163–2170 (2011)
11. Wang, X., Yang, C., Zhou, J.: Clustering aggregation by probability accumulation. Pattern Recognition 42(5), 668–675 (2009)
12. Yi, J., Yang, T., Jin, R., Jain, A.K., Mahdavi, M.: Robust ensemble clustering by matrix completion. In: 2012 IEEE 12th International Conference on Data Mining (ICDM), pp. 1176–1181. IEEE (2012)

A Novel Recursive Kernel-Based Algorithm for Robust Pattern Classification

José Daniel A. Santos[1], César Lincoln C. Mattos[2], and Guilherme A. Barreto[2]

[1] Federal Institute of Education, Science and Technology of Ceará,
Department of Industry, Maracanaú, Ceará, Brazil
`jdaniel@ifce.edu.br`
[2] Federal University of Ceará, Department of Teleinformatics Engineering,
Center of Technology, Campus of Pici, Fortaleza, Ceará, Brazil
`cesarlincoln@terra.com.br, gbarreto@ufc.br`

Abstract. Kernel methods comprise a class of machine learning algorithms that utilize Mercer kernels for producing nonlinear versions of conventional linear learning algorithms. This kernelizing approach has been applied, for example, to the famed least mean squares (LMS) [1] algorithm to give rise to the *kernel least mean squares* (KLMS) algorithm [2]. However, a major drawback of the LMS algorithm (and also of its kernelized version) is the performance degradation in scenarios with outliers. Bearing this in mind, we introduce instead a kernel classifier based on the least mean M-estimate (LMM) algorithm [3] which is a robust variant of the LMS algorithm based on M-estimation techniques. The proposed *Kernel LMM* (KLMM) algorithm is evaluated in pattern classification tasks with outliers using both synthetic and real-world datasets. The obtained results indicate the superiority of the proposed approach over the standard KLMS algorithm.

Keywords: Labelling errors, outliers, kernel methods, M-estimation, robust pattern recognition.

1 Introduction

Kernel methods have solid mathematical bases and, by possessing the property of mapping nonlinear data to high-dimensional linear space, they have obtained considerable experimental successes. As a consequence, kernel methods such as support vector machines (SVM) [4], least squares support vector machines (LS-SVM) [5], Gaussian processes [6] and regularization networks [7] have found wide applicability, most notably in classification and regression tasks.

However, the aforementioned algorithms are usually derived in batch mode, requiring a significant amount of operations and memory for data storage. This hinders the use of kernel methods in many real-world applications that require online computations. In this regard, the celebrated least mean squares (LMS) algorithm is a popular and practical algorithm used extensively in signal processing, communications, and control. This algorithm provides an efficient online solution for estimating the parameters of linear supervised learning methods.

E. Corchado et al. (Eds.): IDEAL 2014, LNCS 8669, pp. 150–157, 2014.

The LMS algorithm minimizes the mean squared error (MSE) by updating the weight vector in the negative direction of the instantaneous gradient of the MSE with respect to the weight vector. It has been demonstrated that the LMS algorithm is optimal in H_∞ sense since it tolerates small disturbances, such as measurement noise, parameter drifting and modelling errors [8,9]. However, when the disturbances are not small (e.g. presence of impulsive noise) the performance of the LMS algorithm deteriorates considerably [3].

It worth emphasizing that the aforementioned studies on the robustness of the LMS algorithm have been ascertained for regression-like tasks, typically found in the signal processing, communications, and control domains. In the present paper, we focus instead on pattern classification tasks contaminated with outliers, in particular those resulting from label noise[1], as surveyed in [10].

The Kernel Least Mean Squares (KLMS) [2] was proposed as a kernelized version of the LMS algorithm. However, a kernel classifier based on the KLMS are not expected to be robust to outliers, since in essence it also tries to minimize the MSE. Bearing this in mind, in this paper we introduce a kernel classifier based on the least mean M-estimate (LMM) algorithm [3], which is inherently robust to outliers since it is based on M-estimation methods. The proposed algorithm can be understood in two ways: either as a robust version of the KLMS algorithm or as a kernelized version of the LMM algorithm. We compare the performance of the proposed approach with those achieved by the KLMS and Kernel Adatron (KAdatron) [11] algorithms in benchmarking pattern classification problems contaminated with outliers.

The remainder of the paper is organized as follows. In Section 2 all applied algorithms are described, in Section 3 we present and discuss the obtained experimental results with both artificial and real datasets, and in Section 4 the final comments are posed.

2   Methods and Algorithms

Given a sequence of N input-output pairs $\{(\boldsymbol{x}_i, y_i)\}_i^N \in \mathbb{R}^D \times \mathbb{R}$, the corresponding output of a generic classifier is estimated as $\hat{y}_i = f(\boldsymbol{x}_i)$, where $f(\cdot)$ is some function which describes the model and $\hat{y}_i \in \mathbb{R}$ is the estimated output.

In the linear case, the model function can be written as $f(\boldsymbol{x}_i) = \boldsymbol{w}_i^T \boldsymbol{x}_i$, where $\boldsymbol{w}_i \in \mathbb{R}^D$ is a vector of weights. The LMS algorithm [1] has been one of the most popular recursive approaches applied to obtain a solution that minimizes the MSE (Mean Squared Error) of linear models:

$$J_{\mathrm{MSE}}(\boldsymbol{w}_i) = \mathbb{E}\{e_i^2\} = \mathbb{E}\{(y_i - \boldsymbol{w}_i^T \boldsymbol{x}_i)^2\}, \tag{1}$$

where $\mathbb{E}\{\cdot\}$ is the expectation operator and $e_i = y_i - \boldsymbol{w}_i^T \boldsymbol{x}_i$ is the error for the i-th iteration. Minimization of the Eq. (1) is obtained by taking its gradient

---

[1] This type of outlier may result either from mistakes during labelling the data points (e.g. misjudgment of a specialist) or from typing errors during creation of data storage files (e.g. by striking an incorrect key on a keyboard).

with respect to the weights: $\frac{\partial J_{\text{MSE}}(\boldsymbol{w}_i)}{\partial \boldsymbol{w}_i} = -2\mathbb{E}\{e_i \boldsymbol{x}_i\}$. The recursive algorithm is calculated updating $\boldsymbol{w}_i$ at each iteration in the negative direction of this gradient, which involves approximating $\mathbb{E}\{e_i \boldsymbol{x}_i\}$ by its instantaneous value $e_i \boldsymbol{x}_i$:

$$\boldsymbol{w}_{i+1} = \boldsymbol{w}_i - \mu \frac{\partial J_{\text{MSE}}(\boldsymbol{w}_i)}{\partial \boldsymbol{w}_i} = \boldsymbol{w}_i + \mu e_i \boldsymbol{x}_i, \tag{2}$$

where μ is a learning step which controls the convergence rate.

By applying a nonlinear transformation to the input data, it is possible to obtain a nonlinear classifier from the same error function in Eq. (1). In a kernel context, the KLMS algorithm operates on the feature space obtained by applying a mapping $\Phi(\cdot)$ to the inputs, generating a new sequence of input-output pairs $\{(\Phi(\boldsymbol{x}_i), y_i)\}_{i=1}^{N}$ [2]. Weight updating is similar to the LMS Eq. (2):

$$\boldsymbol{w}_{i+1} = \boldsymbol{w}_i + \mu e_i \Phi(\boldsymbol{x}_i). \tag{3}$$

Considering $\boldsymbol{w}_0 = \boldsymbol{0}$, where $\boldsymbol{0}$ is the null-vector, after N iterations we get

$$\boldsymbol{w}_N = \mu \sum_{i=1}^{N-1} e_i \Phi(\boldsymbol{x}_i), \tag{4}$$

$$\hat{y}_N = \boldsymbol{w}_N^T \Phi(\boldsymbol{x}_N) = \mu \sum_{i=1}^{N-1} e_i \kappa(\boldsymbol{x}_i, \boldsymbol{x}_N), \tag{5}$$

where $\kappa(\boldsymbol{x}_i, \boldsymbol{x}_j) = \Phi(\boldsymbol{x}_i)^T \Phi(\boldsymbol{x}_j)$ is a positive-definite kernel function. It should be noted that only Eq. (5) is needed both for training and testing. Although the values of the weight vector do not need to be computed, the *a priori* errors $e_i, i \in \{1, \cdots N\}$, and the training inputs $\boldsymbol{x}_i, i \in \{1, \cdots N\}$, must be maintained for prediction.

In order to add robustness to the standard LMS algorithm one can use the M-estimation framework introduced by Huber [12]. This was done by Zou *et al.* [13], who proposed the least mean M-estimate (LMM) as a robust variant of the LMS algorithm. In this paper we introduce a kernelized version of the LMM algorithm, to be called the *kernel LMM* (KLMM) algorithm. The KLMM algorithm is derived from a more general objective function than the MSE:

$$J_{\text{KLMM}}(\boldsymbol{w}_i) = \mathbb{E}\{\rho(e_i)\} = \mathbb{E}\{\rho(y_i - \boldsymbol{w}_i^T \boldsymbol{x}_i)\}, \tag{6}$$

where $\rho(\cdot)$ is the M-estimate function [12]. The function $\rho(\cdot)$ computes the contribution of each error e_i to the objective function $J_{\text{KLMM}}(\boldsymbol{w}_i)$. Note that when $\rho(u) = u^2$, the function $J_{\text{KLMM}}(\boldsymbol{w}_i)$ reduces to the MSE function $J_{\text{MSE}}(\boldsymbol{w}_i)$.

Weight updating in KLMM is given by

$$\boldsymbol{w}_{i+1} = \boldsymbol{w}_i - \mu \frac{\partial J_{\text{KLMM}}(\boldsymbol{w}_i)}{\partial \boldsymbol{w}_i} = \boldsymbol{w}_i + \mu q(e_i) e_i \Phi(\boldsymbol{x}_i), \tag{7}$$

where $q(e_i) = \frac{1}{e_i} \frac{\partial \rho(e_i)}{\partial e_i}$ is the weight function. The output of the KLMM algorithm is similar to that of the KLMS (Eq. 5):

$$\boldsymbol{w}_N = \mu \sum_{i=1}^{N-1} q(e_i) e_i \Phi(\boldsymbol{x}_i), \tag{8}$$

$$\hat{y}_N = \boldsymbol{w}_N^T \Phi(\boldsymbol{x}_N) = \mu \sum_{i=1}^{N-1} q(e_i) e_i \kappa(\boldsymbol{x}_i, \boldsymbol{x}_N). \tag{9}$$

In the present paper the bisquare M-estimate function is considered [14]:

$$\rho(e) = \begin{cases} \frac{\xi^2}{6} \left\{ 1 - \left[1 - \left(\frac{e}{\xi}\right)^2 \right]^3 \right\}, & |e| < \xi \\ \frac{\xi^2}{6}, & \text{c.c.} \end{cases} \tag{10}$$

$$q(e) = \begin{cases} \left[1 - \left(\frac{e}{\xi}\right)^2 \right]^2, & |e| < \xi \\ 0, & \text{c.c.} \end{cases} \tag{11}$$

where ξ is a threshold parameter specified to diminish the impact large errors in the estimation process. Smaller values of ξ produce more resistance to outliers, but at the expense of lower efficiency when the errors are normally distributed.

In [13] it is suggested that the parameter ξ could be adjusted through a technique called ATS (Adaptive Threshold Selection). Let the error e_i have a Gaussian distribution possibly corrupted with some impulsive noise. The error variance σ_i^2 at iteration i is estimated by the following robust estimator:

$$\hat{\sigma}_i^2 = \lambda \hat{\sigma}_{i-1}^2 + c(1 - \lambda)\text{med}(A_i), \tag{12}$$

where λ is a forgetting factor with value close but not equal to 1, med($\cdot$) is the median operator, $A_i = \{e_i^2, e_{i-1}^2, \cdots, e_{i-N_w+1}^2\}$, N_w is the fixed window length for the median operation and $c = 1.483(1 + 5/(N_w - 1))$ is the estimator's correction factor. In this paper the values $\lambda = 0.98$ and $N_w = 14$ were applied.

Given a rejection probability of $Pr\{|e_i| > \xi_i\} = \text{erfc}\left(\frac{\xi_i}{\sqrt{2}\hat{\sigma}_i}\right)$, where $\text{erfc}(x) = \frac{2}{\sqrt{\pi}} \int_x^\infty \exp(-t^2)dt$ is the the complementary error function, the threshold ξ_i can be calculated. For example, for the probabilities 0.05, 0.025 and 0.01, we have $\xi_i = 1.96\hat{\sigma}_i$, $\xi_i = 2.24\hat{\sigma}_i$ and $\xi_i = 2.576\hat{\sigma}_i$, respectively. The last value was chosen for the experiments in this paper.

The last estimation algorithm to be described, the Kernel Adatron (KAdatron) [11], is an on-line algorithm for training linear perceptron-like classifiers by providing a procedure that emulates SVM without resorting to any quadratic programming toolboxes. The first step is to initialize the Lagrange multipliers ($\alpha_i = 0, \forall i = 1, \ldots, N$). So, we use the data-dependent representation $\{(\boldsymbol{x}_i, y_i)\}_{i=1}^N$ and the kernel function to compute

$$z_i = \sum_{j=1}^N \alpha_j y_j \kappa(\boldsymbol{x}_i, \boldsymbol{x}_j) \quad \text{and} \quad \gamma_i = y_i z_i. \tag{13}$$

Let $\delta\alpha_i = \mu(1 - \gamma_i)$ be the proposed change to α_i, whose adaptation rule is

$$\alpha_i = \begin{cases} \alpha_i + \delta\alpha_i, & (\alpha_i + \delta\alpha_i) > 0 \\ 0, & \text{otherwise.} \end{cases} \tag{14}$$

If the maximum number of training epochs has been exceeded then stop, otherwise return to the procedure from the Eq. (13). Finally, the output estimation for a new input $\boldsymbol{x}_*$ can be written as:

$$\hat{y}_* = \sum_{i \in SV} y_i \alpha_i^o \kappa(\boldsymbol{x}_*, \boldsymbol{x}_i), \tag{15}$$

where α_i^o is the solution of KAdatron algorithm and SV represents the index set of support vectors. For all algorithms evaluated in this paper (i.e. KLMS, KLMM and KAdatron) we use the gaussian kernel function

$$\kappa(\mathbf{x}_i, \mathbf{x}_j) = \exp\left\{ -\frac{\|\mathbf{x}_i - \mathbf{x}_j\|^2}{2\sigma^2} \right\}, \tag{16}$$

where σ is the width of the kernel and is a hyperparameter to be optimized.

3   Experimental Results and Discussion

The experimental results were separated into two groups: one with artificial two-dimensional data for the sake of visualization of the decision regions obtained by each classifier; and another with four real-world datasets (Vertebral Column, Pima Indian Diabetes, Breast Cancer and Ionosphere)[2], for analyzing classifier's performance due to the presence of outliers.

The first experiments involved 1000 two-dimensional samples from two classes (red and black). All samples are used for training, since the goal is to visualize the final position of the decision curves, not to compute recognition rates. An increasing number of outliers labelled as belonging to the red class was added at each experiment. Outliers were intentionally added at the lower right corner of the figures, far from the region originally associated with samples from the red class. The obtained decision curves are shown in Figure 1.

It may be observed that the KLMS- and KAdatron-based classifiers are strongly affected by the addition of outliers during training, modifying their decision curves to include those samples. In the other hand, the KLMM-based classifier is practically insensitive to the presence of mislabeled patterns, even after the addition of 15% of outliers.

In the second group of experiments, with the real-world datasets, 80% of the data was used for training and the remaining 20% for testing. Furthermore, the addition of outliers was done by deliberately mislabeling a percentage of samples from a given class. The percentage of mislabeled samples covered 0%, 5%, 10%, 20% and 30% of the training samples of a given class.

The Vertebral Column dataset was configured as a two-class problem since we removed the samples from the Disk Hernia class and considered only the ones from Normal and Spondylolisthesis classes. During the experiments, randomly chosen samples from the Spondylolisthesis class had their labels changed to the Normal class. For the Pima Indian Diabetes, Breast Cancer and Ionosphere

---

[2] Freely available at `http://archive.ics.uci.edu/ml`

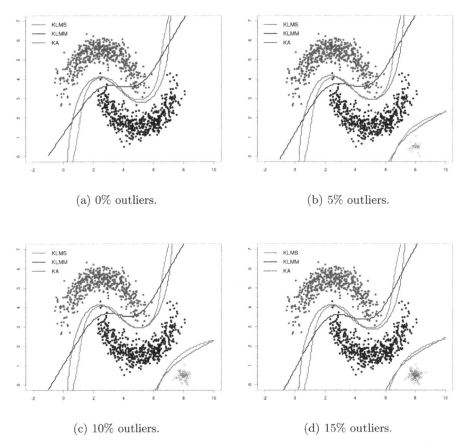

(a) 0% outliers.   (b) 5% outliers.

(c) 10% outliers.   (d) 15% outliers.

Fig. 1. Modifications observed in the decision curves as a consequence of the presence of an increasing number of outliers in the data

datasets, the mislabeled samples (outliers) were generated by changing the labels of samples randomly selected from the class with largest number of samples. All the algorithms were trained for one single pass over the training set, since we aimed to evaluate their online nature. The values of the learning step and kernel width parameters used in the experiments are presented in Table 1 and the results averaged after 100 repetitions are presented in Table 2.

As expected, for all datasets the KLMS- and KAdatron-based classifiers achieved good results only for the scenarios without outliers. The performance of those learning deteriorated with the increasing number of mislabeled patterns. This phenomenon is particularly observed for the Vertebral Column and Ionosphere datasets. Only for the Breast Cancer dataset, the KAdatron-based classifier has achieved higher rates than the KLMM-based classifier. The performance of the KLMS-based classifier was poor in all scenarios with high percentages of outliers (20% and 30%). For some datasets (i.e. Vertebral Column, Breast

Table 1. Values of the parameters used in the experiments

Dataset	KLMS	KLMM	KA
Vertebral Column	$\eta = 0.60; \sigma^2 = 2.00$	$\eta = 0.60; \sigma^2 = 2.00$	$\eta = 0.60; \sigma^2 = 2.00$
Diabetes	$\eta = 0.20; \sigma^2 = 10.00$	$\eta = 0.20; \sigma^2 = 10.00$	$\eta = 0.10; \sigma^2 = 5.00$
Breast Cancer	$\eta = 0.50; \sigma^2 = 0.50$	$\eta = 0.10; \sigma^2 = 0.25$	$\eta = 0.10; \sigma^2 = 0.50$
Ionosphere	$\eta = 0.75; \sigma^2 = 0.50$	$\eta = 0.10; \sigma^2 = 5.00$	$\eta = 0.75; \sigma^2 = 0.50$

Table 2. Classification results without and with outliers during training step

Vertebral Column dataset					
	0%	5%	10%	20%	30%
KLMS	93.76 ± 3.32	92.58 ± 3.59	89.56 ± 5.94	84.26 ± 9.10	78.78 ± 10.62
KLMM	90.64 ± 4.00	90.02 ± 3.49	90.70 ± 3.40	89.68 ± 3.78	87.86 ± 5.16
KAdatron	93.36 ± 3.93	91.78 ± 4.76	90.42 ± 4.98	84.70 ± 8.62	77.52 ± 11.06
Pima Indian Diabetes dataset					
KLMS	76.16 ± 3.62	75.33 ± 3.67	73.69 ± 4.35	71.07 ± 6.64	66.49 ± 9.55
KLMM	75.85 ± 4.33	76.31 ± 3.08	75.76 ± 3.49	73.96 ± 4.50	69.25 ± 8.16
KAdatron	76.83 ± 3.18	76.36 ± 2.89	74.97 ± 3.39	72.84 ± 4.69	67.61 ± 6.89
Breast Cancer dataset					
KLMS	97.23 ± 1.32	96.84 ± 1.67	96.03 ± 1.77	93.07 ± 2.86	85.89 ± 5.08
KLMM	94.75 ± 1.84	94.49 ± 2.01	94.77 ± 1.88	94.74 ± 2.23	95.36 ± 1.96
KAdatron	96.41 ± 1.75	96.90 ± 1.62	96.65 ± 1.55	95.80 ± 1.82	93.14 ± 2.96
Ionosphere dataset					
KLMS	90.66 ± 2.81	88.52 ± 5.58	86.18 ± 8.26	80.52 ± 11.10	73.98 ± 13.64
KLMM	87.17 ± 3.44	86.77 ± 3.45	86.96 ± 3.96	85.69 ± 3.95	84.49 ± 4.46
KAdatron	89.73 ± 3.72	88.83 ± 5.09	85.77 ± 7.55	81.45 ± 10.39	77.18 ± 12.81

Cancer and Ionosphere) for 30% of outliers, the average classification rate of the KLMM-based classifier is approximately 10 points higher than that achieved by the KLMS-based classifier. As a final remark, it is important to point out a remarkable feature of the proposed KLMM-based classifier: the obtained values for the standard deviation of the classification rate tend to be much smaller than those achieved for the other evaluated kernel classifiers, especially for scenarios with high number of outliers.

4   Conclusions

In this paper the problem of pattern classification in the presence of outliers was tackled by the proposal of a robust online kernel-based classifier. The proposed kernel classifier incorporates ideas from the M-estimation theory and from the celebrated LMS algorithm. The proposed method was compared to the original KLMS and the KAdatron algorithms through several experiments of binary classification where an increasing number of outliers (mislabeled patterns) were added during training. Both synthetic and real-world datasets were used. The obtained results made it clear the deterioration of performance of the KLMS- and KAdatron-based classifier as the number of outliers increase. The proposed KLMM-based classifier achieved better overall results than the other two kernel classifiers.

Acknowledgments. The authors thank the financial support of FUNCAP (Fundação Cearense de Apoio ao Desenvolvimento Científico e Tecnológico) and NUTEC (Núcleo de Tecnologia Industrial do Ceará).

References

1. Widrow, B., Winter, R.: Neural nets for adaptive filtering and adaptive pattern recognition. IEEE Computer 21(3), 25–39 (1988)
2. Liu, W., Pokharel, P., Principe, J.: The kernel least-mean-square algorithm. IEEE Transactions on Signal Processing 56(2), 543–554 (2008)
3. Chan, S.C., Zhou, Y.: On the performance analysis of the least mean M-estimate and normalized least mean M-estimate algorithms with gaussian inputs and additive gaussian and contaminated gaussian noises. Journal of Signal Processing Systems 80(1), 81–103 (2010)
4. Vapnik, V.: The Nature of Statistical Learning Theory. Springer (1995)
5. Suykens, J.A.K., Gestel, V.T., DeBrabanter, J., DeMoor, B., Vandewalle, J.: Least Squares Support Vector Machines. World Scientific (2002)
6. Rasmussen, C.E., Williams, C.K.I.: Gaussian Processes for Machine Learning. MIT Press, Cambridge (2006)
7. Girosi, F., Jones, M., Poggio, T.: Regularization theory and neural networks architectures. Neural Computation 7, 219–269 (1995)
8. Bolzern, P., Colaneri, P., De Nicolao, G.: H_∞-robustness of adaptive filters against measurement noise and parameter drift. Automatica 35(9), 1509–1520 (1999)
9. Hassibi, B., Sayed, A.H., Kailath, T.: H_∞ optimality of the LMS algorithm algorithm. IEEE Transactions on Signal Processing 44(2), 267–280 (1996)
10. Frénay, B., Verleysen, M.: Classification in the presence of label noise: a survey. IEEE Transactions on Neural Networks and Learning Systems 25(5), 845–869 (2014)
11. Friess, T.T., Cristianini, N., Campbell, C.: The kernel Adatron algorithm: A fast and simple learning procedure for support vector machines. In: Proceedings of the 15th International Conference of Machine Learning (ICML 1998), pp. 188–196 (1998)
12. Huber, P.J.: Robust estimation of a location parameter. Annals of Mathematical Statistics 35(1), 73–101 (1964)
13. Zou, Y., Chan, S.C., Ng, T.S.: Least mean M-estimate algorithms for robust adaptive filtering in impulsive noise. IEEE Transactions on Circuits and Systems II 47(12), 1564–1569 (2000)
14. de Paula Barros, A.L.B., Barreto, G.A.: Improving the classification performance of optimal linear associative memory in the presence of outliers. In: Rojas, I., Joya, G., Gabestany, J. (eds.) IWANN 2013, Part I. LNCS, vol. 7902, pp. 622–632. Springer, Heidelberg (2013)

Multi-Objective Genetic Algorithms for Sparse Least Square Support Vector Machines

Danilo Avilar Silva and Ajalmar Rêgo Rocha Neto

Instituto Federal do Ceará, IFCE, Brasil,
Programa de Pós-Graduação em Engenharia de Telecomunicações, PPGET
daniloavilar@gmail.com, ajalmar@ifce.edu.br

Abstract. This paper introduces a new approach to building sparse least square support vector machines (LSSVM) based on multi-objective genetic algorithms (GAs) for classification tasks. LSSVM classifiers are an alternative to SVM ones due to the training process of LSSVM classifiers only requires to solve a linear equation system instead of a quadratic programming optimization problem. However, the lost of sparseness in the Lagrange multipliers vector (i.e. the solution) is a significant drawback which comes out with theses classifiers. In order to overcome this lack of sparseness, we propose a multi-objective GA approach to leave a few support vectors out of the solution without affecting the classifier's accuracy and even improving it. The main idea is to leave out outliers, non-relevant patterns or those ones which can be corrupted with noise and thus prevent classifiers to achieve higher accuracies along with a reduced set of support vectors. We point out that the resulting sparse LSSVM classifiers achieve equivalent (in some cases, superior) performances than standard full-set LSSVM classifiers over real data sets. Differently from previous works, genetic algorithms are used in this work to obtain sparseness not to find out the optimal values of the LSSVM hyper-parameters.

Keywords: Least Square Support Vector Machines, Pruning Methods, Genetic Algorithms.

1 Introduction

There are many works based on both Least Square Support Vector Machines (LSSVM) [9] and Evolutionary Computation (EC), especially Genetic Algorithms (GAs). GAs are mostly applied to optimize the LSSVM kernel parameters, as well as the classifier parameters [12,5]. Genetic Algorithms (GAs) are optimization methods and therefore they can be used to generate useful solutions to search problems. Due to the underlying features of GAs some optimization problems can be solved without the assumption of linearity, differentiability, continuity or convexity of the objective function. Unfortunately, these desired characteristics are not found in several mathematical methods when applied to the same kind of problems. GAs as a meta-heuristic method are also used to deal with classification tasks even with LSSVM.

E. Corchado et al. (Eds.): IDEAL 2014, LNCS 8669, pp. 158–166, 2014.
© Springer International Publishing Switzerland 2014

A theoretical advantage of kernel methods such as Support Vector Machines [11] concerns the empirical and structural risk minimization which balances the complexity of the model against its success at fitting the training data, along with the production of sparse solutions [7]. Support Vector Machines (SVM) are the most popular kernel methods. The LSSVM is an alternative to the standard SVM formulation [11]. A solution for the LSSVM is achieved by solving linear KKT systems[1] in a least square sense. In fact, the solution follows directly from solving a linear equation system, instead of a QP optimization problem. On the one hand, it is in general easier and less computationally intensive to solve a linear system than a QP problem. On the other hand, the resulting solution is far from sparse, in the sense that it is common to have all training samples being used as SVs.

To handle the lack of sparseness in SVM and LSSVM solutions, several *reduced set* (RS) and pruning methods have been proposed, respectively. These methods comprise a bunch of techniques aiming at simplifying the internal structure of those models, while keeping the decision boundaries as similar as possible to the original ones. They are very useful in reducing the computational complexity of the original models, since they speed up the decision process by reducing the number of SVs. They are particularly important for handling large datasets, when a great number of data samples may be selected as support vectors, either by pruning less important SVs [3,4] or by constructing a smaller set of training examples [6,1], often with minimal impact on performance.

In order to combine the aforementioned advantages of LSSVM classifiers and genetic algorithms, this work aims at putting both of them to work together for reducing the number of support vectors and also achieve equivalent (in some cases, superior) performances than standard full-set LSSVM classifiers. Our proposal finds out sparse classifiers based on a multi-objective genetic algorithm which takes into account both the accuracy of classification and the support vector pruning rate. In order to do this, we also propose a new multi-objective fitness function which incorporates a cost of pruning in its formulation. Differently from previous works, genetic algorithms are used in this work to obtain sparseness not to find out the optimal values of the kernel and classifier parameters.

The remaining part of this paper is organized as follows. In Section 2 we review the fundamentals of the LSSVM classifiers. In Section 3 we briefly present some methods for obtaining sparse LSSVM classifiers. In Section 4, we introduce genetic algorithms which are necessary to understanding of our proposal in Section 5. We present our simulations and conclusions in sections 6 and 7, respectively.

2 LSSVM Classifiers

The formulation of the primal problem for the LSSVM [9] is given by

$$\min_{\mathbf{w},\xi_i}\left\{\frac{1}{2}\mathbf{w}^T\mathbf{w} + \gamma\frac{1}{2}\sum_{i=1}^{L}\xi_i^2\right\}, \tag{1}$$

$$\text{subject to} \quad y_i[(\mathbf{w}^T\mathbf{x}_i) + b] = 1 - \xi_i, i = 1,\ldots,L$$

---

[1] Karush-Kuhn-Tucker systems.

where γ is a positive cost parameter, as well as the slack variables $\{\xi_i\}_{i=1}^{L}$ can assume negative values.

The Lagrangian function for the LSSVM is then written as

$$L(\mathbf{w}, b, \boldsymbol{\xi}, \boldsymbol{\alpha}) = \frac{1}{2}\mathbf{w}^T\mathbf{w} + \gamma\frac{1}{2}\sum_{i=1}^{L}\xi_i^2 - \sum_{i=1}^{L}\alpha_i(y_i(\mathbf{x}_i^T\mathbf{w} + b) - 1 + \xi_i), \qquad (2)$$

where $\{\alpha_i\}_{i=1}^{L}$ are the Lagrange multipliers. The conditions for optimality can be given by the partial derivatives, namely, $\frac{\partial L(\mathbf{w},b,\boldsymbol{\xi},\boldsymbol{\alpha})}{\partial \mathbf{w}} = 0$, $\frac{\partial L(\mathbf{w},b,\boldsymbol{\xi},\boldsymbol{\alpha})}{\partial \alpha_i} = 0$, $\frac{\partial L(\mathbf{w},b,\boldsymbol{\xi},\boldsymbol{\alpha})}{\partial b} = 0$ and $\frac{\partial L(\mathbf{w},b,\boldsymbol{\xi},\boldsymbol{\alpha})}{\partial \xi_i} = 0$, such that $\mathbf{w} = \sum_{i=1}^{L}\alpha_i y_i \mathbf{x}_i$, $\sum_{i=1}^{L}\alpha_i y_i = 0$, $y_i(\mathbf{x}_i^T\mathbf{w} + b) - 1 + \xi_i = 0$ and $\alpha_i = \gamma\xi_i$, respectively.

Thus, based on these conditions, one can formulate a linear system $\mathbf{Ax} = \mathbf{B}$ in order to represent this problem as

$$\left[\begin{array}{c|c} 0 & \mathbf{y}^T \\ \hline \mathbf{y} & \boldsymbol{\Omega} + \gamma^{-1}\mathbf{I} \end{array}\right] \left[\begin{array}{c} b \\ \boldsymbol{\alpha} \end{array}\right] = \left[\begin{array}{c} 0 \\ \mathbf{1} \end{array}\right] \qquad (3)$$

where $\boldsymbol{\Omega} \in \mathbb{R}^{L \times L}$ is a matrix whose entries are given by $\Omega_{i,j} = y_i y_j \mathbf{x}_i^T \mathbf{x}_j$, $i, j = 1, \ldots, L$. In addition, $\mathbf{y} = [y_1 \quad \cdots \quad y_L]^T$ and the symbol $\mathbf{1}$ denotes a vector of ones with dimension L. The solution of this linear system can be computed by direct inversion of matrix $\mathbf{A}$, i.e., $\mathbf{x} = \mathbf{A}^{-1}\mathbf{B}$. Furthermore, one should use $\Omega_{i,j} = y_i y_j K(\mathbf{x}_i, \mathbf{x}_j)$ for nonlinear kernels.

3 Sparse Classifiers

In this section, we present Pruning LSSVM and IP-LSSVM methods used to obtain sparse classifiers. Both of them reduce the number of support vectors based on the values of Lagrange multipliers.

3.1 Pruning LSSVM

Pruning LSSVM was proposed by Suykens in 2000 [8]. In this method, support vectors and then their respective input vectors are eliminated according to the absolute value of their Lagrange multipliers. The process is carried out recursively, with gradual vector elimination at each iteration, until a stop criterion is reached, which is usually associated with decrease in performance on a validation set. Vectors are eliminated by setting the corresponding Lagrange multipliers to zero, without any change in matrix dimensions. The resolution of the current linear system, for each new reduced set, is needed at each iteration, and the reduced set is selected from the best iteration. This is a multi-step method, since the linear system needs to be solved many times until the convergence criterion is reached.

3.2   IP-LSSVM

IP-LSSVM [1] uses a criterion in which patterns close to the separating surface and far from the support hyperplanes are very likely to become support vectors. In fact, since patterns with $\alpha_i \gg 0$ are likely to become support vectors, the margin limits are located closer to the separating surface than the support hyperplanes. As we can see, that idea applied in IP-LSSVM training is based on SVM classifiers working. According to these arguments, the new criteria proposed in IP-LSSVM is to leave those patterns out of the support vector set whether the lagrange multiplier $\alpha_i \geq 0$, $\alpha_i < 0$ or $\alpha_i \ll 0$. Therefore, support vector set has patterns where $\alpha_i \gg 0$.

The proposed criteria is applied to IP-LSSVM in order to eliminate non-relevant columns of the original matrix $\mathbf{A}$ and to build a non-squared reduced matrix $\mathbf{A}_2$ to be used a posteriori. The eliminated columns correspond to the least relevant vectors for the classification problem, selected according to their Lagrange multiplier values. The rows of $\mathbf{A}$ are not removed, because its elimination would lead to a loss of labeling information and in performance [10].

4   Genetic Algorithms

Genetic algorithm [12] is a search meta-heuristic method inspired by natural evolution, such as inheritance, mutation, natural selection, and crossover. This meta-heuristic can be used to generate useful solutions to optimization problems. Due to the characteristics of GA methods, it is easier to solve few kind of problems by GA than other mathematical methods which do have to rely on the assumption of linearity, differentiability, continuity, or convexity of the objective function.

In a genetic algorithm, a population of candidate individuals (or solutions) to an optimization problem is evolved toward better solutions by natural selection, i.e., a fitness function. In this population, each individual has a set of genes (gene vector), named chromosome, which can be changed by mutation or combined generation-by-generation with other one to build new individuals by reproduction processes which use crossover. The most common way of representing solutions is in a binary format, i.e., strings of 0s and 1s.

Standard implementation of genetic algorithm for natural selection is the roulette wheel. After the selection process, the selected individuals are used as input to other genetic operators: crossover and mutation. Crossover operator combines two strings of chromosomes, but mutation modifies a few bits in a single chromosome.

5   Proposal: Multi-Objective Genetic Algorithm for Sparse LSSVM (MOGAS-LSSVM)

Our proposal, named MOGAS-LSSVM, is based on a multi-objective genetic algorithm which aims at both improving (maximizing) the classifier's accuracy over the training (or validation) dataset and reducing (minimizing) the number of support vectors. Each individual or chromosome in our simulations is represented by a binary vector of genes where each one is set to either "zero" (false) whether a certain pattern will be used as a support vector in the training process or "one" (true) otherwise. Our scalarization-based fitness function [2] is a balance between the value of the resulting accuracy of a classifier when we take into account the genes that were set to "one" and the proportion of support vector pruned to the number of training patterns. In this approach, each individual also has as many support vectors as the number of genes set up to "one", except if some of them receive a zero value of its Lagrange multiplier.

5.1   Individuals or Chromosomes

In order to avoid misunderstanding, we present simple examples of how to construct individuals using MOGAS-LSSVM. Let us consider the matrix **A** of LSSSVM linear system for a very small training set only with four patterns. In our proposal, for all the patterns in the support vectors set, we have to set up a gene vector to $[1 \ 1 \ 1 \ 1]$. For a individual without the fourth pattern in the SV set, we need to set up a gene vector to $[1 \ 1 \ 1 \ 0]$. This means that the fifth column has to be eliminated.

Generally speaking, in order to eliminate a certain pattern from the support vector set which is described by the i-th gene (g_i) in the vector $\mathbf{g} = [g_1 \ g_2 \ g_3 \ g_4]$, it is necessary to remove the $i + 1$ column. It is worth emphasizing again that we must not remove the first column in order to avoid lost of performance.

5.2   Fitness Function for MOGAS-LSSVM

With respect to individual modeling, MOGAS-LSSSVM classifier tries to both maximize the accuracy and minimize the number of support vectors, but this optimization process takes in account a certain cost θ of pruning patterns which belong to the support vector set. This can be achieved as follows

$$fitness = 1 - \left\{ \frac{(1 - accuracy) + ((1 - \theta) * reduction)}{2 - \theta} \right\} \tag{4}$$

where $\theta \in [0, 1]$, accuracy is the hit rate over training (validation) dataset and $reduction = \frac{\#training \ patterns \ pruned}{\#training \ patterns}$.

It is easy to see that the values of the fitness function lies in $[0, 1]$, since whether the cost θ is high (one), so values of fitness function are only depending on accuracy; whether the cost θ is low (zero), so values of fitness function are still depending on accuracy but in direct proportion to the reduction. Whenever $\theta = 1$, our MOGAS-LSSVM classifier is guided only by accuracy and then is single-objective.

5.3   MOGAS-LSSVM Algorithm

MOGAS-LSSVM algorithm for training a classifier can be described as follows.

1. Initialize $t = 0$, where t stands for the generation;
2. Generate initial population $P(t)$, i.e., sets of genes and build their related matrices $\{\mathbf{A}_i\}_{i=1}^s$, where s is the number of individuals at the generation t, randomly;
3. For each individual i in $P(t)$
4.   Solve the LSSVM linear $(\mathbf{x}_i = (\mathbf{A}_i^T \mathbf{A}_i)^{-1} \mathbf{A}_i^T \mathbf{b}_i)$;
5.   Evaluate the fitness function, i.e., the function presented in Eq. (4);
6. While $t \leq t_{max}$, where t_{max} is the maximum number of generations
7.   Select individuals i and their matrices $\mathbf{A}_i$;
8.   Apply crossover operation to selected individuals;
9.   Apply mutation operation to selected individuals;
10.   Compute $t = t + 1$;
11.   Build a matrix $\mathbf{A}_i$ for each individual in $P(t + 1)$.
12.   Evaluate training set accuracies for each $\mathbf{A}_i$;
13. Select the best individual or solution, i.e., $\mathbf{A}^o$.

6   Simulations and Discussion

In this work, we carried out tests with real-world benchmarking datasets. Four UCI Machine Learning datasets were used for evaluation, namely, Diabetes (PID), Haberman (HAB), Breast Cancer (BCW) and Vertebral Column Pathologies (VCP) with 768, 306, 683 and 310 patterns, respectively. Initially, the GA randomly creates a population of 120 feasible solutions. At each generation, we carried out an elitist ranking selector that clones top 10% of individuals. Besides that, 80% of new individuals comes from applying the crossover operator and then mutation one is used to complete 30 new individuals. In our simulations, the cost of reduction θ was zero, due to our goal is to obtain most reduction of SVs. All the parameters of LSSVM and GAs were tuned by trial and error.

In Table 1 and Table 2, we report performance metrics (mean value and standard deviation of the recognition rate) on testing dataset averaged over 30 independent runs. We also show the average number of SVs ($\#SVs$), the number of training patterns ($\#TP$) as well as the values of the parameter γ (LS-SVM) and the reduction obtained. We perform two different kind of simulations, the first one with results presented in Table 1 is based on splitting our data set into two others, training and testing. In this configuration 80% and 20% of the patterns were selected at random for training and testing at each independent run. The second kind of simulation is based on splitting our data set into three other ones: training, validation and testing, see Table 2. In this configuration 60%, 20% and 20% of the patterns were selected at random for training, validation and testing, at each independent run. Accuracies in Eq. (4) for the first and second type of simulations are evaluated over training and validation dataset, respectively.

By analyzing these table, one can easily conclude that the performances of the reduced-set classifiers (MOGAS-LSSVM) were equivalent to those achieved by the full-set classifiers. In some cases, as shown in this table for the VCP and Pima Diabetes, the performances of the reduced-set classifiers were even better. It is worth mentioning that, as expected, the multi-objective proposal achieves a

Table 1. Results for the LSSVM, IP-LSSVM and MOGAS-LSSVM classifiers. The cost of reduction for MOGAS-LSSVM is zero ($\theta = 0$).

Dataset	Model	γ	Accuracy	# TP	# SVs	Reduction
VCP	LSSVM	0.05	81.2 ± 4.9	248	248.0	–
VCP	IP-LSSVM	0.05	75.4 ± 8.6	248	184.0	25.8%
VCP	MOGAS-LSSVM	0.05	84.3 ± 4.5	248	53.0	78.6%
HAB	LSSVM	0.04	73.8 ± 5.0	245	245.0	–
HAB	IP-LSSVM	0.04	70.6 ± 4.9	245	223.0	9.0%
HAB	MOGAS-LSSVM	0.04	75.7 ± 4.6	245	57.2	76.6%
BCW	LSSVM	0.04	96.7 ± 1.1	546	546.0	–
BCW	IP-LSSVM	0.04	96.8 ± 1.1	546	284.0	48.0%
BCW	MOGAS-LSSVM	0.04	96.5 ± 1.1	546	146.8	73.1%
PID	LSSVM	0.04	75.8 ± 2.7	614	614.0	–
PID	IP-LSSVM	0.04	74.0 ± 5.3	614	503.0	18.1%
PID	MOGAS-LSSVM	0.04	77.7 ± 2.7	614	199.6	67.5%

Table 2. Results for the LSSVM, P-LSSVM, and MOGAS-LSSVM classifiers. The cost of reduction for MOGAS-LSSVM is zero ($\theta = 0$).

Dataset	Model	γ	Accuracy	# TP	# SVs	Reduction
VCP	LSSVM	0.05	80.1 ± 3.2	186	186.0	–
VCP	P-LSSVM	0.05	80.0 ± 4.7	186	120.5	35.2%
VCP	MOGAS-LSSVM	0.05	81.3 ± 5.3	186	29.4	84.2%
HAB	LSSVM	0.04	73.7 ± 3.6	184	184.0	–
HAB	P-LSSVM	0.04	74.2 ± 6.1	184	94.0	48.9%
HAB	MOGAS-LSSVM	0.04	75.1 ± 7.1	184	23.1	87.5%
BCW	LSSVM	0.04	96.9 ± 0.8	410	410.0	–
BCW	P-LSSVM	0.04	96.3 ± 1.5	410	210.3	48.7%
BCW	MOGAS-LSSVM	0.04	96.0 ± 1.6	410	91.0	77.8%
PID	LSSVM	0.04	75.8 ± 1.5	461	461.0	–
PID	P-LSSVM	0.04	75.6 ± 3.3	461	230.5	50.0%
PID	MOGAS-LSSVM	0.04	76.9 ± 3.9	461	113.6	75.4%

much better reduction in terms of support vectors when compared to the single-objective ($\theta = 1$). We can also point out that the mean of support vectors for MOGAS-LSSVM with $\theta = 0$ is lower than the means for LSSVM and indeed MOGAS-LSSVM with $\theta = 1$ for each size of training dataset, see Fig. 1

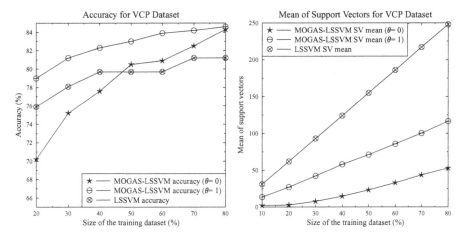

Fig. 1. Figure on left presents accuracy versus size of the training dataset and then figure on right presents mean of SVs versus size of the training dataset

7   Conclusions

In this work, we present a proposal called MOGAS-LSSVM which is based on a multi-objective genetic algorithms and guided by both the training (or validation) accuracy and the reduction of support vectors. It is also proposed, in this work, a way of balancing the accuracy and reduction amounts as a new fitness function. Based on the obtained results, MOGAS-LSSVM was able to reduce the number of support vectors and to maintain similar or even higher classifier's performance. Therefore, MOGAS-LSSVM classifier is attractive when one seeks a competitive classifier with large datasets and limited computing resources.

References

1. Carvalho, B.P.R., Braga, A.P.: IP-LSSVM: A two-step sparse classifier. Pattern Recognition Letters 30, 1507–1515 (2009)
2. Dubois-Lacoste, J., López-Ibáñez, M., Stützle, T.: Combining two search paradigms for multi-objective optimization: Two-phase and pareto local search. In: Talbi, E.-G. (ed.) Hybrid Metaheuristics. SCI, vol. 434, pp. 97–117. Springer, Heidelberg (2013)
3. Geebelen, D., Suykens, J.A.K., Vandewalle, J.: Reducing the number of support vectors of SVM classifiers using the smoothed separable case approximation. IEEE Trans. on Neural Networks and Learning Systems 23(4), 682–688 (2012)
4. Li, Y., Lin, C., Zhang, W.: Improved sparse least-squares support vector machine classifiers. Neurocomputing 69, 1655–1658 (2006)
5. Mustafa, M., Sulaiman, M., Shareef, H., Khalid, S.: Reactive power tracing in pool-based power system utilising the hybrid genetic algorithm and least squares SVM. IET Generation, Transmission & Distribution 6(2), 133–141 (2012)
6. Peres, R., Pedreira, C.E.: Generalized risk zone: Selecting observations for classification. IEEE Pattern Analysis and Machine Intelligence 31(7), 1331–1337 (2009)

7. Steinwart, I.: Sparseness of support vector machines. Journal of Machine Learning Research 4, 1071–1105 (2003)
8. Suykens, J.A.K., Lukas, L., Vandewalle, J.: Sparse least squares support vector machine classifiers. In: Proceedings of the 8th European Symposium on Artificial Neural Networks (ESANN 2000), pp. 37–42 (2000)
9. Suykens, J.A.K., Vandewalle, J.: Least squares support vector machine classifiers. Neural Processing Letters 9(3), 293–300 (1999)
10. Valyon, J., Horvath, G.: A sparse least squares support vector machine classifier. In: Proceedings of IEEE IJCNN, vol. 1, pp. 543–548 (2004)
11. Vapnik, V.N.: Statistical Learning Theory. Wiley-Interscience (1998)
12. Yu, L., Chen, H., Wang, S., Lai, K.K.: Evolving least squares SVM for stock market trend mining. IEEE Trans. on Evolutionary Computation 13(1), 87–102 (2009)

Pixel Classification and Heuristics
for Facial Feature Localization

Heitor B. Chrisóstomo, José E.B. Maia, and Thelmo P. de Araujo

Universidade Estadual do Ceará, UECE
Av. Paranjana, 1700, Fortaleza, CE, Brazil, 60740-903
{bc.heitor,thelmo.dearaujo}@gmail.com, jose.maia@uece.br

Abstract. In his work, we use a broad set of pixel features of low computational cost—which includes first order gray-level parameters, second order textural features, moment invariant features, multi-scale features, and frequency domain features—for pixel classification based on facial feature localization. A Radial Basis Function Neural Network performs the classification into three regions of interest. Morphological filters and intrinsic geometric properties of the human face are combined into a post-processing heuristic to finish the feature localization. We present the results, which are qualitative and quantitative satisfactory.

Keywords: Facial feature localization, pixel classification, face segmentation.

1 Introduction

Facial feature localization problem is treated in this work in two phases: a first one, in which face segmentation is done by pixel classification, followed by a second phase that applies heuristics based on morphological filters and some intrinsic geometrical properties of the human face.

Image segmentation is a process of partitioning a given image into regions of interest satisfying certain criteria. It is one of the fundamental processes in the visual recognition chain, making the generation of a compact description of an image (such as contours, connected regions, etc.) possible, which is much more manageable than the whole image. Most of the image segmentation methods can be broadly grouped into boundary-based (or edge-based) techniques, region based techniques, and hybrid methods. They all rely on two basic pixel neighborhood properties: discontinuity and similarity. Pixel discontinuity gives rise to boundary-based methods, whereas pixel similarity gives rise to region-based methods. This work explores region-based methods. Facial image segmentation is a hard problem to solve due to the large number of variations in image appearance, such as pose variations, image orientations, illumination conditions, and facial expressions. For an additional detailed survey of the different techniques, we direct the reader to the literature [1–4, 6].

All the above mentioned approaches may be applied to facial image segmentation. Nevertheless, specific knowledge about the application domain is useful to improve algorithm performance. One well-known way to improve facial image segmentation takes advantage of the locations of face elements, their proportions, the distances between them, and uses templates to guide the algorithm through its tasks of face localization,

E. Corchado et al. (Eds.): IDEAL 2014, LNCS 8669, pp. 167–174, 2014.

segmentation, recognition, and identification. In this work a weak form of template is used in the last step of the algorithm to label the regions of interest.

Our goal in this research is not to extract the exact shapes of facial features, but to locate regions corresponding to facial features such as eyes, eyebrows, mouth, nose, and ears. Facial image segmentation is treated as a classification problem and solved by supervised machine learning techniques. More specifically, we start with a large number of features (102) of various categories in the literature of image segmentation (texture, edges, DCT, Gabor, etc.) and then select features via PCA. The segmentation is based on classification of pixels which is performed by a Radial Basis Function Neural Network (RBFNN). We also present the formal definition of all features used in the research. Pixel classification on gray-level images, however, is not enough for a fine facial feature localization. Therefore, post-processing heuristics based on morphological filters and intrinsic geometrical properties of frontal faces.

The rest of the paper is organized as follows: Section 2 presents the segmentation methodology in detail, including the formal definition of each feature, and the post-processing heuristic. Section 3 provides data description, illustrates the implementation procedure, and analyzes the result obtained by the proposed approach. Finally, Section 4 concludes the work of this paper.

2   Methodology

2.1   General View

We first prepare the data for 10 runs of the experiment by dividing it randomly into two sets: a training set (80%) and a testing set (20%). Figure 1 presents the steps followed in each run. The left-hand part of Figure 1 shows the training phase and the right-hand part of the figure shows the testing phase.

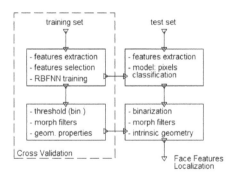

Fig. 1. General view of the method used

Following the literature, 102 features usually used for image segmentation are selected to be in the initial set for the training phase. Since the method is based on pixel classification, features are computed using a window W around each image pixel. How-ever, the best feature set may vary for each window size, therefore the window size has to be also selected. For this, we adopted an iterative process: for each window size in an exploratory range, a filter type procedure (explained in Section 2) selects the best feature set. The procedure is validated by training a RBFNN and using cross-validation with the pixel classification accuracy as the performance metric. After this phase, port-processing parameters are adjusted: binarization threshold, morphological filter, and facial geometric properties.

Data is balanced before entering the RBFNN using a method proposed by Fu et al. [4] to determine cost factors [1] automatically. K-means algorithm is used in the non-supervised phase of the RBFNN [1] to compute the centers c_i of the Gaussian functions. The distances to the B nearest neighbors are averaged to get the radii of each Gaussian function [9]: $\sigma_i = \frac{1}{B} \sum_{j=1}^{B} \| c_i - c_j \|$. In this work, we use 39 neurons in the RBFNN and $B = 7$.

In the testing phase, the relevant features are extracted from the test image and the segmentation algorithm is applied. Post-processing finishes the feature localization.

2.2 Feature Extraction

In this work, we apply a simple PCA technique to select the q most relevant features among $p = 102$ features previously chosen for their low computational cost. The orig-inal features are divided into five categories: 3 first order gray-level parameters, 64 second order textural features, 7 moment invariant features, 8 multi-scale features, and 20 frequency domain features.

The first order gray-level and the multi-scale features are defined in Table 1. The summations in the computation of gray-level mean and variance are done in a window W of size $w \times w$, for $w = 3, 5, 7, 9$. The two multi-scale features use Gaussian kernel

$$G_\sigma(x, y) = \frac{1}{2\pi\sigma^2} \exp\left(\frac{x^2 + y^2}{2\sigma^2}\right) ,$$

with 4 different radii ($\sigma = 0.5, 1.0, 2.0, 4.0$) [11], to perform convolution with gray-level intensity in a 5×5 window.

16 second order textural measures are computed accordingly to the formulas in Table 1 using co-occurrence matrices $P(i, j, d, \theta)$ [6] for 4 different angles θ ($0°$, $45°$, $90°$, and $135°$) with $d = 1$ pixel distance, and 4 window sizes $w = 3, 5, 7, 9$. In these features we used the following notation:

- $p(i, j) = P(i, j, d, \theta)/R$, where $R = 2w(w - 1)$, for $\theta = 0°, 90°$; and $R = 2w(w - 2) + 2$, for $\theta = 45°, 135°$;
- N_g is the number of gray levels in the images;
- μ_x, σ_x, μ_y and σ_y are the mean and variance of the marginal probabilities p_x and p_y, respectively;
- $p^+(k) = \sum_{\substack{i=1 \\ i+j=k,\, k=2,...,2N_g}}^{N_g} \sum_{j=1}^{N_g} p(i, j)$ and $p^-(k) = \sum_{\substack{i=1 \\ |i-j|=k,\, k=0,...,N_g-1}}^{N_g} \sum_{j=1}^{N_g} p(i, j)$;

Table 1. First and second order textural features and multi-scale features

#	Feature	Formula	#	Feature	Formula
1	Gray-level intensity	$I(x,y)$	11	Sum average	$f_6 = \sum\limits_{i=2}^{2N_g} i\, p^+(i)$
2	Gray-level mean	$\mu_{gl} = \dfrac{1}{w^2} \sum\limits_{(x,y)\in W} I(x,y)$	12	Sum variance	$f_7 = \sum\limits_{i=2}^{2N_g} (i - f_6)^2\, p^+(i)$
3	Gray-level variance	$\sigma_{gl}^2 = \sum\limits_{(x,y)\in W} \dfrac{(I(x,y)-\mu_{gl})^2}{w^2-1}$	13	Sum entropy	$f_8 = - \sum\limits_{i=2}^{2N_g} p^+(i)\log(p^+(i))$
4	Sq. gradient of L	$\left(I \otimes \dfrac{\partial G_\sigma}{\partial x}\right)^2 + \left(I \otimes \dfrac{\partial G_\sigma}{\partial y}\right)^2$	14	Entropy	$f_9 = - \sum\limits_{i=1}^{N_g}\sum\limits_{j=1}^{N_g} p(i,j)\log(p(i,j))$
5	Laplacian of L	$I \otimes \dfrac{\partial^2 G_\sigma}{\partial x \partial x} + I \otimes \dfrac{\partial^2 G_\sigma}{\partial y \partial y}$	15	Difference variance	$f_{10} = \sum\limits_{i=0}^{N_g-1} (i - \mu_{x-y})^2\, p^-(i)$
6	Angular 2nd moment	$f_1 = \sum\limits_{i=1}^{N_g}\sum\limits_{j=1}^{N_g} (p(i,j))^2$	16	Difference entropy	$f_{11} = - \sum\limits_{i=0}^{N_g-1} p^-(i)\log(p^-(i))$
7	Contrast	$f_2 = \sum\limits_{n=0}^{N_g-1} n^2 p^-(n)$	17	IMC 1	$f_{12} = \dfrac{f_9 - H_{xy1}}{\max\{H_x, H_y\}}$
8	Correlation	$f_3 = \sum\limits_{i=1}^{N_g}\sum\limits_{j=1}^{N_g} \dfrac{(ij)p(i,j)}{\sigma_x\sigma_y} - \mu_x\mu_y$	18	IMC 2	$f_{13} = \left(1 - e^{-2(H_{xy2}-f_9)}\right)^{\frac{1}{2}}$
9	Var. of sum of squares	$f_4 = \sum\limits_{i=1}^{N_g}\sum\limits_{j=1}^{N_g} (i-\mu)^2 p(i,j)$	19	Homogeneity	$\sum\limits_{j=0}^{N_g-1} \dfrac{1}{j^2+1}\, p^-(j)$
10	Inverse diff. moment	$f_5 = \sum\limits_{i=1}^{N_g}\sum\limits_{j=1}^{N_g} \dfrac{1}{1+(i-j)^2}\, p(i,j)$	20	Energy	$E = \sum\limits_{i=0}^{2N_g} (p^+(i))^2 \sum\limits_{j=0}^{N_g-1} (p^-(j))^2$
21	Sum & diff. entropy	$H = - \sum\limits_{i=2}^{2N_g} p^+(i)\log(p^+(i)) - \sum\limits_{j=0}^{N_g-1} p^-(i)\log(p^-(i))$			

- $H_x = - \sum_{i=1}^{N_g} p_x(i) \log(p_x(i))$ and $H_y = - \sum_{j=1}^{N_g} p_y(j) \log(p_y(j))$;
- $H_{xy1} = - \sum_{i=1}^{N_g} \sum_{j=1}^{N_g} p(i,j) \log(p_x(i)\, p_y(j))$;
- $H_{xy2} = - \sum_{i=1}^{N_g} \sum_{j=1}^{N_g} p_x(i)\, p_y(j) \log(p_x(i)\, p_y(j))$.

The 7 moment invariant features [7] in Table 2 are also calculated in a window W of size $w \times w$, for $w = 3, 5, 7, 9$. For $p, q \in \mathbb{N}$, regular and central moments are defined, respectively, by

$$\eta_{pq} = \sum_{x\in W} \sum_{y\in W} x^p y^q I(x,y) \quad \text{and} \quad m_{pq} = \sum_{x\in W} \sum_{y\in W} (x - \bar{x})^p (y - \bar{y})^q I(x,y),$$

where $\bar{x} = \frac{\eta_{10}}{\eta_{00}}$ and $\bar{y} = \frac{\eta_{01}}{\eta_{00}}$; which are normalized for scale invariance [7] by doing

$$\mu_{pq} = \frac{m_{pq}}{m_{00}^\gamma}, \quad \text{where } \gamma = \frac{p+q}{2} + 1.$$

Two types of frequency domain features are considered in Table 2: discrete cosine transform (DCT) and Gabor wavelets [5]. Because 2D DCT is not a convolution, we use a convolution-like heuristic to generate DCT features for each image pixel. A $N \times N = 4 \times 4$ DCT kernel window slides over the image generating 16 features for each image

Table 2. Moment invariant, DCT, and Gabor features

#	Feature	Formula	#	Feature	Formula
22	Moment invariant 1	$\phi_1 = \mu_{20} + \mu_{02}$	24	Moment invariant 3	$\phi_3 = (\mu_{30} - 3\mu_{12})^2 + (3\mu_{21} - \mu_{03})^2$
23	Moment invariant 2	$\phi_2 = (\mu_{20} - \mu_{02})^2 + 4\mu_{11}^2$	25	Moment invariant 4	$\phi_4 = (\mu_{30} + \mu_{12})^2 + (\mu_{21} + \mu_{03})^2$
26	Moment invariant 5	$\phi_5 = (\mu_{30} - 3\mu_{12})(\mu_{30} + \mu_{12})[(\mu_{30} + \mu_{12})^2 - 3(\mu_{21} + \mu_{03})^2] + (3\mu_{21} - \mu_{03})(\mu_{21} + \mu_{03})[3(\mu_{30} + \mu_{12})^2 - (\mu_{21} + \mu_{03})^2]$			
27	Moment invariant 6	$\phi_6 = (\mu_{20} - \mu_{02})[(\mu_{30} + \mu_{12})^2 - (\mu_{21} + \mu_{03})^2] + 4\mu_{11}(\mu_{30} + \mu_{12})(\mu_{21} + \mu_{03})$			
28	Moment invariant 7	$\phi_7 = (3\mu_{21} - \mu_{03})(\mu_{30} + \mu_{12})[(\mu_{30} + \mu_{12})^2 - 3(\mu_{21} + \mu_{03})^2] - (\mu_{30} - 3\mu_{03})(\mu_{21} + \mu_{03})[3(\mu_{30} + \mu_{12})^2 - (\mu_{21} + \mu_{03})^2]$			
29	2D DCT	$F(u, v) = \sum_{x=0}^{N-1} \sum_{y=0}^{N-1} \cos\left(\frac{\pi}{N}(x + \frac{1}{2})u\right) \cos\left(\frac{\pi}{N}(y + \frac{1}{2})v\right) I(x, y)$			
30	real Gabor filter	$G(x, y; \sigma, \omega, \theta) = \frac{1}{2\pi\sigma^2} \exp\left(-\frac{\pi(x^2 + y^2)}{\sigma^2}\right) \cos(2\pi\omega(x\cos\theta + y\sin\theta))$			

pixel. The first feature (corresponding to the top left kernel window pixel) is discarded for it represents the gray-level average, a feature already considered.

In order to reduce the number of Gabor features, we varied the Gabor filter parameters and applied the proposed scheme to select only five Gabor filters: 3 15×15 masks ($\theta = 0°$, $\sigma = 14$, $\lambda = 50$; $\theta = 90°$, $\sigma = 14$, $\lambda = 50$; and $\theta = 90°$, $\sigma = 14$, $\lambda = 100$) and 2 21×21 masks ($\lambda = 50$ and $\lambda = 100$, both with $\theta = 45°$ and $\sigma = 14$).

For each of the 116 $m \times m$ gray-scale images, an $m^2 \times p$ image feature matrix is computed and its columns are normalized to have its elements in the $[0, 1]$ range. In order to apply PCA, a matrix X is composed by appending all r training image feature matrices, resulting in an $n \times p$ matrix, with $n = rm^2$.

2.3 Feature Selection

Jolliffe [8] describes three main types of method for feature selection that use principal components (PCs). Considering the first q PCs out of p PCs, the first method associates $p - q$ original features with each of the last $p - q$ PCs. This can be done once or iteratively, using, in the latter case, the remaining q original features and so on. "The reasoning behind this method is that small eigenvalues correspond to near-constant relationships among a subset of variables. If one of the variables involved in such a relationship is deleted (a fairly choice for deletion is the variable with the highest coefficient in absolute value in the relevant PC) little information is lost" [8, p. 138].

A second method associates a set of $p - q$ original variables en bloc with the last PCs, choosing this variables for deletion. In the third method, q original variables are associated with each of the first q PCs. In this work, we use the latter method (whose reasoning is complementary to the first one), which is described in details as follows.

Singular value decomposition (SVD) is applied to X to obtain $X = U\Sigma V^T$, where U is an $n \times p$ matrix, Σ is a $p \times p$ diagonal matrix with the singular values in decreasing order, and V is a $p \times p$ orthogonal matrix, whose columns are the orthonormal right singular vectors of X (which are the eigenvectors of the correlation matrix $X^T X$).

The q first eigenvectors are chosen to achieve a cumulative percentage of total variation of $E_c\%$ (which is achieved with slightly more than q original variables [8]), and the selection of the q most representative original features proceeds as follows [3]:

The greatest (in absolute value) coordinate, say v_{k1}, of the first eigenvector $\mathbf{v}_1 = [v_{11} \quad \ldots \quad v_{p1}]^T$ corresponds to the original feature that has the greatest load on the first principal direction, i.e., the k-th original feature is the one which has greatest loading on the (first) dominant principal direction.

The greatest (in absolute value) coordinate, say v_{j2}, of the second eigenvector $\mathbf{v}_2$ indicates that the j-th feature is the second most relevant of the original features. If the j-th original feature has already been chosen, the next greatest coordinate is chosen. The selection proceeds until the q most relevant original features have been selected.

Note that this way of feature selection helps to reduce—but not necessarily avoiding [2]—information redundancy. This is because, if there are groups of highly related original variables, the method will select only one variable from the group [8].

Variable nof in Table 3 describes the number of features q necessary to achieve each of 4 percentage levels of accumulated energy by window size $w \times w$.

Table 3. Accuracy statistics: in each cell, nof is the number of features necessary to achieve energy level; μ is the accuracy mean; σ is the accuracy standard deviation; fpr is the false positive rate; and fnr is the false negative rate

Energy		Window size $w \times w$			Energy		Window size $w \times w$		
level(%)		3×3	5×5	7×7	9×9	level(%)		3×3	5×5	7×7	9×9

Energy level(%)		3×3	5×5	7×7	9×9	Energy level(%)		3×3	5×5	7×7	9×9
75	nof	18	30	29	28	85	nof	24	40	40	38
	μ	0.832	0.830	0.838	0.765		μ	0.829	0.830	0.841	0.834
	σ	0.0141	0.00981	0.00632	0.107		σ	0.0255	0.0108	0.00500	0.00675
	fpr	0.0451	0.0467	0.0406	0.108		fpr	0.0523	0.0510	0.0442	0.0438
	fnr	0.123	0.123	0.120	0.128		fnr	0.119	0.119	0.115	0.122
80	nof	21	35	**34**	33	90	nof	29	47	47	45
	μ	0.755	0.831	**0.841**	0.833		μ	0.842	0.834	0.828	0.842
	σ	0.117	0.0099	**0.00555**	0.00851		σ	0.0120	0.00907	0.0125	0.00842
	fpr	0.108	0.0491	**0.0450**	0.0442		fpr	0.0539	0.0443	0.0415	0.0406
	fnr	0.128	0.120	**0.114**	0.122		fnr	0.104	0.121	0.131	0.118

2.4 Post-processing Heuristic

A three-step post-processing heuristic completes the work of the RBFNN. In the first step, a simple threshold technique is used to get a binary version of each image obtained after classification (Figure 2 left). In the second step, morphological filters localize the most populated connected components, and their centroids are computed. These centroids are the location candidates for the facial features. In the third step, we apply some intrinsic geometric properties between the face features to complete their localization (right block on Figure 2). This heuristic is still in development at this time. Examples of the expected results are shown in Section 3.

3   Results and Discussion

In our experiments, 116 images from FERET database [10] were used. 75% of them were randomly selected to the training set and the remaining 29 images composed the testing set. In order to select the more appropriate window size and accumulated energy level (and, hence, the number of features), a 4-fold cross validation scheme was applied to the training set. These experiments generated Table 3. Normalization was applied to the training set and the training set means are used in the testing set.

Using Table 3, we chose to work with a 7×7 window with 80% of accumulated energy. The 34 features selected by the PCA scheme are: feature 11 (with $\theta = 0°$); 19 ($\theta = 45°$); 5 ($\sigma = 0.5$); 4 ($\sigma = 2.0$); feature 24; 16 ($\theta = 45°$); 18 ($\theta = 90°$); 19 ($\theta = 0°$); 29 ($u = 2, v = 3$); 5 ($\sigma = 2.0$); feature 26; 5 ($\sigma = 1.0$); 29 ($u = 3, v = 3$); 15 ($\theta = 0°$); 29 ($u = 3, v = 1$); 29 ($u = 3, v = 2$); 29 ($u = 3, v = 0$); 18 ($\theta = 45°$); 17 ($\theta = 45°$); 15 ($\theta = 90°$); 10 ($\theta = 90°$); 10 ($\theta = 135°$); feature 3; 4 ($\sigma = 4.0$); feature 27; 14 ($\theta = 90°$); 5 ($\sigma = 4.0$); 14 ($\theta = 45°$); 7 ($\theta = 90°$); feature 23; feature 25; 7 ($\theta = 0°$); 4 ($\sigma = 1.0$); and 7 ($\theta = 135°$).

Average results (10 executions) for the testing set are: accuracy mean $\mu = 0.8493$, st.dev. $\sigma = 0.0231$, and false positive and negative rates $fpr = 0.0414$, $fnr = 0.1192$.

The importance of false positive and negative rates rests upon the need to post-process the resulting image after classification in order to localize the face features. High false positive rates mean that many pixel positions were left as candidates for feature localization. High false negative rates, by their turn, result in many missing pixels that would be used for feature localization. One may notice that the observed $fnr = 0.1192$ is more than the double of the $fpr = 0.0414$.

Figure 2 (left) shows four typical gray-scale images resulting from the classification phase. The central block shows the binary images after the morphological filter is applied. One may notice that few regions in the images are left to be candidates to the final localization process, which becomes much easier for this same reason.

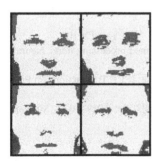  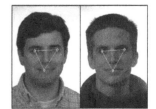

Fig. 2. Images after classification (left), after post-processing (center), and after face feature localization (right)

The coordinates of the localized facial features are mapped back to the original image, shown in the right block of Figure 2, in which one may see typical expected results after the last step of face feature localization.

4   Conclusion

We approach the complex problem of facial feature localization in gray-scale face images by first classifying pixels for image segmentation, followed by heuristics based on morphological filters and human face intrinsic geometry. Our main contribution in this work is the selection and recommendation of the 34 features (see Section 3) selected among 102 most used features in the literature when dealing with the difficult tasks of frontal face segmentation and facial feature localization.

Results have being shown promising. A clear limitation is that only frontal face images were used in the experiments. Nevertheless, frontal face images have their due applications.

For the next steps in this research, we intend to validate the procedure with a larger and more diversified set of images and use facial feature localization obtained here as input for facial expression interpretation.

References

1. Bishop, C.M.: Pattern Recognition and Machine Learning. Springer, New York (2006)
2. Cohen, I., Xiang, Q.T., Zhou, S., Sean, X., Thomas, Z., Huang, T.S.: Feature selection using principal feature analysis (2002)
3. Dunteman, G.H.: Principal Components Analysis. Sage University Paper series on Quantitative Appl. in the Social Sciences, No. 07-069. Sage, Newbury Park (1989)
4. Fu, X., Wang, L., Chua, K.S., Chu, F.: Training RBF neural networks on unbalanced data. In: Proceedings of the 9th International Conference on Neural Information Processing (ICONIP 2002), vol. 2, pp. 1016–1020 (2002)
5. Gonzalez, R.C., Woods, R.E.: Digital Image Processing, 3rd edn. Prentice-Hall, Inc., Upper Saddle River (2006)
6. Halarick, R.M., Shanmugam, K., Dinstein, I.: Textural features for image classification. IEEE Transactions on Systems, Man, and Cybernetics 3(6), 610–621 (1973)
7. Hu, M.-K.: Visual pattern recognition by moment invariants. IRE Transactions on Information Theory 8(2), 179–187 (1962)
8. Jolliffe, I.T.: Principal Component Analysis, 2nd edn. Springer, New York (2010)
9. Moody, J., Darken, C.J.: Fast learning in networks of locally-tuned processing units. Neural Comput. 1(2), 281–294 (1989)
10. Phillips, P.J.: The facial recognition technology FERET database. IEEE Transactions on Pattern Analysis and Machine Intelligence 22 (2004)
11. Witkin, A.P.: Scale-space filtering. In: Proceedings IJCAI 1983, pp. 1019–1022. Morgan Kaufmann Publishers Inc., San Francisco (1983)

A New Appearance Signature
for Real Time Person Re-identification

Mayssa Frikha, Emna Fendri, and Mohamed Hammami

Sfax University, Faculty of Science,
Road Sokra Km 3 BP 1171, 3000 Sfax, Tunisia
frikha.mayssa@hotmail.fr,
fendri.msf@gnet.tn,
mohamed.hammami@fss.rnu.tn

Abstract. Appearance based person re-identification attracts the attention of researchers and presents an active research area for intelligent video surveillance systems. In this paper, we propose a new approach for person re-identification in multi-camera systems. This approach consists in computing a new person signature by extracting a texture descriptor, not from the entire body, but only from stripes selected automatically. In addition, in this work, unlike existing solutions using gray leveled body, we propose to compute the texture descriptor from HSV colored body. Our approach has been compared to state-of-the-art methods using the highly challenging VIPeR dataset. We prove from this comparative study, that the proposed approach improves both, time and quality performances of person re-identification.

Keywords: video surveillance, person re-identification, co-occurrence matrix, SFFS, body stripes selection, complexity.

1 Introduction

Person re-identification is one of the most active research topics in computer vision. It consists in observing a person in one camera's fold of view and re-identifying that same person again in the same camera or another camera view with different acquisition conditions at different times and locations. Since target persons do not change their clothes when they transit over different cameras areas, their appearances seem to present their most reliable information. However, appearance-based person re-identification presents a highly challenging problem due to variations that can affect the human appearance. The most common variations concern human poses, camera's viewpoints, acquisition conditions, camera's characteristics and lighting conditions.

Typically, person re-identification approaches have focused on extracting visual characteristics highlighting the important aspects of a person's appearance. This task concerns the body parts description and part-based body model.

The former focuses on building a descriptor based on low level features such as color, texture, shape or combinations thereof. Color features are the most

E. Corchado et al. (Eds.): IDEAL 2014, LNCS 8669, pp. 175–182, 2014.
© Springer International Publishing Switzerland 2014

adopted features as they are viewpoint invariant and can be sensed at a far distance from the camera. Color histogram is widely used with different color spaces such as RGB [17][5], HSV [5][18], or combinations thereof [7] but no one has been proven the best. However, histograms do not contain any information on the spatial distribution of colors on the body. An alternative way is to use the spatial relationship between pixels such as Spatiogram [5], Color/Path-Length [17], Vertical Feature [11] and, Regional Histogram [5]. Their drawback is they are sensitive to the lighting variations and color response of the sensor [4][12].

Several recent methods insist on the importance of texture features as they provide high information details and characteristic based on clothing textures such as Co-occurrence Matrix [5] and Local Binary Patterns [4][10][18], or as a filter response such as Haar Wavelet [1], Gabor filters and Schmid filters [5]. Their advantage is that they are not affected by the camera sensor variations. Generally, these features are extracted from gray-level image and do not retain any information about the body appearance color.

On the other hand, shape features are rarely employed [4][5][18], given that the person is a highly articulated object, his body shape changes dramatically according to the person's pose and movement. Finally, hybrid methods combine different features [2][6][8][13]. Unfortunately, these methods are more expensive in term of computation cost that they are impractical in real-time scenarios.

Concerning the part-based body model, low-level features can be extracted either from the whole body image [5][17] or from body parts instead. The former solution is faster and economical in terms of computional cost but less discriminating. The latter is based on a simple report that clothing appearance of the body parts follows the body structure (e.g. upper body is different from lower body). Therefore, it divides body in different stripes describing each body part separately to capture visually distinct areas of the appearance. In literature, the stripes number is chosen empirically [5][11] or based on human anatomy. Generally, the division of the body can be done in 3 stripes [4][2] corresponding to the head, torso and legs, or, 6 equal-sized horizontal stripes [7][8][12][13][19] corresponding to the head, upper/lower torso, upper/lower legs, and chooses. We can notify that, although this second solution is more complex (in terms of memory space and processing time), it is highly more interesting in term of re-identification accuracy because it provides a more detailed decomposition of the body. Nevertheless, real-time performance of person re-identification task is an important need for wide area security applications.

In this paper, we propose a new appearance signature that is based on a texture feature extracted from colored body image. In addition, unlike state-of-the-art methods, we propose to automatically select the most relevant body stripes. We will prove that carefully designed visual appearance modeling can provide a significant improvement of re-identification accuracy, processing time and memory requirement in order to deal with real-time application.

The remainder of the paper is organized as follows. Section 2 describes the proposed approach. Section 3 provides the experimental results. Section 4 concludes the paper.

2   Proposed Approach

In this section we detail the proposed approach described in fig. 1. First we extract texture features from colored body stripes. Then we automatically select the salient appearance stripes.

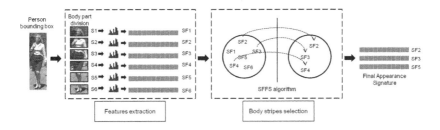

Fig. 1. Proposed approach for determining our appearance signature

2.1   Feature Extraction

Inspired by re-identification results using Co-occurrence Matrix [5][6] and the importance of color information to model people appearance, we extract Co-occurrence Matrix from six HSV colored body stripes. Co-occurrence Matrix [9] of a colored image I, of size $N \times N$, represents the frequency of a pixel with value i is adjacent to the pixel with value j as can be seen in equation 1.

$$CM_k(i,j,k) = \sum_{x=1}^{N} \sum_{y=1}^{N} \begin{cases} 1, & if\, I(x,y,k) = i \ and \ I(x+\Delta_x, y+\Delta_y, k) = j \, . \\ 0, & otherwise \, . \end{cases}$$

(1)

where k denoted the color component (H, S, V), Δ_x and Δ_y are the offset, we calculate CM according to the direct neighbor and by considering 4 directions $(0°, 45°, 90°,$ and $135°)$. The 4 adjacency matrices are summed then converted into a vector. The effectiveness can be explained by using the sum of the matrices which encodes the distributions of the intensities and totally ignores the relative position of neighboring pixels. This method is more invariant to pose and viewpoint variations. Also, HSV color space is very close to the way the human brain perceives color and it's known by the separation of the image intensity from the color information which is more robust to lighting variations. In order to capture the distinct areas of the appearance, we extract CM from six body stripes.

2.2   Body Stripes Selection

In uncontrolled acquisition conditions scenarios, some body stripes are more informative and stable to capture the salient appearance characteristics. Our contribution is to automatically select the salient body stripes while discarding the noisy ones, instead of using the whole body stripes as the state-of-the-art methods. The advantage of this selection is to reduce the impact of partial occlusions, pose and viewpoint variations, to improve the processing time and memory requirement in order to deal with real time applications. We have adopted to Sequential Forward Feature Selection Algorithm (SFFS) [16] for its effectiveness. SFFS is based on a heuristic approach to automatically reduce the initial set of stripes without specifying the final stripes number.

Algorithm SFFS starts from an empty selected stripes set $\Lambda = \oslash$ and selected stripes number $k = 0$.

Step 1: Select the best strip s^+ (forward step) that maximise the Correct Re-identification Rate (CRR) by the equation 2 and update the selected stripes set $\Lambda_k = \Lambda_k \cup s^+$ and the selected stripes number $k = k + 1$.

$$s^+ = arg\ max_{s \notin \Lambda_k} CRR\left(\Lambda_k \cup s\right) . \tag{2}$$

Step 2: Select the worst strip s^- (backward step) from the selected stripes set that minimise the CRR by the equation 3.

$$s^- = arg\ max_{s \in \Lambda_k} CRR\left(\Lambda_k \backslash s\right) . \tag{3}$$

Step 3: If $CRR\left(\Lambda_k \backslash s^-\right) \succ CRR\left(\Lambda_k\right)$ then updates the selected stripes set by eliminating the worst strip s^- formulated by $\Lambda_{k-1} = \Lambda_k \backslash s^-$, $k = k - 1$ and go back to step 2 else go back to step 1.

The whole process will stops if there is no further improvement in the Correct Re-identification Rate.

3   Experimental Results

The proposed appearance modeling has been evaluated on the challenging benchmark VIPeR [8]. We present two series of experiments: the first concerns the features extraction and the second concerns body stripes selection. Then, we compared our method with the state-of-the-art methods.

3.1   Presentation of VIPeR Database and Experimental Setup

VIPeR dataset [8] contains 632 person image pairs taken from two non overlapping camera views (camera A and camera B). Each image is scaled to 64×128 pixels. The appearance exhibit significant variation in pose and viewpoint, illumination conditions with the presence of occlusions. Most of the matched image pairs contain a viewpoint change of about $90°$ or $180°$, as can be seen in fig. 2.

We randomly selected 316 image pairs for testing set, and the rest are used for training set. This procedure was repeated 10 times as proposed in [8]. Considering images from cam B as the gallery set, and images from cam A as the probe

Fig. 2. VIPeR dataset [8]: People images captured from different viewpoints

set, each image of the probe set is matched with gallery set. For evaluation, we use the most common metrics are the CMC (Cumulative Match Characteristic) curve and Rank-1 accuracy. We report the average performance over 10 trials.

3.2   Evaluation of Our Appearance Signature Parameters

The re-identification process of a query person Q from a gallery set Γ based on our appearance can be formulated as the sum of the similarity between features vector of the corresponding selected body stripes as can be seen in equation 4.

$$ P = \underset{P_i}{arg\,min} \left[\sum_{j=1}^{SN} D\left(SF_{P_i}^{j}, SF_{Q}^{j} \right) \right], P_i \in \Gamma . \tag{4} $$

Where $\Gamma = \{P_1, \ldots, P_N\}$ is a gallery of N appearance signatures, $D\,(.,.)$ is the Bhattacharyya distance [3] which is the most adopted [2][6][8], SN is the selected stripes number, and SF^j is the features vector of the j-th body strip.

Firstly, a comparative study will allow us to validate the performance of Co-occurence Matrix against 3 well designed features in term of re-identification accuracy, processing time and memory requirement. As color features, we opted for Color histogram and Spatiogram. Spatiogram [14] represents an extension of color histogram that incorporates spatial information by calculating the spatial mean and covariance for each bin. As texture feature, we opted for Local Binary Patterns [15] which encodes the local structure around each pixel by generating a binary code. We use a model consisting of 8 points with a radius of 1 pixel. For each descriptor, different experiments were performed by varying the color space. We have chosen the most adopted color spaces, i.e. RGB, HSV, and YCbCr. Each color component is quantified into 32 bins.

Table 1 shows rank-1 accuracy (%) for each descriptor. Similar performance is achieved using color features in RGB space and Co-occurrence Matrix in HSV space. The significant improvement for Co-occurrence Matrix, get up to 4% rank-1 compared with gray level, is explained by its extraction from colored body image allowing the incorporation of color information with texture information.

Table 2 presents a comparison of the complexity cost for the three best features. Comparing with Color histogram and spatiogram, we can see that the

Table 1. Rank-1 accuracy (%) for each descriptor in different color spaces on VIPeR dataset

Descriptors	Color spaces			
	RGB	YCrCb	HSV	Gray level
Color Histogram	**6.17**	5.57	4.94	-
Spatiogram	**6.14**	6.08	4.91	-
Co-occurrence Matrix	2.69	4.97	**5.85**	1.71
Local Binary Patterns	1.74	**2.31**	2.15	1.30

Table 2. Comparison of computational cost for the descriptors on VIPeR dataset

	Features Extraction Time	Similarity Measurement Time	Size of Features Vector per strip
Color Histogram	0.0057 s	0.6261 s	32^3
Spatiogram	0.0097 s	2.8640 s	$32^3 * (1 + 2 + 4)$
Co-occurrence Matrix	0.0143 s	0.0618 s	$32^2 * 3$

average of features extraction time for Co-occurrence Matrix is slightly slow (over than 2 times), but it is the faster (over than 10 times) in term of average similarity measurement time and memory requirement. Thus, in order to respect real-time constraints, Co-occurrence Matrix in HSV space presents an efficient tradeoff between computational cost and re-identification accuracy.

Table 3. Matching Rates (%) at rank r with and without stripes selection on VIPeR

	Matching Rates	
# stripes	r=1	r=5
6 Stripes	9.84	19.75
$\{B_2, B_3, B_4\}$	11.65	28.48

Secondly, after applying the SFFS algorithm selected stripes are $\{B_2, B_3, B_4\}$ corresponding to the upper torso, lower torso, and upper legs which confirms our intuition (c.f. fig. 1). These selected stripes present the area of the body that provides maximum distinguishing appearance information. Table 3 presents matchings rates at rank 1 and 5 with and without stripes selection. Stripes selection get up to 2% rank-1 and 9% rank-5 re-identification rates which prove that some body stripes are more invariant to pose and viewpoint changes. Also, this selection exhibits an important reducing of the computational cost to the half which is very interesting for real-time scenarios.

Table 4. VIPeR dataset: matching rates (%) at rank r with 316 persons

Methods	r=1	r=5
SDALF (hybrid method) [2]	19.87	38
Proposed appearance signature	18.39	33.73
Spatiogram HSV [6]	7.69	16.27
Gray level Co-occurrence Matrix [5]	2.88	8.23

3.3   Comparison with Recent State-of-the-Art Methods

We compared our method with recent state-of-the-art methods SDALF [2], Spatiogram HSV [6] and Gray level Co-occurrence Matrix [5] as can be seen in table 4 using the same experimental conditions of [2] to make a fair comparison.

Our method outperforms Spatiogram HSV [6] get up to 10% rank-1 and 17% rank-5 correct re-identification rates, and, Gray Level Co-occurrence Matrix [5] get up to 16% rank-1 and 25% rank-5. Compared with SDALF, our method is slightly lower at rank 1 and 5. It should be noted that the computation time of SDALF (hybrid method: two color features and texture feature) is extremely expensive and it does not recommend for further real-time constraints. The advantage of our method can be explained by extracting a texture feature from colored body in order to incorporate the color information with the texture information, and, by eliminating the noisy body stripes to only capture the salient appearance details. The proposed appearance modeling proves to be robust enough to deal with partial occlusions, pose and viewpoint variations.

4   Conclusion

This paper introduces a novel appearance signature for real-time person re-identification. The proposed approach consists in extracting co-occurrence matrix from colored HSV body for 3 automatically selected body stripes corresponding to upper torso, lower torso, and upper legs. We have present two novel strategies for person re-identification by extracting texture feature from colored body image and by automatically selecting the most reliable and invariant body stripes using the SFFS algorithm. Further, we evaluated our method on the challenging VIPeR dataset. Compared to the state-of-the-art methods, we proved that the proposed approach improves person re-identification rate, processing time and memory requirement.

References

1. Bak, S., Corvee, E., Brémond, F., Thonnat, M.: Person Reidentification Using Haar-based and DCD-based Signature. In: Workshop on Activity Monitoring by Multi-Camera Surveillance Systems (2010)
2. Bazzani, L., Cristani, M., Murino, V.: Symmetry-Driven Accumulation of Local Features for Human Characterization and Re-identification. Comput. Vis. Image Underst. (2013)

3. Bhattacharyya, A.: On a Measure of Divergence Between two Statistical Populations Defined by Their Probability Distribution. Bull. Calcutta. Math. Soc. (1943)
4. Bialkowski, A., Denman, S., Sridharan, S., Fookes, C., Lucey, P.: A Database for Person Re-identification in Multi Camera Surveillance Networks. In: International Conference on Digital Image Computing Techniques and Applications (2012)
5. Derbel, A., Ben Jemaa, Y., Canals, R., Emile, B., Treuillet, S., Ben Hamadou, A.: Comparative Study Between Color Texture and Shape Descriptors for Multi-Camera Pedestrians Identification. In: International Conference on Image Processing Theory, Tools and Applications, Istanbul (2012)
6. Derbel, A., Ben Jemaa, Y., Treuillet, S., Emile, B., Canals, R., Ben Hamadou, A.: Robust Descriptors Fusion for Pedestrians' Re-identification and Tracking Across a Camera Network. In: International Conference on Computer Vision Theory and Applications, Spain (2013)
7. Du, Y., Ai, H., Lao, S.: Evaluation of Color Spaces for Person Re-identification. In: 21st International Conference on Pattern Recognition, Japan (2012)
8. Gray, D., Tao, H.: Viewpoint Invariant Pedestrian Recognition with an Ensemble of Localized Features. In: Forsyth, D., Torr, P., Zisserman, A. (eds.) ECCV 2008, Part I. LNCS, vol. 5302, pp. 262–275. Springer, Heidelberg (2008)
9. Haralick, R.M., Shanmugam, K., Dinstein, I.: Textural Features for Image Classification. IEEE Trans. Syst. Man Cybernet. (1973)
10. Hirzer, M., Roth, P.M., Bischof, H.: Person Re-identification by Efficient Impostor-Based Metric Learning. In: 9th International Conference on Advanced Video and Signal-Based Surveillance (2012)
11. Ilyas, A., Scuturici, M., Miguet, S.: A Combined Motion and Appearance Model for Human Tracking in Multiple Cameras Environment. In: International Conference on Engineering and Technology, Pakistan (2010)
12. Kuo, C.H., Khamis, S., Shet, V.: Person Re-identification Using Semantic Color Names and Rankboost. In: IEEE Workshop on Applications of Computer Vision (2013)
13. Liu, C., Gong, S., Loy, C.C., Lin, X.: Person Re-identification: What Features are Important? In: 1st International Workshop on Re-Identification (2012)
14. O'Conaire, C., O'Connor, N.E., Smeaton, A.F.: An Improved Spatiogram Similarity Measure for Robust Object Localisation. In: IEEE International Conference on Acoustics, Speech, and Signal Processing (2007)
15. Ojala, T., Pietikäinen, M., Harwood, D.: Performance Evaluation of Texture Measures with Classification based on Kullback Discrimination of Distributions. In: 12th International Conference on Pattern Recognition (1994)
16. Pudil, P., Novovicova, J., Kittler, J.: Floating Search Methods in Feature Selection. In: 12th International Conference on Pattern Recognition (1994)
17. Truong Cong, D.-N., Khoudour, L., Achard, C.: People Reacquisition across Multiple Cameras with Disjoint Views. In: Elmoataz, A., Lezoray, O., Nouboud, F., Mammass, D., Meunier, J. (eds.) ICISP 2010. LNCS, vol. 6134, pp. 488–495. Springer, Heidelberg (2010)
18. Yang, Z., Jin, L., Tao, D.: A Comparative Study of Several Feature Extraction Methods for Person Re-identification. In: Zheng, W.-S., Sun, Z., Wang, Y., Chen, X., Yuen, P.C., Lai, J. (eds.) CCBR 2012. LNCS, vol. 7701, pp. 268–277. Springer, Heidelberg (2012)
19. Zheng, W., Gong, S., Xiang, T.: Re-identification by Relative Distance Comparison. IEEE Trans. on Pattern Anal. Mach. Intell. (2012)

A New Hand Posture Recognizer
Based on Hybrid Wavelet Network Including
a Fuzzy Decision Support System

Tahani Bouchrika, Olfa Jemai, Mourad Zaied, and Chokri Ben Amar

REsearch Groups in Intelligent Machines (REGIM-Lab), University of Sfax,
National Engineering School of Sfax BP 1173, 3038 Sfax, Tunisia
tahani.bouchrika@ieee.org
http://members.regim.org/tahani08

Abstract. In this paper we present a novel hand posture recognizer based on wavelet network learnt by fast wavelet transform (FWN) including a fuzzy decision support system (FDSS). Our contribution in this paper resides in proposing a new classification way for the FWN classifier. The FWN having an hybrid architecture (using as activation functions both wavelet and scaling ones) provides hybrid weight vectors when approximating an image. The FWN classification phase was achieved by computing simple distances between test and training weight vectors. Those latter are composed of two types of coefficients that are not in the same value range which may influence on the distances computing. This can cause wrong recognitions. So, to overcome this lacuna, a new classification strategy is proposed. It operates a human reasoning mode employing a FDSS to calculate similarity degrees between test and training images. Comparisons with other works are presented and discussed. Obtained results have shown that the new hand posture recognizer performs better than previously established ones. . . .

Keywords: hand posture recognition, hybrid fast wavelet network, fuzzy decision support system.

1 Introduction

Hand gesture recognition is an active area of research in the vision community because of its extensive applications in virtual reality, sign language recognition, computer games etc. . . Hand gestures can be divided into two categories: static gestures[1] and dynamic ones[2]. The static gesture recognition methods extract features from a single frame to be classified, while the dynamic gesture recognition ones use the movement relationship between successive frames to determine hand paths. In this paper we focused on the problem of static hand gesture recognition. So, we have created a new hand posture recognizer based on an hybrid fast wavelet network including a fuzzy decision support system (HFWN-FDSS). Hand gesture recognition can be considered as a classification problem. So, to accomplish this task, authors employed a neural network (NN) classifier in[3,4], a

E. Corchado et al. (Eds.): IDEAL 2014, LNCS 8669, pp. 183–190, 2014.

hidden markov model (HMM) classifier in[5,6], an hybrid wavelet network learnt by fast wavelet transform (HFWN) in[7,8] and a fuzzy-rough (FR) approach in[9]. In[8], given results, proved that the HFWN classifier performs better than the classical wavelet network[10], the NN and the HMM. So, we still have to compare our new approach to the HFWN classifier and to the FR approach. The architecture and the learning algorithm of the HFWN were explained in[11]. It was a result of a sequence of our research group works[10,11,12] trying to overcome shortcomings met in the literature[13]. To prove the robustness of the HFWN classifier, we have used it in[14,15] to recognize 2D and 3D faces and in[16] to classify images. Besides, this classifier was employed in[17] and [18] in order to recognize driver eyes states to inhibit the hypovigilance. Although the HFWN proved its effectiveness in many applications [22,23,24], it suffers from a shortcoming in it classification phase: because of its hybrid aspect, obtained weight vectors are divided into two types that are the approximation and the detail coefficients. So, to make classification, distances will be computed between test and training weight vector images. The values of the two types of coefficients are generally not in the same value range which may influence on distance values that can cause wrong recognitions. From here, as a continuity of wich was performed[10,11,12], we had the idea of working on this point by proposing a new classification strategy which will be detailed in the section 2.2(a). The rest of the paper is organized as follows: Section 2 outlines the proposed approach for hand posture recognition. Section 3 presents the experimental results with the aim of illustrating the effectiveness of the proposed method. In section 4, we end up with conclusion and perspectives for future works.

2   Overview of the Proposed Approach to Recognize Statistic Hand Gestures

The solution which we present to recognize static hand gestures can be divided into two main parts. The first part presents the learning phase which ensures the approximation of hand posture training samples by an HFWN. The second part is the classification phase that includes a FDSS in order to calculate similarity degrees between query images and learning ones.

2.1   Approximation with HFWN

In this part, as it is shown in Fig.1, we proceed by the approximation of every element of the learning base by a HFWN to produce a data signature. The resulting signature is constituted of wavelets and scaling functions (g_i) and their coefficients (α_i). Obtained information will be used to match a query sample with all the samples in the learning base. The learning base is composed of hand posture classes. Each class consists of samples X_{ij}, $(i \leq n, j \leq m)$ with n presents the number of images and m is the number of classes. An HFWN is optimized for each example X_{ij} and stored in a database containing all network models. Obtained network models will be used in the classification phase which will be explained in Section 2.2.

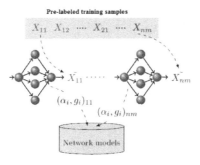

Pre-labeled training samples

X_{11} X_{12} $\cdots$ X_{21} $\cdots$ X_{nm}

$\tilde{X}_{11}$ $\cdots$ $\tilde{X}_{nm}$

$(\alpha_i, g_i)_{11}$

$(\alpha_i, g_i)_{nm}$

Network models

Fig. 1. Training phase

2.2 Classification

In the classification stage, the sample to be classifed is projected on all wavelet networks of the learning samples. After each projection, the family of activation functions remains unchanged (that of the learning samples), while new weights (δ_i), are computed. Coefficient vectors (δ_i and α_i) will be used to calculate similarity degrees between the query image and the learning ones. This step is ensured using fuzzy measures.

a-Computing Similarity Degrees

Computing similarity degree of two images was performed in [8,16] by calculating distances between the two coefficient vectors (δ_i and α_i). Activation functions used to produce coefficient vectors are of two types (scaling and wavelet functions). So, provided coefficients are necessary of two types : the approximation coefficients (a_k) produced by scaling functions and detail ones (d_k) resulti ng from the wavelet functions use (with k is the number of coefficients in each coefficient vector). Values of the two types of coefficients are generally not in the same value range which may influence on the measurement of distances. From this point we had decided of decorticating each vector into two ones containing separately the approximation and the detail coefficients. Then, to classify an image, we calculate distances between vectors containing the same type of coefficients. Resulting distances, will present the entry of a FDSS(Fig.2). This latter will compute similarity degree between the two images in order to ensure the decision-making phase (deciding to which class belongs a posture).

Distances between a query image coefficients and a training image ones (dsa_i and dsd_i) are calculated as follows:

$$dsa_i = \sqrt{\sum_{j=1}^{k}(V_{aq}(j) - V_{ai}(j))^2} \tag{1}$$

$$dsd_i = \sqrt{\sum_{j=1}^{k}(V_{dq}(j) - V_{di}(j))^2} \tag{2}$$

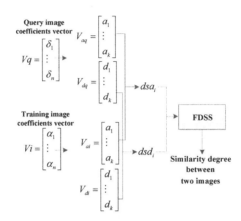

Fig. 2. Computing similarity degree between two images

With:

k = Number of approximation coefficients(AC) and detail coefficients(DC).

dsa_i = Distance between AC vectors of the query image and an image i.

dsd_i = Distance between DC vectors of the query image and an image i.

V_{aq} = AC vector of the query image.

V_{ai} = AC vector of an image i.

V_{dq}= DC vector of the query image.

V_{di} = DC vector of an image i.

Once all similarity degrees are calculated between the posture image to be recognized and the training ones, our classifier will be able to decide to which class belongs the query image. The training class having the largest similarity degree with the query sample will be considered as the result of the classification phase.

b-Fuzzy Decision Support System

The internal configuration of decision support systems including the fuzzy logic is generally composed of three parts that are: Fuzzification, Fuzzy Inference, and Defuzzification.

Fuzzification. It presents a symbolic/digital conversion. It defines membership functions for different variables, particularly for input variables. In this phase, real variables are passed into linguistic ones (fuzzy variables) that can be processed by inferences. The triangular and trapezoidal membership functions are generally chosen. In this work, the triangular shape is retained [19] because of its good given results and its simplicity of implementation.

Fuzzy inference. This step expresses the relationship between the input data and the output variable by linguistic rules. There are several methods of inference: MAX-MIN, MAX-PROD, SUM-PROD. The employed method in this work is the MAX-MIN one because it is simple to implement. The Fuzzy inference rules that manage the FDSS are summarized in the Table 1.

Table 1. Fuzzy inference table relative to the global similarity values

dsd_i		dsa_i		
		Low	Medium	Large
	Low	Very Large	Large	Medium
	Medium	Large	Medium	Low
	Large	Medium	Low	Very Low

As example, the ith rule can be formulated as follows:
If dsa_i is Large **And If** dsd_i is Large,
Then,
The global similarity value of the image i is Very Low.

Defuzzification. The defuzzification transforms the fuzzy decision provided by the inference methods into numerical value. The center of gravity method [20] is used in this work to determine the global degree of similarity.

3   Experimental Results

To highlight the robustness of the HFWN-FDSS, and for reasons of comparison, experiments were performed using the same hand posture base employed in [8]. So, we have evaluated global recognition rates using the same evaluation protocol. Thus, we have used six hand posture (HP_i) classes. Samples from hand posture classes are shown in Fig.3.

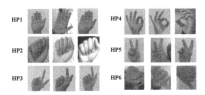

Fig. 3. Samples from the hand posture base

As it is shown in Fig.3, hand postures are performed by different subjects under varying illumination conditions, different angles and different distances. The obtained average recognition rate(RR) applying the HFWN-FDSS on this base was about 95,33%, while the global RR obtained in [8] was 93,16%. Table 2 shows RR of each hand posture using respectively the HFWN classifier and the HFWN-FDSS. These evaluations were achieved employing 100 samples for each hand posture.

Comparison results show that the HFWN-FDSS performs better than the one employed in [8].

Table 2. Hand posture recognition rates using the HFWN and the HFWN-FDSS

Hand postures	RR(HFWN)	RR(HFWN-FDSS)
HP1	94%	96%
HP2	91%	92%
HP3	89%	92%
HP4	97%	98%
HP5	93%	96%
HP6	95%	98%

To be surer about the effectiveness of our approach, we have compared its recognition performances to the FR approach ones [9] employing a well konwn dataset. This latter is the Jochen Triesh dataset[21], a widely used base in the domain of hand gesture recognition. It is composed of ten classes of hand postures performed by 24 different persons against light and dark backgrounds. Samples from this dataset are shown in Fig.4.

Fig. 4. Samples from the Triesh dataset

Table 3 presents obtained RR employing the FR approach, the Support Vector Machine (SVM) which was reported in[9] and our new recognizer. RR were obtained employing the same evaluation protocol detailed in[9]. So considering that N is the number of subsets chosen in[9], two RR were calculated by varing the number of training samples (NTS).

Table 3. Recognition performances of the FR, the SVM and the HFWN-FDSS employing Triesh dataset

NTS	Recognition rates		
	FR[9]	SVM[9]	HFWN-FDSS
120(N=4)	95.83%	94.40%	97.64%
240(N=2)	98.75%	97.91%	98.30%

Testing our approach on images of the Triesh dataset gives results better than those obtained in[9].

4   Conclusion and Future Works

This paper is a contribution on our part in the domain of static hand gesture recognition by creating new hand posture recognizer. The architecture of the new recognizer is based on an HFWN and a FDSS. The novelty in this paper resides in creating a new classification method for the HFWN using fuzzy measures to calculate similarities between query samples and training ones. Results of comparisons with hand shape recognizers in the literature have shown that ours performs better than other ones. In the aim of improving this work, we intend firstly to extend our work by testing it on other datasets. Secondly, we have the intention to compare it to other famous hand posture recognizers in the literature. Finally, we are thinking of extending our approach by employing it to recognize dynamic hand gestures.

Acknowledgments. The authors would like to acknowledge the financial support of this work by grants from General Direction of Scientific Research (DGRST), Tunisia, under the ARUB program.

References

1. Kelly, D., McDonald, J., Markham, C.: A person independent system for recognition of hand postures used in sign language. Pattern Recognition Letters 31, 1359–1368 (2010)
2. Han, J., Awad, G., Sutherland, A.: Modelling and segmenting subunits for sign language recognition based on hand motion analysis. Pattern Recognition Letters 30, 623–633 (2009)
3. Stephan, J., Khudayer, S.: Gesture Recognition for Human-Computer Interaction (HCI). International Journal of Advancements in Computing Technology 02 (2010)
4. Stergiopoulou, E., Papamarkos, N.: Hand gesture recognition using a neural network shape fitting technique. Engineering Applications of Artificial Intelligence 22, 1141–1158 (2009)
5. Chang-Yi, K., Chin-Shyurng, F.: A Human-Machine Interaction Technique: Hand Gesture Recognition Based on Hidden Markov Models with Trajectory of Hand Motion. Control Engineering and Information Science 15, 3739–3743 (2011)
6. Mitra, S., Acharya, T.: Gesture Recognition: A Survey. IEEE Transaction on Systems, Man, and Cybernetics-part C: Application and Reviews 37 (2007)
7. Bouchrika, T., Zaied, M., Jemai, O., Ben Amar, C.: Ordering computers by hand gestures recognition based on wavelet networks. In: Int. Conf. on Communications, Computing and Control Applications Proceedings, Marseilles, France, December 06-08, pp. 1–6 (2012), doi:10.1109/CCCA.2012.6417911
8. Bouchrika, T., Zaied, M., Jemai, O., Ben Amar, C.: Neural solutions to interact with computers by hand gesture recognition. MTAP: Multimedia Tools and Applications (2013), doi:10.1007/s11042-013-1557-y
9. Kumar, P.P., Vadakkepat, P., Ai Poh, L.: Hand posture and face recognition using a fuzzy-rough approach. International Journal of Humanoid Robotics 7(3), 331–356 (2010)

10. Jemai, O., Zaied, M., Ben Amar, C., Alimi, A.M.: Pyramidal Hybrid Approach: Wavelet Network with OLS Algorithm Based-Image Classification. IJWMIP: International Journal of Wavelets, Multiresolution and Information Processing 9(1), 111–130 (2011)

11. Jemai, O., Zaied, M., Ben Amar, C., Alimi, A.M.: Fast Learning algorithm of wavelet network based on Fast Wavelet Transform. Int. J. Pattern Recognition and Artificial Intelligence 25(8), 1297–1319 (2011)

12. Zaied, M., Jemai, O., Ben Amar, C.: Training of the Beta wavelet networks by the frames theory: Application to face recognition. In: The International Workshops on Image Processing Theory, Tools and Applications, Tunisia, pp. 165–170 (November 2008)

13. Zhang, Q., Benveniste, A.: Wavelet networks. IEEE Transactions on Neural Networks 3, 889–898 (1992)

14. Zaied, M., Said, S., Jemai, O., Ben Amar, C.: A novel approach for face recognition based on fast learning algorithm and wavelet network theory. IJWMIP: International Journal of Wavelets, Multiresolution and Information Processing 9(6), 923–945 (2011)

15. Said, S., Ben Amor, B., Zaied, M., Ben Amar, C., Daoudi, M.: Fast and efficient 3D face recognition using wavelet networks. In: International Conference on Image Processing-ICIP, Egypt, pp. 4153–4156 (2010)

16. Jemai, O., Zaied, M., Ben Amar, C., Alimi, A.M.: FBWN: an architecture of Fast Beta Wavelet Networks for Image Classification. In: 2010 IEEE World Congress on Computational Intelligence, the 2010 International Joint Conference on Neural Networks, CCIB, Barcelona, Spain, 1953-, CCIB, Barcelona, Spain, July 18-23, pp. 1953–1960 (2010)

17. Teyeb, I., Jemai, O., Bouchrika, T., Ben Amar, C.: Detecting Driver Drowsiness Using Eyes Recognition System Based on Wavelet Network. In: 5th International Conference on Web and Information Technologies, Tunisia, Hammamet, pp. 245–254 (May 2013)

18. Jemai, O., Teyeb, I., Bouchrika, T., Ben Amar, C.: A Novel Approach for Drowsy Driver Detection Using Eyes Recognition System Based on Wavelet Network. IJES: International Journal of Recent Contributions from Engineering, Science & IT 1, 46–52 (2013)

19. Murshid, A.M., Loan, S.A.: Architectural design of fuzzy inference processor using triangular-shaped membership function. In: IEEE Conference on Open Systems (ICOS), pp. 16–20 (September 2011)

20. Werner, V.L., Etienne, E.K.: Defuzzification: criteria and classification. Fuzzy Sets and Systems 108, 159–178 (1999)

21. Triesch, J., Malsburg, C., Marcel, S.: Hand posture and gesture datasets: Jochen Triesch static hand posture database,
http://www.idiap.ch/resources/gestures/

22. Zaied, M., Mohamed, R., Ben Amar, C.: Power tool for content-based image retrieval using multiresolution wavelet network modeling and dynamic histograms. International REview on Computers and Software (IRECOS) 7(4) (2012)

23. Ejbali, R., Zaied, M., Ben Amar, C.: Wavelet network for recognition system of Arabic word. International Journal of Speech Technology 13(3), 163–174 (2010)

24. Zaied, M., Ben Amar, C., Alimi, A.M.: Beta Wavelet Networks for Face Recognition. Journal of Decision Systems, JDS 14(1-2), 109–122 (2005)

Sim-EA: An Evolutionary Algorithm
Based on Problem Similarity

Krzysztof Michalak

Department of Information Technologies,
Institute of Business Informatics,
Wroclaw University of Economics, Wroclaw, Poland
krzysztof.michalak@ue.wroc.pl

Abstract. In this paper a new evolutionary algorithm Sim-EA is presented. This algorithm is designed to tackle several instances of an optimization problem at once based on an assumption that it might be beneficial to share information between solutions of similar instances. The Sim-EA algorithm utilizes the concept of multipopulation optimization. Each subpopulation is assigned to solve one of the instances which are similar to each other. Problem instance similarity is expressed numerically and the value representing similarity of any pair of instances is used for controlling specimen migration between subpopulations tackling these two particular instances.

Keywords: multipopulation algorithms, evolutionary optimization, combinatorial optimization, travelling salesman problem.

1 Introduction

This paper proposes an evolutionary algorithm Sim-EA which is designed to solve multiple similar instances of an optimization problem by utilizing the idea of a multipopulation evolutionary algorithm.

Probably the most common motivation for employing multipopulation evolutionary algorithms is diversity preservation. Multipopulation evolutionary algorithms are commonly used for such types of problems for which converging to a single optimum is not good enough. This situation is typical to multimodal problems in which there are many equally good or almost equally good solutions with different parameters. It is usually desirable to find many such solutions to allow the decision-maker to choose the best option. Techniques such as species conservation [7], algorithms based on an island model [1] and small-world topologies [5] were applied in the literature to multimodal problems.

Another area in which multipopulation algorithms are often used is multiobjective optimization. Because in multiobjective optimization solutions are evaluated using several different criteria it is not desirable to select only one solution arbitrarily. Instead, an optimization algorithm is usually expected to return an entire Pareto front of nondominated solutions. In [9] parallel approaches to multiobjective optimization are reviewed.

E. Corchado et al. (Eds.): IDEAL 2014, LNCS 8669, pp. 191–198, 2014.

The third type of problems in which multipopulation algorithms are commonly used are dynamic optimization problems. In this type of problems it is undesirable for the population to converge to a single optimum because when the environment changes it is very hard to restore diversity in the population and to start searching for a new optimum. Multipopulation algorithms proposed for this type of problems include forking genetic algorithms (FGAs) [11], Shifting Balance GA (SBGA) [3], the multinational GA (MGA) [12] and Self-Adaptive Differential Evolution algorithm (jDE) [2].

In this paper the multipopulation approach is used for a different purpose: to organize information exchange between simultaneous attempts of solving a set of similar instances of an optimization problem.

The migration scheme influences the information exchange between populations and thus influences the behaviour of the algorithm. In the algorithm proposed in this paper subpopulations are explicitly assigned to different instances of the optimization problem and thus are required to converge to different optima. A similarity measure is used to control the migration of specimens between populations so as to promote information exchange between similar subproblems.

The rest of this paper is structured as follows. In Section 2 the Sim-EA algorithm is described. Section 3 describes the experimental setup and presents the results of the experiments. Section 4 concludes the paper.

2   Algorithm Description

The Sim-EA algorithm proposed in this paper is based on the idea of a multipopulation evolutionary algorithm. The overview of the Sim-EA algorithm is presented in Algorithm 1. The parameters of the algorithm are: N_{gen} - the number of generations, N_{pop} - the size of each subpopulation, N_{prob} - the number of problem instances, N_{imig} - the number of migrated specimens.

In the Sim-EA algorithm subpopulations are assigned to different instances of a given optimization problem. It is assumed that a certain similarity measure can be used to quantitatively describe the similarity of problem instances. Denote the number of problem instances solved simultaneously by N_{prob}. We assume, that a similarity matrix $S_{[N_{prob} \times N_{prob}]}$ is given or can be calculated in the preprocessing phase of the algorithm. For example, if the algorithm solves 20 instances of the Travelling Salesman Problem involving K cities we have 20 cost matrices $C_{[K \times K]}^{(1)}, C_{[K \times K]}^{(2)}, \ldots, C_{[K \times K]}^{(20)}$ that contain distances between the cities (or some other travel cost measure). The similarity between instances i and j, where $i, j \in \{1, \ldots, 20\}$ is simply the similarity of the cost matrices $C_{[K \times K]}^{(i)}$ and $C_{[K \times K]}^{(j)}$. The similarity of such two matrices can be, for example, calculated as $S_{i,j} = -\sum_{p=1}^{K} \sum_{q=1}^{K} (C_{p,q}^{(i)} - C_{p,q}^{(j)})^2$. A similarity matrix is subsequently used to control migration of specimens between populations. Obviously, many migration strategies are possible. Preliminary research shown that one of the best performing ones is to migrate N_{imig} best specimens from one subpopulation which is working on the most similar problem instance. In the Sim-EA algorithm the migration to population P_d is performed as follows. The N_{imig} best specimens are

Algorithm 1. The overview of the Sim-EA algorithm

IN:

    N_{gen} - the number of generations
    N_{pop} - the size of each subpopulation
    N_{prob} - the number of problem instances
    N_{imig} - the number of migrated specimens

Calculate the problem instance similarity matrix $S_{[N_{prob} \times N_{prob}]}$

Initialize subpopulations $P_1, P_2, \ldots, P_{N_{prob}}$.

for $g = 1, \ldots, N_{gen}$ **do**
    Apply genetic operators
    for $d = 1, \ldots, N_{prob}$ **do**
        $s = \operatorname*{argmax}_{t}(S_{d,t})$
        $P'_d =$ the N_{imig} best specimens from P_s
    end for
    for $d = 1, \ldots, N_{prob}$ **do**
        for $x \in P'_d$ **do**
            $P'_d = P'_d - \{x\}$
            w = the weakest specimen in P_d
            $P_d = P_d - \{w\}$
            b = BinaryTournament(w, x)
            $P_d = P_d \cup \{b\}$
        end for
    end for
    Apply genetic operators
    for $d = 1, \ldots, N_{prob}$ **do**
        e = the best specimen in P_d
        $P_d = \text{Select}(P_d \backslash \{e\}, N_{pop} - 1)$
        $P_d = P_d \cup \{e\}$
    end for
end for

selected from the population P_s which solves the most similar problem instance and placed in a set P'_d. After all the sets P'_d, where $d = 1, \ldots, N_{prob}$ are selected the immigrants are merged into respective populations (i.e. each P'_d is merged into the respective P_d). The merging phase is performed using the binary tournament [8] procedure in which each immigrant is compared to the current weakest specimen in the existing population P_d. If the immigrant wins the tournament it replaces the weakest specimen in the population P_d. Genetic operators are applied before and after the migration phase. The aim of the second application of genetic operators is to allow the information from migrated specimens to be incorporated into the target population before the selection step. The selection phase can be performed using any selection procedure such as a roulette wheel selection or a binary tournament. In the proposed algorithm the elitism is used, i.e. the best specimen in each subpopulation P_d is always promoted to the next generation.

3   Experiments and Results

The experiments were performed on a single-objective version of the Travelling Salesman Problem (TSP) [6]. Because in this problem solutions are represented as permutations the Inver-Over genetic operator [10] was used. In the selection phase the binary tournament selection procedure was used. Evolutionary algorithms, especially those dealing with combinatorial optimization are often augmented with local search procedures. In this paper the 2-opt local search [13] was used to improve the quality of the results. Parameters of the Sim-EA algorithm used in the experiments are summarized in Table 1.

In order to verify the assumption that it is beneficial to perform migration based on problem instance similarity, three different strategies were compared: **1-nearest-N-best** (N_{imig} best specimens are migrated from the nearest population), **1-uniform-N-best** (N_{imig} best specimens are migrated from one population which is selected randomly with uniform probability distribution among populations) and **none** (no migration is performed).

Table 1. Parameters of the Sim-EA algorithm used in the experiments

Parameter name	Value
Number of subproblems (N_{prob})	20
Problem size (the number of cities, K)	50
Number of generations (N_{gen})	200
Population size (N_{pop})	100
Random inverse rate for the inver-over operator (δ_i)	0.02

3.1   Test Problem Definition

A set of $N_{prob} = 20$ instances of the TSP with size $K = 50$ each was prepared as follows. The first cost matrix $C^{(1)}_{[K \times K]}$ was randomly initialized by drawing the elements above the diagonal from the uniform probability distribution $U[0, 100]$. To obtain a symmetric matrix the elements above the diagonal were copied symmetrically below the diagonal of the matrix. Obviously, the elements on the diagonal were all set to 0.

The remaining cost matrices $C^{(2)}_{[K \times K]}, \ldots, C^{(20)}_{[K \times K]}$ were generated iteratively. The $C^{(j)}_{[K \times K]}$ matrix was generated from the $C^{(j-1)}_{[K \times K]}$ by replacing $1/N_{pop}$ (i.e. $1/20 = 5\%$) of the non-diagonal elements by random values drawn from the uniform probability distribution $U[0, 100]$. Symmetry of the cost matrix was preserved by changing both $C^{(j)}_{m,n}$ and $C^{(j)}_{n,m}$ to the same value.

Clearly, this procedure ensures that the cost matrices $C^{(i)}_{[K \times K]}$ and $C^{(j)}_{[K \times K]}$ are more different for larger differences $|i - j|$. Note, that in the procedure described above no special attention was paid to satisfy the triangle inequality $C_{p,q}^{(i)} + C_{q,r}^{(i)} \geq C_{p,r}^{(i)}$. While this inequality holds in all metric spaces this is not a strict requirement for the TSP problem, because in various applications the cost matrix may represent a non-metric quantity, such as ticket costs, travel risks etc..

3.2   Results

During the experiments 30 iterations of the test were performed for each of the three migration strategies. From the 30 runs median values of the objective function of the best specimen obtained in each run were calculated. Table 2 summarizes median values obtained by each of the migration strategies for each of the subproblems. Obviously, lower values are better (lower travel costs). For each subproblem the best of the values obtained using the three strategies is marked in bold in Table 2.

Table 2. Median values of the travel cost obtained by each of the migration strategies for each of the subproblems

Subproblem	1-nearest N-best	1-uniform N-best	none	Subproblem	1-nearest N-best	1-uniform N-best	none
1	**171.3077**	176.1562	176.0278	11	**219.3591**	221.2456	221.6789
2	**179.3029**	182.5553	183.6177	12	**223.8487**	226.7204	227.2988
3	**190.1753**	193.9355	195.0356	13	211.1367	**210.3828**	214.3309
4	**194.6214**	198.8933	201.1124	14	**212.1491**	214.721	214.4843
5	**186.2072**	192.4362	192.5718	15	**222.9707**	223.0239	224.8621
6	**180.9736**	186.2206	187.4909	16	**228.5528**	230.5679	229.6890
7	**188.1285**	191.1580	192.2368	17	**263.6930**	266.6074	268.6715
8	**193.7582**	194.4849	198.1625	18	**262.3891**	263.0081	266.0693
9	**202.9813**	206.9935	209.6731	19	**257.2304**	258.7912	260.6183
10	**223.4921**	225.8248	227.4927	20	**249.1812**	252.2827	253.0103

With the exception of the subproblem #13 the 1-nearest-N-best migration strategy gave the best results in the tests. The 1-uniform-N-best migration strategy outperformed the algorithm with no migration in the case of all subproblems except #1, #14 and #16. Clearly, there is some advantage in migrating specimens between subproblems even at random, but the migration based on problem instance similarity has the advantage over other tested approaches.

In order to verify the significance of the results statistical testing was performed. Because the normality of distribution of the measured values cannot be guaranteed the Wilcoxon rank test [14] which does not assume normality was used. This test was recommended in a recent review article [4] which analyzed various methods of statistical testing of the results given by metaheuristic methods. Table 3 summarizes the statistical tests. For each of the subproblems the p-values are given for the null hypothesis that the 1-nearest-N-best strategy gives worse results than each of the two remaining strategies (1-uniform-N-best and none). In Table 3 the interpretation of the p-values is also given. The interpretation is "signif." if the median value obtained by the 1-nearest-N-best strategy is lower than the other one and the p-value is ≤ 0.05. If the median value obtained by the 1-nearest-N-best strategy is lower than the other one, but the p-value is larger than 0.05 the interpretation is "insignif.". If the median value obtained by the 1-nearest-N-best strategy is higher than the other one the interpretation is "worse".

Table 3. The p-values for the null hypothesis that the 1-nearest-N-best strategy gives worse results than each of the two remaining strategies obtained in the statistical verification of the experimental results. Low p-values (≤ 0.05) indicate that 1-nearest-N-best strategy is significantly better than the strategy to which it is compared.

Sub-prob-lem	vs. none		vs. 1-uniform-N-best		Sub-prob-lem	vs. none		vs. 1-uniform-N-best	
	p-value	interp.	p-value	interp.		p-value	interp.	p-value	interp.
1	0.0001891	signif.	0.00017423	signif.	11	0.0082167	signif.	0.11093	insignif.
2	0.00061564	signif.	0.0054597	signif.	12	0.00096266	signif.	0.0014839	signif.
3	0.00052872	signif.	0.0082167	signif.	13	0.057096	insignif.	0.97539	worse
4	3.1123e-005	signif.	2.163e-005	signif.	14	0.013194	signif.	0.17138	insignif.
5	0.0001057	signif.	0.0003065	signif.	15	0.020671	signif.	0.89364	insignif.
6	6.3391e-006	signif.	8.4661e-006	signif.	16	0.40483	insignif.	0.13059	insignif.
7	1.2381e-005	signif.	0.0064242	signif.	17	0.00048969	signif.	0.031603	signif.
8	0.00066392	signif.	0.036826	signif.	18	0.0019646	signif.	0.4908	insignif.
9	9.3157e-006	signif.	0.00035888	signif.	19	0.0024147	signif.	0.17791	insignif.
10	0.0017088	signif.	0.0046818	signif.	20	0.00083071	signif.	0.0046818	signif.

Clearly, the 1-nearest-N-best strategy significantly outperforms the others in most cases. For six subproblems the obtained results are better than for the 1-uniform-N-best strategy, but without statistical significance. For the subproblem #13 the results obtained using the 1-uniform-N-best strategy are better than the results produced by the 1-nearest-N-best strategy.

The dynamic behaviour of the algorithm is presented in Figure 1. In this figure the median values of the best specimen cost calculated over all 30 runs are presented for subproblems #1, #10 and #20. The subproblem #13 for which the 1-nearest-N-best strategy performed worse than the 1-uniform-N-best strategy is also presented. Note, that the graphs in Figure 1 are plotted against total calculation time. The total calculation time includes the time used for calculating the elements of the similarity matrix $S_{[N_{prob} \times N_{prob}]}$. Obviously, this calculation has to be performed only in the case of the 1-nearest-N-best strategy. The figures present the final half of the evolution because in the first half the values change significantly which makes the figures much less readable.

From the figures it can be seen that even if the calculation time of the similarity matrix $S_{[N_{prob} \times N_{prob}]}$ is taken into account the 1-nearest-N-best strategy outperforms the others. Also, even though for the subproblem #13 the 1-uniform-N-best strategy gives the best results the difference between this strategy and 1-nearest-N-best is very small.

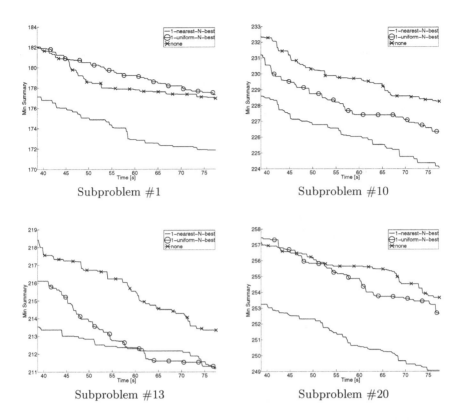

Fig. 1. Median values of the best specimen cost calculated over all 30 runs for sub-problems #1, #10, #13 and #20

4   Conclusion

In this paper a new algorithm Sim-EA is proposed. This algorithm uses a multi-population approach to solve several similar instances of an optimization problem simultaneously. A similarity measure is used to determine to what extent are the subproblems similar. This similarity measure is used in the 1-nearest-N-best strategy to select one population which solves the most similar subproblem for the migration of specimens.

The presented algorithm was tested on a set of 20 Travelling Salesman Problem instances with cost matrices that have varying degree of similarity to each other. The experiments show that the 1-nearest-N-best migration strategy which uses the similarity measure defined in this paper is performing better than the 1-uniform-N-best strategy which selects the source for the migration randomly. Without migration the results are even further deteriorated.

Further work may include experiments with updating the similarity measure during the algorithm runtime based on information discovered during the optimization process.

References

1. Bessaou, M., Petrowski, A., Siarry, P.: Island model cooperating with speciation for multimodal optimization. In: Deb, K., Rudolph, G., Lutton, E., Merelo, J.J., Schoenauer, M., Schwefel, H.-P., Yao, X. (eds.) PPSN 2000. LNCS, vol. 1917, pp. 437–446. Springer, Heidelberg (2000)
2. Brest, J., Zamuda, A., Boskovic, B., Maucec, M.S., Zumer, V.: Dynamic optimization using self-adaptive differential evolution. In: IEEE Congress on Evolutionary Computation, pp. 415–422. IEEE (2009)
3. Chen, J., Wineberg, M.: Enhancement of the shifting balance genetic algorithm for highly multimodal problems. In: Proceedings of the 2004 IEEE Congress on Evolutionary Computation, pp. 744–751. IEEE Press, Portland (2004)
4. Derrac, J., et al.: A practical tutorial on the use of nonparametric statistical tests as a methodology for comparing evolutionary and swarm intelligence algorithms. Swarm and Evolutionary Computation 1(1), 3–18 (2011)
5. Giacobini, M., Preuß, M., Tomassini, M.: Effects of scale-free and small-world topologies on binary coded self-adaptive CEA. In: Gottlieb, J., Raidl, G.R. (eds.) EvoCOP 2006. LNCS, vol. 3906, pp. 86–98. Springer, Heidelberg (2006)
6. Gutin, G., Punnen, A. (eds.): The Traveling Salesman Problem and Its Variations. Combinatorial Optimization. Springer (2007)
7. Li, J.P., Balazs, M.E., Parks, G.T., Clarkson, P.J.: A species conserving genetic algorithm for multimodal function optimization. Evol. Comput. 10(3), 207–234 (2002)
8. Miller, B.L., Goldberg, D.E.: Genetic algorithms, tournament selection, and the effects of noise. Complex Systems 9, 193–212 (1995)
9. Talbi, E.-G., Mostaghim, S., Okabe, T., Ishibuchi, H., Rudolph, G., Coello Coello, C.A.: Parallel approaches for multiobjective optimization. In: Branke, J., Deb, K., Miettinen, K., Słowiński, R. (eds.) Multiobjective Optimization. LNCS, vol. 5252, pp. 349–372. Springer, Heidelberg (2008)
10. Tao, G., Michalewicz, Z.: Inver-over operator for the TSP. In: Eiben, A.E., Bäck, T., Schoenauer, M., Schwefel, H.-P. (eds.) PPSN 1998. LNCS, vol. 1498, pp. 803–812. Springer, Heidelberg (1998)
11. Tsutsui, S., Fujimoto, Y., Ghosh, A.: Forking genetic algorithms: Gas with search space division schemes. Evolutionary Computation 5(1), 61–80 (1997)
12. Ursem, R.K.: Multinational GA Optimization Techniques in Dynamic Environments. In: Whitley, D., Goldberg, D., Paz, C.E., Spector, L., Parmee, I., Beyer, H.G. (eds.) Genetic and Evolutionary Computation Conference, pp. 19–26. Morgan Kaufmann (2000)
13. Watson, J., Ross, C., Eisele, V., Denton, J., Bins, J., Guerra, C., Whitley, L.D., Howe, A.: The traveling salesrep problem, edge assembly crossover, and 2-opt. In: Eiben, A.E., Bäck, T., Schoenauer, M., Schwefel, H.-P. (eds.) PPSN 1998. LNCS, vol. 1498, pp. 823–832. Springer, Heidelberg (1998)
14. Wilcoxon, F.: Individual comparisons by ranking methods. Biometrics Bulletin 1(6), 80–83 (1945)

Multiobjective Dynamic Constrained Evolutionary Algorithm for Control of a Multi-segment Articulated Manipulator

Krzysztof Michalak[1], Patryk Filipiak[2], and Piotr Lipinski[2]

[1] Department of Information Technologies,
Institute of Business Informatics,
Wroclaw University of Economics, Wroclaw, Poland
krzysztof.michalak@ue.wroc.pl
[2] Computational Intelligence Research Group,
Institute of Computer Science,
University of Wroclaw, Wroclaw, Poland
{patryk.filipiak,piotr.lipinski}@cs.ii.uni.wroc.pl

Abstract. In this paper a multiobjective dynamic constrained evolutionary algorithm is proposed for control of a multi-segment articulated manipulator. The algorithm is tested in simulated dynamic environments with moving obstacles. The algorithm does not require previous training - a feasible sequence of movements is found and maintained based on a population of candidate movements. The population is evolved using typical evolutionary operators as well as several new ones that are dedicated for the manipulator control task. The algorithm is shown to handle manipulators with up to 100 segments. The increased maneuverability of the manipulator with 100 segments is well utilized by the algorithm. The results obtained for such manipulator are better than for the 10-segment one which is computationally easier to handle.

Keywords: inverse kinematics, multiobjective evolutionary optimization, constrained problems.

1 Introduction

Inverse Kinematics (IK) is the problem of finding such configuration of an articulated robotic arm that satisfies certain constraints concerning the position and the orientation of its end effector. Applications of IK are very frequent in contemporary robotics, e.g. in steering of industrial planar robots [5], automatization of medical steerable needles [6], performing an optical motion capture [1] or robotic posture control and collision avoidance [8]. Although IK can be expressed in the algebraic form, it is highly inefficient to solve it explicitly. It was stated in [9] that finding the desired pose for the popular case of IK problem with 6 degrees of freedom is equivalent to solving a 16th order polynomial equation. In order to alleviate this difficulty, a number of numerical approaches were proposed instead.

E. Corchado et al. (Eds.): IDEAL 2014, LNCS 8669, pp. 199–206, 2014.

2   Problem Statement

In this paper we consider a multi-segment articulated manipulator consisting of N_s segments that is mounted at a given point $O = [x_o, y_o]$. The segments are connected by joints $J_1, \ldots, J_{N_s}$ (the first joint being attached at the starting point O). The manipulator itself is described by a list of segment lengths $\{l_1, \ldots, l_{N_s}\}$. The position of the manipulator is determined by a list of relative angles $\{\alpha_1, \ldots, \alpha_{N_s}\}$, $\alpha_i \in (-\pi, \pi)$ (i.e. angles relative to previous segment orientation). In this paper we assume, that $\alpha_i = 0$ represents a segment pointing in the same direction as the previous one. The endpoint of the manipulator is expected to reach and remain as close as possible to a given target point $T = [x_t, y_t]$. The environment in which the manipulator operates includes a set of N_o obstacles $\{O_i\}_{i=1,\ldots,N_o}$. Each obstacle O_i is a convex polygon with M_i vertices. At no time t the manipulator may intersect any of the obstacles. Obviously, the manipulator has to move to reach the target point T and to avoid any obstacles. We assume that the movement of the manipulator is defined by setting values of angles between manipulator segments at discrete time instants. Therefore, the sequence of moves that the manipulator performs during all the N_t time steps of the entire simulation can be represented as: $\{\alpha_1(t), \ldots, \alpha_{N_s}(t)\}_{t=0,\ldots,N_t}$. A movement between time instants t and $t+1$ is performed as a linear change of all the angles: $\alpha_j(t + \delta) = \alpha_j(t) \cdot (1 - \delta) + \alpha_j(t+1) \cdot \delta$, for $j = 1, \ldots, N_s$, $\delta \in [0, 1]$. At each time instant $t = 0, \ldots, N_{t-1}$ the algorithm has to calculate a set of angles $\{\alpha_1(t+1), \ldots, \alpha_{N_s}(t+1)\}$ for the time instant $t+1$ based on the current set of angles $\{\alpha_1(t), \ldots, \alpha_{N_s}(t)\}$ in such a way that the endpoint E of the manipulator remains possibly close to the target point T and the manipulator does not intersect any obstacles during the interval $[t, t+1]$.

3   Evolutionary Algorithm

At each time instant $t = 0, \ldots, N_{t-1}$ the evolutionary algorithm tries to find a new set of angles for the manipulator $\{\alpha_1(t+1), \ldots, \alpha_{N_s}(t+1)\}$ for the time instant $t+1$ based on the current set of angles $\{\alpha_1(t), \ldots, \alpha_{N_s}(t)\}$. The genotype of each specimen represents a candidate set of angles for the time instant $t+1$. The evolutionary algorithm proposed in this paper contains both some elements typical to evolutionary algorithms used for constrained optimization and some elements typical to dynamic optimization. For dealing with constraints the algorithm uses two mechanisms used in the Infeasibility Driven Evolutionary Algorithm (IDEA) [10]: a violation measure is used as one of the objectives and a fraction of the population is reserved for infeasible specimens. When solving dynamic optimization problems the loss of diversity is often an issue. To remedy this, random immigrants are added to every generation as proposed by [7].

Objectives and Constraints
The algorithm minimizes three objectives f_1, f_2 and f_3 under two constraints g_1 and g_2. All objectives and constraints are calculated for N_δ simulation steps from time t to $t+1$. The n-th step ($n = 1, \ldots, N_\delta$) corresponds to time $t + \frac{n-1}{N_\delta - 1}$,

where: t - the time instant for which the evolutionary algorithm calculates the transition to $t + 1$.

$\mathbf{f_1}$: The first objective is **the distance between the manipulator end-point E and the target point T**. If there are no obstacles between E and T the Euclidean distance is used. If the $\overline{ET}$ line segment crosses an obstacle at points C_1 and C_2 the distance is calculated as: $f_1 = d_E(E, T) - d_E(C_1, C_2) + d_O$ where d_E denotes Euclidean distance and d_O is the shorter of two paths between C_1 and C_2 around the obstacle. The values for all simulation steps $n = 1, \ldots, N_\delta$ are averaged with weights equal to $\frac{n}{N_\delta}$. Therefore, a higher selective pressure is put on minimizing the distance from E to T at the end of the time interval $[t, t + 1]$. This is intended to give the manipulator some freedom to adjust between time instants t and $t + 1$, while promoting convergence to T towards the end of the time interval $[t, t + 1]$.

$\mathbf{f_2}$: The second objective is **a measure of displacement of the manipulator between time instants t and $t + 1$**: $f_2 = \sum_{k=1}^{N_s} \left[(x_{j_k}(t + 1) - x_{j_k}(t))^2 + (y_{j_k}(t + 1) - y_{j_k}(t))^2 \right]$, where: $x_{j_k}(t)$, $y_{j_k}(t)$ - the coordinates of the $k - th$ joint of the manipulator calculated for angles at the time instant t. Minimizing this objective is intended to limit the occurence of rapid or violent movements of the manipulator.

$\mathbf{f_3}$: The third objective is **a violation measure** proposed in [10] which represents how much the constraints are violated.

The constraints represent collisions with obstacles and self-intersections of the manipulator. Both g_1 and g_2 have to be 0 for the specimen to be feasible. Infeasible specimens have $g_1 > 0$ or $g_2 > 0$.

$\mathbf{g_1}$: The first constraint is **a measure of intersection with obstacles**. If the manipulator does not intersect with a given obstacle then the contribution of this obstacle to the g_1 constraint is 0. Otherwise, for each pair of intersection points C_1, C_2 the length of the shorter of the paths connecting C_1 and C_2 on the circumference of the obstacle is added to g_1. If the endpoint of the manipulator is inside the obstacle the length of the arm inside the obstacle is added.

$\mathbf{g_2}$: The second constraint is **a measure of self-intersections of the manipulator**. This measure is calculated as the sum of $\frac{1}{j \cdot k}$ for those j and k for which the manipulator segments $\overline{J_i J_{i+1}}$ and $\overline{J_k J_{k+1}}$ intersect.

Values of both constraints are summed for all simulations steps $n = 1, \ldots, N_\delta$.

Operators

The evolutionary algorithm proposed in this paper uses two genetic operators typically used for real-valued chromosomes: the SBX crossover described in [2] and the polynomial mutation operator introduced in [4]. Typical genetic operators mentioned above treat the set of angles between manipulator segments as just an array of real numbers. Additionally, three other operators are proposed in this paper that are dedicated for the task of articulated manipulator control.

Single Joint Mutation. If the polynomial mutation operator changes the value of an angle α_i positions of the segments that are placed after α_i change

significantly. This effect may cause a mutated manipulator position to become
infeasible, especially in crowded environments. To mitigate this problem a second mutation operator was designed. The new operator uses the polynomial
mutation operator to mutate individual angles. If an angle α_i is mutated to α_i' a
correction is performed by calculating the change of the i-th angle $\delta_i = \alpha_i' - \alpha_i$
and by turning the joints that follow the mutated one in the opposite direciton:
$\forall j > i : \alpha_j = \alpha_j - \delta_i$. This correction is intended to limit the influence of mu-
tating one angle α_i on the entire part of the manipulator from joint J_i to the
endpoint E.

The Unfold-3 operator. This operator is intended to help the manipulator
straighten by replacing any three segments by two if possible. It is designed to
be used for manipulators in which all the segments are equal to a given constant
length L. By definition this operator is only applied when there exist two joints
J_i and J_{i+3} for which $d_E(J_i, J_{i+3}) \leq 2L$, where $d_E(\cdot)$ is the Euclidean distance.
The operator sets the angles α_i and α_{i+1} so that J_{i+2} and J_{i+3} are placed at
the same positions as J_{i+3} and J_{i+4} were before the operator was applied. Then,
all the angles that follow are corrected: $\forall i + 3 \leq j < N_s : \alpha_j = \alpha_{j+1}$. The last
segment of the manipulator has no preceding position to which to refer, so the
endpoint of the manipulator i directed towards the target point T.

The RepairSelfIntersections Operator
This operator tries to untangle the manipulator if self-intersections are present.
First, a set of intersecting pairs of manipulator segments is identified:

$$I = \left\{ \langle j, k \rangle : \overline{J_j J_{j+1}} \text{ and } \overline{J_k J_{k+1}} \text{ intersect} \right\} . \tag{1}$$

Based on the set I the first and the last joint in the entangled part of the
manipulator are identified:

$$j_f = min \left\{ j : \exists k : \langle j, k \rangle \in I \right\} + 1 , \tag{2}$$

$$j_l = max \left\{ k : \exists j : \langle j, k \rangle \in I \right\} . \tag{3}$$

One index j_{fix} is randomly selected from the range $j_f, \ldots, j_l$ with uniform
probability. The angle $\alpha_{j_{fix}}$ is modified by setting $\alpha_{j_{fix}} = \eta \alpha_{j_{fix}}$, where η is a
random number drawn from the $U[0, 1]$ distribution. This makes the segment
$\overline{J_{j_{fix}} J_{j_{fix}+1}}$ closer to pointing straight with respect to the previous segment
$\overline{J_{j_{fix}-1} J_{j_{fix}}}$.

3.1   The Main Loop

The main algorithm loop is presented in Algorithm 1. This main loop of the
algorithm is executed for every time instant $t = 0, \ldots, N_t - 1$. The population
P is initialized only once at the first time instant $t = 0$ before entering the main
loop. The parameters that affect the execution of the algorithm are: N_{gen} - the
number of generations, N_{pop} - population size, N_{rnd} - the number of random

immigrants, N_f - the number of feasible specimens to select, and N_{inf} - number of infeasible specimens to select.

Algorithm 1. The main algorithm loop.

---

for $g = 1 \rightarrow N_{gen}$ **do**
  $P_{offspring} = \emptyset$
  $P_{mate} = \text{SelectMatingPool}(P)$
  for $i = 1 \rightarrow N_{pop}/2$ **do**
    $\langle O_1, O_2 \rangle = \text{Crossover}(P_{mate}[2 * i - 1], P_{mate}[2 * i])$
    $\text{Mutate}(O_1); \quad \text{Mutate}(O_2)$
    $\text{MutateOneJoint}(O_1); \quad \text{MutateOneJoint}(O_2)$
    $\text{RepairSelfIntersections}(O_1); \quad \text{RepairSelfIntersections}(O_2)$
    $P_{offspring} = P_{offspring} \cup \{O_1, O_2, \}$
  end for
  $P_{rnd} = \text{InitPopulation}(N_{rnd})$
  $\text{RepairSelfIntersections}(P_{rnd})$
  $P_{U3} = \text{Unfold-3}(P_{offspring}) \cup \text{Unfold-3}(P_{rnd})$
  $P_{av} = \text{AvoidObstacles}(P_{offspring}) \cup \text{AvoidObstacles}(P_{rnd}) \cup \text{AvoidObstacles}(P_{U3})$
  $\text{RepairSelfIntersections}(P_{av})$
  $\text{Evaluate}(P_{offspring} \cup P_{rnd} \cup P_{U3} \cup P_{av})$
  $P = P \cup P_{offspring} \cup P_{rnd} \cup P_{U3} \cup P_{av}$
  $\langle P_f, P_{inf} \rangle = Split(P)$
  $\text{Rank}(P_f); \quad \text{Rank}(P_{inf})$
  $P = P_{inf}[1 : N_{inf}] \cup P_f[1 : N_f]$
end for

Apart from the operators described in the "Operators" section the algorithm uses the following procedures.

SelectMatingPool. Selects a mating pool using a binary tournament.

Crossover. Generates offspring by applying the crossover operator with probability P_{cross} to a pair of parents or by copying them as they are with probability $1 - P_{cross}$.

InitPopulation. Returns a given number of new specimens with randomly initialized genotypes. It is used to initialize a new population at time $t = 0$ and for generating random immigrants.

AvoidObstacles. Modifies those candidate solutions that intersect obstacles. Moves the solution outside the obstacle by adjusting the last angle before the obstacle, so that the manipulator segment becomes tangent to an ϵ-envelope of the obstacle.

Evaluate. Calculates values of the f_1, f_2 and f_3 objectives as well as the g_1 and g_2 constraints.

Split. Splits the population to feasible and infeasible specimens (those having $g_1 = 0$ and $g_2 = 0$ and those having $g_1 > 0$ or $g_2 > 0$ respectively).

Rank. Ranks a set of specimens using nondominated sorting and then crowding distance sorting used in the NSGA-II algorithm [3]. The set of feasible specimens P_f and the set of infeasible specimens P_{inf} are ranked separately.

4   Experiments and Results

The proposed algorithm was tested on four scenarios involving different obstacle courses with different movement types and different number of manipulator segments. In all tests the number of time instants was $N_t = 100$. The first three scenarios involved manipulators with $N_s = 10$ segments while the last scenario involved a manipulator with $N_s = 100$ segments. All the scenarios were tested for the number of generations $N_{gen} = 50, 100, 200$ and population sizes $N_{pop} = 50, 100, 200$. For each pair of parameters 10 iterations were performed. Visualizations of these scenarios are given in Figure 1.

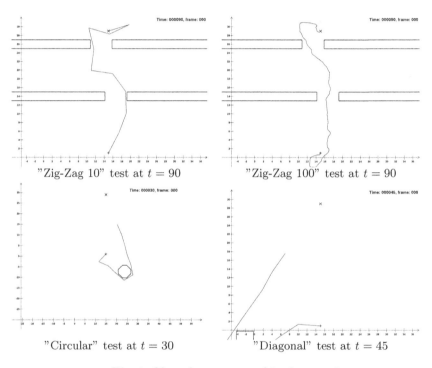

"Zig-Zag 10" test at $t = 90$ "Zig-Zag 100" test at $t = 90$

"Circular" test at $t = 30$ "Diagonal" test at $t = 45$

Fig. 1. Obstacle courses used in the experiments

The parameters of the evolutionary algorithm were set as follows: crossover probability $P_{cross} = 0.9$, crossover distribution index $\eta_{cross} = 15$, mutation probability $P_{mut} = 0.1$, mutation distribution index $\eta_{mut} = 20$, percentage of random immigrants $N_{rnd}/N_{pop} = 20\%$, percentage of infeasible solutions $N_{inf}/N_{pop} = 20\%$. The effectiveness of the algorithm in keeping the manipulator endpoint E close to the target point T was measured by calculating the average Euclidean distance $d_E(E, T)$ and the fraction q of time instants during which $d_E(E, T) < 0.5$. In Table 1 the average Euclidean distance $d_E(E, T)$ between the manipulator endpoint E and the target point T and the fraction q of time instants during which $d_E(E, T) < 0.5$ are given.

Table 1. The average Euclidean distance $d_E(E,T)$ between the manipulator endpoint E and the target point T and the fraction q of time instants during which $d_E(E,T) <$ 0.5 obtained in the tests

Circular		$N_{pop} = 50$		$N_{pop} = 100$		$N_{pop} = 200$	
		$d_E(E,T)$	q	$d_E(E,T)$	q	$d_E(E,T)$	q
	50	10.64	28.69%	10.55	27.44%	10.97	26%
N_{gen}	100	9.43	34.46%	10.67	28.37%	9.74	32.09%
	200	9.96	29.69%	9.32	33.15%	8.41	32.53%

Diagonal		$N_{pop} = 50$		$N_{pop} = 100$		$N_{pop} = 200$	
		$d_E(E,T)$	q	$d_E(E,T)$	q	$d_E(E,T)$	q
	50	5.47	57.76%	4.57	60.98%	4.33	62.02%
N_{gen}	100	4.84	58.34%	4.50	61.16%	4.19	62.44%
	200	4.88	56.44%	4.38	61.57%	4.23	61.55%

Zig-Zag 10		$N_{pop} = 50$		$N_{pop} = 100$		$N_{pop} = 200$	
		$d_E(E,T)$	q	$d_E(E,T)$	q	$d_E(E,T)$	q
	50	5.00	16.6%	3.95	28.59%	3.42	33.35%
N_{gen}	100	4.44	24.93%	3.32	38.19%	2.61	48.14%
	200	2.91	44.23%	2.94	41.94%	2.30	54.46%

Zig-Zag 100		$N_{pop} = 50$		$N_{pop} = 100$		$N_{pop} = 200$	
		$d_E(E,T)$	q	$d_E(E,T)$	q	$d_E(E,T)$	q
	50	4.40	24.53%	4.01	29.98%	3.04	38.87%
N_{gen}	100	4.74	31.26%	2.87	42.84%	2.29	52.07%
	200	3.12	44.30%	2.82	48.05%	1.78	61.03%

Comparison of values from Table 1 shows that the manipulator with 100 segments is handled effectively by the proposed algorithm. The average distance from the target $d_E(E,T)$ obtained for the "Zig-Zag 100" test is very similar to that obtained for the "Zig-Zag 10" test. The fraction q of time instants during which $d_E(E,T) < 0.5$ is higher for the "Zig-Zag 100" test.

In two problems "Circular" and "Diagonal" it might be beneficial for the algorithm to backtrack and "go around" the obstacle. For these problems increasing the values of the parameters (the population size N_{pop} and the number of generations N_{gen}) provides only a moderate improvement in solution quality. This may suggest that the algorithm should include elements of planning. For example it could be beneficial to optimize a sequence of moves for several time instants, not just one movement at a time. Increasing the values of the N_{pop} and N_{gen} parameters improves the results significantly in the "Zig-Zag 10" and "Zig-Zag 100" tests which feature more obstacles and a narrower path to the target. This effect may be caused by a large number of specimens required to find a feasible movement in such crowded space.

5   Conclusions

In this paper a multiobjective dynamic constrained evolutionary algorithm was proposed for control of a multi-segment articulated manipulator. The main advantage of the presented algorithm is that it does not require any tuning for the manipulator parameters nor for any particular environment. The test with 100-segment manipulator shows, that the algorithm is able to utilize the increased maneuverability of this manipulator compared to the one with 10 segments. In an environment with moving obstacles the algorithm managed to keep the end-point of the 100-segment manipulator closer on average to the target point T than in the case of the 10-segment one and for a larger fraction of time.

Further work may include an elaboration of a similar method for 3D space. This would require, at the very least, defining the distance around an obstacle in a way applicable to 3D and a modification of evaluation of constraint g_1 (intersections with obstacles). From the point of view of development of intelligent methods it may be useful to employ AI planning which would allow going around obstacles in directions opposite to where the target point is.

References

1. Aristidou, A., Lasenby, J.: Motion capture with constrained inverse kinematics for real-time hand tracking. In: Proceedings of the IEEE International Symposium on Communications, Control and Signal Processing (ISCCSP 2010), pp. 1–5 (2010)
2. Deb, K., Agarwal, R.: Simulated binary crossover for continuous search space. Complex Systems 9(2), 115–148 (1995)
3. Deb, K., et al.: A fast and elitist multiobjective genetic algorithm: NSGA-II. IEEE Transactions on Evolutionary Computation 6, 182–197 (2002)
4. Deb, K., Goyal, M.: A combined genetic adaptive search (GeneAS) for engineering design. Computer Science and Informatics 26, 30–45 (1996)
5. Dong, H., et al.: Workspace density and inverse kinematics for planar serial revolute manipulators. Mechanism and Machine Theory 70, 508–522 (2013)
6. Duindam, V., Xu, J., Alterovitz, R., Sastry, S., Goldberg, K.: Three-dimensional motion planning algorithms for steerable needles using inverse kinematics. The International Journal of Robotics Research 29(7), 789–800 (2010)
7. Grefenstette, J.: Genetic algorithms for changing environments. In: Parallel Problem Solving from Nature 2, pp. 137–144. Elsevier (1992)
8. Kallmann, M.: Analytical inverse kinematics with body posture control. Computer Animation and Virtual Worlds 19(2), 79–91 (2008)
9. Pieper, D.L.: The kinematics of robots under computer control. Phd thesis, Stanford University (1968)
10. Singh, H.K., Isaacs, A., Ray, T., Smith, W.: Infeasibility driven evolutionary algorithm (IDEA) for engineering design optimization. In: Wobcke, W., Zhang, M. (eds.) AI 2008. LNCS (LNAI), vol. 5360, pp. 104–115. Springer, Heidelberg (2008)

Parameter Dependence in Cumulative Selection

David H. Glass

School of Computing and Mathematics, University of Ulster,
Newtownabbey, Co. Antrim, BT37 0QB, UK
dh.glass@ulster.ac.uk

Abstract. Cumulative selection is a powerful process in which small changes accumulate over time because of their selective advantage. It is central to a gradualist approach to evolution, the validity of which has been called into question by proponents of alternative approaches to evolution. An important question in this context concerns how the efficiency of cumulative selection depends on various parameters. This dependence is investigated as parameters are varied in a simple problem where the goal is to find a target string starting with a randomly generated guess. The efficiency is found to be extremely sensitive to values of population size, mutation rate and string length. Unless the mutation rate is sufficiently close to a value where the number of generations is a minimum, the number of generations required to reach the target is much higher if it can be reached at all.

1 Introduction

From Darwin originally proposed the idea to the present day, gradualism has been an important concept in evolutionary theory [1]. Roughly speaking, the idea is that evolution proceeds by the accumulation of small changes over time. Richard Dawkins discusses the process in a particularly clear way in his popular book *The Blind Watchmaker* [2], where he illustrates the power of cumulative selection by showing how a target string can be reached from a randomly generated initial guess in a small number or steps. He also discusses the relevance of this to the question of how complex entities such as the vertebrate eye might have evolved, a topic which has also been explored using a computer model [3]. Despite the fact that gradualism has many prominent defenders, many questions have been raised in the evolutionary literature. For example, research has been carried out on whether gradualist or alternative models fit fossil data better [4] and computational models have been used to argue that compositional evolution, which involves the combination of pre-adapted genetic material, can provide evolutionary adaptation that is not possible in the gradualist framework [5].

This paper explores a gradualist approach in the context of a simple search problem, obtaining a target string from a random initial guess. Related to gradualism, Dawkins makes the more specific claim that cumulative selection works no matter how small the steps (and so how many steps there are) or how small the selective advantage of each step. This raises the question, however, as to how such changes might affect the efficiency of cumulative selection. A related question is how the efficiency scales with changes in mutation rates or other parameters. For example, do small changes in the parameters give rise to correspondingly small changes in efficiency?

E. Corchado et al. (Eds.): IDEAL 2014, LNCS 8669, pp. 207–214, 2014.

This paper attempts to begin to address some of these questions by considering two models. The first approach is similar to that used by Dawkins, while the second involves a more standard genetic algorithm. Although the first approach is very simple, it turns out to be instructive in terms of how the number of generations depends on the parameters and the results obtained anticipate those from the second approach to some extent. Furthermore, since a genetic algorithm approach is used, it should be noted that the sensitivity of genetic algorithms to various parameters is well-known and that a lot of research has been undertaken to determine how to select optimal parameters (see for example [6]). Hence, the goal here is not to address this general problem in genetic algorithms, but to investigate its significance for gradualism in a simple evolutionary scenario. Including a predefined target string is, of course, unrealistic from an evolutionary point of view, but as in Dawkins' case the goal here is just to investigate cumulative selection, not to model all the aspects of evolution. Considerable effort has been given to simulating other aspects of evolution (see for example [7–9]), but the focus in this paper is much narrower.

2   The Weasel Program

In *The Blind Watchmaker* [2], Richard Dawkins describes a computer program that finds the specified target

```
METHINKS IT IS LIKE A WEASEL
```

by starting with a randomly generated string of the same length and using a process which involves repeated copying with mutations and selection. While Dawkins did not describe his algorithm in detail, general features of his model are captured here including:

- a pre-specified target string,
- a randomly chosen string as the initial guess which is then used to generate a new population of strings,
- for each generation the string that most closely resembles the target is selected,
- the selected string is used to generate the population for the next generation,
- mutation is involved in the process of generating the new population.

2.1   The Model

In addition to these general features, it is necessary to specify exactly what is meant by 'most closely resembles the target' and also how mutations occur. As far as the former is concerned, this is essentially a question about the fitness function. Here it is simply taken to be the number of correct characters and so the selected string will be the string (or one of the strings) with most characters correct. Other options are possible. An obvious example would be that instead of having a fitness function that assigns one to a correct character and zero to an incorrect one, there could be degrees of fitness for characters so that, for example, a would be closer to b than to c. This approach will not be explored here.

The issue of mutations is more complicated. First, when a mutation of a particular character occurs, what kind of mutation should it be? Here it is simply replaced with another character at random (which could in fact be the same character). Once again other options are possible. For example, mutations could be restricted so that a character can only be replaced by one that is sufficiently close to it or, alternatively, be more likely to be replaced by characters that are close to it and less likely to be replaced by characters that are further away. This kind of approach would clearly be compatible with the alternative fitness function mentioned above.

A further consideration for mutations concerns how to determine which characters should be mutated. Clearly there should be some probability for mutations to occur, but should this only apply to incorrect characters in the selected string with correct characters kept fixed, or should it apply to all characters equally, or should there be a bias so that incorrect characters are more likely to mutate? The approach adopted here is to have two different mutation rates, one for the correct characters m_1 and one for the incorrect characters m_2. To avoid confusion the former will be referred to as the correct character mutation (CCM) rate and the latter simply as the mutation rate. A CCM rate of zero, $m_1 = 0$, means that correct characters do not mutate, i.e. they are kept fixed. Clearly, $m_1 = m_2$ corresponds to the special case where there is no bias between correct and incorrect characters so that both are equally likely to mutate.

2.2 Results

Following Dawkins, the set of characters is confined to 27, consisting of the 26 letters of the alphabet and a space, and the target sequence also remains the same. Results are presented in Fig. 1 for a range of values of the selection rate and mutation rate. For all of these calculations the population size was fixed to ten. For each point, the calculation was repeated 1000 times (10,000 for some points at a CCM rate of 0.06) and the average number of generations taken. Not surprisingly, for any given value of the mutation rate, the number of generations is lowest when the CCM rate is equal to zero and the number of generations increases with the CCM rate. By comparing each of the curves it is clear that for higher values of the CCM rate the number of generations increases much more dramatically for low values of the mutation rate.

Fig. 2 explores the dependence on the CCM rate further. For fixed values of the difference between the mutation and CCM rates, the dependence on the CCM rate increases dramatically above a CCM rate of about 0.04. Although not shown on the graph, results obtained averaging over 1000 cases show that the number of generations required when the mutation and CCM rates are equal with a value of 0.06 is approximately 660,000.

The most realistic case is to have the same mutation rates for correct and incorrect characters since correct characters should have an advantage due to the selection process, not due to differential rates of mutation. The results for this case (i.e. a value of 0 in Fig. 2) indicate a minimum in the number of generations at a CCM rate of 0.028, where the number of generations required to reach the target is approximately 730. A positive or negative change in the mutation rate away from this minimum can dramatically increase the number of generations. This is also a key feature in the more detailed genetic algorithm approach considered below.

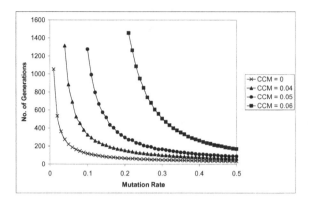

Fig. 1. The number of generations required to reach the target plotted as a function of the mutation rate for four different values of the correct character mutation rate (CCM)

3   A Genetic Algorithm Approach

The model considered above is very simplistic in a number of respects, particularly with regard to the selection procedure since just one string is selected from a given generation to produce the subsequent generation. Here a genetic algorithm [10] is used where a probabilistic approach is adopted to select pairs of strings, based on their fitness, for the production of the next generation.

3.1   The Model

Various decisions need to be made in the development of the algorithm. First, how is the initial population of strings to be selected? The most obvious approach would be to select N strings randomly. The problem with this approach, as some experiments demonstrated, is that if N is large enough for a given string length, then the target string can be reached even if the mutation rate is zero. This choice would be appropriate if the goal were simply to find the target as efficiently as possible, but it is unrealistic and of little use if the goal is to investigate dependence on mutation rate. Instead, a single random guess is chosen and the remaining $N - 1$ guesses are copied from it with each character having a one percent chance of undergoing mutation.

The fitness function is simply taken to be the proportion of characters matching the target. Clearly, other options are available here, but this function seems to be an obvious way of representing gradualism in evolution. Once the fitness of each string has been determined, these values are then normalized and treated as probabilities for the purposes of selection. In the selection process, $N/2$ pairs of strings are selected with replacement from the population based on their respective probabilities, each pair being used to generate a new pair. In the generation of a new string, each character is selected from one of its parents with equal probability, although there is also a probability of a mutation occurring.

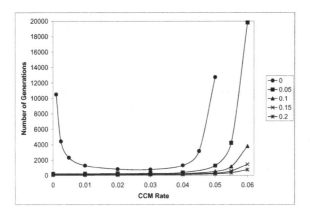

Fig. 2. The number of generations required to reach the target plotted as a function of the correct character mutation rate (CCM). The different curves correspond to differences between the mutation rate (of incorrect characters) and CCM, where the latter is assumed to be smaller.

There is clearly a certain amount of arbitrariness in some of these choices. It is important to point out that they have not been made on the basis of finding the most efficient way to reach the target, but rather to try to make realistic evolutionary choices at least as far as the confines of the model permit. And, of course, it would certainly be possible to select different fitness functions, selection procedures, etc. to see how it affects the results. In summary, the core of the algorithm is given below:

read mutation rate, population size N, target string, initial guess
set count to one
initialize the population of strings by copying the initial guess with random errors
repeat
    for $i = 1$ to N **do**
        set fitness of string to the proportion of characters matching the target
    end for
    set probability of string to the normalized value of the fitness
    for $i = 1$ to $N/2$ **do**
        select with replacement two strings from the population based on their probabilities
        generate two new strings for the next generation (each character has equal probability of being taken from each parent and a probability of being mutated)
    end for
    increment count
until one of the strings matches the target

3.2   Results

Results are presented in Fig. 3 for population sizes of $N = 250$, $N = 500$ and $N = 1000$. For each point, the calculation was repeated 100 times and the average number of generations taken. In each figure results are presented for a range of lengths of the target sequence. For example, in Fig. 3, where $N = 250$, results are presented for target lengths L of 30, 40 and 50. In the previous section the target string was 28 characters in length and so corresponds roughly with $L = 30$ in this section.

A common feature of all the results is that for a given string length L there is a minimum in the number of generations required to reach the target as a function of the mutation rate. This feature is discussed further below, but it is worth noting how this relates to the dependence of the number of generations required to reach the target on L. Not surprisingly, for fixed values of N and the mutation rate the number of generations increases with L, but it is clear from the results that there is no simple relationship between the two. To illustrate how the dependence can change dramatically with just a small change in the mutation rate consider results for $L = 30$ and $L = 50$ when $N = 500$. The ratio of the number of generations at $L = 50$ to the number of generations at $L = 30$ is approximately three when the mutation rate is 0.005, whereas it is about 39 when the mutation rate is 0.007.

Fig. 4 presents results for a particular trial rather than averaging over a large number of trials. The results are for mutation rates of 0.001, 0.005, 0.007 and 0.02 when $N = 500$ and $L = 50$. For each mutation rate results are presented for a typical case. It can be seen that for the lowest mutation rate of 0.001, which is below the mutation rate of the minimum in Fig. 3(b), slow and steady progress is made to the target in just over 5,000 generations. For mutation rates of 0.005 and 0.007 a fitness of around 0.9 is acquired quite quickly, within about 1,000 generations. Beyond this point, the curves follow different trajectories with the curve at a mutation rate of 0.005, which is close to the minimum in Fig. 3(b), converging to the target quite quickly in 1,760 generations while the curve at a mutation rate of 0.007, which is above the minimum in Fig. 3(b), fluctuates around 0.9 for quite a while before reaching the target in just over 9,000 generations.

For a mutation rate of 0.02 the fitness increases quickly to start with, but once it reaches a certain level it fluctuates around it and does not make any more progress towards the target. It fluctuates around a fitness of about 0.68, which corresponds to 34 correct characters out of 50. This can be continued for many more generations (several million) with no change to the behaviour.

4 Conclusions

The goal in this paper has been to investigate how the efficiency of cumulative selection depends on population size N, string length L and the mutation rate in the context of finding a specified target string starting with a randomly generated guess. Both the simple and genetic algorithm models illustrate that the efficiency of cumulative selection can be extremely sensitive to the values of N, L and the mutation rate. Both also exhibit a characteristic minimum in the number of generations required as a function of the mutation rate for fixed values of N and L. As L increases this minimum becomes much more pronounced and, as a consequence, if the mutation rate is not close to the minimum the number of generations can be extremely high. In fact, in some cases a mutation rate slightly above the minimum results in the target not being reached at all.

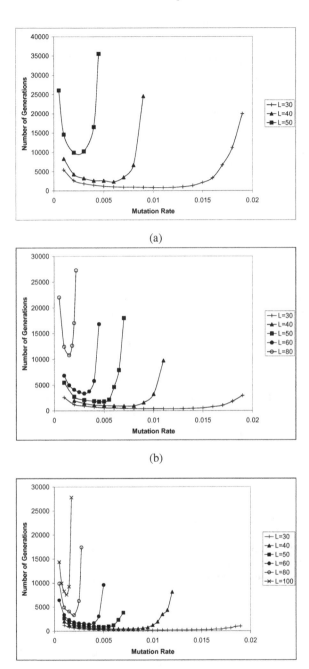

(a)

(b)

(c)

Fig. 3. Results for a) $N = 250$, b) $N = 500$ and c) $N = 1000$

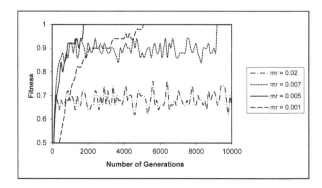

Fig. 4. The fitness plotted every 100th generation for three values of the mutation rate

Some initial work has been carried out on an alternative fitness function and similar behaviour is observed, but there are clearly many other directions for further research. Extending the results to consider larger population sizes, longer strings, other fitness functions, alternative selection procedures and alternative ways of generating the initial population of strings are some obvious examples. This work has only treated a limited case and so future work will focus on investigating models that are more realistic from an evolutionary point of view.

Acknowledgments. The author would like to thank Rory Mullan and Karen McAuley for discussions based on related work carried out in undergraduate projects and also Dr. Mark McCartney and anonymous reviewers for helpful suggestions.

References

1. Kutschera, U., Niklas, K.J.: The modern theory of biological evolution: an expanded synthesis. Naturwissenschaften 91, 255–276 (2004)
2. Dawkins, R.: The Blind Watchmaker. Penguin Books, London (1983)
3. Nilsson, D.E., Pelger, S.: A pessimistic estimate of the time required for an eye to evolve. Proceedings of the Royal Society of London B 256, 53–58 (1994)
4. Hunt, G.: Gradual or pulsed evolution: when should punctuational explanations be preferred? Paleobiology 34, 360–377 (2008)
5. Watson, R.A.: Compositional Evolution. MIT Press (2006)
6. Myers, R., Hancock, E.R.: Empirical modelling of genetic algorithms. Evolutionary Computation 9, 461–493 (2001)
7. Ray, T.S.: An approach to the synthesis of life. In: Langton, C., Taylor, C., Farmer, J.D., Rasmussen, S. (eds.) Artificial Life II. Addison-Wesley, Redwood City (1991)
8. Lenski, R.E., Ofria, C., Pennock, R.T., Adami, C.: The evoultionary origin of complex features. Nature 423, 139–144 (2003)
9. Auerbach, J.E., Bongard, J.C.: Environmental influence on the evolution of morphological complexity in machines. PLoS Computational Biology 10, e1003399 (2014)
10. Mitchell, M.: An Introduction to Genetic Algorithms. MIT Press (1998)

Explanatory Inference under Uncertainty

David H. Glass and Mark McCartney

School of Computing and Mathematics, University of Ulster,
Newtownabbey, Co. Antrim, BT37 0QB, UK
{dh.glass,m.mccartney}@ulster.ac.uk

Abstract. This paper investigates the performance of explanatory or abductive inference in certain hypothesis selection tasks. The strategy is to use various measures of explanatory power to compare competing hypotheses and then make an inference to the best explanation. Computer simulations are used to compare the accuracy of such approaches with a standard approach when uncertainty is present and when several causal scenarios occur including one where the conditions for explaining away are met. Results show that some explanatory approaches can perform well and in certain scenarios they perform much better than the standard approach.

1   Introduction

A prominent approach to explanation in the artificial intelligence literature makes use of probability theory. In particular, probabilistic networks and related approaches provide a suitable context for finding explanations and for explanatory inference, especially when interpreted causally (see for example [1–4]). In general, when explanation has been used for inference in such networks an explanation is selected that maximizes the posterior probability of all the non-evidence nodes (Most Probable Explanation, MPE) or a subset of them (Maximum A Posteriori, MAP). Considerable effort has gone into approximation methods to find such explanations (see [5, 6]).

However, the idea that posterior probability should be maximized has also been questioned and alternative approaches proposed [2, 7]. The current paper extends [8] by considering uncertainty in likelihoods as well as prior probabilities and the effect of including causal relationships including intercausal relationships. This last point is important because intercausal relationships such as explaining away are important in explanatory reasoning and have been studied in detail in the context of Bayesian networks and qualitative probabilistic networks [9, 10]. For this reason, their effect on inference is of interest.

2   Measures of Explanatory Power

A variety of measures for selecting the best explanatory hypothesis have been proposed in the literature. Here a number of the probabilistic measures considered in [8], which are presented in table 1, will be used. For each measure the

E. Corchado et al. (Eds.): IDEAL 2014, LNCS 8669, pp. 215–222, 2014.
© Springer International Publishing Switzerland 2014

idea is to select the explanatory hypothesis, H, which has the highest value of the measure for evidence, E.

Table 1. Measures for selecting an explanatory hypothesis. These measures were used in [8].

Name	Abbreviation	Measure
Most Probable Explanation	MPE	$Pr(H|E)$
Overlap Coherence Measure	OCM	$\frac{Pr(H \wedge E)}{Pr(H \vee E)}$
Maximum Likelihood	ML	$Pr(E|H)$
Likelihood Ratio	LR	$log\left[\frac{Pr(E|H)}{Pr(E|\neg H)}\right]$
Difference Measure	DIFF	$Pr(H|E) - Pr(H)$

Posterior probability is the ideal as far as inference is concerned. At least, this is true when no uncertainty is present in the prior probabilities (see section 3). However, as has frequently been pointed out, posterior probability has weaknesses as a measure of explanatory power. Nevertheless, the MPE approach, which simply selects the explanation with maximum posterior probability given the evidence, can be considered as the benchmark against which other approaches are to be compared in terms of their performance.

Table 2. Further measures for selecting an explanatory hypothesis discussed in [11]

Name	Measure
ED	$Pr(E|H) - Pr(E)$
EC	$Pr(E|H) - Pr(E|\neg H)$
EP	$\frac{Pr(E|H) - Pr(E)}{Pr(E|H) + Pr(E)}$
EM	$log\left[\frac{Pr(E|H)}{Pr(E)}\right]$
$\mathcal{E}$	$\frac{Pr(H|E) - Pr(H|\neg E)}{Pr(H|E) + Pr(H|\neg E)}$

In recent work, Schupbach has considered a number of further measures of explanatory power in addition to the OCM measure in table 1 [11]. These measures are presented in table 2. Like the likelihood ratio and difference measures in table 1, all the measures in table 2 are confirmation measures which means that they are positive (negative) if there is a positive (negative) probabilistic dependence between H and E and zero if they are probabilistically independent (i.e. $Pr(E|H) = Pr(E)$). It is easy to show that for evidence E measures ED, EM, EP and $\mathcal{E}$ will rank two hypotheses H_1 and H_2 in the same way as ML. For this reason, only the EC measure from table 2 will be included in this study.

3 Methodology

In light of the above discussion the approaches to be compared are MPE, OCM, ML, LR, DIFF and EC. In order to do so computer simulations have been carried out as in [8]. The idea is to generate hypotheses H_i, each with a corresponding prior probability and a likelihood for a particular piece of evidence E. The goal is to determine which hypothesis provides the best explanation according to each of the various approaches and to see whether it turns out to be the one that has been designated as the actual hypothesis. By repeating the process numerous times it is possible to determine how successful each approach is on average at identifying the actual hypothesis.

This approach was adopted for the case of mutually exclusive hypotheses in [8], which essentially amounts to treating the hypotheses as separate values of a single hypothesis variable. In most realistic explanatory scenarios, however, there is more than one hypothesis variable. A simple case with two independent binary hypothesis variables, H_1 and H_2, is explored in this paper. In terms of priors, each hypothesis variable is treated as in the mutually exclusive case, but likelihoods are now assigned to each combination of values of the hypothesis variables. Inference is carried out by determining which joint setting of values of the hypothesis variables maximizes a given measure.

The basic algorithm, which is based on that used in [8] is as follows:

1. Randomly assign prior probabilities summing to one for each hypothesis variable H_1 and H_2.
2. Randomly assign a likelihood $Pr(E|H_1, H_2)$ to each combination of values of the hypothesis variables.
3. Randomly select a value for each of the hypothesis variables using their prior probability distributions and designate the joint setting of these values as the actual hypothesis H_A.
4. Select whether E or $\neg E$ occurs using the likelihood of H_A so that there is a probability $Pr(E|H_A)$ of E occurring and $1 - Pr(E|H_A)$ of $\neg E$ occurring.
5. For each approach (MPE, OCM, ML, LR, DIFF and EC), if E occurs, identify which hypothesis (i.e. which joint setting of the hypothesis variables) provides the best explanation of E, and similarly if $\neg E$ occurs.
6. For each approach, if the hypothesis identified in step 5 matches the actual hypothesis, count this as a success, otherwise count it as a fail.
7. Repeat steps 1 to 6 to obtain the accuracy for each approach.

In order to allow for uncertainty in the model, subjective probabilities can be introduced in addition to the objective probabilities generated in steps 1 and 2. The idea is that there is an objective prior probability distribution over the hypotheses and objective likelihoods for each hypothesis. In the presence of uncertainty, however, the reasoner estimates these probabilities and so in step 5 replaces them with subjective probabilities. In terms of subjective priors, the same approach as in [8] is used where step 5 is replaced with:

5a. Introduce a random error to the prior probabilities assigned in step 1 by adding a number sampled from a Gaussian distribution with mean zero and a specified standard deviation, provided the resulting probability lies in the interval $[0, 1]$. If it does not, the process should be repeated until it does.
5b. Normalize the probabilities resulting from the previous step to ensure they sum to one.
5c. For each approach, if E occurs, identify which hypothesis provides the best explanation of E, and similarly if $\neg E$ occurs.

4   Simulations and Results

The results presented in the subsequent sections make use of the algorithms in section 3 by using 10^6 repetitions for each point in the figures to ensure accurate values are obtained.

Before considering cases where the hypotheses are overlapping, some results are presented for mutually exclusive hypotheses and so extend the results presented in [8]. Figs. 1(a) and 1(b) present results for two complementary scenarios, complete uncertainty in all likelihoods and complete uncertainty in all priors respectively. Not surprisingly, the accuracy is lower in both cases than it is when no uncertainty is present. The general pattern, however, between Fig. 1(a) and results presented in [8] is very similar with MPE performing best, followed by OCM and ML performing worst. Since ML only compares likelihoods its performance is no better than a random guess in this case. The trend is reversed when there is complete uncertainty in priors. Now ML performs best while MPE performs worst, although the difference between the approaches is not as great in this case.

In the results that follow the hypotheses are not assumed to be mutually exclusive. As noted earlier, with two hypothesis variables, it is possible to explore causal relationships and different kinds of intercausal relationships. A lot of work has been undertaken in the study of probabilistic causality, but here a state of a hypothesis variable, H_i will be considered as a cause of the evidence E if $Pr(E|H_i) > Pr(E|\neg H_i)$. Given independence the following expressions hold for H_1 (and similarly for H_2):

$$Pr(E|H_1) = Pr(E|H_1, H_2)Pr(H_2) + Pr(E|H_1, \neg H_2)Pr(\neg H_2) \qquad (1)$$

$$Pr(E|\neg H_1) = Pr(E|\neg H_1, H_2)Pr(H_2) + Pr(E|\neg H_1, \neg H_2)Pr(\neg H_2) \qquad (2)$$

When likelihoods are randomly selected for each of the four combined hypothesis states, if the largest value is assigned to $Pr(E|H_1, H_2)$ and the smallest to $Pr(E|\neg H_1, \neg H_2)$, this guarantees that $Pr(E|H_1) > Pr(E|\neg H_1)$ and

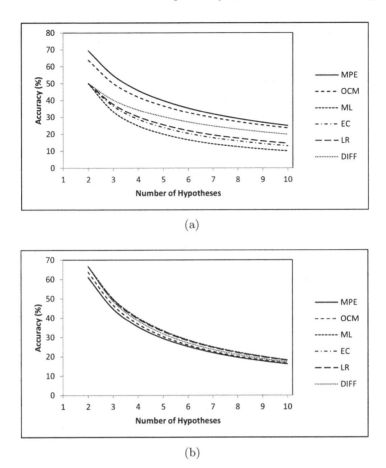

(a)

(b)

Fig. 1. Accuracy against number of hypotheses when there is (a) complete uncertainty in all likelihoods and (b) complete uncertainty in all priors

$Pr(E|H_2) > Pr(E|\neg H_2)$ and so H_1 and H_2 can be considered as causes of E. The hypothesis state $\neg H_1 \& \neg H_2$ can then be considered as a catch-all hypothesis since it invokes no causes.

Two kinds of intercausal relationships discussed by Wellman and Henrion can also be considered [10]. The first of these, which they call *negative product synergy* describes the relationship between causes in contexts where *explaining away* can occur. The idea is that even though two causes of a given piece of evidence may be marginally independent, learning that one occurred reduces the probability of the other. The second is *positive product synergy* and occurs in cases where learning that one cause occurred increases the probability of the other cause having occurred as well. This would apply in contexts where two causes, although marginally independent, both need to be present to produce the effect. The condition for negative product synergy between two causes H_1 and H_2 is:

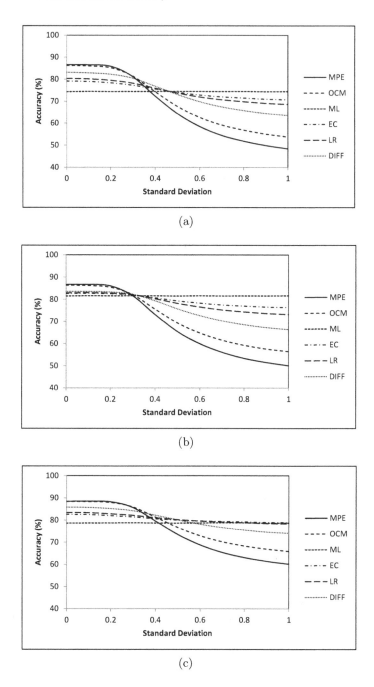

Fig. 2. Accuracy against standard deviation when priors for both causes H_1 and H_2 are selected from the interval $[0.8,1]$ in (a) the general case, (b) the case of negative product synergy, i.e. explaining away, and (c) the case of positive product synergy

$$Pr(E|H_1, H_2) \times Pr(E|\neg H_1, \neg H_2) < Pr(E|H_1, \neg H_2) \times Pr(E|\neg H_1, H_2) \quad (3)$$

and the expression for positive product synergy is obtained by replacing $<$ with $>$. Negative product synergy can be implemented in the simulations by selecting $Pr(E|\neg H_1, \neg H_2)$ randomly from the interval $[0, \alpha]$, where

$$\alpha = \frac{Pr(E|H_1, \neg H_2)Pr(E|\neg H_1, H_2)}{Pr(E|H_1, H_2)}$$

and positive product synergy by selecting $Pr(E|H_1, H_2)$ randomly from the interval $[\beta, 1]$, where

$$\beta = \frac{Pr(E|H_1, \neg H_2)Pr(E|\neg H_1, H_2)}{Pr(E|\neg H_1, \neg H_2)}.$$

To illustrate the influence of these causal relationships a scenario is considered where the prior probability of each cause, H_1 and H_2, lies in the interval $[0.8,1]$. Since causes of E are considered, accuracies are obtained by averaging over cases where E occurs. Results are presented for accuracy against increasing uncertainty in the priors for the general case where H_1 and H_2 are causes (Fig. 2(a)), for the case where negative product synergy is included (Fig. 2(b)) and for the case where positive product synergy is included (Fig. 2(c)).

As in other cases where uncertainty is present in the priors, MPE performs best for low values of uncertainty, but it can be outperformed by other approaches for high uncertainty. This is particular evident in this case since MPE performs much worse when the standard deviation is above 0.4. In the general case (Fig. 2(a)), averaging over all values of uncertainty considered, there is very little to separate ML, EC and LR and their performance is much better than MPE. In the case of negative product synergy, the results for MPE, OCM and DIFF are very little different compared to the general case, but ML, EC and LR all improve with ML performing best overall. It is noticeable that even for low values of uncertainty MPE has very little benefit over the other approaches. In the case of positive product synergy, MPE again performs worst overall although the advantage of the best approaches (ML and EC) over MPE is not as pronounced in this case.

Overall, these results show that various explanatory measures can have a substantial advantage over MPE. This is particularly true when there is a lot of uncertainty in the prior probabilities and in cases where explaining away can occur.

5   Conclusion

Explanatory inference has been simulated by using a number of measures of explanatory power and comparing their performance at hypothesis selection against the standard approach of selecting the explanatory hypothesis with maximum posterior probability (MPE). When two hypothesis variables and causal relationships are included, the performance of MPE is worse than all other approaches in some cases. In particular, when causes have high prior probability and explaining away can occur MPE performs very poorly especially when compared with ML, LR and EC. Clearly, more work needs to be done to delineate when such advantages over MPE occur, but the simulation results presented here do show that at least some explanatory approaches to inference can perform well and that in some cases they can perform much better than MPE.

Acknowledgments. This publication was made possible through the support of a grant from the John Templeton Foundation. The opinions expressed in this publication are those of the authors and do not necessarily reflect the views of the John Templeton Foundation.

References

1. Shimony, A.E.: Explanation, irrelevance and statistical independence. In: Proceedings of AAAI 1991, pp. 482–487 (1991)
2. Chajewska, U., Halpern, J.Y.: Defining explanation in probabilistic systems. In: Proceedings of the 13th Conference on Uncertainty in AI, pp. 62–71 (1997)
3. Halpern, J.Y., Pearl, J.: Causes and explanations: A structural-model approach. Part ii: Explanations. British Journal for the Philosophy of Science 56, 889–911 (2005)
4. Lacave, C., Luque, M., Diez, F.: Explanation of bayesian networks and influence diagrams in elvira. IEEE Transactions on Systems, Man, and Cybernetics, Part B: Cybernetics 37, 952–965 (2007)
5. Gamez, J.A.: Abductive inference in bayesian networks: A review. In: Gamez, J.A., Moral, S., Salmeron, A. (eds.) Advances in Bayesian Networks. STUDFUZZ, vol. 146, pp. 101–120. Springer, Heidelberg (2004)
6. Kwisthout, J.: Most probable explanations in bayesian networks: Complexity and tractability. International Journal of Approximate Reasoning 52, 1452–1469 (2011)
7. Yuan, C., Lim, H., Lu, T.: Most relevant explanation in bayesian networks. Journal of Artificial Intelligence Research 42, 309–352 (2011)
8. Glass, D.H.: Inference to the best explanation: does it track truth? Synthese 185, 411–427 (2012)
9. Pearl, J.: Probabilistic Reasoning in Intelligent Systems. Morgan Kaufman, San Mateo (1988)
10. Wellman, M.P., Henrion, M.: Explaining 'explaining away'. IEEE Trans. Pattern Anal. Mach. Intell. 15, 287–292 (1993)
11. Schupbach, J.N.: Comparing probabilistic measures of explanatory power. Philosophy of Science 78, 813–829 (2011)

A Novel Ego-Centered Academic Community Detection Approach via Factor Graph Model

Yusheng Jia, Yang Gao, Wanqi Yang, Jing Huo, and Yinghuan Shi

State Key Laboratory for Novel Software Technology, Nanjing University, China, 210023

Abstract. Extracting ego-centered community from large social networks is a practical problem in many real applications. Previous work for ego-centered community detection usually focuses on topological properties of networks, which ignore the actual social attributes between nodes. In this paper, we formalize the ego-centered community detection problem in a unified factor graph model and employ a parameter learning algorithm to estimate the topic-level social influence, the social relationship strength between nodes as well as community structures of networks. Based on the unified model we can obtain more meaningful ego community compared with traditional methods. Experimental results on co-author network demonstrate the effectiveness and efficiency of the proposed approach.

Keywords: Ego-centered Community, Community Detection, Factor Graph Model, Social Networks.

1 Introduction

Recently, quite a few methods [1][2][3] have been proposed for the macro-level community detection. In a practical application, the selection of experts for evaluating technology projects usually requires the community information according to one specific expert [4]. Therefore, the community detection task is not only to be modeled as a community detection problem globally from the aspect of the whole network. Ego network models [5] are utilized to represent the social network from ego perspective. Obviously, different experts have the specific research areas as well as the evaluating criteria. Thus, it is necessary to measure the social influence from different research areas. In this paper, we try to address the two major issues which are related to the social influence analysis, including 1) how to quantify the strength of social influence, and 2) how to differentiate the social influence from different research areas. The key point in ego-centered community detection is how to well quantify the influence among researchers by simultaneously leveraging relationship factor, social influence and community structures.

E. Corchado et al. (Eds.): IDEAL 2014, LNCS 8669, pp. 223–230, 2014.

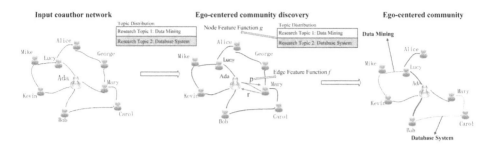

Fig. 1. An illustration of the ego-centered community detection using the co-author networks

Fig.1 illustrates the goal of our work. The input: a co-author network and an expert, (Ada is the ego in Fig.1). As shown in the left of the figure, we can first obtain the research areas distribution of the different experts, and then use ego-centered community detection method to measure the degree that the ego expert is influenced by other experts on each research topic as well as social relationship strength between them. The right of the figure shows the output, where the result reflects not only the closest relationship, but also the mutual influences between them on different research topics.

Our proposed approach for community detection has the following pipeline: (1) we first we get the research areas information by using Latent Dirichlet Allocation (LDA) [8][9] model to analysis the co-author network. (2) Ego Community Factor Graph Model (ECFG) combines the influential factors with the community structure to model the relationship of the networks. By finding the most relevant nodes which connected with the ego node based on the quantitative analysis, we can find out the ego community from the large social networks.

2   Related Work

Recent studies show that there has been a great interest in community detection. Passarella et al. [5] proposed a constructive algorithm to generate the ego networks. The algorithm complemented with an analytical model which considered both the structural properties in the anthropology literature and the properties of their contact process.

Clauset [10] proposed an algorithm to find the local optimal community structures in large networks starting from ego vertex. The method used the greedy strategy to explore the network one vertex at a time where the added vertex maximizes the local community value.

Luo et al. [11] optimized and extended the local community definition, which introduced the connection density measurement to avoid the phenomena of outliers. A greedy strategy and two local search approaches (KL-like and the add-all) are also employed to improve the precision.

However, the discovered communities extracted with traditional methods contain many outliers and with high recall and low accuracy. The main problem is that the discovered communities can't reflect the real state of community structures accurately in a large network which only consider the network anthropology structure and ignore the content information associated with each node.

3 Methodology

3.1 LDA for Research Topic Modeling

In order to estimate the distribution of experts' research topics, a popularly generative probabilistic model, LDA [6], is used to collect discrete data from the co-author networks. Specifically, in the co-author networks, every expert focuses on his own research areas. Formally, for each node $v \in V$ and research topics z, a vector $\theta_v \in \mathbb{R}^T$ ($\sum_z \theta_{vz} = 1$) express the distribution of research topics. Each element represents the importance of node v on topic z. Finally, we use LDA to initialize the research topic distribution of each expert node in the co-author network.

3.2 Ego Community Factor Graph (ECFG) Model

Parameters: The factor graph model [15] contains a set of observed variable $\{v_i\}_{i=1}^N$ represents the input network, a set of hidden vectors $\{y_i\}_{i=1}^N$ which corresponds to the N nodes of the network. A virtual node u_c is introduced for expressing ego community c_e and non-ego community $\bar{c}_e$. Fig. 2 shows a typical example of an ECFG. The network consists of four nodes $\{v_1, ..., v_4\}$, which have corresponding hidden tors $Y = \{y_1, ..., y_4\}$. Two latent variables C=2 are introduced for communities. Specifically, two kinds of feature functions are defined for the ECFG model.

Node Feature Function: $g(v_i, y_i, e, z)$ is a feature function defined on node v_i, which estimate how likely node v_i belongs to the ego community c_e. The node feature function is defined to measure the relationship strength between node v_i and ego node e. Meanwhile, the topic-level social influence should also be taken into account [16]. Thus, the node feature function g can be defined as follows:

$$g(v_i, y_i, e, z) = \begin{cases} \dfrac{w_{iy_i^z}^z}{\sum_{j \in NB(i)}\left(w_{ij}^z + w_{ji}^z\right)} \cdot \dfrac{a_{i\delta_n} \sum_{t \in \delta(i,e)} a_{\delta_t \delta_{t-1}}}{n \sum_{j \in NB(i)} a_{ij}} y_i^z & \neq i \\[4mm] \dfrac{\sum_{j \in NB(i)} w_{ji}^z}{\sum_{j \in NB(i)}\left(w_{ij}^z + w_{ji}^z\right)} y_i^z & = i \end{cases} \tag{1}$$

where NB(i) represents the neighboring nodes of node v_i, $w_{ij}^z = \theta_j^z a_{ij}$, θ_j^z is calculated by LDA which indicates the influential node v_j on specific topic z, and a_{ij} is equal to the edge weight. Then w_{ij}^z is denoted as the topical similarity and association strength between node v_i and v_j. Moreover, $\delta_{(i,e)}$ denotes the set of points for shortest path from ego node to node v_i. δ_n represents the nth node of the shortest path and n is the number of edges of the shortest path. The first part of the feature function evaluates the influential factor while the second part measures the relationship strength with the ego node.

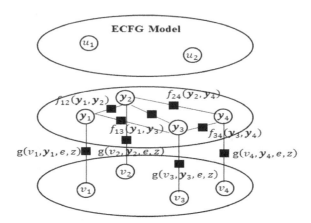

Fig. 2. Graphical representation of the ego-centered community detection via factor graph model

Edge Feature Function: $f_{ij}(y_i, y_j)$ is an edge feature function defined on the edge of the network, describing how likely two connected nodes belong to the same community. One of the key factors, modularity [8], is a significance description of the community structure. The higher value of the modularity corresponds to a better division of a network [17]. Formally, we defined the edge feature function f as follows:

$$f_{ij}(y_i, y_j) = \begin{cases} e^{a_{ij} - \frac{k_i k_j}{2m}} & y_i = y_j \\ 0 & y_i \neq y_j \end{cases} \tag{2}$$

where $k_i = \sum_{j \in O(i)} a_{ij}$ the sum weight of isout-edges of node v_i and $m = \sum_{i,j} a_{ij}$ is the sum of weights of all edges in the network. The defined edge feature function can guarantee the structure of community.

Joint Distribution: Usually, by combining all the feature functions we can maximize an objective likelihood function to construct the model. The definition of joint objective function as follows:

$$P(v, Y) = \frac{1}{Z} \prod_{i=1}^{N} \prod_{z=1}^{T} g(v_i, y_i, e, z) \prod_{i=1}^{N} \prod_{j=1}^{N} f_{ij}(y_i, y_j) \tag{3}$$

where $v = [v_1, ..., v_N]$ and $Y = [y_1, ..., y_N]$ denote all observed and hidden variables respectively. g is node feature function and f is edge feature function. Z is a normalizing factor. As defined by the Eq. (3), maximizing the likelihood function $P(v, Y)$ can obtain the most relevant nodes of the given ego node corresponding to the ego-centered community.

Ego Community Affinity Propagation (ECAP) Learning Method: A Topical Affinity Propagation (TAP) algorithm [16] was proposed based on sum-product [15] approach to evaluate the topic-specific message passing process. However, as for the ego-centered community detection problem, we should also consider the relationship strength and community structure factors. Thus, we update the parameters and propose the ECAP method. The algorithm is summarized in Alogirthm1.

Specifically, two types of variables p_{ije}^z and c_{ije}^z are introduced to explain the ego-centered community discovery task. Here, p_{ije}^z implies how likely user v_i affects v_j belonging to the ego community u_e on the specific research areas z, while c_{ije}^z indicates that how likely user v_j is affected by v_i belonging to ego community u_e in the research areas z. Then, the update rules for the variables as follows [16]:

$$p_{ije}^z = a_{ije}^z + b_{ije}^z - max_{k \in NB(j)}\{a_{ike}^z + b_{ike}^z + c_{ike}^z\} \tag{4}$$

$$c_{jje}^z = max_{k \in NB(j)} min\{p_{kje}^z, 0\} \tag{5}$$

$$c_{ije}^z = min\left(max\{p_{jje}^z, 0\}, -min\{p_{jje}^z, 0\} - \max_{k \in NB(j\{i\})} min\{p_{kje}^z, 0\}\right)$$

$$i \in NB(j) \tag{6}$$

where $NB(j)$ represents the neighboring nodes of node j, a_{ije}^z and b_{ije}^z are the logarithm of the normalized node feature function and edge feature function, respectively. In detail,

$$a_{ije}^z = log \frac{g(v_i, y_i, e, z)}{\Sigma_{k \in NB(i) \cup \{i\}} g(v_i, y_i, e, z)} \tag{7}$$

$$b_{ije}^z = log \frac{f_{ij}(y_i, y_j)}{\Sigma_{k \in NB(i) \cup \{i\} \cup \{j\}} f_{ij}(y_i, y_j)} \tag{8}$$

The measurement of the relationship strength and topic-level social influence can unified with the ego community score which define based on the learned variables p_{ije}^z and c_{jje}^z using a sigmoid function [16] as follows:

$$\mu_{ije}^z = \frac{1}{1 + e^{-(p_{ije}^z + c_{ije}^z)}} \tag{9}$$

Algorithm 1. The ECAP Learning Algorithm

Input: $G=(V, E)$, ego node e and research topic distribution $\{\theta_v\}_{v \in V}$
Output: ego community graphs
1.   Calculate the node feature function $g(v_i, y_i, e, z)$;
2.   Calculate the edge feature function $f_{ij}(y_i, y_j)$;
3.   Calculate a_{ij}^e, b_{ij}^e according to Eq. (7) , Eq. (8);
4.   **repeat**
5.   **for each** edge-ego pair (e_{ij}, e) update p_{ije}^z according to Eq.(4)
6.   **for each** node-ego pair (v_j, e) update c_{jje}^z according to Eq.(5)
7.   **for each** edge-ego pair (e_{ij}, e) update c_{ije}^z according to Eq. (6)
8.   **until** convergence;
9.   **for each** node v_i do
10.   **for each** neighboring node $s \in NB(i) \cup \{i\}$ compute μ_{ije}^z according to Eq. (9)
11.   Generate ego community graph according to $\{\mu_{ije}^z\}$

4   Experiments and Results

4.1   Data Sets and Experiment Settings

We perform our experiments on a coauthor network extracted from an academic search system Arnetminer.org (http://arnetminer.org) [18]. The coauthor network contains 640,134 authors and 1,554,643 co-author relations. Firstly, we apply LDA to model the topic distributions of both the authors and the corresponding papers. About 7 author-specific research topic distributions to each author evaluated from the network. The ECAP learning algorithm is implemented using C++ programming language and Visual Studio 2012 IDE tool. The experiment performed on Windows 2007 with Intel(R) Core(TM) i5 CPU and 4GB memory.

4.2   Performance Analysis

Case Study: To analysis the results, we choose one expert, Prof. Jiawei Han, as the ego node. As Table.1 shows the ego node Prof. Jiawei Han mainly focuses on research topics including "data mining" and "database system". Table.2 evaluates the example list of researchers who are mostly influenced by the ego node. Fig.3 represents the discovered two communities on each topic. The community which includes the ego node represents the ego community of Prof. Jiawei Han. Then, by combining the ego community on two different topics we can finally obtain the ego community of Prof. Jiawei Han. To analysis, only 10 most representative experts of the community are showed in the table. Obviously, the nodes which mostly influenced by the ego node are included in the ego community. This implies the social influence factor can also be measured when detecting the communities.

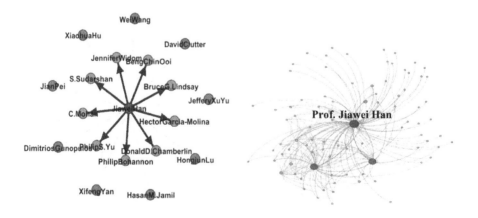

Fig. 3. Ego-centered Community Detection Results on two different topics

Fig. 4. The ego community of Prof. Jiawei Han

Table 1. Top 10 self-representative experts on 7 different research topics

Research Topics	Representative Experts
Data Mining	Jiawei Han, Jian Pei, HeikkiMannila, Wei Wang, Philip S. Yu, Jeffery Xu Yu, Qiang Yang, EamonnJ,Keogh, Christos Faloutsos, Martin Ester
Database System	Gerhard Weikum, Michael Stonebraker, Jennifer Widom, Philip S.Yu, Richard T.Snodgrass, Jiawei Han, Joseph M.Hellerstein, WeiSun, V.S.Subrahamanian
Machine Learning	Pat Langley, Bernhard Scholkopf, Michael I.Jordan, Alex Waibel, Daphne Koller, Yoram Singer, SatinderP.Singh, Manfred K.Warmuth, Floriana Esposito, Trevor Darrell
Bayesian Network	Daphne Koller, David Heckerman, Henri Prade, Floriana Esposito, Michael I. Jordan, Dilier Dubois, LluisGodo, Philippe Smets, Finn Verner Jensen, Paul R.Cohen
Information Retrieval	James P.Callan, OphirFrieder, Gerard Salton, Zheng Chen, Alan F.Smeaton, MouniaLalmas, Chris Buckley, Amanda Spink,W.Bruce Croft, Yi Zhang
Web Services	Yan Wang, Fabio Casati, ZakariaMaamar, Patrick C.K.Hung, Jen-Yao Chung, Ying Li, Liang-Jie Zhang, BoualemBenatallah, Jun-Jang Jeng, Claude Godart
Semantic Web	Dieter Fensel, Stefan Decker, York Sure, Wolfgang Nejdl, James A.Hendler, Enrico Motta, Steffen Staab, Ian Horrocks, Rudi Studer, Carole A.Goble

Table 2. Example of the influence analysis for the coauthor dataset

Ego Researcher	Influenced Researchers
Jiawei Han	David Clutter, Hasan M.Jamil, K.P.Unnikrishnan, RamasamyUthurusamy, Shiwei Tang

The extracted ego community (see Fig.3) evaluates the community structure as well as the social influence between experts on specific topics. The experimental results demonstrate the effectiveness of ECAP of ego-centered community identification on the co-author data sets. Finally, the ego community is extracted from the co-author network shown in Fig.4.

5   Conclusions

In this paper, we study the problem of ego-centered social community analysis. We formally define the problems and propose a unified factor graph model, ECGF, to detect the ego community. Based on the node feature function and edge feature function, we can evaluate the research topic specific social influence, social relationship strength and community structures jointly. Unlike the previous methods which only measure the topological properties of the network, the proposed learning algorithm considers factors more reasonable and practical. Therefore, the detected community can reflect the real state of the social relationships rather than less practical network clusters. Experimental results on co-author network data set demonstrate that the proposed approach can effectively find the ego-centered communities.

Acknowledgments. The work was support by NSFC (61035003, 61175042, 61321491, and 61305068), Jiangsu 973 (BK2011005), Jiangsu NSF (BK20130581), the Program for New Century Excellent Talents in University (NCET-10-0476) and Jiangsu Clinical Medicine Special Program (BL2013033) and the Graduate Research Innovation Program of Jiangsu, China (CXZZ13 0055). Also, this work was partially supported by Collaborative Innovation Center of Novel Software Technology and Industrialization.

References

1. Fortunato, S.: Community detection in graphs. Physics Reports 486(3), 75–174 (2010)
2. Clauset, A.: Finding local community structure in networks. Physical Review E 72(2), 026132 (2005)
3. Newman, M.E.J., Girvan, M.: Finding and evaluating community structure in networks. Physical Review E 69(2), 026113 (2004)
4. Kerzner, H.R.: Project management: a systems approach to planning, scheduling, and controlling. John Wiley & Sons (2013)
5. Passarella, A., Dunbar, R.I.M., Conti, M., et al.: Ego network models for Future Internet social networking environments. Computer Communications 35(18), 2201–2217 (2012)
6. Blei, D.M., Ng, A.Y., Jordan, M.I.: Latent dirichlet allocation. The Journal of Machine Learning Research 3, 993–1022 (2003)
7. Hoffman, M.D., Blei, D.M., Bach, F.R.: Online Learning for Latent Dirichlet Allocation. NIPS 2(3), 5 (2010)
8. Lancichinetti, A., Fortunato, S.: Limits of modularity maximization in community detection. Physical Review E 84(6), 066122 (2011)
9. Newman, M.E.J.: Communities, modules and large-scale structure in networks. Nature Physics 8(1), 25–31 (2012)
10. Clauset, A.: Finding local community structure in networks. Physical Review E 72(2), 026132 (2005)
11. Luo, F., Wang, J.Z., Promislow, E.: Exploring local community structures in large networks. Web Intelligence and Agent Systems 6(4), 387–400 (2008)
12. Wang, C., Blei, D.M.: Collaborative topic modeling for recommending scientific articles. In: Proceedings of the 17th ACM SIGKDD International Conference on Knowledge Discovery and Data Mining, pp. 448–456. ACM (2011)
13. Hong, L., Davison, B.D.: Empirical study of topic modeling in twitter. In: Proceedings of the First Workshop on Social Media Analytics, pp. 80–88. ACM (2010)
14. Blei, D.M.: Probabilistic topic models. Communications of the ACM 55(4), 77–84 (2012)
15. Kschischang, F.R., Frey, B.J., Loeliger, H.A.: Factor graphs and the sum-product algorithm. IEEE Transactions on Information Theory 47(2), 498–519 (2001)
16. Tang, J., Sun, J., Wang, C., et al.: Social influence analysis in large-scale networks. In: Proceedings of the 15th ACM SIGKDD International Conference on Knowledge Discovery and Data Mining, pp. 807–816. ACM (2009)
17. Yang, Z., Tang, J., Li, J., Yang, W.: Social Community Analysis via Factor Graph Model. IEEE Intelligent Systems 26(3), 58–65 (2011)

Intelligent Promotions Recommendation System for Instaprom Platform

Marcos Martín Pozo[1], José Antonio Iglesias[2], and Agapito Ismael Ledezma[3]

Universidad Carlos III de Madrid, Leganés (Madrid), Spain
marcos.martin.pozo.delgado@gmail.com, {jiglesia,ledezma}@inf.uc3m.es

Abstract. The customized marketing is an increasing area where users are progressively demanding and saturated of massive advertising, which has a really low success rate and even discourage the purchase. Furthermore, another important issue is the smash hit of mobile applications in the most known platforms (Android and iPhone), with millions of downloads worldwide. Instaprom is a platform that joins both concepts in a mobile application available for Android and iPhone; it retrieves interesting instant promotions being close to the user but without invading the user's e-mail. Nowadays, the platform sends promotions based on the customized preferences by the user inside the application, although the intelligent system proposed in this paper will provide a new approach for creating intelligent recommendations using similar users promotions and the navigation in the application information.

Keywords: marketing, artificial intelligence, machine learning, recommendation systems, marketing segmentation.

1 Introduction

Over the last years, smartphones on the one hand and promotions applications on the other hand are increasing in size. The first expansion is due to the progress and decreasing costs of technology and the offers applications emerge to combat the effects of economic crisis. Instaprom is a platform that joins the necessity of offers (good for the shopper, which saves money, and for the seller, which frees stock) with the power of smartphones in a mobile application available for Android and iPhone that allows to the shops add offers easily and flexibly and permits to the shoppers receive the offers immediately.

In this paper CRISP-DM methodology [15] –the most used methodology for Data Mining tasks [6]– is applied to create an intelligent recommendation system for Instaprom.

This paper is structured in the following sections: Section 2 describes the CRISP-DM methodology, Section 3 analyzes the datasets and their preprocessing, Section 4 explains the proposed models, Section 5 evaluates these models and finally in Section 6 conclusions and future research are described.

E. Corchado et al. (Eds.): IDEAL 2014, LNCS 8669, pp. 231–238, 2014.

2   Methodology

The methodology applied in the approach proposed in this paper is CRISP-DM, the most used methodology in Data Mining [6]. This methodology has 6 steps in an iterative process which are detailed as follows.

First of all, the business has to be known the context of the process and the objectives that we want to achieve. Secondly, we need to know how many datasets we have and how these datasets are because they are the main point in Data Mining. Data preprocessing is the third step and the most time-consuming because real data are incomplete and inconsistent. Then, the models are designed using Data Mining techniques. The fifth step is to evaluate the generated models. Finally, the sixth step is to deploy the system in a production environment.

Instaprom platform is the business in our case so we want to get the most relevant promotions which can be interesting for each user. This is achieved comparing the available promotions with the previous favorite promotions and with the promotions that the most similar users like.

The available data are the user profile introduced in the mobile application, the most loved promotions for the user and similar users. Data are described with more details in the next section.

Finally, steps 3-5 are analyzed in greater details in next sections, and deployment details have less relevance to the Data Mining process.

3   Preprocessing

This project uses heterogeneous data from varied sources, and therefore each one has to be processed differently. These data have been obtained for two months by recording Android application information.

Firstly, promotions text is used to find similar promotions to the most valuated by the user. These promotions are stored in a database, but they must be processed in order to compare them correctly. This process consists in applying text mining to promotions fields (title, body and conditions). The text mining process applied is 1) *tokenize* the text (to split the text in words), 2) convert all characters to lower case, 3) remove stop words (words without relevance as prepositions and conjunctions with some specific of domain words as 'promotion'), 4) apply a stemmer (to convert words to their root, removing genre, number and conjugation ambiguity) and 5) applying TF-IDF (technique that assigns a relevance numeric value to each word in function of the frequency in the text and the inverse of the frequency in other documents).

Secondly, user navigation is recorded capturing events in Android application. For each event the following data are captured: name of the event, date, user, promotion, shop, commercial area and geolocalization. This task allows how to know the way in which the user uses the application (relevant to find users that use the applications for the same purpose) and also permits to know who is interested in each promotion. This interest is calculated assigning a score to each event and adding the event score to the promotions for the user every time the user triggers the event. Events score has been assigned in function of the relevance of each event and it is shown in Table 1.

Finally, application user profile is used to find similar users and recommend them their most valuated promotions. This information consists of gender, year of birth, province and interests. All the fields are optional, except interests, that must be at least three. This information is stored in a database, but it is necessary to extract and clean it. In addition, this profile information is merged with user events. A different dataset which has been considered in this research is the division of these events in sessions, considering a session as a sequence of same user events with less than 10 minutes between an event and the next one. Ultimately, profile information and sessions are merged in another dataset.

Table 1. Score of each user event in the application

Event	Description	Score
Promotion	User enters in a promotion	20
Conditions	User views the promotion conditions	10
QR	User views the promotion QR code (purchase is assumed)	80
Shop	User views store information	30
Like	User likes promotion	50
Not like	User unchecks that he likes promotion	-50
Preshare	User enters in share option	10
Share	User share promotion	100
Directory	User enters in shop directory	0
CA	User enters in commercial area	0
Preferences	User enters in preferences	0
About	User enters in about application section	0
List	User enters in promotions list	0
Back	User presses back button	0
E-mail	User enters in send e-mail to shop option	0
Call	User enters in call to shop option	0
Web	User enters in view shop web option	0
Contact	User enters in application contact option	0
Help	User enters in help option	0
Play	User starts promotions reception	0
Pausa	User pauses promotions reception	0
Reload	User reload promotions list	0

4   Models

This project uses different techniques to process the datasets. These techniques are renowned and massively used. For clustering tasks EM, K-means, Cobweb and Furthest First are used. For classification tasks C4.5 is used because decision trees permit to analyze results in an affordable manner. Finally, a less known adaptive technique is used to process the users events sequences. Many other techniques have been studied, but they cannot be applied due to time restrictions: dynamic bayessian networks [12,13], ontologies [9,14], fuzzy logic, simulation [2], artificial neural networks [2], genetic algorithms [10,11] and user models [1,3,4,5,7,17].

On the one hand, text mining is used to preprocess promotions text and afterwards get a similarity measure among them. This process is applied to several datasets: promotion title, body, conditions and all fields merged in an unique dataset. To analyze the best dataset, and because analyzing TF-IDF results is intricate and complex, clustering and classification techniques are used. Firstly, EM is used to cluster the TF-IDF processed promotions because using Weka it gets automatically the number of clusters using cross validation [16] and because its mathematical base is suitable for the numerical nature of TF-IDF results. Next, C4.5 is applied to clusters to analyze the goodness of results.

On the other hand, user application and navigation profile are used to find similar users and to recommend their most valuated promotions. In a first iteration, clustering were used to both profiles. Thus, EM, K-means, Cobweb and Furthest First were applied on several datasets: profile, sessions, and sessions with profile. In a last iteration an adaptive clustering technique was used to user events sequences: this technique stores events sequences of a maximum length in a *trie* (a special type of tree suitable for information retrieval) and then it uses the cosine distance to cluster users. This technique uses also a special algorithm in which the behavior of the users can evolve, for more details see [8].

The before techniques constitute the intelligent recommendation system. After this processing a strict filter is used to remove the promotions that don't match the user gender, age, location or interests. This procedure is schematized in Figure 1.

5   Evaluation

Evaluation is essential in a data mining project in order to validate the created models: without a correct evaluation it is not possible to guarantee that the models work properly. In this research a complete evaluation couldn't be accomplished because data collected was not large enough. Nevertheless, bad models were rejected and remaining models seemed to be adequate, although a re-evaluation with more data is necessary to guarantee the fitness of the models.

Content-based recommendation models are validated analyzing the decision trees and the precision obtained by C4.5 on EM clusters. In a first iteration stop words were not removed from promotions fields, but despite getting more precision (due to overfitting) the models obtained are worse. For this reason this should be validated with more data. Precision obtained by C4.5 for each dataset is shown in Figure 2. In this chart the number of clusters that EM generated is shown in yellow color inside bars. In this chart we can notice that the dataset with more precision is when we use all fields of the promotions. The dataset that only contains conditions also has a high precision, but analyzing the decision tree we can see its incoherence, due to the short text of conditions (even there

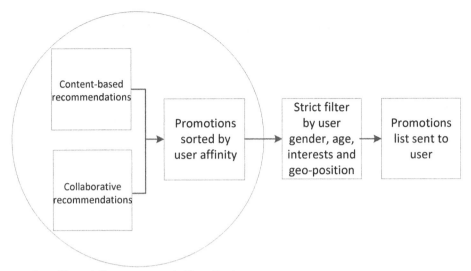

Intelligent Recommendation System

Fig. 1. Intelligent recommendation model

are promotions without conditions). Then, we can say that the best model for content-based recommendation is to use all fields of the promotion removing stop words.

Collaborative recommendation using traditional clustering techniques are validated similarly: clusters are classified by C4.5 and precision and trees coherence are analyzed. In a first iteration data types were incorrect, dealing numeric values as nominal with several possible values. Again, in some cases the precision of this iteration is greater than the precision with correct types due to overfitting, but obviously with correct data types the models are more robust (it can be seen analyzing decision trees) and with more data this would be validated. Precision gotten for each dataset is shown in Figure 3 (hierarchical techniques results are not shown because they were very low), and again number of the clusters that EM generated is shown in yellow color inside bars. In this chart we can notice that precision is high in all datasets, though if we consider sessions the precision is worse. This is because there are few sessions, and they are short. Between profile and sessions with profile the difference is not large, but analyzing sessions with profile decision trees we noticed that user events are almost not present, and they appear in the bottom of trees, where the relevance is low. Finally, in all models K-means is a bit better than EM, but EM calculates the number of clusters automatically, therefore, we determine that the best model for profile clustering is EM using the dataset in which the user profiles are stored.

Finally, since traditional clustering techniques were not adequate for user events profile, it was analyzed using a different technique. This technique generates event sequences of a maximum length (for instance with maximum length 3 it generates sequences of length 1, 2 and 3), then it generates a *trie* with these

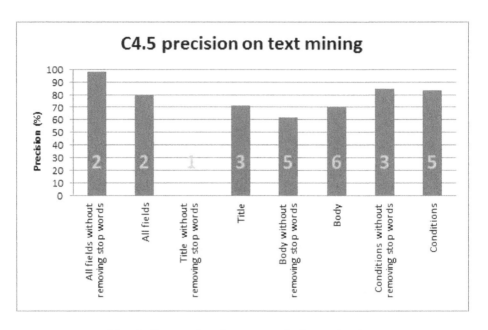

Fig. 2. Content-based recommendation evaluation

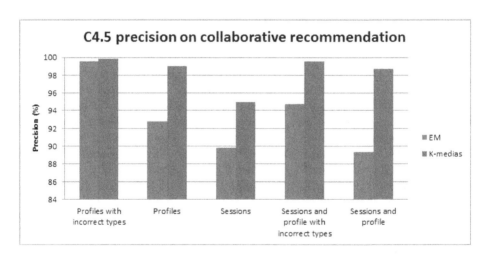

Fig. 3. Collaborative recommendation evaluation

sequences and finally it uses cosine distance to generate clusters. For more details see [8]. This technique is evaluated extracting the sequence of highest χ^2 of each user and then analyzing the average χ^2 and the occurrences of each sequence. This analysis was done with sequences of maximum length 3, 4 and 5; higher maximum lengths were not analyzed because results are similar with these maximum lengths and higher maximum length won't change them. Average χ^2 is

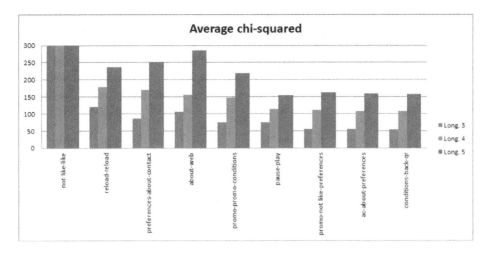

Fig. 4. Average Chi-squared for best sequences of each user

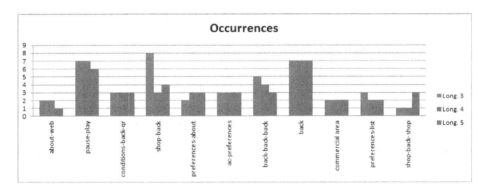

Fig. 5. Occurrences of best sequences of each user

shown in Figure 4 and occurrences are shown in Figure 5. We can appreciate that results hardly change varying maximum length. The best sequences are those with higher average χ^2 and more occurrences because they are the cluster prototypes. These sequences are *pause-play, conditions-back-qr* and *about-web*.

In conclusion, the best model for content-based recommendations is text mining using all fields of the promotions and the best model for collaborative recommendations are combining EM for profile recommendations with *tries* technique for navigation profile recommendations.

6 Conclusions and Future Research

In this project an intelligent recommendation system for Instaprom platform has been designed and evaluated. CRISP-DM methodology has been used for this purpose, and as result several techniques and datasets have been proposed.

In spite of these models are promising, more data are required to completely validate them. A set of validation users would be helpful too in order to guarantee the validity of the recommendation system.

Future Research will be about this validation with more data and cross validation, though other techniques can also be evaluated. According to the research detailed in Section 3, fuzzy logic, artificial neural networks, dynamic bayesian networks, simulation and genetic algorithms are the most promising.

References

1. Bouneffouf, D.: Mobile Recommender Systems Methods: An Overview. CoRR (2013)
2. Çağil, G., Erdem, M.B.: An intelligent simulation model of online consumer behavior. J. Intell. Manuf. 23(4), 1015–1022 (2012)
3. Davidsson, C.: Mobile Application Recommender System (2010)
4. Gavalas, D., et al.: Mobile recommender systems in tourism. Journal of Network and Computer Applications (2013)
5. Godoy, D., Schiaffino, S.N., Amandi, A.: Integrating user modeling approaches into a framework for recommender agents (2010)
6. KdNuggets: Data Mining Methodology (2007), http://www.kdnuggets.com/polls/2007/data_mining_methodology.htm
7. Lakiotaki, K., et al.: Multicriteria User Modeling in Recommender Systems. IEEE Intelligent Systems 26(2), 64–76 (2011)
8. Iglesias, J.A., Angelov, P., Ledezma, A., Sanchis, A.: Evolving classification of agents behaviors: a general approach. In: Evolving Systems. Springer (2010)
9. Liu, W., et al.: Ontology-Based User Modeling for E-Commerce System. In: Third International Conference on Pervasive Computing and Applications, ICPCA 2008, pp. 260–263 (2008)
10. Martínez-López, F.J., Casillas, J.: Marketing Intelligent Systems for consumer behaviour modelling by a descriptive induction approach based on Genetic Fuzzy Systems. Industrial Marketing Management (2009)
11. Martínez-López, F.J., Casillas, J.: Mining uncertain data with multiobjective genetic fuzzy systems to be applied in consumer behaviour modelling. Expert Syst. Appl. 36(2), 1645–1659 (2009)
12. Pearl, J.: Probabilistic reasoning in intelligent systems: networks of plausible inference (1998)
13. Prinzie, A., Poel, D.: Modeling complex longitudinal consumer behavior with Dynamic Bayesian networks: an Acquisition Pattern Analysis application. J. Intell. Inf. Syst., 283–304 (2011)
14. Rodríguez Rodríguez, A., Iglesias García, N., Quinteiro-González, J.M.: Modelling the psychographic behaviour of users using ontologies in web marketing services. In: Moreno-Díaz, R., Pichler, F., Quesada-Arencibia, A. (eds.) EUROCAST 2011, Part I. LNCS, vol. 6927, pp. 121–128. Springer, Heidelberg (2012)
15. Turban, E., Sharda, R., Delen, D.: Decision Support and Business Intelligence Systems (2010)
16. Weka doc.: http://weka.sourceforge.net/doc.dev/weka/clusterers/EM.html
17. Zhang, L., Chen, S., Hu, Q.: Dynamic Shape Modeling of Consumers' Daily Load Based on Data Mining. In: Li, X., Wang, S., Dong, Z.Y. (eds.) ADMA 2005. LNCS (LNAI), vol. 3584, pp. 712–719. Springer, Heidelberg (2005)

Kernel K-Means Low Rank Approximation for Spectral Clustering and Diffusion Maps

Carlos M. Alaíz, Ángela Fernández, Yvonne Gala, and José R. Dorronsoro[*]

Dpto. Ingeniería Informática & Inst. Ingeniería del Conocimiento,
Universidad Autónoma de Madrid, 28049 Madrid, Spain
{carlos.alaiz,a.fernandez,jose.dorronsoro}@uam.es,
yvonne.gala@estudiante.uam.es

Abstract. Spectral Clustering and Diffusion Maps are currently the leading methods for advanced clustering or dimensionality reduction. However, they require the eigenanalysis of a sample's graph Laplacian L, something very costly for moderately sized samples and prohibitive for very large ones. We propose to build a low rank approximation to L using essentially the centroids obtained applying kernel K-means over the similarity matrix. We call this approach kernel KASP (kKASP) as it follows the KASP procedure of Yan *et al.* but coupling centroid selection with the local geometry defined by the similarity matrix. As we shall see, kKASP's reconstructions are competitive with KASP's ones, particularly in the low rank range.

Keywords: Nyström Extension, Kernel K-Means, Diffusion Maps, Subsampling.

1 Introduction

Spectral Clustering (SC,[8]) and Diffusion Maps (DM,[5]) are probably the leading current methods to perform advanced clustering or dimensionality reduction. Among their many appeals stand out an elegant underlying theory, the possibility of relatively simple implementations, their application in an unsupervised setting and the very good results they often give. However, they require in principle the eigenanalysis of the graph's Laplacian (SC) or a graph's Markov matrix (DM), that have the same dimensionality as the sample size. Thus, a first drawback is the cost of the eigenanalysis required, with complexity $O(N^3)$ (with N the sample size), too expensive for moderately sized samples and impossible for very large ones. A second drawback is the difficulty of computing the SC or DM projections of new patterns, as these projections are eigenvector components.

There is a large body of literature dealing with these two issues. The Nyström extension [4] is the standard approach to compute the spectral coordinates of a new pattern. To deal with costs, a common approach is to subsample the original

---

[*] With partial support from Spain's grant TIN2010-21575-C02-01 and the UAM–ADIC Chair for Machine Learning.

E. Corchado et al. (Eds.): IDEAL 2014, LNCS 8669, pp. 239–246, 2014.

patterns retaining a small subset that is used to define a first embedding, which is then extended to the entire sample. Among the state of the art methods is the KASP procedure in [13], where standard K-means is used to build a set of representative centroids over which spectral clustering is done. Our proposal, kKASP, builds on it. Related to this and also relevant to this work are methods that build low rank approximations to the similarity matrix [7]; this has been also done for kernel-based learning [12]. An important recent contribution is also [2], which not only suggests subsampling procedures but also defines a metric to compare the quality of a Nyström based reconstruction, that we will use here.

As suggested in [13], a drawback of subsampling methods is that they often do not exploit the information that the similarity matrix provides. While several approaches can be followed to build this matrix, the standard one is to use a kernel to define the similarity between two patterns. In turn, a powerful and well known extension of K-means is kernel K-means (kKM; [10]). Thus a natural idea that we pursue here is to replace standard K-means as used in the already mentioned KASP procedure with kernel K-means. Hence, in our proposal, that we call kernel KASP (kKASP), instead of working with Euclidean pattern distances in the sample, we project its patterns using a kernel (which can be the one used to build the similarity matrix that we want to subsample) into a high dimensional space where K-means is performed. Although we will not pursue it here, we point out the deeper relationship between kKM and normalized cut spectral clustering; indeed they are essentially equivalent [6].

While in principle the projection kernel could be decoupled from that of the graph's similarity matrix, to use the same kernel for both tasks is a very natural choice in SC and DM. In fact, doing so not only simplifies computations but links patterns' similarity with distances in projected space. Moreover, assuming as we shall do, Gaussian kernels, it is easy to see that a high similarity between two patterns is equivalent to them being close in projected space. Thus, deriving the subsample from the kKM centroids, links explicitly pattern sampling with the graph structure upon which SC and DM are built and, as we shall see, yields accurate low rank matrix reconstructions of the graph's Laplacian (or the Markov matrix P) they use. In summary our kKASP model is a simple and natural kernel variant of KASP that, to the best of our knowledge, has not been considered before. While our kKASP model can be used to work out SC/DM clustering or dimensionality reduction on its entirely, we will limit ourselves here to compare the quality of kKASP's low rank reconstruction with that of random subsampling and KASP. For this we use the metric proposed in [2] and, as we shall see, kKM-based subsampling yields a reconstruction of the Markov matrix P competitive with KASP and even slightly better in some cases.

The paper is organized as follows. In Sect. 2 we briefly review SC and DMs; while we recall that they will not be further used, we stress that our goal is to achieve a good approximation to the Markov matrix that is at the heart of both methods and, thus, their brief description. We shall also describe in this section Nyström's extension and the quality measure proposed in [2]. In Sect. 3 we review kernel K-means (kKM) and describe our kKASP approach. Since kKM centroids are known only implicitly but we actually need a subsample, we will approximate kKM centroids with the closest sample point of their clusters. Section 4 contains the experimental results and the paper ends with a short discussion and pointers to further work.

2   SC, DM and Nyström's Formula

As previously introduced, Spectral Clustering (SC,[8,9,11]) and Diffusion Maps (DM,[5]) are manifold learning methods for clustering and dimensionality reduction that have received recently a very large attention. In its first versions, SC looked for clusters in a sample $S = \{x_1, \ldots, x_N\}$ by building an appropriate local similarity matrix W in such a way that the clusters were the connected components of the weighted graph G defined by W. These were then identified analyzing the eigenstructure of a suitable normalization of the graph's Laplacian $L = D - W$ with D the diagonal degree matrix (i.e., $D_{ii} = d_i = \sum_j w_{ij}$). For instance, working as we do here with the random walk Laplacian $L_{\mathrm{rw}} = D^{-1}L = I - D^{-1}W = I - P$, the number C of connected components of G (and hence, clusters) is given by the multiplicity of the smallest eigenvalue (equal to 0) of L_{rw} and the clusters are found applying K-means over the spectral projections $v(x_i) = (v_i^1, \ldots, v_i^m)^\top$ of a sample point x_i, where m is a suitably chosen projection dimension, $\{v^p\}_{p=0}^{N-1}$ are the right eigenvectors of L_{rw} or, equivalently, P, and we drop the first eigenvalue $\lambda_0 = 0$ and its constant eigenvector $(1, \ldots, 1)^\top$. Notice that for this to work it is no longer required the similarity to be local, as the graph G could be indeed connected, something that will happen if the similarity matrix verifies $w_{ij} > 0$. Moreover, the SC coordinates $(v_i^1, \ldots, v_i^m)^\top$ can also be used for dimensionality reduction purposes [3].

Diffusion Maps add two changes to SC. First, we select an α value, $0 \leq \alpha \leq 1$, to normalize W into $W^{(\alpha)}$ with $w_{ij}^{(\alpha)} = w_{ij}/d_i^\alpha d_j^\alpha$, and define then the Markov probability matrix $P^{(\alpha)} = (D^{(\alpha)})^{-1}W^{(\alpha)}$ on the graph G. As shown in [5], α reflects the role of the underlying and unknown sample density. For instance, when $\alpha = 0$, we go back to the previously discussed normalized graph Laplacian. A better choice could be $\alpha = 1$ for which it can be shown that defining the similarity matrix through a radial kernel, the infinitesimal generator of the diffusion process coincides with the manifold's Laplace–Beltrami operator and, hence, captures the geometry of the sample's underlying manifold. In any case, we drop from now on the α parameter and write just P instead of $P^{(\alpha)}$.

The second addition is the definition of a diffusion distance [5] for t steps

$$D^t(x_i, x_j)^2 = \int_z \left(p^t(x_i, z) - p^t(x_j, z)\right)^2 \frac{dz}{\phi(z)},$$

where ϕ is the stationary distribution of P. The quantity $D^t(x_i, x_j)$ will be small if $p^t(x_i, z) \simeq p^t(x_j, z)$ for all z, i.e., there is large probability of going from x_i to x_j and vice-versa. Now it turns out that $D^t(x_i, x_j)^2$ can be expressed as $D^t(x_i, x_j)^2 = \sum_{k=1}^{N-1} \lambda_k^{2t}(v_i^k - v_j^k)^2$, with v^k again the right eigenvectors of P and we drop v^0. Notice that the eigenvalues λ_k of P that DM uses are $\lambda_k = 1 - \hat{\lambda}_k$, with $\hat{\lambda}_k$ the eigenvalues of L that are used in SC. Setting now $\Psi^t(x_i) = (\lambda_1^t v_i^1, \ldots, \lambda_{N-1}^t v_i^{N-1})^\top$, $D^t(x_i, x_j)$ is just the Euclidean distance between $\Psi^t(x_i)$ and $\Psi^t(x_j)$. Since λ_j decrease, dimensionality reduction can be easily achieved by simply retaining the first m components of Ψ and dropping the others that correspond to small eigenvalues. Moreover, as the diffusion distance is approximated by Euclidean distance in $\mathbb{R}^m$, it is natural to use the K-means algorithm if clustering is sought. In summary, DM lends itself very naturally to dimensionality reduction and clustering.

SC and DM share two drawbacks. The first one is the eigenanalysis they require, which will be very costly for moderate size matrices and impossible for large ones. The second is the difficulty of computing the SC or DM projections of new, unseen patterns. Both can be dealt with using Nyström extension [2,4]. Assume a symmetric positive semidefinite kernel $a(x_i, x_j)$ and its kernel matrix A, i.e., $a_{ij} = a(x_i, x_j)$ for which we assume the eigendecomposition $AU = U\Lambda$ with U orthonormal. Then we can define the Nyström extension to a new pattern x as the approximation $\tilde{u}^k(x)$ to the true $u^k(x)$ given by

$$\tilde{u}^k(x) = \frac{1}{\lambda_k^u} \sum_{j=1}^{N} a(x, x_j) u_j^k. \tag{1}$$

To extend this to the asymmetric cases of SC and DM, recall that we have $P = D^{-1}W$. If we set $A = D^{1/2}PD^{-1/2} = D^{-1/2}WD^{-1/2}$, then A is symmetric, we have again $AU = U\Lambda$ and setting now $V = D^{-1/2}U$ we have $PV = V\Lambda$ and applying the extension (1) of A's eigenvectors u is easy to see that $\tilde{v}^k(x) = \frac{1}{\lambda_k} \sum_{j=1}^{N} P(x, x_j) v_j^k$.

Nyström extension can be used to build a low rank approximation first to W and then to P. Consider the representations

$$W = \begin{pmatrix} \tilde{W} & B^\top \\ B & C \end{pmatrix}, \quad D = \begin{pmatrix} \tilde{D} & 0 \\ 0 & \hat{D} \end{pmatrix}, \quad P = D^{-1}W = \begin{pmatrix} \tilde{P} & B'_P \\ B_P & C_P \end{pmatrix},$$

with $\tilde{W}$ the $K \times K$ similarity of a K pattern subsample $\tilde{S}$, $B_P = \hat{D}^{-1}B$, $C_P = \hat{D}^{-1}C$, $B'_P = \tilde{D}^{-1}B^\top$ and $\tilde{P} = \tilde{D}^{-1}\tilde{W}$. Let $W = U\Lambda U^\top$ and $\tilde{W} = \tilde{U}\tilde{\Lambda}\tilde{U}^\top$ be the eigendecompositions of W and $\tilde{W}$. It is easy to see that (1) yields $\mathcal{U} = \begin{pmatrix} \tilde{U} \\ B\tilde{U}\tilde{\Lambda}^{-1} \end{pmatrix}$ as the approximation to the first K eigenvectors U of W which, in turn, we can use to build the $\tilde{S}$-based approximations W' and $P' = D^{-1}W'$ to the entire W and P, namely

$$W' = \mathcal{U}\tilde{\Lambda}\mathcal{U}^\top = \begin{pmatrix} \tilde{W} & B^\top \\ B & B\tilde{W}^{-1}B^\top \end{pmatrix}, \quad P' = \begin{pmatrix} \tilde{P} & B'_P \\ B_P & B_P\tilde{P}^{-1}B'_P \end{pmatrix},$$

where in W', $C' = B\tilde{W}^{-1}B^\top$, and then for P' we have $C_P = \hat{D}^{-1}C' = \hat{D}^{-1}B\tilde{W}^{-1}\tilde{D}\tilde{D}^{-1}B^\top = B_P\tilde{P}^{-1}B'_P$. We call P' to the Nyström's approximation to the full Markov matrix P and we shall use the Frobenius norm of $P - P'$ (or, equivalently, $L - L'$), given by

$$d_F(L, L') = d_F(P, P') = \|P - P'\|_F = \|C_P - B_P\tilde{P}^{-1}B'_P\|_F,$$

to define the reconstruction error for comparing different ways of selecting $\tilde{S}$.

Algorithm 1. kKASP Algorithm.

Input: $S = (x_1, \ldots, x_N)$; K, the subsample size.

1 Apply kernel K-means on S and select K pseudo-centroids $\tilde{S}^K = \{z_1, \ldots, z_K\}$;

2 Perform the eigenanalysis of the matrix P^K associated to $\tilde{S}^K$;

3 Compute Nyström extensions $\tilde{V}^K$ as described in Sect. 2;

4 If desired, perform dimensionality reduction on the $\tilde{V}^K$ and clustering;

---

3 Kernel KASP

Standard (Euclidean distance-based) K-means proceeds by choosing K initial centroids $\{C_k^0\}_{k=1}^K$, associating sample patterns x_p to their nearest centroid and building a first set of clusters $\{\mathcal{C}_k^0\}_{k=1}^K$, with $x_p \in \mathcal{C}_k^0$ if $k = \arg\min_\ell \|x_p - C_\ell^0\|$. The new centroids C_k^1 are now the means of the $\mathcal{C}_k^0$ which, in turn, are used as before to define a new set of clusters $\mathcal{C}_k^1$ and of centroids C_k^2. We proceed iteratively until no sample point changes its centroids. At each step i we have

$$\sum_{x_p \in \mathcal{C}_k^i} \|x_p - C_k^{i+1}\|^2 = \min_Z \sum_{x_p \in \mathcal{C}_k^i} \|x_p - Z\|^2, \quad \forall k.$$

Thus, the iterations progressively minimize the within cluster sum of squares $\sum_{k=1}^K \sum_{x_p \in \mathcal{C}_k^i} \|x_p - C_k^i\|^2$. The basic K-means can be enhanced in a kernel setting replacing the original sample patterns x by non linear extensions $\Phi(x)$ where, if Φ corresponds to a reproducing kernel $\mathcal{K}$, we do not have to work explicitly with the $\Phi(x)$ extensions, since the distances $\|\Phi(x_p) - C_k^i\|^2$ are given by

$$\|\Phi(x_p) - C_k^i\|^2 = \Phi(x_p) \cdot \Phi(x_p) + C_k^i \cdot C_k^i - 2\Phi(x_p) \cdot C_k^i$$

$$= \mathcal{K}(x_p, x_p) + \frac{1}{|\mathcal{C}_k^i|^2} \sum_{x_q, x_r \in \mathcal{C}_k^i} \mathcal{K}(x_q, x_r) - \frac{2}{|\mathcal{C}_k^i|} \sum_{x_q \in \mathcal{C}_k^i} \mathcal{K}(x_p, x_q).$$

Therefore, the previous Euclidean K-means procedure extends straightforwardly to a kernel setting [10].

As mentioned, we follow the KASP procedure in [13] but applying kernel K-means with the similarity W as the kernel matrix. We call our procedure kernel KASP (kKASP). Notice that in kKASP the centroids are only known implicitly. Since we need concrete subsample patterns, we shall retain K pseudo-centroids, i.e., the sample points x_p^k in the final clusters $\mathcal{C}_k$ closest to the C^k centroids. Algorithm 1 summarizes our procedure.

The complexity analysis of the preceding is easy. The cost of the first step is $O(KNI)$, with I the number of iterations, plus the cost $O(N^2)$ of pre-computing the similarity matrix; the eigenanalysis of P^K has a $O(K^3)$ cost and that of the Nyström extensions is $O(KN)$. This is obviously more efficient than a straight SC or DM approach over the entire sample, whose eigenanalysis of the complete singularity matrix would have a complexity of $O(N^3)$.

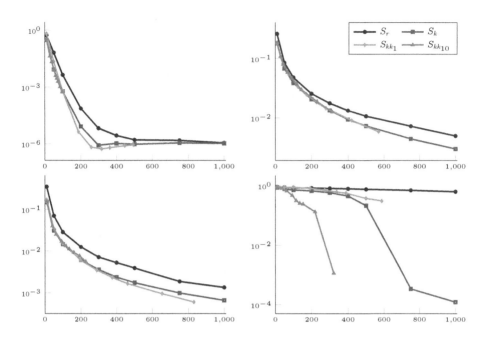

Fig. 1. Reconstruction error (median) in function of the number of patterns for *Fishbowl* (top left), *Musk* (top right), *Digits* (bottom left) and *Image* (bottom right)

4   Numerical Experiments

In this section we compare experimentally different approaches to select a subsample over which the Nyström extension is applied. Recall that the distance $d_F(P, P')$ is used as a quality measure, where $P = D^{-1}W$ is the transition probability matrix of SC (and that of DM with $\alpha = 0$). The similarity W is defined over a Gaussian kernel where we have set the width parameter σ as the 10% percentile of all the distances. We consider four approaches: a random selection of patterns (S_r), a KASP selection using the pseudo-centroids (S_k) and two kKASP selections using a kernel with two different values of σ: the percentiles 1% and 10%. With the first one (S_{kk_1}) the model is more local and thus it provides more clusters, whereas in the second, $S_{kk_{10}}$, the kernel matrix is precisely the similarity matrix W defined above. We consider different subsample sizes, namely 10, 50, 100, 200, 300, 400, 500, 750 and 1,000. For S_r and S_k these are the final sizes but, for S_{kk_1} and $S_{kk_{10}}$, the kernel K–means implementation we use can collapse some of the clusters to 0 patterns; thus the subsample can be smaller.

We consider four examples. The first one is the synthetic fish-bowl of [2], where we generate uniform random points on the unit sphere, and then remove all points over a height of 0.7; we build this way 10,000 three-dimensional sample patterns. The other three are classification problems that come from the UCI database [1] and are also used in [13]: the musk problem (with 6,598 patterns and 166 features representing musk and non-musk molecules), the pen-based

Table 1. Summary of the results for S_k and S_{kk_1}. For each dataset and subsample size, the first two rows include medians and deviations of the reconstruction errors for S_k and S_{kk_1}, and the third row medians and deviations of the final sizes of S_{kk_1}.

	Size	50	100	200	500	1,000
Fishb.	Error S_k	$8.3\text{E}^{-3}\pm5.9\text{E}^{-4}$	$5.9\text{E}^{-4}\pm1.4\text{E}^{-4}$	$8.2\text{E}^{-6}\pm4.0\text{E}^{-6}$	$9.5\text{E}^{-7}\pm4.0\text{E}^{-7}$	$1.1\text{E}^{-6}\pm1.6\text{E}^{-7}$
	Error S_{kk_1}	$2.2\text{E}^{-2}\pm4.0\text{E}^{-3}$	$6.4\text{E}^{-4}\pm1.1\text{E}^{-4}$	$4.3\text{E}^{-6}\pm1.4\text{E}^{-6}$	$6.3\text{E}^{-7}\pm1.7\text{E}^{-7}$	$8.8\text{E}^{-7}\pm2.5\text{E}^{-7}$
	Size S_{kk_1}	50 ± 0	99 ± 0	186 ± 2	363 ± 6	493 ± 7
Musk	Error S_k	$6.8\text{E}^{-2}\pm1.5\text{E}^{-3}$	$3.9\text{E}^{-2}\pm1.5\text{E}^{-3}$	$2.1\text{E}^{-2}\pm6.9\text{E}^{-4}$	$7.3\text{E}^{-3}\pm3.3\text{E}^{-4}$	$3.0\text{E}^{-3}\pm9.4\text{E}^{-5}$
	Error S_{kk_1}	$8.2\text{E}^{-2}\pm4.9\text{E}^{-3}$	$5.6\text{E}^{-2}\pm2.3\text{E}^{-3}$	$3.4\text{E}^{-2}\pm1.7\text{E}^{-3}$	$1.5\text{E}^{-2}\pm6.2\text{E}^{-4}$	$5.8\text{E}^{-3}\pm1.9\text{E}^{-4}$
	Size S_{kk_1}	42 ± 2	73 ± 3	124 ± 6	279 ± 6	570 ± 8
Digits	Error S_k	$3.1\text{E}^{-2}\pm1.1\text{E}^{-3}$	$1.5\text{E}^{-2}\pm6.9\text{E}^{-4}$	$6.0\text{E}^{-3}\pm3.8\text{E}^{-4}$	$1.7\text{E}^{-3}\pm8.0\text{E}^{-5}$	$6.6\text{E}^{-4}\pm5.9\text{E}^{-5}$
	Error S_{kk_1}	$3.4\text{E}^{-2}\pm1.1\text{E}^{-3}$	$1.5\text{E}^{-2}\pm7.5\text{E}^{-4}$	$6.1\text{E}^{-3}\pm1.5\text{E}^{-4}$	$1.7\text{E}^{-3}\pm6.6\text{E}^{-5}$	$6.0\text{E}^{-4}\pm2.6\text{E}^{-5}$
	Size S_{kk_1}	50 ± 0	100 ± 0	196 ± 1	460 ± 4	831 ± 7
Image	Error S_k	$7.8\text{E}^{-1}\pm1.6\text{E}^{-2}$	$7.3\text{E}^{-1}\pm2.0\text{E}^{-4}$	$6.7\text{E}^{-1}\pm3.6\text{E}^{-2}$	$2.1\text{E}^{-1}\pm6.7\text{E}^{-2}$	$1.2\text{E}^{-4}\pm4.2\text{E}^{-6}$
	Error S_{kk_1}	$8.9\text{E}^{-1}\pm1.0\text{E}^{-4}$	$8.9\text{E}^{-1}\pm4.0\text{E}^{-4}$	$8.2\text{E}^{-1}\pm3.4\text{E}^{-2}$	$5.7\text{E}^{-1}\pm6.5\text{E}^{-2}$	$3.1\text{E}^{-1}\pm1.0\text{E}^{-1}$
	Size S_{kk_1}	50 ± 0	100 ± 0	192 ± 2	389 ± 5	587 ± 7

handwritten digits recognition problem (with $10,992$ instances of dimension 16) and the image segmentation problem (with $2,310$ randomized instances of dimension 19 from a database of 7 outdoor images).

Given K-means dependence on the initial centroids, we repeat each experiment 25 times starting at random different C_k^0, and Fig. 1 shows for each requested subsample the medians of the subsample sizes obtained and of the reconstruction errors; we use the median and median absolute deviation as robust variants of the average and standard deviation in account of kKM's dependence on its initialization. Obviously this error will decrease as the sizes grow (being 0 for a full sample). As can be seen, S_{kk_1} performs slightly better than S_k (with the same subsample size), except in the image problem, where S_k's error descends quickly with more than 500 patterns (i.e., a relatively large 21.5% of sample size), whereas the decrease of S_{kk_1}'s error is much slower. $S_{kk_{10}}$ gives a good performance for low ranks (similar to S_k and S_{kk_1} and even better for Image) but it quickly collapses centroids and the subsample size attainable is thus smaller. As expected, S_r is the weakest performer, as it does not use any information about the problem. Table 1 gives a more detailed comparison between S_k and S_{kk_1}, including also the deviations and reinforcing the previous discussion. The header gives the number of S_k centroids whereas the last line one of each example those of S_{kk_1}, which are usually smaller. This has to be considered when comparing the S_k and S_{kk_1} reconstruction errors (e.g., the *Image* error of S_{kk_1} in the $1,000$ centroid column should be compared with the S_k error for 500 centroids).

5   Discussion and Conclusions

In this work we have introduced kKASP, an extension of the K-means-based approximate spectral clustering procedure proposed in [13] to a kernel framework that couples the kernel clustering matrix with the graph's similarity matrix and uses the resulting pseudo-centroids to compute a low rank L' approximation to the graph's random walk Laplacian L. We have compared the approximations of kKASP with that of KASP over four datasets, obtaining promising results in terms of the reconstruction error between L and L'.

In any case this is a first step as the main goals of SC and DM are dimensionality reduction and clustering, for which the reconstruction error is just an initial metric that has to be refined. In fact, while this error decreases as the rank of L' grows, the region of interest is that of low rank values, where reconstruction error will be larger. Therefore, other quality measurements closer to the problem at hand should be used, such as for example confusion matrices comparing a full SC or DM clustering with their low rank counterparts. Moreover, from a DM point of view our results are given for the $\alpha = 0$ case, while $\alpha = 1$ is often a better choice. We are currently working on these and other related issues.

References

1. Bache, K., Lichman, M.: UCI machine learning repository (2013), http://archive.ics.uci.edu/ml
2. Belabbas, M., Wolfe, P.: Spectral methods in machine learning and new strategies for very large datasets. Proceedings of the National Academy of Sciences 106(2), 369–374 (2009)
3. Belkin, M., Nyogi, P.: Laplacian Eigenmaps for dimensionality reduction and data representation. Neural Computation 15(6), 1373–1396 (2003)
4. Bengio, Y., Delalleau, O., Roux, N.L., Paiement, J., Vincent, P., Ouimet, M.: Learning eigenfunctions links Spectral Embedding and Kernel PCA. Neural Computation 16(10), 2197–2219 (2004)
5. Coifman, R., Lafon, S.: Diffusion Maps. Applied and Computational Harmonic Analysis 21(1), 5–30 (2006)
6. Dhillon, I., Guan, Y., Kulis, B.: Kernel k-means: Spectral clustering and normalized cuts. In: Proceedings of the 10th ACM SIGKDD International Conference on Knowledge Discovery and Data Mining, pp. 551–556. ACM, New York (2004)
7. Fowlkes, C., Belongie, S., Chung, F., Malik, J.: Spectral grouping using the nyström method. IEEE Trans. Pattern Anal. Mach. Intell. 26(2), 214–225 (2004)
8. Luxburg, U.: A tutorial on spectral clustering. Statistics and Computing 17(4), 395–416 (2007)
9. Ng, A., Jordan, M., Weiss, Y.: On spectral clustering: Analysis and an algorithm. In: Advances in Neural Information Processing Systems, pp. 849–856. MIT Press (2001)
10. Schölkopf, B., Smola, A., Müller, K.R.: Nonlinear component analysis as a kernel eigenvalue problem. Neural Comput. 10(5), 1299–1319 (1998)
11. Shi, J., Malik, J.: Normalized cuts and image segmentation. IEEE Transactions on Pattern Analysis and Machine Intelligence 22(6), 888–905 (2000)
12. Williams, C., Seeger, M.: Using the nyström method to speed up kernel machines. In: Advances in Neural Information Processing Systems 13, pp. 682–688. MIT Press (2001)
13. Yan, D., Huang, L., Jordan, M.: Fast approximate spectral clustering. In: Proceedings of the 15th ACM SIGKDD International Conference on Knowledge Discovery and Data Mining, pp. 907–916. ACM, New York (2009)

Linear Regression Fisher Discrimination Dictionary Learning for Hyperspectral Image Classification

Liang Chen[1], Ming Yang[1,*], Cheng Deng[1], and Hujun Yin[2]

[1] School of Computer Science and Technology,
Nanjing Normal University, Nanjing, P.R. China
[2] School of Electrical and Electronic Engineering,
The University of Manchester, Manchester, M139PL, UK
myang@njnu.edu.cn

Abstract. In this paper, we propose a novel dictionary learning method for hyperspectral image classification. The proposed method, linear regression Fisher discrimination dictionary learning (LRFDDL), obtains a more discriminative dictionary and a classifier by incorporating linear regression term and the Fisher discrimination into the objective function during training. The linear regression term makes predicted and actual labels as close as possible; while the Fisher discrimination is imposed on the sparse codes so that they have small with-class scatters but large between-class scatters. Experiments show that LRFDDL significantly improves the performances of hyperspectral image classification.

Keywords: Dictionary Learning, Linear Regression, Fisher Discrimination, Hyperspectral Image Classification.

1 Introduction

Hyperspectral image (HSI) classification has become a research hotspot for HSI processing, where pixels are given labels based on their spectral characteristics. The methods of HSI classification can be divided into three categories: supervised [1], unsupervised [2] and semi-supervised [3]. The recent HSI classification techniques utilize the spatial information to improve the performances of classification and can be found in [4-5].

Recently, sparse coding has been adopted to solve the problems in computer vision and image analysis [6-7], and it has also been successfully applied in HSI classification [8-11], relying on the fact that an unknown hyperspectral pixel can be represented by a few training samples (atoms) from a given dictionary and classified based on the sparse code. The dictionaries in [8-11] consist of training samples collected randomly from the image of interest. However, a more robust dictionary can be learned from training samples [12-13]. The dictionary learning (DL) methods in [12-13] only take the reconstructive error into consideration, so they generate

---

* Corresponding author.

E. Corchado et al. (Eds.): IDEAL 2014, LNCS 8669, pp. 247–254, 2014.

reconstructive sparse codes. But for the task of classification, it would be better to use discriminative sparse codes.

DL aims to learn a dictionary from training samples, which can better represent or code the samples. Recently, dictionary learning methods pay attention on the discrimination of the learned dictionary for classification. Several dictionary learning methods incorporate discriminative terms into the objective function, e.g. [14-17]. The discrimination criteria include linear predictive classification error [14], Fisher discrimination criterion [15], label consistent [16] and bilinear discrimination [17].

We here propose a novel dictionary learning method, linear regression Fisher discrimination dictionary learning (LRFDDL), which is able to generate a discriminative dictionary and a classifier. The linear regression term makes predicted and actual labels as close as possible, while the Fisher discrimination criterion is imposed on the sparse codes so that they have small within-class scatters but large between-class scatters. Small with-class scatters make the sparse codes belonged to the same class be more compact and large between-class scatters keep the sparse codes of different classes apart. Thus the sparse codes can be separated easily. Besides, the discrimination of the learned dictionary can be represented by the sparse codes. In this manor, we obtain a discriminative dictionary and sparse codes. Then we utilize the learned dictionary and classifier to improve the performances of hyperspectral image classification.

The rest of this paper is organized as follows. Section 2 gives a brief review of the related work. Section 3 provides detailed explanations of our method. Experimental results are given in section 4. Section 5 gives related conclusions.

2   Related Work

Ref. [10] proposed a classification method based on sparse representation for hyperspectral image classification with a few labeled samples. Given a test pixel y, the algorithm of the method is as follows:

i) Sparsely code y over the dictionary A via l_1-norm minimization

$$\hat{x} = \arg \min_x \left\{ \| y - Ax \|_2^2 + \lambda \| x \|_1 \right\} \tag{1}$$

where λ is a constant scalar.

ii) Classification

$$class(y) = \arg \min_i \left\{ e_i \right\} \tag{2}$$

where $e_i = \| y - A_i \hat{x}_i \|_2$.

Dictionary learning techniques [12-13] are applied in hyperspectral image classification. But these dictionary learning methods in [12-13] only take the reconstructive errors into consideration, which are not suitable for classification. Hence, we propose a novel dictionary learning method, which generates a discriminative dictionary as well as a classifier for classification in section 3.

3   Linear Regression Fisher Discrimination Dictionary Learning for Hyperspectral Image Classification

3.1   Linear Regression Fisher Discrimination Dictionary Learning (LRFDDL)

In this session, we propose a novel dictionary learning model, named linear regression Fisher discrimination dictionary learning (LRFDDL). We first give some notations below.

Denote by $A = [A_1, A_2, ..., A_c] \in R^{d \times n}$ the set of training samples, where A_i is the training samples of class i. Dictionary learning aims to learn a dictionary $D = [D_1, D_2, ..., D_c] \in R^{d \times k}$. Denote by $X = [X_1, X_2, ..., X_c] \in R^{k \times n}$ the coding coefficient matrix of A coded over D.

Our LRFDDL model is expressed as follows

$$J_{(D,X,W)} = \arg\min_{(D,X,W)} \left\{ \|A - DX\|_F^2 + \lambda \|X\|_1 + \alpha f(X) + \beta t(X) \right\} \qquad (3)$$

where the first term is the reconstruction error; the second term is the sparsity regularization term; the third term is the Fisher discrimination imposed on the coefficient matrix X; the forth term is the linear regression term. Next, we shall explain the last two terms in details.

Linear Regression Term. We aim to learn a more suitable classifier for classification. Hence, we hope the predictive label based on the sparse code is the same with the actual label as much as possible. So as in [14, 16], we include the classification error as a term in the objective function for dictionary learning. Here we use a linear predictive classifier $f(x; W) = Wx$, so $t(X)$ is defined as

$$t(X) = \|H - WX\|_F^2 \qquad (4)$$

where $H = [h_1 ... h_n] \in R^{c \times n}$ are the class labels of the input samples A. $h_i = [0, 0 ... 1 ... 0, 0]^T \in R^c$ is a label vector corresponding to an input sample y_i, where the non-zero position indicates the class of y_i.

Fisher Discrimination Term. The discrimination of the learned dictionary can be represented by the sparse codes, so we can make sparse codes be discriminative. Motivated by Ref. [15], we impose Fisher discrimination criterion on coding coefficient matrix X to make the sparse codes and dictionary be discriminative by minimizing the with-class scatters and maximizing the between-class scatters.

Let $S_W(X)$ and $S_B(X)$ denote the with-class scatter matrix and the between-class scatter matrix of X, they are defined as:

$$S_w(X) = \sum_{i=1}^{C} \sum_{x_j \in X_i} (x_j - m_i)(x_j - m_i)^T$$
$$S_B(X) = \sum_{i=1}^{C} n_i (m_i - m)(m_i - m)^T$$

(5)

where m_i and m are the mean vector of X_i and X respectively, and n_i is the number of samples in class A_i.

We adopt the Fisher discrimination term $f(X)$ similar to Ref. [15] as follows:

$$f(X) = tr(S_W(X)) - tr(S_B(X)) + \eta \|X\|_F^2$$

(6)

Note that $\|X\|_F^2$ is to make $f(X)$ convex and η is set as 1 according to [15].

In summary, the objective function of LRFDDL is expressed as:

$$J_{(D,X,W)} = \underset{(D,X,W)}{\arg\min} \left\{ \begin{array}{l} \|A - DX\|_F^2 + \lambda\|X\|_1 + \alpha\left(tr(S_W(X) - S_B(X)) + \eta\|X\|_F^2\right) \\ + \beta\|H - WX\|_F^2 \end{array} \right\}$$

(7)

Optimization. The problem in Eq. (7) could be solved by iteratively optimizing the dictionary D, sparse codes X and classifier parameters W.

Firstly, we fix D and W, and optimize X. Here we compute X_i class by class, and the objective function in Eq.(7) is reduced to:

$$J_{(X_i)} = \underset{(X_i)}{\arg\min} \left\{ \begin{array}{l} \|A_i - DX_i\|_F^2 + \lambda\|X_i\|_1 + \alpha\left(\|X_i - M_i\|_F^2 - \sum_{j=1}^{c}\|M_j - M\|_F^2 + \eta\|X_i\|_F^2\right) \\ + \beta\|H_i - WX_i\|_F^2 \end{array} \right\}$$

(8)

where $M_j = [m_j, m_j, \ldots, m_j] \in R^{k \times n_j}$ and $M = [m, m, \ldots, m] \in R^{k \times n_j}$ (m_j and m are the mean vectors of X_i and X, respectively). The Iterative Projection Method (IPM) in [18] is employed to solve Eq.(8).

Secondly, we fix X and W, and optimize D. We also update D_i class by class. The objective function in Eq.(7) is reduced to:

$$J_{(D_i)} = \arg\min_{(D_i)} \left\| A - D_i X^i - \sum\nolimits_{j=1, j \neq i}^{c} D_j X^j \right\|_F^2 \qquad (9)$$

We adopt the method proposed in [19] to solve the above problem.

Finally, we fix X and D, and optimize W. We use the ridge regression model [20] and obtain the following solution:

$$W = HX^T \left(XX^T + \lambda_1 I \right)^{-1} \qquad (10)$$

3.2   Hyperspectral Image Classification Based on LRFDDL

Given a learned dictionary D and the classification parameters W, classification of an unlabeled pixel y follows the following procedures.

i) obtain the sparse code $\hat{x}$ over D by solving the optimization problem in Eq.(1);

ii) classify y via

$$class(y) = \arg\max_i l_i \qquad (11)$$

where $l = W\hat{x}$.

Next, we apply the method on hyperspectral images to verify the effectiveness of the method.

4   Experiments and Results

To verify the effectiveness of the proposed method, we test the proposed method on two hyperspectral image sets, the University of Pavia and the Center of Pavia images. They are urban images acquired by the Reflective Optics System Imaging Spectrometer (ROSIS). The ROSIS sensor generates 115 spectral bands ranging from 0.43 to 0.86 um and has a spatial resolution of 1.3m per pixel.

4.1   The University of Pavia Images

The University of Pavia Images consist of 610×340 pixels, each having 103 bands, with the 12 most noisy bands removed, as shown in Fig. 1. There are nine ground-truth classes of interest. We chose 200 pixels for each class for training and the rest for testing. We repeated the procedure 10 times and the average results for different methods are reported. The overall accuracy (OA) is the ratio between correctly classi-fied test pixels and the total number of test pixels, and the kappa coefficient (κ) is a robust measure of the degree of agreement.

Fig. 1. For the University of Pavia images: (a) Original the University of Pavia images and (b) Ground truth map

The classification results are summarized in Table 1. From this table, we can find that LRFDDL performed the best with the highest classification accuracy: 83.74%. It can be seen that LRFDDL improved at least 2% over l_1 and OMP which chose training samples as a dictionary. The OA of LRFDDL is 5.25% higher than DKSVD and 0.63% higher than FDDL. The results show that LRFDDL is effective.

Table 1. Classification accuracy (%) for the University of Pavia images

	SVM	l_1	OMP	DM	DKSVD	FDDL	**LRFDDL**
OA	82.55%	81.62%	77.55%	82.95%	78.49%	83.11%	**83.74%**
K	0.7741	0.7616	0.7057	0.7783	0.7172	0.7803	**0.7894**

4.2   The Center of Pavia Images

The center of Pavia images contain 1096×492 pixels, each having 102 spectral bands after 13 noisy bands are removed. In our experiment, we chose a subset of the images, which contains 480×492 pixels, as shown in Fig.2. There are nine ground-truth classes of interests. We chose 200 pixels for each class for training and the rest for testing. We repeated the procedure 10 times and the average results for different methods are reported.

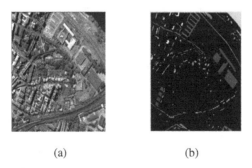

(a)   (b)

Fig. 2. For the Center of Pavia images:(a) Original the Center of Pavia images and (b) Ground truth map

The classification results are summarized in Table 2. From this table, we can find that the LRFDDL achieved the highest classification accuracy: 94.57%. Over l_1 and OMP, LRFDDL improved 0.86% and 5.27%. DM, which generates reconstructive sparse codes, was also inferior to LRFDDL. The OA of LRFDDL is 4.51% higher than DKSVD and 0.43% higher than FDDL. The results show that LRFDDL improves the performances of hyperspectral image classification.

Table 1. Classification accuracy(%) for the Center of Pavia images

	SVM	l_1	OMP	DM	DKSVD	FDDL	**LRFDDL**
OA	93.81%	93.71%	89.30%	93.86%	90.06%	94.14%	**94.57%**
K	0.9226	0.9214	0.8669	0.9233	0.8760	0.9267	**0.9319**

5   Conclusions

A novel dictionary learning method, termed linear regression Fisher discrimination dictionary learning (LRFDDL), is proposed for hyperspectral image classification. LRFDDL aims to obtain both a discriminative dictionary and a classifier via the linear regression term and the Fisher discrimination criterion. The linear regression makes predicted and actual labels as close as possible, making the dictionary and classifier more suitable for classification; while the Fisher discrimination criterion imposed on the sparse code makes the dictionary and sparse codes more discriminative. Promising results of the proposed method are shown on two hyperspectral image sets.

Acknowledgments. This work is supported in part by National Natural Science Foundation of China under Grant No.61272222. Natural Science Foundation of Jiangsu Province of China under Grant No.BK2011782, and Key (Major) Program of Natural Science Foundation of Jiangsu Province of China under No. BK2011005.

References

1. Melgani, F., Bruzzone, L.: Classification of Hyperspectral Remote Sensing Images with Support Vector Machines. IEEE Transactions on Geoscience Remote Sensing 42(8), 1778–1790 (2004)
2. Zhong, Y., Zhang, L., Huang, B., Li, P.: An unsupervised artificial immune classifier for multi/hyper-spectral remote sensing image. IEEE Transactions on Geoscience Remote Sensing 44(2), 420–431 (2006)
3. Yang, L.X., et al.: Semi-Supervised Hyperspectral Image Classification Using Spatio-Spectral Laplacian Support Vector Machine. IEEE Geoscience Remote Sensing Letters 11(3), 651–655 (2014)
4. Kang, X., Li, S., Benediktsson, J.A.: Spectral–Spatial Hyperspectral Image Classification With Edge-Preserving Filtering. IEEE Transactions on Geoscience Remote Sensing 52(5), 2666–2677 (2014)

5. Li, W., Prasad, S., Fowler, J.E.: Hyperspectral Image Classification Using Gaussian Mixture Models and Markov Random Field. IEEE Geoscience Remote Sensing Letters 11(1), 153–157 (2014)
6. Yang, J., Wright, J., Huang, T., Ma, Y.: Image Super resolution as Sparse Representation of Raw Patches. In: IEEE Conference on Computer Vision and Pattern Recognition (2008)
7. Wright, J., Yang, M., Ganesh, A., Sastry, S., Ma, Y.: Robust Face Recognition via Sparse Representation. IEEE Transactions on Pattern Analysis and Machine Intelligence 31(2), 210–227 (2009)
8. Chen, Y., Nasrabadi, N.M., Tran, T.D.: Hyperspectral Image Classification via Kernel Sparse Representation. IEEE Transactions on Geoscience Remote Sensing 51(1), 217–231 (2013)
9. Zhang, H., et al.: A Nonlocal Weighted Joint Sparse Representation Classification Method for Hyperspectral Imagery. IEEE Journal of Selected Topics in Applied Earth Observations and Remote Sensing pp(99), 1–10 (2013)
10. Haq, S.U., et al.: A Fast and Robust Sparse Approach for Hyperspectral Data Classification Using a Few Labeled Samples. IEEE Transactions on Geoscience Remote Sensing 50(6), 2287–2302 (2012)
11. Sun, L., et al.: A Novel Supervised Method for Hyperspectral Image Classification with Spectral-Spatial Constraints. Chinese Journal of Electronics 23(1), 135–141 (2014)
12. Castrodad, A., et al.: Learning Discriminative Sparse Representations for Modeling, Source Separation, and Mapping of Hyperspectral Imagery. IEEE Transactions on Geoscience and Remote Sensing 49(11), 4263–4281 (2011)
13. Zhao, Y., et al.: Hyperspectral Imagery Super-resolution by Sparse Representation and Spectral Regularization. EURASIP Journal on Advances in Signal Processing 2011(1), 1–10 (2011)
14. Zhang, Q., Li, B.: Discriminative K-SVD for Dictionary Learning in Face Recognition. In: IEEE Conference on Computer Vision and Pattern Recognition, pp. 2691–2698 (2010)
15. Yang, M., et al.: Fisher Discrimination Dictionary Learning for Sparse Representation. In: IEEE International Conference on Computer Vision, pp. 543–550 (2011)
16. Jiang, Z., Lin, Z., Davis, L.: Label Consistent K-SVD: Learning a Discriminative Dictionary for Recognition. IEEE Transactions on Pattern Analysis and Machine Intelligence 35(11), 2651–2664 (2013)
17. Liu, H.-D., et al.: Bilinear Discriminative Dictionary Learning for Face Recognition. Pattern Recognition 47(5), 1835–1845 (2014)
18. Rosasco, L., et al.: Iterative Projection Methods for Structured Sparsity Regularization. MIT Technical Reports
19. Yang, M., et al.: Metaface Learning for Sparse Representation based Face Recognition. In: IEEE International Conference on Image Processing, pp. 1601–1604 (2010)
20. Golub, G., Hansen, P., O'leary, D.: Tikhonov regularization and total least squares. SIAM Journal on Matrix Analysis and Applications 21(1), 185–194 (1999)

Ensemble-Distributed Approach in Classification Problem Solution for Intrusion Detection Systems

Vladimir Bukhtoyarov[1] and Vadim Zhukov[2]

[1] Siberian State Aerospace University, Department of Information Technologies Security, Krasnoyarsky Rabochy Av. 31, 660014 Krasnoyarsk, Russia
vladber@list.ru
[2] Siberian State Aerospace University, Department of Information Technologies Security, Krasnoyarsky Rabochy Av. 31, 660014 Krasnoyarsk, Russia
vadimzhukov@mail.ru

Abstract. Network activity has become an essential part of daily life of almost any individual or company. At the same time the number of various network threats and attacks in private and corporate networks is constantly increasing. Therefore, the development of effective methods of intrusion detection is an urgent problem nowadays. In the paper the basic scheme and main steps of the novel ensemble-distributed approach are proposed. This approach can be used to solve a wide range of classification problems. Its scheme is well suited for the problem of intrusion detection in computer networks. Unlike traditional ensemble approaches the proposed approach provides partial obtaining of adaptive solutions by individual classifiers without an ensemble classifier. The proposed approach has been used to solve some test problems. The results are presented in the article. The approach was also tested on a data set KDD Cup '99 and the results confirm the high efficiency of the proposed scheme of ensemble-distributed classification. In comparison with the traditional approaches for distributed intrusion detection systems there is a significant reduction (about 10%) of information flows between distributed individual classifiers and a centralized ensemble classifier. Possible ways of approach improving and possible applications of the proposed collective-distributed scheme are presented in the final part of the article.

Keywords: information security, intrusion detection, ensemble approach, classification.

1   Introduction

Nowadays intrusion detection systems (IDS) as automated information systems for the detection of information security incidents have become an essential part of comprehensive solution for information security problems in automated systems with network infrastructure. Typically intrusion detection systems are software and hardware facilities which perform the functions of automated detection of threats in information systems. There are some requirements for such information security systems. One of them is the following: an intrusion detection system must perform the

E. Corchado et al. (Eds.): IDEAL 2014, LNCS 8669, pp. 255–265, 2014.

analysis of data collected for the purpose of intrusion detection using both signature-based techniques and heuristic methods.

The methods based on data mining technologies (DMT) are some of the most promising and effective techniques for the development of algorithmic support in intrusion detection systems. These methods include artificial neural networks, fuzzy and neural-fuzzy systems, evolutionary algorithms, immune and multi-agent systems [1-3]. The analysis of the use of such techniques as parts of intrusion detection systems and data mining problems as a whole shows that most efficient use of DMT can be achieved by applying several DMTs in a single system. The improvement of problem solution quality (for selected performance criteria, such as the number of errors of the first and second order) is achieved due to a number of factors, such as:

- achieving of synergetic effect by several DMTs. Ensembles of DMTs may be composed of technologies of the same type and technologies of different types;
- the scheme of distributed data analysis with individual predictors and the scheme of their individual solutions integration are in good agreement with the structures of modern automated systems in which they are used.

However, permanent desire to improve the quality of IDS, the increasing number of problems, the increasing demands for performance and the need to develop and to use parallel computing systems lead to the necessity of developing new approaches. A neural network ensemble approach has been a promising and popular approach used to solve classification problems lately. The development of this approach was initiated in the article of Hansen and Salomon [4]. Before that the collective approach using other techniques for classification was applied [5].

In this article we focus on the classification problems and the development and analysis of appropriate neural network ensemble methods for network IDS. We suppose that the improvement of the techniques used to combine experts (participants of the ensemble) decisions may be one of the ways for further development of the ensemble methods for solving problems of classification. Generally, there are many strategies for combining the individual classifier's decisions. In this paper we propose a new three-level approach, which extends the idea of a stacked generalized method developed by Wolpert [6]. In the proposed approach the trained classifiers with continuous outputs placed on the second level are used for the aggregation of individual experts' decisions. Unlike the traditional ensemble approaches the proposed approach can provide adaptive solutions obtaining by individual classifiers in some cases without recourse to the ensemble classifier. Parameters that allow to adjust the frequency of the ensemble classifier involvement are based on the characteristics of a specific problem. In general, any other appropriate technologies, such as fuzzy logic, neural fuzzy and tree classifiers can be used as classifiers. Furthermore one of the fundamental ideas of the proposed approach is the assumption of the possibility of classifiers ensembles formation on the second level. In this case the automated procedure of forming of such effective groups is constructed, and then the solutions of these groups are integrated in the third-level aggregator. It is partially assumed that this approach will improve the generalization capability of the classifier, reduce the influence of noise in training sample by the use of ensembles and the final aggregator. In addition to this attempts are made to improve the adaptability of signatures detection for specific nodes due to the fact that some of the solutions are left for the individual classifiers located in them. To implement these ideas we have proposed the approach described below and explored it on a set of tasks

The proposed approach is described in detail in the last part of Section 2.

Section 3 is devoted to a statistical investigation of the performance of the proposed approach. Finally, we discuss the results of our statistical investigation.

2   Technique Overview

2.1   Ensemble-Distributed Approach for Detecting Incidents of Information Security

In some problems of information security incidents for distributed systems it is possible to achieve better performance by partial solving the problem without evaluation of the solutions using all DMT in the ensemble. In some cases a distinct classifier (agent) efficiently performs detection. A "distinct classifier" is an individual classifier which is placed in a separate node of network infrastructure but participates in the ensemble. One of the ways to improve the efficiency of neural network-based IDS is to develop distributed neural network-based IDS. Such distributed systems increase the load (amount of traffic) in computer networks but allow to avoid some lacks of host-based IDS. In particular the computational load associated with the work of intrusion detection systems is uniformly distributed over the network. As a result, servers load decrease significantly and performance of IDS increases. The distributed scheme allows IDS to work adequately (with good performance) in today's fast speed computer networks. The distributed intrusion detection scheme with the use of neural networks requires a significant adaptation of the neural network approach because it is not quite clear how to distribute neural networks over computer networks nodes.

In this paper we propose the neural network ensemble approach as a basis for distributed intrusion detection systems. This approach assumes that "agents" (relatively simple neural network classifiers) are functioning in each node of the computer network. In most cases these agents independently identify anomalous activity and detect intrusions. In case of "boundary" situations (a single classifier "doubts the solution") defined by a prespecified threshold value the agent obtains the solution from the ensemble stored on the server.

2.2   The Main Steps of Developing the Ensemble-Distributed IDS

Designing of Neural Networks Ensemble Members. The purpose of this phase is to form a set of sufficiently effective neural networks which in most cases can operate as individual agents,that is, they can independently detect the anomalous activities/intrusions. Designing of structures and training of neural networks are performed using above-described methods similarly to the case when neural networks are used in the host-based IDS. In general the total number of neural networks is assumed to be equal to the number of nodes in the computer network in which we suppose to place agents. If there are too many nodes (and correspondingly too many neural networks) computer network nodes can be combined into homogeneous groups according to some criteria. Then a single neural network is designed for each group and a priori knowledge specific for each group may be considered.

We applied a probabilistic approach to design individual neural network classifiers in our experiments [7]. This probabilistic approach called PGNS (Probability based Generator of neural Networks Structures) extends the ideas of EDA-like probability-based genetic algorithm. The method is described in details in [7].

Neural Networks Ensemble Design. The purpose of this stage is to create a neural network ensemble classifier on the basis of neural networks formed at the previous stage. The ensemble of neural networks and the way of evaluating the ensemble classifier solutions are formed with ensuring sufficient intrusion detection efficiency using any appropriate method. There are many ways to design neural network ensembles. We employed the approach called Genetic Programming based ENsembling (GPEN) [7]. The GPEN method automatically constructs a program using the genetic programming operators. This program shows how to combine the component networks predictions in order to get a good ensemble prediction. The GPEN also selects those networks which provide predictions to be taken into consideration by including them into the set of input variables of the program.

Later the neural networks ensemble classifier can be modified in accordance with changing conditions of operation (for example, the emergence of new types of attacks, etc.) using the algorithms that are applied to its construction. The parameters of the neural network ensemble classifier are stored on the server or distributed on several main nodes of computer networks.

Distribution of Ensemble Members on Network Nodes and Independent Operation of the Individual Agents. Obviously the calculation of the neural networks ensemble solution requires more time and computational resources in comparison with the evaluation of the solution using a single neural network from the ensemble. Besides, each neural network which forms the ensemble in most cases provides sufficient intrusion detection efficiency. Therefore single neural networks formed during the first stage are supposed to work partially independently (as "independent agents") in nodes of a computer network . At step 3 neural networks are distributed among nodes in the computer network either in an arbitrary order or with a priori information if it was used at the first stage. Thereafter a single neural network operates individually in each node or an individual agent-network appeals to the neural network ensemble classifier.

Appeal to the Neural Network Ensemble Classifier. Appeal to the neural network ensemble classifier occurs when a single agent (classifier) is "not confident" in its prediction. The threshold of "confidence" is given as a numerical value that characterizes the range of values at the output of a single neural network classifier. This threshold characterizes the emergence of the uncertainty situation, that is, the situation when an agent can't identify a class for specific pattern with sufficient level of confidence. The idea is that if such situations emerge then a single neural network classifier should appeal to the ensemble classifier and use its prediction to provide a "highly-confident" classification. Because of some neural networks ensemble classifier properties discussed above in these situations there are more reasons to trust the ensemble classifier solution than the solution obtained using a single agent. The threshold value which characterizes the emergence of uncertainty should be selected on the basis of specific implementation of the neural networks classifiers.

We developed the following scheme to determine whether the individual agents should operate independently or they should appeal to the ensemble classifier. At first individual classifiers solution and its confidence level ξ are calculated. Then confidence level value ξ is compared with a predetermined threshold value Δ. Finally on the basis of this comparison the decision about the necessity to appeal to the ensemble classifier for ensemble decision-making is taken:

- if $\xi \leq \Delta$ then it is necessary to appeal to an ensemble classifier for converging on a common solution;
- else the pattern is classified by the individual neural network without appeal to the ensemble classifier.

The main point here is to find the way to calculate confidence level value ξ for each individual classifier. We propose a two-factor estimate, which is based on consideration of the following two parameters:

1) Degree of certainty ξ_i of i-th individual classifier about correctness of its decision ("individual degree of certainty"). Obviously there are various ways to evaluate individual degree of certainty for different types of DMT and in certain cases it is difficult to estimate ξ_i - for example, for decision trees. As ensembles of artificial neural networks are considered in our study then the individual degree of confidence can be calculated using the signal level at the output of a neuron corresponding to the class determined by the individual neural network classifier for the pattern. Two ways to evaluate the individual degree of certainty were proposed:

1.1) Individual degree of certainty is evaluated without taking into account the output of the remaining neurons in the output layer:

$$\xi_i = \max_{j=1, output\_layer\_size} \left(S_{i,j} \right). \tag{1}$$

Here $S_{i,j}$ is the value of the j-th neuron of the output layer in the i-th neural network (individual classifier). This way to estimate the individual degree of certainty does not take into account the possibility of forming sufficiently close signals on several neurons in the output layer of the classifier neural network.

1.2) The second way to evaluate the individual degree of certainty was proposed as an alternative approach. According to this formula we calculate the relative level of the signal at the output of a neuron corresponding to the class determined by the individual classifier:

$$\xi_i = \max_{j=1, output\_layer\_size} \left(S_{i,j} \right) \times \frac{1}{\sum_{k=1}^{output\_layer\_size} S_{i,k}} \tag{2}$$

Since the output values of the neurons are normalized, the values ξ_i of the individual degrees are in the range $[0;1]$.

2) ρ_i is the individual degree of confidence in the i-th individual classifier. This value determines how effective this individual classifier is. In our study the individual degree of confidence has been viewed as a value directly proportional to the number of true classified patterns in the test sample:

$$\rho_i = \frac{N_{cc}^i}{N_{ts}}. \tag{3}$$

Here N_{cc}^i is the number of patterns from the test sample which are correctly classified by i-th classifier, N_{ts} is the total number of patterns in the test sample. Individual degree of confidence ρ_i calculated with this formula is in the range $[0;1]$. Other approaches to evaluate the degree of confidence in the individual classifier can be used and it is the direction of further research.

The total value of confidence level ξ is calculated according the following two-factor formula:

$$\xi = \xi_i \times \rho_i \tag{4}.$$

Obviously the values of confidence level are in the range $[0;1]$. The proposed approach for determining the degree of confidence does not imply the need for further calculations of the value obtained for the sample points of individual and collective classification. It is remarkable that there is no need to perform additional high-cost calculations to obtain classifiers solutions in two-factor estimate procedure described above.

Values ρ_i are calculated while designing initial neural networks for the ensemble. Anyway this procedure includes the stage of calculating classification error of individual neural classifiers. As for ξ_i values they are simply obtained on the output layer in case of using neural networks as individual classifiers. Thus the cost of additional calculations can be considered negligible and this fact is proved by evaluation of the time of the proposed approach work on test problems. The choice of the threshold to appeal to the neural network ensemble classifier is one of the areas for further research.

Adaptation of Individual Agents and the Whole Neural Networks Ensemble Classifier. It is obvious that the proposed ensemble scheme for IDS should not lose one of the major advantages of systems based on neural networks: adaptability and learning ability during IDS functioning. Improvement of an information exchange mechanism is one of the directions for further research in this area.

Statistical experiments were carried out using test samples to study the applicability and effectiveness of the approach described above. In the statistical study we investigate the parameters which allow to empirically prove that the proposed ensemble-distributed scheme makes it possible to decrease the total number of requests to ensemble classifiers and individual classifiers are able to classify a part of patterns independently.

It is obvious that the predetermined threshold value Δ affects the number of appeals to the ensemble classifier. Appeal rate can be adjusted by changing this value: the higher the value is the more often the ensemble classifiers will be appealed, and vice versa. The way to determine the optimal value in terms of minimizing the classification error is the subject for further research. Requirements for information exchange intensity between distributed individual classifiers and the ensemble classifier can also be considered while tuning threshold value Δ.

In this study the results are obtained with the thresholds defined in the preliminary studies of the approach effectiveness. The description of the conditions and results of statistical experiments are given in the next section of the article.

3 Numerical Experiments

3.1 PROBE Attacks Detection

The proposed ensemble-distributed approach was tested on a dataset "KDD'99 Cup", also hosted in the Machine Learning Repository. The objective was to detect PROBE attacks. To evaluate the approach effectiveness all patterns relevant to PROBE attacks were marked as referring to the first class, the other patterns were marked as belonging to the second class. We used the following attributes in our experimental study: 1, 3, 5, 8, 33, 35, 37, 40. The choice of these attributes has been made empirically on the basis of the analysis of related works and their description can be found in [8]. Development and description of a formal procedure for selecting the attributes for the proposed approach are the directions for further research.

Artificial neural networks with multilayer perceptron architecture were used as basic individual classifiers. The same scheme of statistical study as for the problems discussed above was used to assess effectiveness. In order to objectively evaluate the effectiveness of the approach, the results were compared with the results on this task published for other approaches and collected in [9]. General parameters of the algorithms which we utilized to design individual neural networks are presented in Table 1.

The comparison results are shown in Table 2.

Table 1. General parameters of the algorithms

		PGNS	Genetic algorithm (ANN learning)	GPEN
Number of generations (maxima)		200	50	20
Number of individuals		50	50	50
Mutation		-	Average	Average
Selection type		Tournament	Tournament	Tournament

Table 2. Probe Attacks Detection Results

Classification techniques	Detection rate, %	False Positive rate, %
PSO-RF	99.9	0.029
PART (C4.5)	99.6	0.1
BayesNet	98.5	1
SMO (SVM)	84.3	3.8
Random Forest	99.8	0.1
Logistic	84.3	3.4
Bagging	99.6	0.1
Jrip	99.5	0.1
NBTree	99.6	0.1
Neural Network Ensemble Approach	99.8	0.045
Ensemble-Distributed Approach	99.8	0.05

The results obtained using the proposed approach show that the reliability of detection of PROBE attacks increased in comparison with almost all the competing approaches included in our research. The only approach having higher reliability of detection and a lower number of false positives was PSO-RF approach [9]. It should be taken into account that the parameters which we used for our method were faintly tuned during short preliminary study. They are unlikely to be optimal, and we hope that in future we will be able to get better results by proper tuning of these parameters. It is also worth mentioning that the results for other methods are taken from the third-party open-source, in which the performance characteristics of the methods in terms of required computational resources and time are not given.

3.2   All Attacks Detection

A number of additional experiments were carried out to compare the effectiveness of the proposed approach with other known algorithms for the problem of detection of all kinds of attacks presented in data set "KDD Cup'99". For the experiments the initial data set was divided into bootstrap subsamples: 67% of the records were used as a training sample, the remaining 33% were used as an examining sample. The total number of runs performed to evaluate the effectiveness of approaches was 50. The following methods were considered and implemented as alternative approaches: a single multi-layer perceptron [10], decision trees constructed by C4.5 [11], Bayesian classifier and the classification algorithm based on Hyperspheres [12].

The results of the experiments are shown in Table 3.

Table 3. All Attacks Detection Results

Classification techniques	Detection rate, %	False Positive rate, %
Multilayer Perceptron	81.8	0.7
Trees	79.6	1.1
Bayes	86.2	11.7
Hypersphere	81.0	1.0
GA-based Ensemble Approach	90.8	0.3
Ensemble GASEN Approach	94.8	0.3
Ensemble-distributed Approach	97.2	0.4

ANOVA and Wilcoxon tests were used to verify the statistical significance of the results. Both methods confirmed the statistical significance of the benefits of the proposed approach by maximizing detection rate.

3.3   Threshold and Redistribution of the Load

To investigate the robustness of the ensemble-distributed approach a series of numerical experiments were carried out. Here we apply the proposed approach to detect not only PROBE attacks but all kinds of attacks presented in the "KDD'99 Cup" data set. We also used ANOVA and Wilcoxon tests to verify the statistical significance of the results and 67% and 33% bootstrap sampling.

Each experiment includes the following steps:

1. Design of a neural network ensemble.
2. Selection of the threshold values Δ.
3. Estimation of the distribution of the computational load between the individual classifier and the ensemble classifier. We calculated the relative frequency of solving the problem by individual classifiers and the corresponding value for the ensemble classifier. The data gathered during threshold selection also allowed us to estimate the intermediate values of the distribution of the computational load. The results of experiments are given in Table 4.

Table 4. Threshold Value Research Results

Percentage of patterns classified by individual classifiers, %	Detection rate of ensemble-distributed approach, %
50	68.0
40	78.4
30	84.6
20	91.2
10	96.8
0	97.2

It is seen that if there is 10% load distribution in favour of individual classifiers then the detection rate value is practically not reduced. In the situation where 10% means tens of thousands of queries for data analysis it is an essential value, providing a significant reduction in the load on the network communications.

The proposed ensemble-distributed approach showed statistically indistinguishable results when we used load redistribution in favor of the individual classifiers at 10%. Obviously the problem of setting the optimal value of the threshold requires further investigation. In particular, it seems promising to use the individual threshold for each classifier and to apply formalized algorithms for determining such thresholds. It is also necessary to explore in more detail the overall dynamics depending on changes in classification error threshold and the corresponding redistribution of computational load between the individual classifier and the collective classifiers.

4   Summary

A new approach for intrusion detection in distributed classification-based IDS is presented in this paper. This approach is based on using ensembles of neural networks. The basic scheme of neural network ensembles usage in computer networks concerning intrusion detection is proposed. The main ideas of a neural network ensemble approach for solving classification problems are described. We present the results of an experimental study comparing the effectiveness of the proposed approach and other methods. These results show that the proposed ensemble approach allows us to solve the considered classification problem not worse than some well-known methods, the results of whose work were obtained and published by their authors.

Thus the use of the evolutionary ensemble-distributed approach can effectively solve the problem of information security incidents detection in automated systems. The proposed approach requires further development in the form of creating the environment that allows to simulate the functioning of the IDS on the basis of the use of neural network ensembles. We suppose that open challenges and promising directions for future research are also the following:

- the development and investigation of more effective mechanisms for the exchange of "knowledge" among the individual ensemble members, ensemble members and the ensemble as a whole;

- the formalization of the methodology of selecting a threshold value to decide on the necessity to appeal to a single agent of the neural networks ensemble;

- the development and research of more effective methods for evaluating ensemble decisions for classification problems, modeling problems, etc.

We hope that in the nearest future we will finally provide an effective integrated solution for distributed IDS based on neural network ensembles. Additionally, we plan to further develop our approach for classification problems and use it to solve modern large-scale problems.

References

1. Patcha, A., Park, J.-M.: An Overview of Anomaly Detection Techniques: Existing Solutions and Latest Technological Trends., Computer Networks (2007)
2. Akbar, S., Nageswara Rao, K., Chandulal, J.A.: Implementing Rule based Genetic Algorithm as a Solution for Intrusion Detection System. International Journal of Computer Science and Network Security 11(8), 138–144 (2011)
3. Amalraj Victoire, T., Sakthivel, M.: A Refined Differential Evolution Algorithm Based Fuzzy Classifier for Intrusion Detection. European Journal of Scientific Research 65(2), 246–259 (2011)
4. Hansen, L.K., Salamon, P.: Neural network ensembles. IEEE Transactions on Pattern Analysis and Machine Intelligence 12, 993–1001 (1990)
5. Rastrigin, L.A., Erenstein, R.H.: Method of collective recognition. Energoizdat, Moscow (1981)
6. Wolpert, D.H.: Stacked generalization. Neural Networks 5, 241–259 (1992)
7. Bukhtoyarov, V., Semenkina, O.: Comprehensive evolutionary approach for neural network ensemble automatic design. In: IEEE World Congress on Computational Intelligence 2010, Barcelona, pp. 1640–1645 (2010)
8. Stolfo, S., Fan, W., Lee, W., Prodromidis, A., Chan, P.: Cost-based Modeling for Fraud and Intrusion Detection: Results from the JAM Project. In: Proceedings of the 2000 DARPA Information Survivability Conference and Exposition, DISCEX 2000 (2000)
9. Malik, A.J., Shahzad, W., Khan, F.A.: Binary PSO and random forests algorithm for PROBE attacks detection in a network. In: 2011 IEEE Congress on Evolutionary Computation (CEC), New Orleans, LA, pp. 662–668 (June 2011)
10. Haykin, S.: Neural networks: a comprehensive foundation. Prentice Hall PTR (1994)
11. Quinlan, J.R.: C4.5: programs for machine learning. Morgan Kaufmann (1993)
12. Ong, Y.S., et al.: Classification of adaptive memetic algorithms: a comparative study. IEEE Transactions on Systems, Man, and Cybernetics, Part B: Cybernetics 36(1), 141–152 (2006)

Weight Update Sequence in MLP Networks

Mirosław Kordos[1], Andrzej Rusiecki[2], Tomasz Kamiński[1], and Krzysztof Greń[1]

[1] University of Bielsko-Biala, Department of Mathematics and Computer Science,
Bielsko-Biała, Willowa 2, Poland
mkordos@ath.bielsko.pl
[2] Wroclaw University of Technology, Institute of Computer Engineering, Control and Robotics,
Wrocław, Wybrzeże Wyspiańskiego 27, Poland
andrzej.rusiecki@pwr.edu.pl

Abstract. The advantages of Variable Step Search algorithm - a simple local search-based method of MLP training is that it does not require differentiable error functions, has better convergence properties than backpropagation and lower memory requirements and computational cost than global optimization and second order methods. However, in some applications, the issue of training time reduction becomes very important. In this paper we evaluate several approaches to achieve this reduction.

1 Introduction

The most popular methods used to train MLPs are analytical gradient-based algorithms using error backpropagation (BP). These algorithms include standard gradient descent, resilient backpropagation (RPROP), Quickprop, Levenberg-Marquardt algorithm , several versions of conjugate gradients or the scale conjugate gradients methods [2]. Regular gradient-based algorithms are usually able to obtain satisfiable solutions, which are not necessarily global optima.

Another group of methods consists of global optimization techniques such as genetic algorithms [3], simulated annealing [4] and its variants, particle swarm optimization or tabu search [5]. These global methods are computationally expensive (especially with evolutionary approach) and also do not guarantee better network performance.

In this paper, we consider the third group of methods, namely algorithms based on local search techniques [1,7], in particular Variable Step Search Algorithm (VSS) first described in [9] and then successfully applied in [10,11]. The idea of the method was based on inspection of the learning process, which helped in formulating basic rules to change one weight at time. The VSS supports non-differentiable transfer and error functions, what was crucial in our applications [10,12]. Also minimizing the network training time was crucial and therefore we especially address this issue in this paper.

The main focus of this work is the discussion of how to reduce the computational cost of the method. In the next sections we describe the basic VSS algorithm and its novel variants with various weight update schema including, among others, updating only selected weights, discarding weights for which the update did not reduce the error, and combination of these approaches. Section 4 presents experimental results for 5 classification and 5 regression benchmark tasks, where network performances for the

E. Corchado et al. (Eds.): IDEAL 2014, LNCS 8669, pp. 266–274, 2014.

tested methods and several levels of computational efforts are compared. Based on the simulation results, general conclusions are formulated in the last section.

2 The VSS Algorithm

Because of the need for MLP learning algorithms that work with non-differentiable error functions, and a high computational cost of the known global search methods that can address this issue, we began experimenting with determining the gradient direction numerically instead of analytically. Thus we changed each weight a little and measured how it affected the network error. Then the weight was restored to its original value and the next weight was examined. The gradient direction we obtained was very close to the gradient direction obtained by the backpropagation algorithm, but the computational cost was even higher, due to the need of determining the network error as many times in one epoch as the number of weights. However, we noticed that if changing one weight by dw caused the error decrease and we did not restore the weight to its original value but try to change the next weight in that new point, we obtained much better and faster convergence of the algorithm (the convergence abilities were comparable to those of the Levendberg-Maguardt (LM) algorithm [9]). That is the basic idea of the VSS algorithm. Additional advantage of the VSS algorithm is that it can be applied with any feedforward network structure. It can be directly used to train deep neural architectures [6], eg. with 10 hidden layers (in contrary to backpropagation) and after introducing some modifications, which are described later, it is suitable for big networks, when the training time increases much slower than that of the LM algorithm.

Because of these advantages, especially that we have to use non-differentiable error functions, we recently used the VSS algorithm for hundreds of thousands of extensive tests [9]. Then the problem of decreasing the training time became crucial, and even as little time reduction as 20% was noticeable.

3 Reduction of Computational Cost

To reduce the training time we introduced the following enhancements: individually adjusted changes of each weight, remembering signals in a table and, recently, optimization of weight probing points. The paper focuses on the third enhancement, but before discussing it we shortly introduce the idea of the first and second one, to present a full picture.

Individually Adjusted Weight Changes. The algorithm starts with random weights and the initial $dw(i) = 0.1$ for each weight $w(i)$ in the output layer and $dw(i) = 0.3$ in the hidden layer. Then it individually adjust $dw(i)$ for each weights; $dw(i)$ becomes the value by which the weight changed in the current epoch. When the next epoch begins, the initial guess of the optimal change of each weight is $d1 \cdot dw(i)$, because as the experiments showed, the weights tend to keep the approximate proportion between their changes in two consecutive epochs. For the same reason when we have to reverse the direction of changing a weight, we decrease the step, multiplying the previous step by $0.5 \cdot d1 \cdot dw(i)$ instead $d1 \cdot dw(i)$.

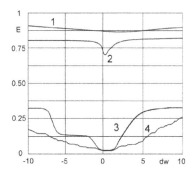 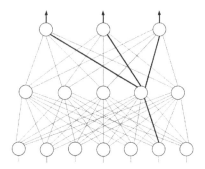

Fig. 1. Left: Typical error surface cross-sections in the direction of: (1) hidden weight at the beginning of the training; (2) output weight at the beginning of the training, (3) output weight at the end of the training, (4) hidden weight at the end of the training. Right: The idea of Signal Table (When the weight between the 4th hidden and 6th input neuron is modified, the signals are propagated only through the connections shown in bold).

Signal Table. As we change only one weight, there is no reason to propagate the signals through the entire network every time. That must be done only once, at the beginning of the training, to fill the "signal table" - an array, which stores the input and the output signals for each neuron for each training vector. Then, when we modify a single weight, we update only the portion of the signal table, which was affected by this change (the bold lines in figure 1-right in case of modifying a hidden layer weight and only one input and output of a single neuron in case of modifying an output layer weight). That allows reducing the computational cost by up to 98%, depending on the network size and structure.

Optimization of Weight Probing Points. The questions are: what is the optimal sequence of particular weight evaluation, in what direction should the search go and what is the optimal precision of the search? Finding the minimum on the error surface in a particular weight direction requires probing the error value in several points. The more points we use, we are able to find the minimum more precisely but on the other hand the computational cost of such a search is higher. There are some methods known from numerical analysis, such as the three-point formula [8], however the problem of minimizing the computational cost while modifying one weight at a time is so specific, that the general methods cannot be successfully applied to it. For that reason, in this paper, we evaluate different strategies (see Algorithms 1 and 2), where the assessment criterion is the network error (MSE) after a given training time expressed by the computational effort (ce):

1. d1 - probing the error value in one point in given direction
2. d1d2 - probing the error value in two points in given direction
3. d1d2d3 - probing the error value in three point in given direction
4. parabola - probing the error in two points and going to the parabola vertex

We tested all the four strategies with "superweights" ("s" in tables 1-3), with weight freezing ("f") and with "superweights" and weight freezing ("sf"). That gave together

Algorithm 1. VSS-d1, VSS-d1d2 and VSS-d1d2d3 algorithm

for $n = 1$ to numberOfEpochs **do**
    for $i = 1$ to numberOfWeights **do**
        $E_0 = E_{min}$
        $w(i,n) = w(i,n-1) + d1 \cdot dw(i)$
        if $E < E_{min}$ **then**
            $E_{min} = E$
            if $d2$ **then**
                $w(i,n) = w(i,n) + d2 \cdot dw(i)$
                if $E < E_{min}$ **then**
                    $E_{min} = E$
                  if $d3$ **then**
                      $w(i,n) = w(i,n) + d3 \cdot dw(i)$
                      if $E < E_{min}$ **then**
                        $E_{min} = E$
                    else
                      $w(i,n) = w(i,n) + d2 \cdot dw(i)$
                  end if
                end if
            else
                $w(i,n) = w(i,n-1) + d1 \cdot dw(i)$
            end if
        end if
        else
            $w(i,n) = w(i,n-1) - 0.5 \cdot d1 \cdot dw(i)$
            if $E < E_{min}$ **then**
                $E_{min} = E$
                if $d2$ **then**
                  $w(i,n) = w(i,n) - 0.5 \cdot d2 \cdot dw(i)$
                if $E < E_{min}$ **then**
                      $E_{min} = E$
                  if $d3$ **then**
                      $w(i,n) = w(i,n) - 0.5 \cdot d3 \cdot dw(i)$
                      if $E < E_{min}$ **then**
                        $E_{min} = E$
                    else
                      $w(i,n) = w(i,n) - 0.5 \cdot d2 \cdot dw(i)$
                  end if
                end if
                else
                  $w(i,n) = w(i,n-1) - 0.5 \cdot d1 \cdot dw(i)$
                end if
            end if
        end if
        end if
        if $E_0 > E_{min}$ **then**
            $dw(i) = w(i,n) - w(i,n-1)$
        else
            $w(i,n) = w(i,n-1)$
            $dw(i) = 0.5 \cdot dw(i)$
        end if
    end for
end for

16 combinations. By superweights we mean the 25% of weights, which when changed resulted in the greatest error reduction. Then the superweights are once again adjusted in the same training cycle before next training cycle begins. There are usually also several weights, which when changed did not cause the error decrease. They can get frozen for the next training cycle, because it is likely that changing them in the next cycle will reduce the error only to a minimal extend (or not at all) and thus the computational effort can be better used for examining other weights in the next training cycle. The experimentally determined optimal $dw1$, $dw1$, $dw3$ values were close to 1.5 and that values we used.

Algorithm 2. VSS-parabola algorithm

---

for $n = 1$ to numberOfEpochs **do**
    for $i = 1$ to numberOfWeights **do**
        $E_0 = E$
        $w_0 = w(i, n)$
        $w_1 = w(i, n) = w(i, n - 1) + d1 \cdot dw(i)$
        if $E_1 = E < E_{min}$ **then**
            $w_2 = w(i, n) = w(i, n) + d2 \cdot dw(i)$
        else
            $w_2 = w(i, n) = w(i, n) - 0.5 \cdot d1 \cdot dw(i)$
        end if
        $E_2 = E$
        $w(i, n) = vertexOfParabola(w_0, E_0, w_1, E_1, w_2, E_2)$
        if $parabolaIsConcave$ **and** $E < E_{min}$ **then**
            $dw(i) = w(i, n) - w(i, n - 1)$
        else
            $w(i, n) = min_E(w_0, w_1, w_2)$
            $dw(i) = w(i, n) - w(i, n - 1)$
        end if
    end for
end for

---

4   Experimental Evaluation

In the experiments we used the following datasets for classification tasks: Iris (4 attributes / 150 vectors / 3 classes), Ionosphere (34/351/2), Climate Simulation Changes (19/540/2), Image Segmentation (19/1050/7), Glass (10/214/6) and the following for regression tasks: Steel (13 attributes / 960 vectors), Yacht Hydrodynamics (7/308), Building (15/1052), Concrete Compression Strength (8/1030), Crime and Communities (8/318) All of them but Steel and Building come from [13]. To perform the experiments we created the software in C#. The datasets and the source code are available from [14].

To make the comparison easy, each dataset was standardized before the training (inputs in classification tasks and input and output in regression tasks). For classification

Table 1. MSE on training set after $ce=1000$

method	iris	ion.	clim.	img.	glass	steel	yacht	bld.	conc.	crime	av-c	av-r
d1	0.764	1.372	0.594	7.033	5.285	0.313	0.321	0.186	0.859	0.327	0.625	0.417
d2	0.800	1.461	1.483	7.248	5.571	0.380	0.311	0.206	0.861	0.336	0.741	0.436
d3	0.854	1.355	1.428	7.339	5.684	0.387	0.296	0.219	0.865	0.359	0.734	0.441
par.	0.984	1.392	1.414	7.420	5.778	0.411	0.358	0.231	0.900	0.361	0.751	0.480
d1,s	0.779	1.343	0.450	7.152	5.238	0.315	0.293	0.179	0.847	0.320	0.610	0.401
d2,s	0.784	1.416	1.441	7.289	5.413	0.397	0.280	0.204	0.857	0.330	0.727	0.425
d3,s	0.890	1.396	1.497	7.297	5.707	0.397	0.276	0.220	0.867	0.359	0.747	0.432
par,s	0.924	1.452	1.371	7.425	5.792	0.455	0.329	0.218	0.882	0.369	0.749	0.472
d1,f	0.784	1.336	0.583	6.910	5.342	0.309	0.306	0.178	0.856	0.322	0.620	0.405
d2,f	0.784	1.560	1.591	7.519	5.471	0.371	0.305	0.194	0.862	0.333	0.765	0.427
d3,f	0.803	1.452	1.431	7.520	5.781	0.387	0.284	0.204	0.861	0.362	0.749	0.431
par,f	0.940	1.419	1.480	7.199	5.694	0.455	0.342	0.255	0.875	0.347	0.748	0.487
d1,sf	0.786	1.365	0.438	7.208	5.240	0.304	0.293	0.176	0.848	0.320	**0.613**	**0.397**
d2,sf	0.809	1.384	1.445	7.073	5.558	0.358	0.282	0.187	0.846	0.325	0.724	0.409
d3,sf	0.829	1.414	1.586	7.418	5.598	0.398	0.281	0.199	0.858	0.357	0.754	0.429
par,sf	0.862	1.396	1.273	7.392	5.627	0.452	0.330	0.239	0.873	0.363	0.723	0.478

we used neurons with hyperbolic tangent transfer functions. The number of output neurons was equal to the number of classes and we trained the network so the signal of the neuron corresponding to the current class should be greater than 0.995 (if the signal was greater than 0.995 we assumed that the error made by this neuron is zero, the purpose of that was to prevent unnecessary growth of output layer weights) and the signals of the remaining output neurons should be smaller than -0.995. For regression tasks the output neuron had linear transfer function. For each of the 10 dataset and each of the 16 combinations of parameters we trained the network 100 times starting from random weight values. The average values of the 100 trainings are presented in tables 1-3. The standard deviations were of similar order for each dataset: about $\sigma=0.33$ for $ce=1000$, $\sigma=0.10$ for $ce=2000$ and $\sigma=0.02$ for $ce=4000$, where ce is the computational effort, which is proportional to the training time. The training time could not be reliably measured directly in many cases, because it was frequently only a fraction of second. Thus ce was calculated in the following way: When the network training starts $ce=0$. For changing each weight of an output neuron:

$$ce = ce + 1 + x, \tag{1}$$

For changing each weight of a hidden neuron:

$$ce = ce + 1 + L2 * (L1 + 1) * 0.2 + L2 * 0.8 + x; \tag{2}$$

where $L1$ is the number of neurons in the hidden layer and $L2$ is the number of neurons in the output layer. $L2 * (L1 + 1)$ is the number of signals that must be recalculated in the output layer (for additions and subtractions the cost is 0.2) and $L2$ is the number of transfer functions that must be calculated in the output layer (for calculating hyperbolic tangent the cost is 0.8). For the current layer the cost is $0.2 + 0.8 = 1$. x is the cost the constant operations independent of the weight location within the network structure.

Table 2. MSE on training set after $ce=2000$

method	iris	ion.	clim.	img.	glass	steel	yacht	bld.	conc.	crime	**av-c**	**av-r**
d1	0.613	0.603	0.370	6.124	2.587	0.248	0.185	0.136	0.802	0.295	0.399	0.317
d2	0.564	0.638	0.389	6.489	2.647	0.289	0.165	0.150	0.803	0.296	0.414	0.323
d3	0.583	0.920	0.365	6.726	2.774	0.321	0.149	0.148	0.810	0.316	0.452	0.327
par.	0.690	0.955	0.385	6.730	2.708	0.340	0.186	0.187	0.845	0.343	0.463	0.370
d1,s	0.545	0.585	0.347	6.303	2.413	0.217	0.152	0.122	0.796	0.293	0.390	0.291
d2,s	0.474	0.628	0.372	6.571	2.522	0.264	0.143	0.128	0.804	0.296	0.403	0.301
d3,s	0.517	0.937	0.373	6.666	2.840	0.286	0.139	0.146	0.811	0.312	0.451	0.315
par,s	0.658	0.985	0.387	6.809	2.757	0.454	0.249	0.179	0.834	0.339	0.468	0.413
d1,f	0.588	0.564	0.363	6.121	2.457	0.231	0.162	0.130	0.797	0.291	0.389	0.301
d2,f	0.529	0.658	0.364	6.796	2.607	0.292	0.145	0.143	0.804	0.293	0.419	0.313
d3,f	0.546	0.982	0.333	6.868	2.855	0.325	0.130	0.148	0.804	0.317	0.459	0.320
par,f	0.723	0.967	0.385	6.529	2.768	0.359	0.190	0.195	0.825	0.303	0.462	0.369
d1,sf	0.521	0.556	0.338	6.269	2.410	0.195	0.155	0.115	0.797	0.290	**0.384**	**0.285**
d2,sf	0.491	0.656	0.347	6.361	2.626	0.246	0.141	0.117	0.799	0.295	0.402	0.291
d3,sf	0.519	0.918	0.337	6.763	2.758	0.272	0.145	0.140	0.807	0.316	0.445	0.312
par,sf	0.710	0.927	0.376	6.747	2.766	0.364	0.230	0.189	0.844	0.325	0.463	0.390

Table 3. MSE on training set after $ce=4000$

method	iris	ion.	clim.	img.	glass	steel	yacht	bld.	conc.	crime	**av-c**	**av-r**
d1	0.222	0.470	0.279	2.769	2.207	0.134	0.074	0.073	0.758	0.273	0.242	0.216
d2	0.213	0.467	0.281	2.893	2.301	0.179	0.053	0.079	0.761	0.274	0.248	0.220
d3	0.223	0.456	0.274	3.780	2.289	0.195	0.047	0.082	0.765	0.278	0.272	0.223
par.	0.343	0.509	0.378	3.261	2.490	0.219	0.072	0.095	0.776	0.289	0.288	0.246
d1,s	0.219	0.461	0.264	2.435	2.157	0.122	0.050	0.071	0.754	0.272	0.229	0.203
d2,s	0.207	0.462	0.263	2.742	2.222	0.164	0.050	0.076	0.762	0.275	0.239	0.214
d3,s	0.206	0.451	0.280	3.509	2.257	0.191	0.045	0.077	0.768	0.277	0.262	0.219
par,s	0.383	0.462	0.363	3.101	2.389	0.277	0.067	0.111	0.807	0.297	0.276	0.267
d1,f	0.222	0.432	0.275	2.621	2.216	0.124	0.073	0.071	0.758	0.272	0.234	0.213
d2,f	0.209	0.431	0.293	2.890	2.276	0.171	0.053	0.077	0.764	0.273	0.245	0.217
d3,f	0.202	0.398	0.264	3.679	2.236	0.197	0.042	0.079	0.763	0.280	0.259	0.220
par,f	0.332	0.479	0.359	3.246	2.560	0.234	0.097	0.106	0.817	0.287	0.284	0.267
d1,sf	0.220	0.432	0.229	2.348	2.147	0.113	0.049	0.067	0.750	0.271	**0.219**	**0.196**
d2,sf	0.219	0.440	0.248	2.749	2.279	0.155	0.045	0.071	0.761	0.276	0.238	0.209
d3,sf	0.232	0.398	0.259	3.478	2.232	0.183	0.045	0.077	0.765	0.281	0.255	0.218
par,sf	0.278	0.470	0.358	3.146	2.371	0.251	0.095	0.093	0.777	0.286	0.270	0.260

$x = 0.3$ for determining parabola vertex, $x = 0.2$ for selecting the superweights and $x = 0.1$ for all other methods.

We also made another series of experiments in a 10-fold crossvalidation to find out how the weight update scheme used in the network training influences the final prediction ability. However, it turned out that the way the network reached the predefined MSE value (which was the stopping criteria) did not influenced the prediction accuracy.

The average value column (av-c) in tables 1-3 for classification tasks is the weighted average of the values for particular datasets weighted by the inverse of output neuron numbers - thus it represents the average MSE per one output neuron:

$$av - c = (iris/3 + ion/2 + clim/2 + img/7 + glass/6)/5. \tag{3}$$

For regression it is:

$$av - r = (steel + yacht + bld + 0.5 \cdot conc + crime)/5. \tag{4}$$

5 Conclusions

We presented some improvements of the VSS algorithm, which was our algorithm of choice in the cases that we had to use non-differentiable error functions for big datasets. The conclusions are as follows: it does not make sense to locate the minimum in each weight direction precisely, because the closer to the minimum we are, the error surface gets flatter and the less we can gain (see fig. 1 - left). It better pays off to use the computational effort to modify the subsequent weight. Because the learning trajectory changes its direction rather gradually than suddenly, if changing same weights values in one epoch was very successful it is a good idea to modify them once again before going to the next epoch. On the contrary, if changing some weights did not cause improvement, the weights can be frozen for the subsequent epoch and the computational power can be used to change the more promising weights. The optimal $d1$ value is about 1.5, allowing for gradual step increase as the error surface is getting flatter with the training progress. The additionally obtained time reduction is on average about 30%. Obviously, well written and optimized code is another field of improvement. The 30% may seem not much for one training of a small or medium size network. However, in the case when we have to conduct thousands of experiments, on real-world datasets, it can shorten our work by many hours.

References

1. Aarts, E., Lenstra, J.K.: Local Search in Combinatorial Optimization. John Wiley & Sons, Inc., New York (1997)
2. Du, K.-L., Swamy, M.-N.S.: Neural Networks and Statistical Learning. Springer (2013)
3. Garcia-Pedrajas, N., et al.: An alternative approach for neural network evolution with a genetic algorithm. Neural Networks 19(4), 514–528 (2006)
4. Engel, J.: Teaching Feed-forward Neural Networks by Simulated Annealing. Complex Systems 2, 641–648 (1988)
5. Battiti, R., Tecchiolli, G.: Training Neural Nets with the Reactive Tabu Search. IEEE Trans. on Neural Networks 6, 1185–1200 (1995)
6. Bengio, Y.: Learning Deep Architectures for AI. Foundations and Trends in Machine Learning 2(1), 1–127 (2009)
7. Beliakov, G., Kelarev, A., Yearwood, J.: Derivative-free optimization and neural networks for robust regression. Optimization 61(12), 1467–1490 (2012)

8. Burden, R.L., Douglas Faires, J.: Numerical Analysis, Cengage Learning (2010)
9. Kordos, M., Duch, W.: Variable Step Search Algorithm for Feedforward Networks. Neurocomputing 71(13-15), 2470–2480 (2008)
10. Kordos, M., Rusiecki, A.: Improving MLP Neural Network Performance by Noise Reduction. In: Dediu, A.-H., Martín-Vide, C., Truthe, B., Vega-Rodríguez, M.A. (eds.) TPNC 2013. LNCS, vol. 8273, pp. 133–144. Springer, Heidelberg (2013)
11. Rusiecki, A., Kordos, M., Kamiński, T., Greń, K.: Training Neural Networks on Noisy Data. In: Rutkowski, L., Korytkowski, M., Scherer, R., Tadeusiewicz, R., Zadeh, L.A., Zurada, J.M. (eds.) ICAISC 2014, Part I. LNCS, vol. 8467, pp. 131–142. Springer, Heidelberg (2014)
12. Rusiecki, A.: Robust learning algorithm based on LTA estimator. Neurocomputing 120, 624–632 (2013)
13. Merz, C., Murphy, P.: UCI repository of machine learning databases (2014),
http://www.ics.uci.edu/mlearn/MLRepository.html
14. Source code and datasets used in the paper,
http://www.kordos.com/software/ideal2014.zip

Modeling of Bicomponent Mixing System Used in the Manufacture of Wind Generator Blades

Esteban Jove[1,2], Héctor Aláiz-Moretón[2], José Luis Casteleiro-Roca[1],
Emilio Corchado[3], and José Luis Calvo-Rolle[1]

[1] Departamento de Ingeniería Industrial,
Escuela Universitaria Politécnica, University of A Coruña,
Avda. 19 de febrero s/n, 15495, Ferrol, A Coruña, Spain
esteban.jove@udc.es
[2] Departamento de Ingeniería Eléctrica y de Sistemas y Automática,
Escuela de Ingenierías Industrial e Informática, Universidad de León,
Campus de Vegazana s/n, 24071, León, León, Spain
[3] Departamento de Informática y Automática,
Universidad de Salamanca,
Plaza de la Merced s/n, 37008, Salamanca, Salamanca, Spain

Abstract. The clean energy use has increased during the last years, especially, electricity generation through wind energy. Wind generator blades are usually made by bicomponent mixing machines. With the aim to predict the behavior of this type of manufacturing systems, it has been developed a model that allows to know the performance of a real bicomponent mixing equipment. The novel approach has been obtained by using clustering combined with regression techniques with a dataset obtained during the system operation. Finally, the created model has been tested with very satisfactory results.

Keywords: Clustering, SOM, MLP, SVM, Wind generator.

1 Introduction

There was an increase of the energy consumption and a raise of the enviromental pollution during the last years. This fact involves the expansion of the renewable energy generation [30].

In this context, wind energy solutions has increased over the world. With the aim to improve the competitive all the involved tasks have been and still being improved [17]. One of these improvements is the use of carbon fiber as a base material to manufacture wind generator blades. This compound is obtained by mixing components [28]. They are called bicomponents because these materials have different characteristics when they are separated. However, they react when they are mixed and their properties change significantly.

E. Corchado et al. (Eds.): IDEAL 2014, LNCS 8669, pp. 275–285, 2014.

In the system studied on the present research, one of the components is an epoxy resin, and the order one is a catalyst. Both components are Non-Newtonian fluids, which means that their mechanical properties do not remain constant with variation of efforts received [6]. On this study, a 1 : 1 ratio is required in the final compound to obtain the best properties. The obtained material with the above mentioned components has the advantages of high specific strength and stress, then, it is suitable for the use in wind generator blades [28].

The installation studied is a mixing machine, which have two reservoirs, one by each component, epoxy resin and catalyst. Each tank feeds one pump, and both have to deliver the same flow rate to obtain a constant ratio in the mixing valve.

Due to the characteristics of the fluid and the variation of the pump efficiency, it is a trouble to obtain the needed ratio for the final material. The purpose of this study is to create a model with the aim to predict the behavior of the system. It is obtained by data processing of the information obtained over the real installation when it is in normal operation. Thus, by modeling the system it is possible to predict the flow rate, the ratio and other system variables.

It is possible to split the layout of predictive models in two methods [18]:

- Global models: The model is generated from all available data for training, trying to get the lowest error. Different techniques can be used to generate the model: Multi-Layer Perceptron (Articial Neural Networks, ANN) [2,3,1] and Support Vector Regression (SVR) [12].
- Local models: all the dataset is divided into different clusters, depending on the characteristics of the input data. Typically, in order to make the groups, it can be used algorithms like Self Organization Map (SOM) [4,8] or K-means [10,11]. After that, regression techniques are applied for each cluster.

This paper is organized as follows. It begins with a short description of the case study followed by an explanation of the models to describe the system behaviour. In the next section, results are presented, and finally the conclusions and the acknowledgments are shown.

2   Case of Study

2.1   Mixing Machine Installation

Before system behavior modeling, it is necessary to know the parts of the installation and its operation. As was mentioned in the introduction section, the system aim is mix two components on the same proportion. These components are stored in separate tanks. Both reservoirs feed two pumps in order to supply the required flow for optimum mixture.

Each electric pump is connected to a Variable Frequency Drive (VFD) which control their speeds. The pumped fluids are mixed with a mixing valve that gives the final bicomponent, which is used to make the wind generator blades. Due to the blade form, the output flow will vary depending on the part of the blade under creation. The process does not require a constant flow at the output. An example of the flow set point required is shown on Figure 1. Taken into account that the fluids are Non-Newtonian and the pumps efficiency are variable, the control system is not easy to carry out.

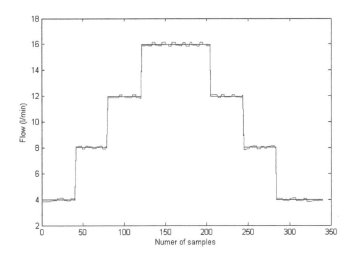

Fig. 1. Scheduled output flow rate

2.2 System Variables

The dataset used on this research was obtained by measuring different system parameters during the plant operation. It is possible to measure the following variables of the system:

- Ratio set point: it represents the required mixture proportion. Its value is 1:1 always on this process.
- Real ratio: it represents real mixture proportion. It is one of the variables to be controlled.
- Flow rate set point: it is the desired flow rate at the output. It depends of the part of the blade under manufacture (cc/min).
- Flow rate A, Flow rate B: they are the flow rate gave by pump A and B (l/min).
- Flow rate A+B: it represents the sum of last two variables. It will vary depending of the flow rate set point value (l/min).
- Pump speed A, Pump speed B: they are the speeds of the pump A and B (rpm) respectively.
- Pressure out pump A, Pressure out pump B: these variables represent the pressure at the output of the pump A and B (bar).
- Pressure out flow meter A, Pressure out flow meter B: they are the pressure measured after each flow meter (bar).

Firstly, it is necessary to classify these empirical data as inputs or outputs. In Figure 2 it is shown the system topology with the inputs and the outputs.

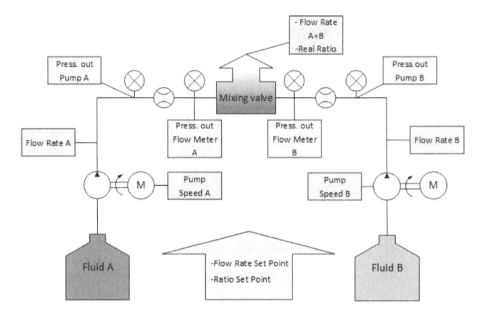

Fig. 2. Process layout

3   Model Approach

As was mentioned above, global and local models have been taken into account with the aim to obtain a good approach performance. There is only one group for the case of global models, and there are several groups when local models are used. Therefore, it is necessary to apply regression techniques on each group. In both cases the modeling process is shown in Figure 3.

The Figure 4 shows in a graphical way the data classification process into N different clusters.

3.1   Obtaining Dataset

The dataset used on this research was obtained by many different tests during the operation process. In Figure 5 it is possible to see the pump speed measured by a sensor while the system is working. This is one of the variables to be modeled. As can be seen in this figure, there are some mistakes on the data acquisition.

The dataset was conditioned by removing the wrong measures for a good modeling. After this task, our dataset was reduced from 9511 to 8549 samples.

3.2   Techniques Considered to Create the Model

In this section are described different techniques used in order to obtain representative models of the system.

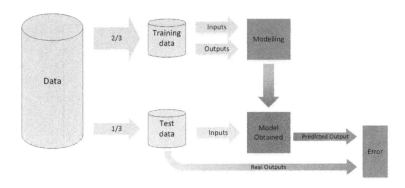

Fig. 3. Modeling process

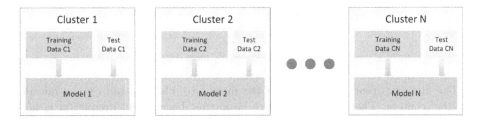

Fig. 4. Obtaining local models

Self-Organization Map (SOM). Self-Organization Map [14] was created to visualize data behaviour on a low dimensional visualization. This technique uses unsupervised learning [16,15]. It is based on an array of nodes which are connected to N inputs by an N-dimensional weight vector. The process is implemented by an iterative online algorithm but it is possible to be implemented by a batch version.

As a result of the process, SOM is expected to capture the geometry of the data. Thus, it is possible to visualize a $2D$ representation. All process allows to obtain an idea of the number of the necessary clusters.

Data Clustering. K-Means Algorithm. Clustering algorithm organizes unlabeled feature vectors into different groups. Each group or cluster contain all data similar with similar features [13]. It is an unsupervised method of data grouping depending on the similarity [20,27]. Many different clustering algorithms have been used and new clustering algorithms continue to appear. Most of these algorithms are based on two clustering methods: agglomerative hierarchical clustering and iterative square-error partitional clustering [19]. The partitional clustering algorithms divide the data into a number of clusters trying to reduce the error function. The number of clusters is normally predefined, but it can be part of the error function [7,23]. The main purpose is to obtain a partition which, for a specific number of clusters, minimizes the square error.

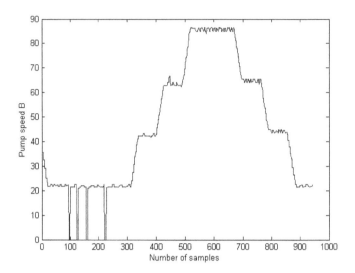

Fig. 5. Pump speed B, pure data

The final grouping will depend of the number of clusters and on the initials clusters centers. The first one is more important because its prediction is not an easy task. K-means partitional clustering algorithm brings good results, over all, if the clusters are closed, well-separated in hyperspace and hyperspherical in shape. It is also able to detect hyper ellipsoidal-shaped clusters and it is computationally effective [7].

Artificial Neural Networks (ANN). Multi-Layer perceptron (MLP)
A multilayer perceptron is a feedforward artificial neural network [26,29]. It is one of the most used ANNs due to its simple structure and its robustness. In spite of this fact, the ANN architecture must be selected properly in order to obtain good results. Multi-Layer Perceptron is composed by one input layer, one or more hidden layers and one output layer. Each layer has neurons, with an activation function. In a typical configuration, all neurons form a layer have the same activation function. This function could be a step, linear, log-sigmoid or tan-sigmoid.

Support Vector Regression (SVR), Least Square Support Vector Regression (LS-SVR) SVR is a modification of the Support Vector Machines (SVM) algorithm, used for classification. The main idea of the SVR is to map the data into a high-dimensional feature space F by a non linear mapping and to do linear regression in this space [5,21].

The approximation of the solution is obtained by solving a linear equations system, and it is similar to SVM in generalization performance terms [22,25]. The use of LS-SVM to regression is known as LS-SVR [9,24]. Its insensitive loss function is replaced by a squared loss function, that makes the Lagrangian by solving a linear KarusKuhn-Tucker KKT.

3.3 Model Selection

After the model training, the Test data was used to calculate the MSE (Mean Squared Error) with the aim to select the best regression technique. The MSE of the global model was compared with the MSE for different number of clusters, to select the optimal configuration.

4 Results

The main objective is to select the best model that predicts the system behaviour.

First of all, there was a study based on SOM analysis to discover if correlation exists between different inputs, reducing the number of the inputs to the final model. The SOM analysis is used too to determine the optimal number of clusters, but, as is showing, the results are not as clear as it was supposed at first. Therefore, the data was divide into clusters with the K-means algorithm, and, for each cluster, different regression techniques were used, selecting the best techniques by MSE calculation.

4.1 SOM Analysis

The neighbor distance, shown in Figure 6, defines the different zones of neurons in SOM structure. The clusters are separated by dark lines, and light areas on the map should correspond to a cluster. The SOM analysis is not definitely clear, and the selection of the number of clusters utilized is a subjective decision.

With the purpose of selection the optimal clusters, the dataset was divided in different number of clusters, from 5 to 10, and the optimal number is selected after an MSE analysis of all procedure. All the results were compared with the MSE for global model.

The SOM input planes for each input of the SOM net, Figure 7, represent the reaction of the net to a specific input. In the figure, the Inputs 3 and 5, the 'Press. out Pump A' and the 'Press. out Flow Meter A', have similar reactions. The same case occurs for Inputs 4 and 6, the same variables for the B side of the system, it is meant that these inputs are correlated each other. The Input 1, the 'Ratio Set Point', has always the same value, 1 : 1 for the bicomponent used in this research. In case of changing this bicomponent, the new 'Ratio Set Point' could be different, but always the same another time. The 'Input 2 - Flow Rate Set Point', is unique and different for the others inputs.

With the last explanation, at final, only three inputs are considered to create the model of the system; these inputs are:

– Input 2 - Flow Rate Set Point.
– Input 3 - Pressure out Pump A.
– Input 4 - Pressure out Pump B.

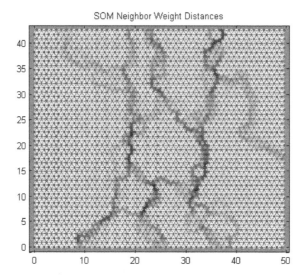

Fig. 6. SOM neighbor distances

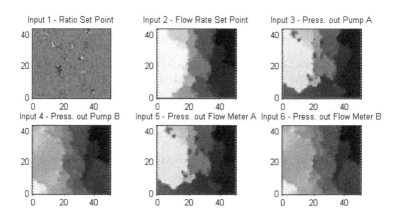

Fig. 7. SOM input planes

4.2 Regression Analysis

As it was explained, the optimal number of clusters was decided after the calculation of the average of the MSE for all the outputs. For each cluster, the two regression techniques (ANN and SVR) were trained. The configuration of the SVR is an automatized task, but the ANN configuration should be selected properly. In this research, the number of neurons in the hidden layer of the network was varied from 3 to 20. The activation function in these neurons was always tan-sigmoid, and the ouput layer neurons had a linear activation function.

The best results for the ANN regression, with 5 neurons in the hidden layer, are shown in Table 1. In this table, the results for all the clusters configuration are presented. It is possible to see that the optimal clusters' number is 6, based on the average MSE. However, if the number of clusters was different for each output, the model performance could be better, but the complexity would be increased more than the improvement achieved.

Table 1. MSE obtained using ANN

	Real ratio	Flow rate A+B	Flow rate A	Flow rate B	Pump speed A	Pump speed B	Average
Global model	0.52260	0.42362	0.10503	0.10373	0.73050	0.84887	0.45572
5 clusters	0.51698	0.37985	0.09366	0.09258	0.59531	0.74568	0.40401
6 clusters	0.42997	0.38097	0.09393	0.09282	0.56373	0.71347	0.37915
7 clusters	0.47842	0.31058	0.07597	0.07511	0.61409	0.79754	0.39195
8 clusters	0.43980	0.45649	0.11283	0.11088	0.96655	1.26514	0.55891
9 clusters	1.27257	0.46047	0.11423	0.11258	0.59370	0.77710	0.55511
10 clusters	0.42680	0.38379	0.09449	0.09315	0.78581	0.96995	0.45900

The results for the SVR regression are shown in Table 2. In this case, the optimal separation of dataset is 9. However, with the SVR occurs the same as before, if the number of clusters was different for each output, the MSE could be better.

Table 2. MSE obtained using SVR

	Real ratio	Flow rate A+B	Flow rate A	Flow rate B	Pump speed A	Pump speed B	Average
5 clusters	0.15511	0.30999	0.07228	0.07466	0.36859	0.41285	0.23225
6 clusters	0.15494	0.23023	0.06358	0.05983	0.42614	0.63239	0.26118
7 clusters	0.16321	0.29375	0.07305	0.07254	0.38717	0.47223	0.24366
8 clusters	0.19468	7.34780	1.85414	1.82709	255.51247	451.48554	119.70362
9 clusters	0.16087	0.23890	0.06001	0.06163	0.32868	0.41169	0.21030
10 clusters	0.15778	0.22213	0.05443	0.06473	0.31153	1.14401	0.32577

5 Conclusions

This research provides an accurate method of modeling a bicomponent mixing system, used in wind generator blades manufacturing.

The achieved model predicts the behaviour of the real system when the inputs are known. This model has been obtained from an empirical dataset. This model approach is based on a hybrid intelligent system by combining different regression techniques on local models.

After several tests, the analysis of the results shows that the best model configuration has 9 clusters. The regression technique employed on each one was SVR. The best average MSE obtained with this configuration was 0.2103.

This kind of analysis could be applied to many different systems with the aim to improve several goals like: efficiency, performance, features of the obtained material, and so on.

Remark that the final hybrid model achieves good results although the system has an important nonlinear nature.

Acknowledgments. This research has been funded by the University of León, and by the Spanish Ministry of Education, Culture and Sport (grant for Collaboration in University Departments).

References

1. Alvarez-Huerta, A., Gonzalez-Miguelez, R., García-Metola, D., Noriega-Gonzalez, A.: Drywell tempeture prediction of a nuclear power plant by means of artificial neural networks. Dyna 86(4), 467–473 (2011)
2. Bishop, C.: Pattern recognition and machine learning (information science and statistics). Springer-Verlag New York, Inc., Secaucus (2006)
3. Calvo-Rolle, J., Casteleiro-Roca, J., Quintián, H., Meizoso-Lopez, M.: A hybrid intelligent system for PID controller using in a steel rolling process. Expert Systems with Applications 40(13), 5188–5196 (2013)
4. Cherif, A., Cardot, H., Boné, R.: SOM time series clustering and prediction with recurrent neural networks. Neurocomput. 74(11), 1936–1944 (2011)
5. Cristianini, N., Shawe-Taylor, J.: An introduction to support Vector Machines and other kernel-based learning methods. Cambridge University Press, New York (2000)
6. Fan, H., Wong, C., Yuen, M.F.: Prediction of material properties of epoxy materials using molecular dynamic simulation. In: 7th International Conference on Thermal, Mechanical and Multiphysics Simulation and Experiments in Micro-Electronics and Micro-Systems, EuroSime 2006, pp. 1–4 (April 2006)
7. Garg, L., Mcclean, S., Meenan, B., Millard, P.: Phase-type survival trees and mixed distribution survival trees for clustering patients' hospital length of stay. Informatica 22(1), 57–72 (2011)
8. Ghaseminezhad, M.H., Karami, A.: A novel self-organizing map (SOM) neural network for discrete groups of data clustering. Appl. Soft Comput. 11(4), 3771–3778 (2011)
9. Guo, Y., Li, X., Bai, G., Ma, J.: Time series prediction method based on LS-SVR with modified gaussian RBF. In: Huang, T., Zeng, Z., Li, C., Leung, C.S. (eds.) ICONIP 2012, Part II. LNCS, vol. 7664, pp. 9–17. Springer, Heidelberg (2012)
10. Jacobs, R., Jordan, M., Nowlan, S., Hinton, G.: Adaptive mixtures of local experts. Neural Comput. 3(1), 79–87 (1991)
11. Jordan, M., Jacobs, R.: Hierarchical mixtures of experts and the EM algorithm. Neural Comput. 6(2), 181–214 (1994)
12. Karasuyama, M., Nakano, R.: Optimizing svr hyperparameters via fast cross-validation using aosvr. In: International Joint Conference on Neural Networks, IJCNN 2007, pp. 1186–1191 (August 2007)
13. Kaski, S., Sinkkonen, J., Klami, A.: Discriminative clustering. Neurocomputing 69(13), 18–41 (2005)

14. Kohonen, T.: The self-organizing map. Proceedings of the IEEE 78(9), 1464–1480 (1990)
15. Kohonen, T.: Exploration of very large databases by self-organizing maps. In: International Conference on Neural Networks, vol. 1, pp. PL1–PL6 (1997)
16. Kohonen, T., Oja, E., Simula, O., Visa, A., Kangas, J.: Engineering applications of the self-organizing map. Proceedings of the IEEE 84(10), 1358–1384 (1996)
17. Li, H., Chen, Z.: Overview of different wind generator systems and their comparisons. IET Renewable Power Generation 2(2), 123–138 (2008)
18. Martínez-Rego, D., Fontenla-Romero, O., Alonso-Betanzos, A.: Efficiency of local models ensembles for time series prediction. Expert Syst. Appl. 38(6), 6884–6894 (2011)
19. Pal, N., Biswas, J.: Cluster validation using graph theoretic concepts. Pattern Recognition 30(6), 847–857 (1997)
20. Qin, A., Suganthan, P.: Enhanced neural gas network for prototype-based clustering. Pattern Recogn. 38(8), 1275–1288 (2005)
21. Steinwart, I., Christmann, A.: Support vector machines, 1st edn. Springer Publishing Company, Incorporated (2008)
22. Suykens, J., Vandewalle, J.: Least squares support vector machine slassifiers. Neural Processing Letters 9(3), 293–300 (1999)
23. Šutienė, K., Makackas, D., Pranevičius, H.: Multistage k-means clustering for scenario tree construction. Informatica 21(1), 123–138 (2010)
24. Wang, L., Wu, J.: Neural network ensemble model using PPR and LS-SVR for stock market forecasting. In: Huang, D.-S., Gan, Y., Bevilacqua, V., Figueroa, J.C. (eds.) ICIC 2011. LNCS, vol. 6838, pp. 1–8. Springer, Heidelberg (2011)
25. Wang, R., Wang, A.-M., Song, Q.: Research on the alkalinity of sintering process based on LS-SVM algorithms. In: Jin, D., Lin, S. (eds.) Advances in CSIE, Vol. 1. AISC, vol. 168, pp. 449–454. Springer, Heidelberg (2012)
26. Wasserman, P.: Advanced methods in neural computing, 1st edn. John Wiley & Sons, Inc., New York (1993)
27. Ye, J., Xiong, T.: Svm versus least squares SVM. Journal of Machine Learning Research - Proceedings Track 2, 644–651 (2007)
28. Young, W.B., Wu, W.H.: Optimization of the skin thickness distribution in the composite wind turbine blade. In: 2011 International Conference on Fluid Power and Mechatronics (FPM), pp. 62–66 (August 2011)
29. Zeng, Z., Wang, J.: Advances in neural network research and applications, 1st edn. Springer Publishing Company, Incorporated (2010)
30. Zuo, Y., Liu, H.: Evaluation on comprehensive benefit of wind power generation and utilization of wind energy. In: 2012 IEEE 3rd International Conference on Software Engineering and Service Science (ICSESS), pp. 635–638 (June 2012)

Branching to Find Feasible Solutions in Unmanned Air Vehicle Mission Planning

Cristian Ramírez-Atencia[1], Gema Bello-Orgaz[1],
Maria D. R-Moreno[2], and David Camacho[1]

[1] Departamento de Ingeniería Informática, Universidad Autónoma de Madrid,
C/Francisco Tomás y Valiente 11, 28049 Madrid, Spain
cristian.ramirez@inv.uam.es, {gema.bello,david.camacho}@uam.es
aida.ii.uam.es
[2] Departamento de Automática, Universidad de Alcalá,
Carretera Madrid Barcelona, km 33 600, 28871 Madrid, Spain
mdolores@aut.uah.es

Abstract. Mission Planning is a classical problem that has been traditionally studied in several cases from Robotics to Space missions. This kind of problems can be extremely difficult in real and dynamic scenarios. This paper provides a first analysis for mission planning to Unmanned Air Vehicles (UAVs), where sensors and other equipment of UAVs to perform a task are modelled based on Temporal Constraint Satisfaction Problems (TCSPs). In this model, a set of resources and temporal constraints are designed to represent the main characteristics (task time, fuel consumption, ...) of this kind of aircrafts. Using this simplified TCSP model, and a Branch and Bound (B&B) search algorithm, a set of feasible solutions will be found trying to minimize the fuel cost, flight time spent and the number of UAVs used in the mission. Finally, some experiments will be carried out to validate both the quality of the solutions found and the spent runtime to found them.

Keywords: unmanned aircraft systems, mission planning, temporal constraint satisfaction problems, branch and bound.

1 Introduction

Unmanned Aircraft Systems (UAS) can take advantage of planning techniques where the application domain can be defined as the process of generating tactical goals for a team of Unmanned Air Vehicles (UAVs). Nowadays, these vehicles are controlled remotely from ground control stations by humans operators who use legacy mission planning systems.

Mission planning for UAS can be defined as the process of planning the locations to visit (waypoints) and the actions that the vehicle can perform (loading/dropping a load, taking videos/pictures, acquiring information), typically over a time period. These planning problems can be solved using different methods such as Mixed-Integer Lineal Programming (MILP) [14], Simulated Annealing [2], Auction algorithms [8], etc. Usually, these methods are the best way

E. Corchado et al. (Eds.): IDEAL 2014, LNCS 8669, pp. 286–294, 2014.

to find the optimal solutions but, as the number of restrictions increase, the complexity grows exponentially because it is a NP-hard problem.

In the literature there are some attempts to implement UAS that achieve mission planning and decision making using temporal action logic (TAL) for reasoning about actions and changes [3], Markov Decision Process (MDP) and dynamic programming algorithms [4], or hybrid partial-order forward-chaining (POFC)[7], among others. Other modern approaches formulate the mission planning problem as a Constraint Satisfaction Problem (CSP), where the tactic mission is modelled and solved using constraint satisfaction techniques.

This work deals with multiple UAVs that must perform one or more tasks in a set of waypoints and specific time windows. The solution plans obtained should fulfill all the constraints given by the different components and capabilities of the UAVs involved over the time periods given. Therefore a Temporal Constraint Satisfaction Problem (TCSP) representation is needed. The approach from [9] is used to model a mission planning problem using Gecode [13] to program the constraints. In this previous work Backtracking (BT) method is applied to find the complete space of solutions, but in many real-life applications it is necessary to find only a good solution, what can be achieved considering a Constraint Satisfaction Optimization Problem (CSOP). For this purpose, in this work a new optimization function has been designed to look for good solutions minimizing the fuel cost, the flight time and the number of UAVs needed. Finally, Branch and Bound (B&B) search is employed for solving this CSOP model.

The rest of the paper is structured as follows: section 2 shows the state of the art in CSPs. Section 3 describes how a Misison is defined in the UAV domain and the modelization of the problem as a TCSP. In Section 4, the objective functions that will be used in the experimental phase are explained in detail. Section 5 explains the experiments performed and the experimental results obtained. Finally, the last section presents the final analysis and conclusions of this work.

2   Constraint Satisfaction Problems

A mission can be described as a set of goals that are achieved by performing some tasks with a group of resources over a period of time. The whole problem can be summed up in finding the correct schedule of resource-task assignments that satisfies the proposed constraints, like a CSP [1]. In a CSP, the states are defined by the values of the variables and the goal test specifies the constraints that the values must obey.

There are many studied methods to search the space of solutions for CSPs, such as BT, Backjumping (BJ) or look-ahead techniques (i.e. Forward Checking (FC)). BT search method solves CSP by incrementally extending a partial solution that specifies consistent values for some of the variables, towards a complete solution, and by repeatedly choosing a value for another variable consistent with the values in the current partial solution.

In many real-life applications it is necessary to find a good solution, and not the complete space of possible solutions. CSOP consists of a standard CSP and an optimization function (objective function) that maps every solution (complete labelling of variables) to a numerical value measuring the quality of the solution.

There are several methods for solving CSOP such as Russian doll search [12], Bucket elimination [11], Genetic algorithms [5] and Swarm intelligence [6]. The most widely used algorithm for finding optimal solutions is called B&B [10]. This algorithm searches for solutions in a depth first manner and behaves like BT except that as soon as a value is assigned to the variable, the value of heuristic function for the labelling is computed. If this value exceeds the bound (initially set to minus or plus infinity given it is a minimization or maximization problem), then the sub-tree under the current partial labelling is pruned immediately. The efficiency of B&B is determined by two factors: the quality of the heuristic function and whether a good bound is found early.

A TCSP is a particular class of CSP where variables represent times (time points, time intervals or durations) and constraints represent sets of allowed temporal relations between them [15]. A UAS mission can be perfectly represented as a set of temporal constraints over the time the tasks in the mission start and end. Besides the temporal constraints, the problem has several constraints imposing the proficiency of the UAVs to perform the tasks.

3   UAV Mission Plan Model Based on TCSPs

A UAV mission can be defined as a number n of tasks to accomplish for a team of UAVs. A task could be exploring a specific area or search for an object in a zone. One or more sensors belonging to a particular UAV, as can be seen in Table 1, may be required to perform a task. Each task must be performed in a specific geographic *area*, in a specific *time interval* and needs an amount of *payloads* to be accomplished.

Table 1. Different task actions considered

Id.	Action	Payload Needed
A1	Taking pictures of a zone	– Camera EO/IR
A2	Taking real-time pictures of a zone	– Camera EO/IR – Communications Equipment
A3	Tracking a zone	– Radar SAR

To perform a mission, there are a number m of UAVs, each one with some specific characteristics: *fuel* consumed, *maximum reachable speed*, *minimum cruise speed*, permission to go to *restricted areas*, and capacities or *payloads* (cameras, radars, communication equipments, ...). Moreover, in each point in time, each UAV is positioned at some specific *coordinates* and is filled with an amount of *fuel*. The main goal to solve the problem is to assign each task with a UAV that is able to perform it, and a start time of the UAV departure to reach the task area in time.

In this approach, the problem domain is modelled as a TCSP where the main variables are the *tasks* and their values will be the *UAVs* that perform each task and their respective *departure times*. There are two additional variables, the *fuel cost* and *distance travelled* for each task, that can be deduced from tasks assignment and UAV characteristics. Further details of this model can be seen at [9], but the main constraints defined in this model are as follows:

- **Temporal** constraints assuring a UAV does not perform two tasks at the same time.
- **Speed** window constraints: the mean cruise speed of a UAV to perform a task is contained in specific speed window v_{max} and v_{min} of the UAV.
- **Payload** constraints: checks whether a UAV carries the corresponding payload to perform a task.
- **Altitude** window constraints: the UAV altitude window must be contained in the altitude window of the area of the performing task.
- **Zone permission** constraints: just UAVs with permissions in restricted areas shall perform tasks developed in restricted areas.
- **Fuel** constraints: the total fuel cost for a UAV in a mission must be smaller than its actual fuel.

4   Optimization Function Description

In order to apply a method for solving CSOP, a new optimization function has been designed. This new function is looking to optimize (minimize) 3 objectives:

- The total **fuel consumed**, i.e the sum of the fuel consumed by each UAV at performing the tasks of the mission.
- The **number of UAVs** used in the mission. A mission performed with a lower number of vehicles is usually better because the remaining vehicles can perform other missions at the same time.
- The total **flight time**, i.e. the sum of the flight time of each UAV at performing the tasks of the mission. We have computed it as the difference between the ending of the last task performed by the UAV and its departure time.

Our model uses weights to map these three objectives into a single cost function, as the similar approach WCOP [16]. This function is computed as the sum of percentage values of these three objectives, as shown in Equation 1. In this sense, in the experimental phase, a comparative assessment of weights for finding feasible solutions of the problem is carried out. To solve the UAVs missions modelled is employed B&B search for minimization implemented by Gecode.

$$f_{cost}(i) = K_F \frac{Fuel(i)}{\max_j Fuel(j)} + K_U \frac{N°UAVs(i)}{\max_j N°UAVs(j)} + K_T \frac{FlightTime(i)}{\max_j FlightTime(j)}$$

$$K_F, K_U, K_T \in [0,1], \qquad K_F + K_U + K_T = 1 \quad (1)$$

5   Experimental Results

5.1   Mission Scenario Description

In this paper, a scenario (from the previously described model) with a group of 9 UAVs to perform a mission of 10 tasks is used for the experimental phase. Each task of the mission collides in time with its two previous tasks, i.e. task 10 collides with tasks 9 and 8; task 9, with tasks 8 and 7, and so on (see Figure 1).

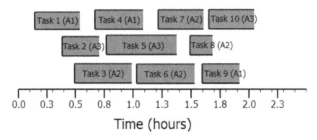

Fig. 1. Scenario perspective with dependency of each task with the two previous tasks

Each task is assigned an action, which type identifier from Table 1 is shown in Figure 1 after each task. The mission described is performed in approximately two hours and involves varied actions in different areas. Each of the 9 UAVs available has different types of payloads for performing the tasks. In this approach, we consider the topology specified in Figure 2.

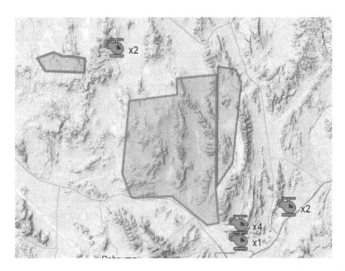

Fig. 2. Topology of the scenario where missions are performed. Coloured areas represent the areas where tasks are performed. Helicopters represent the airports where UAVs are situated at the beginning of the mission.

5.2   Results

Firstly, an analysis of the optimal solution found considering as cost function each one of the objectives individually is carried out. It can be seen in Table 2.

Table 2. Objective values and runtime spent in the search of the optimal solution using cost functions considering individually each objective

Cost function	Flight Time	No. of UAVs	Fuel	Runtime
100% Fuel	22h 8min 13s	4	269.561L	4min 9s
100% No. of UAVs	23h 22min 23s	4	282.003L	8.87s
100% Flight Time	18h 0min 8s	8	284.875L	7min 32s

It can be appreciated when considering cost function 100% Flight Time that, besides the high runtime needed, the optimal solution found has a high number of UAVs and fuel consumption. This could be due to shorter flight times are obtained using UAVs that reach higher speeds but consuming more fuel. Considering this aspect, in this simple approach we have decided to only consider fuel consumption and No. of UAVs for the comparative assessment of optimization function weights, see Table 3.

Table 3. Objective values and runtime spent in the search of the optimal solution using cost functions considering fuel and number of UAVs with different percentages

Cost function	Flight Time	No. of UAVs	Fuel	Runtime
100% Fuel	22h 8min 13s	4	269.561L	4min 9s
90% Fuel + 10% No. of UAVs	22h 8min 13s	4	269.561L	3min 22s
80% Fuel + 20% No. of UAVs	22h 8min 13s	4	269.561L	2min 7s
70% Fuel + 30% No. of UAVs	22h 8min 13s	4	269.561L	1min 39s
60% Fuel + 40% No. of UAVs	22h 8min 13s	4	269.561L	1min 23s
50% Fuel + 50% No. of UAVs	22h 8min 13s	4	269.561L	54.67s
40% Fuel + 60% No. of UAVs	22h 8min 13s	4	269.561L	46.03s
30% Fuel + 70% No. of UAVs	22h 8min 13s	4	269.561L	35.02s
20% Fuel + 80% No. of UAVs	22h 8min 13s	4	269.561L	33.99s
10% Fuel + 90% No. of UAVs	22h 8min 13s	4	269.561L	34.13s
100% No. of UAVs	23h 22min 23s	4	282.003L	8.87s

Analysing results shown in Table 3, it can be appreciated that only considering the fuel consumption in a low percentage, an optimal solution both for the fuel and number of UAVs minimization is reached. Additionally, it takes a better runtime than only considering fuel consumption. For this reason, it can be considered that a cost function of 10% fuel + 90% No. of UAVs is pretty good for searching feasible solutions of the kind of problem solved.

Finally, the runtime spent in the search of feasible solutions and the runtime spent in the search of the entire space of solutions using BT are compared in Figure 3. The time difference observed is very high, as expected. Concretely the BT runtime is higher than B&B in an order of $3 \cdot 10^5$.

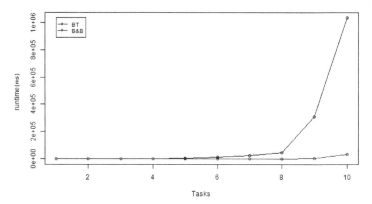

Fig. 3. Runtime spent in the search of the space of solutions with BT and the optimal solution with B&B using the cost function 10% fuel + 90% No. of UAVs

6   Conclusions and Discussion

In this paper, we try to search feasible solutions for a UAV Mission Planning model based on TCSP. The presented approach defines missions as a set of tasks to be performed by several UAVs with some capabilities. The problem is modelled using: (1) temporal constraints to assure that each UAV only performs one task at a time; (2) logical constraints such as the maximum and minimum altitude reachable or restricted zone permissions, and (3) resource constraints, such as the sensors and equipment needed or the fuel consumption.

Concretely, we have designed an optimization function to minimize three objectives: the fuel consumption, the number of UAVs used in the mission and the total flight time of all the UAVs. From the obtained results, we have observed that the flight time does not help in the optimization of the rest of the objectives; so we have considered to put it aside.

Studying the solutions found by several cost functions with different weights for fuel and number of UAVs, we have observed how the runtime spent in the search decrease as the percentage of fuel decreases. Moreover, we show the solutions obtained by these cost functions inside the POF of fuel versus No. of UAVs, observing that for the cost function 10% fuel + 90% number of UAVs we obtain both the optimal solutions obtained with 100% fuel and 100% No. of UAVs, which together with its low runtime makes this cost function pretty good for finding feasible solutions in a reasonable time.

It is important to remark that the results obtained are highly dependant on the proposed scenarios and on the topology of the areas the missions are developed in. So further works should consider different scenarios and topologies, so a more general conclusion would be obtained. Furthermore, we will use a Multiobjective model, such as the Multiobjective Evolutionary Algorithms (MOEAs), and SPEA2 or NSGA-II algorithms; to find the Pareto Optimal Frontier.

As future lines of work, these results need to be compared against other optimization algorithms (such as Tabu Search and Genetic Algorithms, among others) to observe which one is better in terms of optimality of the solutions and runtime spent. Using these new algorithms, new heuristics to reduce the complexity of the problem and adapting our current model, we expect to be able to simulate problems near to real scenarios.

Acknowledgments. This work is supported by the Spanish Ministry of Science and Education under Project Code TIN2010-19872 and Savier Project (Airbus Defence & Space, FUAM-076915). The authors would like to acknowledge the support obtained from Airbus Defence & Space, specially from Savier Open Innovation project members: José Insenser, César Castro and Gemma Blasco.

References

1. Barták, R.: Constraint programming: In pursuit of the holy grail. In: Proceedings of the Week of Doctoral Students, pp. 555–564 (1999)
2. Chiang, W.C., Russell, R.A.: Simulated annealing metaheuristics for the vehicle routing problem with time windows. Annals of Operations Research 63, 3–27 (1996)
3. Doherty, P., Kvarnström, J., Heintz, F.: A temporal logic-based planning and execution monitoring framework for Unmanned Aircraft Systems. Autonomous Agents and Multi-Agent Systems 19(3), 332–377 (2009)
4. Fabiani, P., Fuertes, V., Piquereau, A., Mampey, R., Teichteil-Konigsbuch, F.: Autonomous flight and navigation of VTOL UAVs: from autonomy demonstrations to out-of-sight flights. Aerospace Science and Technology 11(2-3), 183–193 (2007)
5. Fonseca, C., Fleming, P.: Multiobjective optimization and multiple constraint handling with evolutionary algorithms. I. A unified formulation. IEEE Transactions on Systems, Man and Cybernetics 28(1), 26–37 (1998)
6. Gonzalez-Pardo, A., Camacho, D.: A new CSP graph-based representation for ant colony optimization. In: 2013 IEEE Conference on Evolutionary Computation (CEC 2013), vol. 1, pp. 689–696 (2013)
7. Kvarnström, J., Doherty, P.: Automated planning for collaborative UAV systems. In: Control Automation Robotics & Vision, pp. 1078–1085 (December 2010)
8. Leary, S., Deittert, M., Bookless, J.: Constrained UAV mission planning: A comparison of approaches. In: 2011 IEEE International Conference on Computer Vision Workshops (ICCV Workshops), pp. 2002–2009 (November 2011)
9. Ramirez-Atencia, C., Bello-Orgaz, G., R-Moreno, M.D., Camacho, D.: A simple CSP-based model for Unmanned Air Vehicle Mission Planning. In: IEEE International Symposium on INnovations in Intelligent SysTems and Application (2014)
10. Rasmussen, S., Shima, T.: Branch and bound tree search for assigning cooperating uavs to multiple tasks. In: American Control Conference, pp. 6–14 (2006)
11. Rollon, E., Larrosa, J.: Bucket Elimination for Multiobjective optimization problems. Journal of Heuristics 12(4-5), 307–328 (2006)

12. Rollon, E., Larrosa, J.: Multi-objective Russian doll search. In: Proceedings of the National Conference on Artificial Intelligence, vol. 22, p. 249. AAAI Press, MIT Press, Menlo Park, Cambridge (1999, 2007)

13. Schulte, C., Tack, G., Lagerkvist, M.Z.: Modeling and Programming with Gecode (2010), http://www.gecode.org/

14. Schumacher, C., Chandler, P., Pachter, M., Pachter, L.: UAV Task Assignment with Timing Constraints via Mixed-Integer Linear Programming. Tech. rep., DTIC Document (2004)

15. Schwalb, E., Vila, L.: Temporal constraints: A survey. Constraints 3(2-3), 129–149 (1998)

16. Torrens, M., Faltings, B.: Using Soft CSPs for Approximating Pareto-Optimal Solution Sets. In: AAAI Workshop Proceedings Preferences in AI and CP: Symbolic Approaches. AAAI Press (2002)

Towards Data Mart Building
from Social Network for Opinion Analysis

Imen Moalla and Ahlem Nabli

Sfax University, Faculty of Science, Road Sokra Km 3 BP 1171, 3000 Tunisia
MIRACL Laboratory, Sfax, Tunisia
imen.moalla@hotmail.fr,
ahlem.nabli@fsegs.rnu.tn

Abstract. In the recent years, social networks have played a strategic role in the lives of many companies. Therefore several decision makers have worked on these networks for making better decisions. Furthermore, the increased interaction between social networks and web users has lead many companies to use a data warehouse to collect information about their fans. This paper deals with a multidimensional schema construction from unstructured data extracted from social network. This construction is carried out from Facebook page in order to analyze customers' opinions. A real case study has been developed to illustrate the proposed method and confirming that he social network analysis can predict chance of the success of products.

Keywords: Data warehouse, Data Mart, Social Networks analysis, Opinion Analysis, Schema Design.

1   Introduction

Recently, social networks and online communities, such as Twitter and Facebook, have become a powerful source of knowledge being daily accessed by millions of people [2]. Social networks help people to follow breaking news, to keep up with friends or colleagues and to contribute to online debates. Seen the emergence of these social networks, several decision maker's aim to use the abundant information generated by these networks to improve their decisions. Indeed, the main advantage of social networks is to enable companies to operate a new visibility shape on the Internet at a lower cost. Its a way to make them known to different publics and collect information on customers prospective. However, business intelligence provides a solution for companies, which allows to collect, consolidate, model and restore the data of a company offering help to decision makers. The popularity of social networks and high volume of user generated content, especially subjective content caused the heavy demand to adopt sentiment analysis in business applications. Sentiment Analysis helps decision makers to get customer opinion in real-time [3]. This real-time information helps them to design new marketing strategies, improve product features and can predict chances of product failure.

E. Corchado et al. (Eds.): IDEAL 2014, LNCS 8669, pp. 295–302, 2014.

The social network analysis problem took a big importance from the scientific community. In the literature, the data analysis approaches from social networks highlight two areas of research: community analysis and opinions analysis [10]. So, data warehouse can be used to analyze the opinion of customers based on many dimensions. In this context and in order to benefit from existing information in social networks, we are interested in using these data as a source feeding a data Mart schema. In fact, we propose a heuristic to design data Mart schema from Facebook page in order to analyze customers' opinions.

The remainder of this paper is organized as follows: section 2 details the related works. Section 3 describes our method Data Mart construction from the social network Facebook. Opinion analysis is performed in section 4. Finally, section 5 draws conclusion and future works.

2   Related Works

The importance of social networks for decision making process is highlighted in several studies such as [9,10,4,1,7]. For example, Rehman and al. [9] proposed an architecture to extract tweets from Twitter and load them to a data warehouse. Multidimensional cubes resulting can be used for analyzing user's behavior on twitter during event earthquake in Indonesia. But in this work, the authors present schema without detailing how this schema is determined. However, in [10] the authors proposed a Business Intelligence architecture, called OSNBIA (Online Social Networks Business Intelligence Architecture). Therefore, the authors did not explain how the data warehouse schema is performed. In 2011 Kazienko and al. [4] proposed a multidimensional model for social network that allows to capture data about activities and interactions between web users. The multidimensional model of the social network can be used To make a new relationships and to analyze different communication ways. Also, in this study the researchers did not perform an experimental phase to validate their process. A data warehouse schema is proposed by [1] to analyze the big volume of data tweets. The authors suggested using information retrieval approaches to classify the most significant words in the hierarchy level of the dimensions. Moya and al.[7] presented an approach to integrate sentiment data extracted from the web into the corporate data warehouse. In 2013 Mansmann and al.[6] proposed to model data warehouse elements from the dynamic and semi-structured data of Twitter. In addition, they propose to extend the resulting model by including dynamic categories and hierarchies discovered from DM and semantic enrichment methods. In [2] the researchers presented a data analysis framework to discover groups of similar twitter messages posted by users about an event.

The comparative study shows that the data warehouse design methods [9,10,1] are based on the social network twitter except the work of [7] that used an approach from Web. In addition, the multidimensional concepts are generally defined by the designer without presenting how these concepts are determined [7,10,1]. Furthermore all methods operate on the texts except [4] operates on the link. Based on these lakes, we propose a method composed of set steps, applied to Facebook page, which based on heuristics to generate the data Mart schema.

3   Schema Design from Facebook Page

This paper deals with a method to generate a data Mart schema from Facebook page. As depicted in Fig.1, our method encloses four steps: (1) *Data extraction* which involves collecting information about Facebook page, fans and posts. (2) *Data analysis* that includes specific operations to select relevant data and to eliminate inconsistency. (3) *Schema modeling* which suggests multidimensional concepts. (4) *Loading step* that incorporates procedures for charging modeled schema from the second step. Since the data mart is constructed, the decision maker can analyze the opinions of fan page based on MDX query. Before performing our method we start by introducing Facebook page.

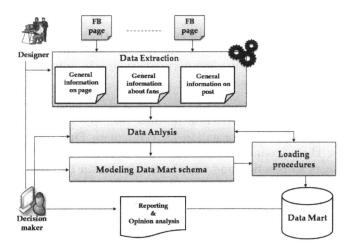

Fig. 1. The proposed method Process

Facebook Page. (FP) is a sort of website integrated into the Facebook community which allows the companies to interact with customers by starting discussions within comments and photos. It has three types of information: general information of the page, information about fans and about posts. In the following sections we give an idea about data extraction and data analysis. After that, we explain the extraction of multidimensional concepts in section 3. Finally, we expose reporting based on opinion analysis query.

3.1   Data Extraction

The first step of our method is to extract data from a FP, so we collect general information of FP, information about fans and information related to posts. In fact, this extraction consists of creating an application enabling the access to the API graph in order to select the access permissions to the data. The general information of FP are numerous, here we elicit some of them: Id (page identifier), name (page name), user name (FP username), category (FP type), number of fans (number of fans in FP), site (link website), etc. We recognize that a post can be photos, links, videos or statutes. So, each one is concerned by some information. For example, a post is defined by the ID (post identifier), name (post name), type(the type of post), created date (date of creation of post), number of likes (number of fans liked post), number of shares (number of fans shared post) etc. We noticed a difference between the data collected on the fans who have a friendship with the admin of FP and who're without friendship with the admin. The data available for the first category are the data that we select their access tokens and which are visible to fans friends. However, for fans not friends we have only public data.

3.2 Data Analysis

Since data are collected we proceed to analyze these data. The analysis concerns the selection of relevant data, the elimination of duplicate and inconsistency data. The decision maker is implied during this step in order to contribute in the selection of relevant data.

Select Relevant Data: The main needs of the decision maker are to determine the client's opinions' and satisfaction towards their products. In fact, client's opinions provide valuable information for companies. First, they help to understand how their products and services are perceived [5]. They yield clues about costumers satisfaction and expectations that can be used to determine their current and future needs and preferences. Second, they may be helpfull in understanding what product dimensions or attributes are important to each clients. Third, client's opinions on the products offered by competitors provide essential information to accomplish a successful competitor analysis.

Eliminating Duplicates and Inconsistency Data: From studied case, we found that, when a post is published many times in the page, Facebook assigns for each post an ID. Therefore, the same post has different ID. In this case, analyzes on the same post are distributed over multiple IDs instead of the same ID. To solve this problem, we decided to group the different ID of the same post published several times and assigned to them the same identifier. Another problem was detected when computing the number of positive comments. Indeed, the information extracted from the comments can be relative of fans or admin; this made us a need to delete the comments of the admin page.

3.3 Modeling Schema

This phase consists in defining the multidimensional schema of the data Mart from the extracted data. This definition consists in determining: facts, measures, dimensions and hierarchies. This definition is based on heuristics adapted from [8].

Heuristics of Fact and Measure Determination: Fact to be analyzed describes the daily activity of events performed in a Facebook page which allow capturing the opinions of fans. Thats why we define the **Fact Opinion**.

The opinions expressed by fans on posts can be positive or negative. The positive opinions on the posts are determined through the following heuristics:

Share-Measure: if the post has an important number of shares then we conclude that there are a number of fans impressed by the product. This allows us to define the measure Number of shares called *Number-shares*.

Likes-Measure: when the post has an important number of likes this means that there are a significant number of fans who liked the product. This allows us to define the measure named Number of likes called *Number-likes*.

Comments-Measure: every post can have a set of comments. Comment can be positive or negative. It is qualified as positive if it contains positive words, which means that the fan is interested by the product. In our context, we determined all the positive words through which we can conclude that the comment is positive. These latter give us to define the measure *Number-comment-positive*. The negative opinions on posts are captured by actions realized by fans on the posts. These actions are deducted when the fans click mask or signal on a post of the page in their newsfeed. So, the measure *Number-comment-negative* is proposed. All the determined measures are depicted in Fig.3.

Heuristics of Dimensions Determination : The extraction of the dimensions is based on a type of object named base object BO which completes the definition of fact. A base object answers the questions: "who", "what", "when" and "where". Every BO defined an axis of analysis which can interest the decision maker. From a FP, the objects which answer these questions for the fact opinion are:

Who declare the opinion ? – Fans declare their opinions.
Where are posted the opinions ? – The opinions are published in page.
What are the opinions ? – Posts encapsulate the opinions.
When the posts have been shared ? – The post has its publication date.

Based on these four questions we obtain four dimensions: the dimensions page, post, fans and date. Due to the lack of space we present only the parameter extraction of Page dimension.

Table 1. Parameters of the page dimension

Tags describing the page	Type concepts	Name of multidimensional concept
⟨id⟩	identifier	Page-ID
⟨name⟩	weak attribute	name
⟨username⟩	weak attribute	user name
⟨website⟩	weak attribute	website
⟨phone⟩	weak attribute	phone
⟨category⟩	Level 2 parameter	category
⟨location⟩	Level 2 parameter	location

Extraction of Dimension Parameters': The dimension Page is determined from the general information of Facebook page. Table 1 presents all the attributes composing the dimension Page (column 1). Based on this information we determine the type of each attribute (weak attribute or parameter) in column 2. Then we define the equivalent multidimensional concept in column 3. Fig.2 depicts a graphical representation of the page and date dimensions.

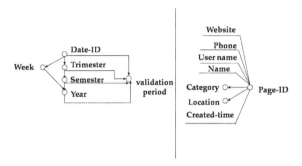

Fig. 2. Hierarchies of Page and Date dimensions

Notes that, a FP can be removed from the Net, so we define a parameter called validation period in which a page is active as shown in Fig 2.

From the previous steps, we generate the data Mart schema modeled in X-DFM structure as shown in Fig 3.

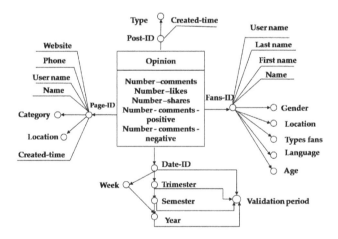

Fig. 3. Data Mart schema generated from Facebook page

The structure of the schema is a graph centered at the fact type node (opinion) which includes all measures (Number-comments, Number-likes, Number-comments-positive, and Number-comments-negative). The fact is relied by four dimensions (Page, Fans, Date, and Post) each of them contents a node key attribute and others node represents dimension parameters (user name, first name, last name, phone, website, created-time).

4   Opinion Analysis

Opinion analysis is a crucial step for communities. The goal is to help decision makers to model strategies and access to relevant information. From our Data Mart constructed we analyzed the number of likes, shares, and the number of positive and negative comments by week, trimester and semester. The decision makers can use of the following queries:

- Analyze the number of comments for posts by week, trimester and semester.
- Analyze the number of likes for posts by week, trimester and semester.
- Analyze the number of shares for posts by week, trimester and semester.
- Analyze the number of positive and negative comments for posts by week, trimester and semester.

We performed a comparison between the measurements taken from the page and actual sales of a company. As shown in Fig .4, we note that the curve for the sharing follows the same shape as the real company sales. So, we may infer that social networks can predict sales trends of a company based on the number of shares.

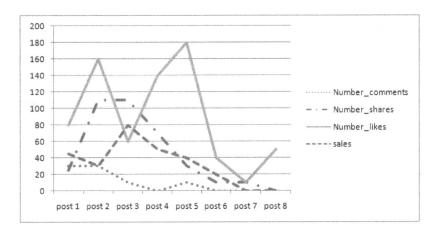

Fig. 4. Best posts for the first semester of 2013

5   Conclusions and Future Work

We have presented in this paper a method to build a data Mart from a social network for opinion analysis. This method uses Facebook page as data source. it's invloved on four steps: the first is data extraction. The second step is the data analysis that consists of making specific operations to select the relevant data and the third step is schema definition that defines the multidimensional concepts of the data Mart. The analysis that we made, we give us an overview on the actual case of the company's sales. Our future orientations consist of studying the possibility of using ontology to better analyses fans opinions. Thus, we will enhance our method to support other types of social networks.

References

1. Bringay, S., Béchet, N., Bouillot, F., Poncelet, P., Roche, M., Teisseire, M.: Towards an On-Line Analysis of Tweets Processing. In: Hameurlain, A., Liddle, S.W., Schewe, K.-D., Zhou, X. (eds.) DEXA 2011, Part II. LNCS, vol. 6861, pp. 154–161. Springer, Heidelberg (2011)
2. Baralis, E., Cerquitelli, T., Chiusano, S., Grimaudo, L., Xiao, X.: Analysis of Twitter Data Using a Multiple-level Clustering Strategy. In: Cuzzocrea, A., Maabout, S. (eds.) MEDI 2013. LNCS, vol. 8216, pp. 13–24. Springer, Heidelberg (2013)
3. Jaganadh, G.: Opinion Mining and Sentiment Analysis, CSI Communications (2012)
4. Kazienko, P., Kukla, E., Musial, K., Kajdanowicz, T., Bródka, P., Gaworecki, J.: A generic model for a multidimensional temporal social network. In: Yonazi, J.J., Sedoyeka, E., Ariwa, E., El-Qawasmeh, E. (eds.) ICeND 2011. CCIS, vol. 171, pp. 1–14. Springer, Heidelberg (2011)
5. Laura, P., Jorge, C.A.: Sentiment Analysis in Business Intelligence: A survey. In: Customer Relationship Management and the Social and Semantic Web. IGI-Global (2011)
6. Mansmann, S., Rehman, N.U., Weiler, A., Scholl, M.H.: Discovering OLAP dimensions in semi-structured data. Information Systems (2013)
7. Moya, L.G., Kudama, S., Aramburu, J., Llavori, R.B.: Integrating Web Feed Opinions into a Corporate Data Warehouse. In: Proceedings of the 2nd International Workshop on Business Intelligence and the WEB BEWEB, New York (2011)
8. Nabli, A.: Thesis: Proposal of an approach of help to the design automated schema of data warehouse (2010)
9. Rehman, N.U., Mansmann, S., Weiler, A., Scholl, M.H.: Building a data warehouse for twitter stream exploration. IEEE/ACM (2012)
10. Santos, P., Souza, F., Times, V., Benevenuto, F.: Towards integrating Online Social Networks and Business Intelligence. In: Proceedings of the IADIS International Conference on Web Based Communities and Social Media (2012)

A Novel Self Suppression Operator Used in TMA

Jungan Chen, ShaoZhong Zhang, and Yutian Liu

Electronic Information Department
Zhejiang Wanli University,
No.8 South Qian Hu Road,
Ningbo, Zhejiang, 315100, China
friendcen21@hotmail.com, {dlut_z88,lyt808}@163.com

Abstract. In V-detector or TMA-OR, the parameters self radius rs or Omin are required to be set by experience. To solve the problem, a novel self suppression operator based on self radius learning mechanism is proposed. The results of experiment show that the proposed algorithm is more effective than V-detector or TMA-OR when KDD and 2-dimensional synthetic data are as the data set.

Keywords: artificial immune system, negative selection, self suppression operator.

1 Introduction

Nowadays, Artificial Immune System (AIS) has been applied to many areas such as computer security, classification, learning and optimization [1]. Negative Selection Algorithm, Clonal Selection Algorithm, Immune Network Algorithm and Danger Theory Algorithm are the main algorithms in AIS [2][3].

A real-valued negative selection algorithm with variable-sized detectors (V-detector Algorithm) applied in abnormal detection is proposed to generate detectors with variable r. A statistical method (naïve estimate) is used to estimate detect coverage in V-detector algorithm[4]. But as reported in Stiboret later work, the performance of V-detector on the KDD Cup 1999 data is unacceptably poor[5]. So a new statistical approach (hypothesis testing) is used to analyze the detector coverage and achieve better performance [6]. In the statistical approach, p is defined as the proportion of covered nonself points, n is defined as the number of detectors.The assumption, np>5, n(1-p)>5 and n>10, are required to be satisfied.When p is set to 90%, n must be set to at least 50. Sometimes the number of detectors do not have to be more than 50, so it is unreasonable and the performance of the algorithm is less effective because the number of detectors affect the detect performance.

Naïve estimate and hypothesis testing are two methods discussed above. When the number of detctors required is less than 50 , hypothesis testing is not the ideal solution. Actually in naïve estimate method, the candidate detector is added to valid detector set only when it is not detected by any of valid detectors, which means that the distance between candidate detector and any of valid detectors is bigger than the match threshold of the related valid detector. This process can maximize the distance

E. Corchado et al. (Eds.): IDEAL 2014, LNCS 8669, pp. 303–308, 2014.
© Springer International Publishing Switzerland 2014

among valid detectors. But it is difficult to find valid detector with the number of valid detectors increasing. At worst, it is possible that there is no other valid detector generated after one valid detector is generated, which leads that naïve estimate method shows unacceptable result on the KDD data. So the distance among valid detectors chosen in naïve estimate method can affect the number of detectors generated.

To choose the appropriate distance among valid detectors and achieve the optimized number of detectors, a parameter overlap rate (Omin) in T-detector Maturation Algorithm (TMA) is proposed to control the distance among detectors [7].But the optimized Omin is required to be set by experience. To solve this problem, a suppression operator called Negative Selection operator (NS operator) proposed in refrence[8] is used in TMA. and there is no parameter Omin in TMA with NS operator.

Later, a self radius learning mechanism is proposed to achieve the adaptive self radius. By combining NS operator and self radius learning mechanism, an augmented TMA called TMA with Adaptive Capability (TMA-AC) is proposed[9]. But TMA-AC shows less effective than TMA-OR when 2-dimensional synthetic data is as the data set.

In this paper, a novel self suppression operator based on the self radius learning mechanism is proposed to suppress the number of detectors and the TMA algorithm with Self Suppression operator (TMA-SS) is put forward.

2   Algorithm

2.1   Match Range Model

$U=\{0,1\}^n$,n is the number of dimensions. The normal set is defined as selves and abnormal set is defined as nonselves. selves∪nonselves=U. selves∩nonselves =Φ.There are two points $x=x_1x_2...x_n$, $y=y_1y_2...y_n$. The Euclidean distance between x and y is:

$$d(x, y) = \sum_{i=1}^{n} (x_i - y_i)^2 \qquad (1)$$

The detector is defined as dct = {<center, selfmin, selfmax > | center ∈ U,   selfmin, selfmax∈N}. center is one point in U. selfmax is the maximized distance between dct.center and selves. selfmin is the minimized distance. The detector set is defined as DCTS. Selfmax and selfmin are calculated by setMatchRange(dct, selves), dct.center∈U, i∈[1, |selves|], $self_i$∈selves :

$$setMatchRa\,nge = \begin{cases} selfmin = min(\{d(sel\,f_i, dct.center)\}) \\ selfmax = max(\{d(sel\,f_i, dct.center)\}) \end{cases} \qquad (2)$$

[selfmin,selfmax] is defined as self area. Others is as nonself area. Suppose there is one point x ∈ U and one detector dct ∈ DCTS. When d(x,dct) ∉ [dct.selfmin, dct.selfmin], x is detected as abnormal.

2.2   Self Radius Learning Mechanism

To learning the appropriate self radius, a property minselfList is added to the detector, which has four properties include center, selfmin, selfmax and minselfList. The property minselfList has three element $\{self_0, self_1, self_2\} \in selves$, which   have the minimized distance with the center of a given detector.

Self radius r_s can be achieved by the equation 4. $\{m,n\} \in \{0,1,2\}$ and $m \neq n$

$$d_{mn} = d(self_m, self_n) \tag{3}$$

$$r_s = \frac{\sum d_{mn}}{3} \tag{4}$$

The average self radius is defined as avgrs. The equation 5 is used to calculate the value of avgrs.

$$avgrs = \begin{cases} r_s, & avgrs = 0 \\ \dfrac{avgrs + r_s}{2}, & others \end{cases} \tag{5}$$

2.3   Self Suppression Operator

The novel self suppression operator is proposed to eliminate those detectors which are recognized by others. The equation 6 is used to decided wehter a given detector dct_x is valid. dct_x will be removed if it is not valid.

$$IsValid(dct_x, DCTS) = \begin{cases} false, & UnValid > 1 \cap dct_x.self\ min > avgrs \\ true, & others \end{cases} \tag{6}$$

$$\begin{aligned} \exists dct_k \in DCTS \\ if(NSMatchAnd(dct_x, dct_k)) \quad UnValid = UnValid + 1 \end{aligned} \tag{7}$$

$$NSMatchAnd = \begin{cases} true, & \forall d_{ab} = 0 \\ false, & others \end{cases} \tag{8}$$

$$d_{ab} = d(dct_x.minselfList.self_a, dct_k.minselfList.self_b), \{a,b\} \in \{0,1,2\} \tag{9}$$

2.4   The Model of Algorithm

The algorithm, called TMA-SS (TMA with Self Suppression operator), is shown in Fig.1. Step 5 is used to decide whether candidate detector is a valid detector according equation 6.Step 10 is used to estimate the detect coverage.

```
1.  Set the desired coverage pc
2.  Generate one candidate detector dct_x randomly
3.  setMatchRange(dct_x,selves)
4.  compute the self radius r_s and average self raddius
avgrs according equation 4,5
5.  if isvalid(dct_x,DCTS) then // equation 6
6.     dct_x is added to detector set DCTS
7.        covered=0
8.  Else
9.        covered ++
10. If covered <1/(1- pc) then goto 2
```

Fig. 1. TMA-SS algorithm model

3　Experiments

For the purpose of comparison, experiments are carried out using KDD and 2-dimensional synthetic data in table 1,which is described in Zhou's paper[10]. In 2-dimensional synthetic data,various shapes over the unit square $[0,1]^2$ are used as the self region. In every shape, there are training data (self data) of 1000 points and test data of 1000 points including both self points and nonself points. As for KDD data, 20 subsets were extracted from the enormous KDD data using a process described in [5]. Self radius and Omin used in TMA-OR are given in table.1. All the results shown in these figures are average of 100 or 20 (see table 1) repeated experiment with coverage rate 99%.

Table 1. Data set and parameters used in experiments

Data set		Parameters		
		r_s	*Omin*	Repeated times
2-dimensional synthetic data	Comb	0.03	0, 0.7	100
	Cross			
	Intersection			
	Pentagram			
	Ring			
	Stripe			
	Triangle			
KDD data		0.05		20

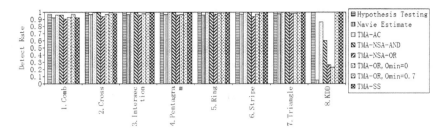

Fig. 2. Detect Rate

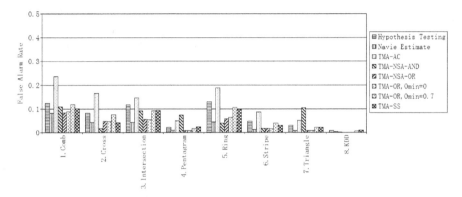

Fig. 3. False Alarm Rate

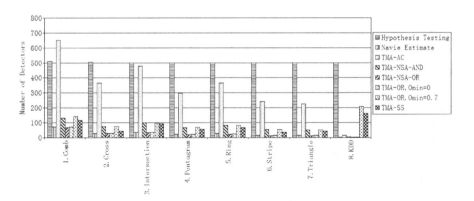

Fig. 4. Number of Detectors

As reference[7], TMA-OR(Omin=0.7) can achieve the best effect. By comparing with TMA-OR(Omin=0.7), TMA-SS get almost the same detect rate (Fig.2) with lower false alarm rate(Fig.3) and smaller number of detectors(Fig.4). So the result of experiment show that TMA-SS is more effective than other algorithm.

4 Conclusion

As the parameters Omin and self radius r_s in TMA-OR are required to be set by experience. To solve the adaptive problem, a novel self suppression operator based on the self radius learning mechanism is proposed, and then an augmented TMA called TMA-SS is proposed. The results of experiment show that the proposed algorithm is more effective than other algorithms.

Acknowledgement. This work is supported by National Natural Science Foundation of China 71071145, Zhejiang Public Welfare Technology Research Project Foundation 2013C31122,Zhejiang Provincial Nature Science Foundation Y1110200, Ningbo Nature Science Foundatio 2012A610010, 2013A610070 , The Science and Technology Innovation Team of Ningbo (Grant No.2013 B82009,2012B82006).Thanks for the assistance received by using KDD Cup 1999 data set [http://kdd.ics.uci.edu/databases / kddcup99/ kddcup99.html] , the 2-dimensional synthetic data set [https:// umdrive. memphis.edu/ zhouji/ www/ vdetector.html].

References

1. Hart, E., Timmis, J.: Application areas of AIS: The past, the present and the future. Journal of Applied Soft Computing 8(1), 191–201 (2008)
2. Timmis, J., et al.: An interdisciplinary perspective on artificial immune systems. Evolutionary Intelligence 1(1), 5–26 (2008)
3. Greensmith, J., Aickelin, U., et al.: Information Fusion for Anomaly Detection with the Dendritic Cell Algorithm. Information Fusion 11(1), 21–34 (2010)
4. Ji, Z., Dasgupta, D.: Real-Valued Negative Selection Algorithm with Variable-Sized Detectors. In: Deb, K., Tari, Z. (eds.) GECCO 2004. LNCS, vol. 3102, pp. 287–298. Springer, Heidelberg (2004)
5. Stibor, T., Timmis, J.I., Eckert, C.: A Comparative Study of Real-Valued Negative Selection to Statistical Anomaly Detection Techniques. In: Jacob, C., Pilat, M.L., Bentley, P.J., Timmis, J.I. (eds.) ICARIS 2005. LNCS, vol. 3627, pp. 262–275. Springer, Heidelberg (2005)
6. Ji, Z., Dasgupta, D.: Estimating the Detector Coverage in a Negative Selection Algorithm. In: Genetic and Evolutionary Computation Conference (2005)
7. Chen, J.: T-detector Maturation Algorithm with Overlap Rate. WSEAS Transactions on Computers 7(8), 1300–1308 (2008)
8. Chen, J.: A novel suppression operator used in optaiNet. BSBT 57, 17–23 (2009)
9. Chen, J., Zhang, Q., Fang, Z.: Improve the Adaptive Capability of TMA-OR. In: Omatu, S., Paz Santana, J.F., González, S.R., Molina, J.M., Bernardos, A.M., Rodríguez, J.M.C. (eds.) Distributed Computing and Artificial Intelligence. AISC, vol. 151, pp. 665–671. Springer, Heidelberg (2012)
10. Ji, Z.: Negative Selection Algorithms: from the Thymus to V-detector. PhD Dissertation, University of Memphis (2006)

Machine Learning Methods for Mortality Prediction of Polytraumatized Patients in Intensive Care Units – Dealing with Imbalanced and High-Dimensional Data

María N. Moreno García[1,*], Javier González Robledo[2], Félix Martín González[2], Fernando Sánchez Hernández[3], and Mercedes Sánchez Barba[4]

[1] Department of Computing and Automation, University of Salamanca, Salamanca, Spain
Department of Computing and Automation, Plaza de los Caídos s/n, 37008 Salamanca
mmg@usal.es
[2] Intensive Care Unit, University Hospital of Salamanca, Salamanca, Spain
[3] School of Nursing and Physiotherapy, University of Salamanca,
Prehospital Emergency Services, Salamanca, Spain
[4] Department of Statistics, University of Salamanca, Salamanca, Spain

Abstract. The aim of this study is the prediction of death of polytraumatized patients based on epidemiological, clinical and health treatment variables by means of data-mining methods. The main problems to be addressed were high dimensionality and imbalanced data. Since the techniques usually used to deal with these drawbacks, as feature selection methods and sampling strategies respectively, did not provided satisfactory results, the aim of the study was to find out the data mining algorithms showing the best behavior in this kind of scenarios. The study was carried out with data from 497 patients diagnosed with severe trauma who were hospitalized in the Intensive Care Unit (ICU) of the University Hospital of Salamanca. The results of the study reveal the better behavior of multiclassifiers as compared with simple classifiers in contexts of high dimensionality and imbalanced datasets, without the need to resort to undersampling and oversampling strategies, which can lead to the loss of valuable data and overfitting problems respectively.

Keywords: Severe trauma, polytrauma, mortality, data mining, classifiers, multiclassifiers.

1 Introduction

Severe trauma is considered to be one of the pathologies with the greatest impact on current society from the point of view of health as well as from the economic perspective. It is the primary cause of mortality of young adults in the world and the most influential as regards the years of potential life lost (YPLL). Regarding the economical aspect, it has been reported that the average economic cost for the treatment of traumatic injury in the United States is greater than the treatment of cancer and cardiovascular diseases [8].

---

* Corresponding author.

E. Corchado et al. (Eds.): IDEAL 2014, LNCS 8669, pp. 309–317, 2014.

The care of polytraumatized patients represents a challenge for current society and, in particular, for health professionals seeking to decrease its negative social and economical impact as well as personal consequences for the patient as much as possible. Current technology affords the possibility of storing huge amounts of medical data as electronic health records (EHRs), which can be processed by advanced techniques, such as data mining algorithms, to obtain useful knowledge on which decisions can be based. However, an important problem to be addressed is the great quantity and variety of variables that is necessary to take into account, from demographic data to clinical variables as well as those related to the healthcare management.

One usual way to deal with that drawback is the application of feature selection methods to know the best attributes for classification in order to use them for building the predictive models, discarding the remainder ones. In most of the cases a better accuracy is achieved when the algorithms work with the selected attributes, however, in the application domain considered in this study these techniques did not provided good results since the accuracy was not improved and in some experiments it became even worse. Two well-known and widely used algorithms whose efficacy has been demonstrated have been applied: CFS (Correlation-based Feature Subset Selection) [6] and a method based on information gain (IG) with respect to the class [7]. On the other hand, an additional problem to be addressed is the treatment of imbalanced data since the number of records belonging to one class is much greater than the number of records of the other class. In this kind of scenarios machine learning algorithms can achieve an acceptable global accuracy but usually the precision for the minority class is low. Oversampling of the minority class records or undersampling of the majority class records are two common approaches to deal with imbalanced datasets, but they have important drawbacks. Undersampling may discard potentially valuable data, while oversampling artificially increases the size of the data set and, as a result, the computational cost of inducing the models. In addition, the replication of existing examples in the minority class causes overfitting problems [9].

The aim of the present study is to apply suitable machine learning algorithms, paying a special attention to multiclassifiers, in order to overcome the mentioned problems. Data mining techniques have been successfully used to infer knowledge in very diverse medical areas; however, in spite of their great interest and promising results, these methods have not yet been exploited in the specific domain of politraumatized patient treatment in intensive care units. In this study they are used to predict the final outcome of these patients.

2   Background

This section is devoted to expose some basic aspects of multiclassifiers that can help to understand their general better behavior against single classifiers. The book of Kuncheva [10] has been taken as reference to develop the contents of sections 2 and 3.

Multiclasifiers combine several individual classifiers induced with different basic methods or obtained from different training datasets with the aim of improving the accuracy of the predictions. The methods for building multiclassifiers can be divided in two groups. The first, such as Bagging [1] and Boosting [3], induce models that merge classifiers with the same learning algorithm, but introducing modifications in the training data set. The second type of methods, named hybrids, such as Stacking [13] and Cascading [4], create new hybrid learning techniques from different base learning algorithms.

Bagging is the acronym for Bootstrap AGGregatING. The method induces a multiclassifier that consists on an ensemble of classifiers built on bootstrap replicates of the training set. Among the different ways of combining the outputs of the classifiers in an ensemble (abstract level, rank level, measurement level...), abstract level is the type used by bagging [10]. Given a set of labels Ω, a set of classifiers D and objects $x \in \mathcal{R}^n$ to be classified, in this approach each classifier Di produces a class label $s_i \in \Omega, i = 1, ..., L$. Thus, for any object $x \in \mathcal{R}^n$ to be classified, the L classifier outputs define a vector $s = [s_1, ..., s_L]^T \in \Omega^L$. The label outputs of the classifiers can be represented as binary vectors $[d_{i,1}, ..., d_{i,c}]^T \in \{0,1\}^c, i = 1, ..., L$, where $d_{i,j} = 1$ if D_i labels x in ω_j, and 0 otherwise, then, the final choice of the class is carried out majority vote. It means that the following rule must be complied.

$$\sum_{i=1}^{L} d_{i,k} = max_{j=1}^{c} \sum_{i=1}^{n} d_{i,j}$$

Majority vote is one of the combination schemes most used in multiclassification. However, unlike other approaches, the abstract level does not provide additional information about the predicted labels, such as probability o correctness.

To apply it some aspect must be considered [10]:

- The number of classifiers, L, is odd.
- The probability of predicting the right class for any $x \in \mathcal{R}^n$ for each classifier is p.
- The classifier outputs are independent, thus the joint probability for any subset of classifiers $A \subseteq D, A = \{D_{i1}, ..., D_{ik}\}$ can be decomposed as

$$P(D_{i1} = s_{i1} ..., D_{ik} = s_{ik}) = P(D_{i1} = s_{i1}) \times ... \times P(D_{ik} = s_{ik})$$
Where s_{ij} is de label output of classifier D_{ij}.

Boosting is a multiclassifier of the same kind of Bagging, however, this method assign weights to the outputs of the induced single classifiers from different training sets (strategies). The weight of a strategy s_i represents the probability that s_i is the most accurate of all of them. In an iterative process, the weights are updated by increasing the weight of strategies with the correct s_i prediction and reducing the weight of strategies with incorrect predictions. In this way the multiclassifier is developed incrementally, adding one classifier at a time. The classifier that joins the ensemble at step k is trained on a data set selectively sampled from the training data set Z. The sampling distribution starts from uniform, and progresses in each k step towards increasing the likelihood of worst classified data points at step k − 1 [10]. This algorithm is called AdaBoost which comes from ADAptive BOOSTing. This algorithm presents the advantage of drive the ensemble training error to zero in very few iterations [10].

3   Methodological Approach

In this study several classification algorithms were applied They were both simple classifiers and multiclassifiers. Two tree induction algorithms were employed, J48 and REPTree. J48 is an advanced version of C4.5 [11], one of the algorithms most used and well known. Both are information gain-based methods and REPTree uses reduced-error pruning with back-fitting. As simple classifiers we also applied a Bayesian network with the K2 search algorithm [2] and a Support Vector Machine (SVM). The multiclassifiers used were Bagging and AdaBoost.

The behavior of classifiers used in an individual way sometimes fails with some training sets, specifically when they consist of a wide variety of heterogeneous data and when the proportion of records of each class is very unequal. In order to address these peculiarities (high dimensionality and imbalanced distribution) characterizing the dataset used in this study some multiclassifiers were applied. Multiclassifiers extend the hypothesis space with respect to single classifiers and achieve a better management of large number of attributes. Another additional advantage of these techniques is the reduction of the overfitting problem, which takes place when the learning process finds a regularity in the data that is distinctive of the training set but cannot be extended to other datasets. Specifically, Adaboost has the capacity to avoid overfitting problems and reduce errors at the same time in spite of the progressively increasing complexity of the induced classifiers [10]. This fact is essential to deal with the imbalanced data problem. As commented before, most of the works in the literature use sampling strategies to deal with these drawbacks. There are some different proposals but they are mainly focused on improving a specific classifier [12] and only a few propose handling several classifiers [5].

In general, the key of the good behavior of classifiers ensemble is the diversity provided by different training sets. Ideally, the training sets should be generated randomly from the distribution of the problem, but in practice only one training set, $Z = \{z_1 \dots z_N\}$, is available. In these cases the bootstrap sampling can be used to generate L training sets. Significant improvements derived from this procedure are achieved mainly when the base classifier is unstable, that is, small changes in the training set should lead to large changes in the classifier output [10]. This is the scenario that takes place when working with imbalanced data.

On the other hand, majority vote properties assure the improvement of the single classifiers results if the outputs were independent and classifiers had the same individual accuracy p. Outputs of Bagging cannot be considered completely independent since the training samples are formed from the same set Z by taking bootstrap replicates, however Bagging improves accuracy of single classifiers due to the bias-variance decomposition of the classification error [10].

4   Case Study

The study included data from 497 patients diagnosed with severe trauma and poly-trauma who were hospitalized in the Intensive Care Unit (ICU) of the University Hospital in Salamanca from 2006 to 2011. The analysis was carried out taking into account 120 attributes grouped into the following categories:

- Epidemiologic: age, sex, kind of accident, origin (prehospital or hospital transfer), personal history.
- Hospital Emergency room: Clinical Data, treatment in the prehospital emergency setting, major changes in evolution during the stay in the Emergency Room and time taken for transfer to the operating room or ICU.
- ICU: Time from hospital arrival to admission at the ICU and origin, clinical data on admission, complications, variables related to mechanical ventilation, evolutionary data and diagnostic.

These attributes are of several types, continuous, discrete, nominal, ordinal and binary. Several data-mining algorithms were applied to obtain models that would allow the prediction of patient "death" from these variables.

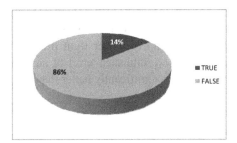

Fig. 1. Percentage of records of the *true* and *false* classes

An important aspect to be considered in the study is the distribution of records belonging to each class ("true" and "false") in the dataset. As showed in figure 1 the proportion of records of each class is quite different (70 records of the "true" class and 427 of the "false" class). This unequal distribution can significantly influence the results of the data-mining algorithms, since high accuracy can be obtained but the rate of correctly classified instances of the minority class may be very low. In these cases, analysis of the results must focus not only on accuracy but also on the precision achieved for the two classes. In addition, the capture of other measures such as the F-measure and the ROC analysis can be useful.

One of the purposes of the work was to check the performance of multiclassifiers against simple classifiers in this context, where the number of attributes is high, and especially in the case in which there is a minority class in the dataset. Accordingly, we applied four simple classification algorithms, two decision trees, J48 and REPT-ree, and Bayes Net and SVM, and two multiclassifiers -Bagging and Adaboost,- using J48, REPTree and SVM in both of them as base classifiers.

5   Results and Discussion

The results obtained are shown in figure 2 where the high accuracy achieved, close to 95% for all the algorithms applied, can be seen. However, the precision of the "true" class (the minority class) was significantly lower. Because of this, an analysis of these results, obtaining F-measures and area under the ROC curve, was carried out in order to examine the behavior of the different classifiers against the problem in hand. Figure 3 shows the result of this analysis.

The study was carried out with a data set containing 497 records and 120 attributes. This large number of attributes may cause a loss of accuracy in the classifiers. Given that the reduction of dimensionality by means of feature selection methods did not lead to an improvement of the accuracy, all the attributes were used as the input of the data-mining algorithms. In all experiments 10-fold cross-validation was applied.

The imbalanced distribution of the records in the dataset elicited a considerable difference in precision between the two classes for most of the algorithms applied. In this context, in general terms multiclassifiers showed better behavior than simple classifiers and, in particular, the best result was provided by Adaboost with J48, which achieved an accuracy of   92.96 %.

Table 1 shows a pair-wise comparison of the classifiers using T-Test that was performed with the aim of evaluating the statistical significance of the results. The notation used is $(x\ y\ z)$ where the value 1 indicates that the classifier in the column is significantly better than the one in the row for x, the same for y and worst for z. The established level of significance was 0.05. T-Tests confirm that Adaboost with J48 is significantly better than the rest of the classifiers.

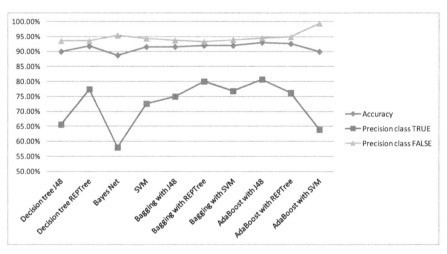

Fig. 2. Results obtained by single and multi-classifiers

Table 1. Statistical significance

	J48	REPTree	Bayes Net	SVM	AdaBoost J48	AdaBoost REPTree	AdaBoost SVM	Bagging J48	Bagging REPTree	Bagging SVM
J48	-	(1/0/0)	(0/0/1)	(0/1/0)	**(1/0/0)**	(1/0/0)	(0/0/1)	(1/0/0)	(1/0/0)	(1/0/0)
REPTree	(0/0/1)	-	(0/0/1)	(0/0/1)	**(1/0/0)**	(0/1/0)	(0/0/1)	(0/1/0)	(1/0/0)	(0/1/0)
Bayes Net	(1/0/0)	(1/0/0)	-	(1/0/0)	**(1/0/0)**	(1/0/0)	(0/1/0)	(1/0/0)	(1/0/0)	(1/0/0)
SVM	(0/1/0)	(1/0/0)	(0/0/1)	-	**(1/0/0)**	(1/0/0)	(1/0/0)	(1/0/0)	(1/0/0)	(1/0/0)
AdaBoost-J48	**(0/0/1)**	**(0/0/1)**	**(0/0/1)**	**(0/0/1)**	-	**(0/0/1)**	**(0/0/1)**	**(0/0/1)**	**(0/0/1)**	**(0/0/1)**
AdaBoost-REPTree	(0/0/1)	(0/1/0)	(0/0/1)	(0/0/1)	**(1/0/0)**	-	(0/0/1)	(0/1/0)	(0/1/0)	(0/1/0)
AdaBoost-SVM	(1/0/0)	(1/0/0)	(0/1/0)	(1/0/0)	**(1/0/0)**	(1/0/0)	-	(1/0/0)	(1/0/0)	(1/0/0)
Bagging -J48	(0/0/1)	(0/1/0)	(0/0/1)	(0/0/1)	**(1/0/0)**	(0/1/0)	(0/0/1)	-	(0/1/0)	(0/0/1)
Bagging -REPTree	(0/0/1)	(0/0/1)	(0/0/1)	(0/0/1)	**(1/0/0)**	(0/1/0)	(0/0/1)	(0/1/0)	-	(0/0/1)
Bagging -SVM	(0/0/1)	(0/1/0)	(0/0/1)	(0/0/1)	**(1/0/0)**	(0/1/0)	(0/0/1)	(0/1/0)	(0/1/0)	-

In order to determine the performance of the algorithms against the different kinds of errors in the context of a dataset with a minority class, an additional analysis was conducted. Besides accuracy, the F-measure and the area under the ROC curve were acquired for all classifiers induced from the imbalanced dataset. Figure 3 shows the values of these metrics. The best results of the F-measure were obtained by the Adaboost multiclassifier with both base classifiers, J48 (92.7%) and RepTree (92.4%). The values of this metric close to 100% indicate that the values of precision and recall are similar and consequently the classifiers are not biased to a specific kind of error. Regarding the ROC area, the best result was provided by Adaboost with J48, furnishing a value of 0.956, followed at some distance by Bagging with RepTree with a value of 0.919. Consequently, it may be concluded that the optimal classifier was that induced by the Adaboost algorithm with J48.

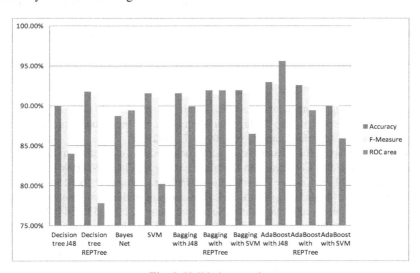

Fig. 3. Validation metrics

In general, the multiclassifiers showed better behavior than the simple classifiers in the case of the imbalanced dataset, as shown in figure 3. Multiclassifiers induce models that merge classifiers with the same learning algorithm, but with different data sets obtained by random resampling, but without using either over-sampling of the records of the minority class or under-sampling of the records of the majority class. These latter approaches are usually applied to address the problem of imbalanced datasets, but they have important drawbacks. Undersampling may discard potentially valuable data, while oversampling artificially increases the size of the data set and, as a result, the computational cost of inducing the models. In addition, the replication of existing examples in the minority class causes overfitting problems [9].

6   Conclusions

This study addresses the problem of predicting death in patients affected by severe trauma and polytrauma, one of the pathologies with the greatest impact in today's society. Several data-mining algorithms were applied to the data from 497 traumatized and polytraumatized patients who required ICU hospitalization. The input variables to the algorithms were both epidemiologic and clinical and were taken at the emergency room and along the stay in the ICU.

The study addresses the problems of managing a large number of attributes and imbalanced distribution of the records, which usually affords greater precision for the majority class than for the minority one. Accordingly, besides accuracy it is necessary to apply some additional validation measures such as the F-measure or the ROC area. For all cases studied, the best results were provided by Adaboost with J48, which achieved an accuracy of 92.96 %; the F-measure was 92.7 and ROC area was 0.956.

In general, the results point to the better behavior of multiclassifiers as compared with simple classifiers in contexts of high dimensionality and imbalanced datasets, without any need to resort to oversampling and undersampling strategies.

References

1. Breiman, L.: Bagging predictors. Machine Learning 24(2), 123–140 (1996)
2. Cooper, G.F., Herskovits, E.: A Bayesian Method for the induction of probabilistic networks from data. Machine Learning 9(3), 09–347 (1992)
3. Freund, Y., Schapire, R.E.: Experiments with a new boosting algorithm. In: Proceedings of the 13th International Conference on Machine Learning, pp. 148–156 (1996)
4. Gama, J., Brazdil, P.: Cascade Generalization. Machine Learning 41(3), 315–343 (2000)
5. Ghazikhani, A., Monsefi, R., Yazdi, H.S.: Ensemble of online neural networks for non-stationary and imbalanced data streams. Neurocomputing 122(25), 535–544 (2013)
6. Hall, M.A.: Correlation-based Feature Selection for Machine Learning. PhD Thesis, University of Waikato, Hamilton, Nueva Zelanda (1999)
7. Hall, M., Frank, E., Holmes, G., Pfahringer, B., Reutemann, P., Witten, I.H.: The WEKA Data Mining Software: An Update. SIGKDD Explorations 11(1), 10–18 (2009)
8. Hemmila, M.R., Jakubus, J.L., Maggio, P.M., et al.: Real money: complications and hospital costs in trauma patients. Surgery 144(2), 307–316 (2008)
9. Hulse, J., Khoshgoftaar, T., Napolitano, A.: Experimental perspectives on learning from imbalanced data. In: Proceedings of the 24th International Conference on Machine Learning, pp. 935–942 (2007)

10. Kuncheva, L.I.: Combining Pattern Classifiers: Methods and Algorithms. John Wiley & Sons (2004)
11. Quinlan, J.R.: C4.5: Programs for Machine Learning. Morgan Kaufmann, San Mateo (1993)
12. Shao, Y.H., Chen, W.J., Zhang, J.J., Wang, Z., Deng, N.Y.: An efficient weighted Lagrangian twin support vector machine for imbalanced data classification. Pattern Recognition 47, 3158–3167 (2014)
13. Wolpert, D.H.: Stacked Generalization. Neural Networks 5(2), 241–259 (1992)

Nonconvex Functions Optimization Using an Estimation of Distribution Algorithm Based on a Multivariate Extension of the Clayton Copula

Harold D. de Mello Jr., André V. Abs da Cruz, and Marley M.B.R. Vellasco

Department of Electrical Engineering
Pontifical Catholic University of Rio de Janeiro
Rio de Janeiro, RJ, Brazil
{harold,andrev,marley}@ele.puc-rio.br

Abstract. This paper presents a copula-based estimation of a distribution algorithm with parameter updating for numeric optimization problems. This model implements an estimation of a distribution algorithm using a multivariate extension of Clayton's bivariate copula (MEC-EDA) to estimate the conditional probability for generating a population of individuals. Moreover, the model uses traditional mutation and elitism operators jointly with a heuristic for a population restarting in the evolutionary process. We show that these approaches improve the overall performance of the optimization compared to other copula-based EDAs.

Keywords: Continuous numeric optimization, evolutionary computation, estimation of distribution algorithms, copulas.

1   Introduction

Numerical optimization is an important task that arises in several different knowledge domains. Evolutionary algorithms are capable of finding good solutions with a lower computational cost for different optimization problems. Nevertheless, conventional evolutionary algorithms have difficulty addressing optimization problems when the number of variables, constraints and goals increases. This difficulty relates partly to the association between the performance and the number of parameters, which must be determined for each problem, and to the inability to extract and use knowledge acquired throughout the search process [1].

One possible approach to overcoming these problems is to use Estimation of Distribution Algorithms (EDAs) [2]. EDAs constitute a class of evolutionary algorithms that construct a probability distribution throughout evolution by analyzing the most promising solutions. This probability distribution is used to generate new individuals instead of using mutation and crossover operators as traditional evolutionary algorithms do.

Different EDA models, which differ in how they estimate the probabilistic model, have been proposed, including but not limited to models based on copula. The

E. Corchado et al. (Eds.): IDEAL 2014, LNCS 8669, pp. 318–326, 2014.
© Springer International Publishing Switzerland 2014

concept of copula was introduced by Sklar (1959) [3] and has been extensively used in finance [4]. More recently, copula theory has been applied to Evolutionary Computation [5]. Copula theory allows any multivariate distribution to be expressed as a function (the copula) of its marginal distributions, which describe the behavior of each individual variable; the copula contains the required information to describe the dependency among all variables and is invariant to non-linear transformation [6].

In the case of multidimensional problems, EDA probabilistic models are built, essentially, from bivariate copula [7] because these represent most parametric copulas. Using bivariate copula makes the estimation of the multivariate system easier. In [8], the use of empirical bivariate Archimedean copulas in EDAs was investigated to optimize n-dimensional problems, while using a constant value for the parameters of the copula.

This study discusses an EDA (MEC-EDA) that is similar to the approach proposed in [8] and applies it to some benchmark functions. Our approach differs in some aspects: the copula parameter is estimated dynamically, using dependency measures; we use information contained in the probability distribution and a classic mutation operator to preserve population diversity; and we use a heuristic to reinitialize the population throughout an elitist evolution. Specifically, MEC-EDA was based on the Clayton's copula because it is simple and no numerical integration is required to compute probabilities. This paper is organized as follows: Section 2 discusses some general aspects regarding copula theory. Section 3 presents copula-based EDAs and our proposed model. Section 4 presents and discusses experimental results on benchmark functions. Section 5 articulates some general conclusions.

2 Copula theory

2.1 Theorems, Properties and Definitions

A function $C(u_1, \dots, u_n): [0, 1]^n \to [0,1]$ is an n-dimensional copula if it satisfies the some basic conditions related to boundary conditions and increasing property [3]. Thus, copula C is a distribution function in $[0,1]^n$ that has marginal uniform functions u_k, $k = 1, \dots, n$ in $(0, 1)$.

According to Sklar's Theorem [3], a joint Cumulative Distribution Function (CDF) F of random variables $X_1, X_2, \dots, X_n$, with continuous marginal distributions $F_1, F_2, \dots, F_n$, respectively, can be characterized by a single n-dimensional dependency function or copula C, such that for all vectors $x \in \overline{\mathbb{R}^n}$:

$$F(x_1, x_2, \dots, x_n) = C\big(F_1(x_1), F_2(x_2), \dots, F_n(x_n)\big) \tag{1}$$

Similarly, for any vector $\mathbf{u} \in [0,1]^n$,

$$C(u_1, u_2, \dots, u_n) = F\big(F_1^{-1}(u_1), F_2^{-1}(u_2), \dots, F_n^{-1}(u_n)\big) \tag{2}$$

where $F_i^{-1}(u) = inf\{x_i \in \mathbb{R}: F_i(x_i) \geq u\}$ for $i = 1, \dots, n$ is the generalized inverse function of the marginal distribution function $u \in (0,1)$.

Copulas as a modelling dependence tool require algorithms that allow generating random variables. They are based on a definition of conditional distribution. If $U_1, \ldots, U_n$ are described by a joint distribution function C, then the conditional distribution U_k, given the values of $U_1, \ldots, U_{k-1}$, can be determined by:

$$C_k(u_k|u_1, \ldots, u_{k-1}) = \frac{\partial^{k-1} C_k(u_1, \ldots, u_k)}{\partial u_1 \cdots \partial u_{k-1}} \Big/ \frac{\partial^{k-1} C_{k-1}(u_1, \ldots, u_{k-1})}{\partial u_1 \cdots \partial u_{k-1}} \tag{3}$$

2.2   Archimedean Copulas

Archimedean copulas can represent highly specialized dependence structures. Nevertheless, this type of copula is not directly derived from Sklar's Theorem. For the bivariate case, this copula holds the following representation:

$$C(u_1, u_2) = \varphi^{-1}\big(\varphi(u_1) + \varphi(u_2)\big) \quad u_1, u_2 \in [0, 1] \tag{4}$$

where $\varphi(\cdot)$ is known as a generator function of the copula, and φ^{-1} is its inverse.

Several Archimedean copula families are described in [3]; most of them depend on a single parameter (θ) that controls the dependency structure. However, this work used only one of them in constructing the EDA (Table 1). In this copula, the perfect dependency among the random variables occurs when $\theta \to \infty$, while $\theta = 1$ indicates total independence among the variables.

Table 1. Clayton's bivariate copula

$C(u_1, u_2)$	$\varphi(t)$	θ
$\left(u_1^{-\theta} + u_2^{-\theta} - 1\right)^{-1/\theta}$	$\theta^{-1}\left(t^{-\theta} - 1\right)$	$[0, \infty]$

An additional advantage of Archimedean copulas is that they can be easily used to generate multivariate distributions from extensions of Archimedean 2-copulas. The simplest method used in this work that is named an exchangeable multivariate Archimedean copula (EAC) nevertheless poses a limitation: the dependency structure between any pair of variables is described by the same parameter θ, regardless of dimension. In this respect, Archimedean constructions (NACs) and pair-copulas models [9] are more flexible.

2.3   Copula Parameter Estimation

The parameter estimation of copulas is usually accomplished using one of the following different approaches: i) maximum likelihood; ii) inference functions for the margins method; or iii) semi-parametric maximum likelihood. An alternative and less computationally intense method that is used in this work is the estimation of moments based on Kendall's tau (τ) [10]. In this method, the relationship between the rank correlation and the copula's θ parameter is used. For Clayton's copula, the following relationships between τ and θ holds:

$$\tau_\theta = \frac{\theta}{\theta+2} \tag{5}$$

3 Copula-Based Estimation of Distribution Algorithms (CEDAs)

EDAs are a class of optimization algorithms whose most important step and bottle-neck is estimating the joint probability distribution associated with the variables from the most promising solutions determined by the evaluation function. This approach is also the aspect that differs most among EDAs. The complexity of this probabilistic model can be classified into the following categories: independent, pairwise dependent, multivariate dependency and mixed models.

3.1 General Copula-Based EDAs

Recently, a new approach to developing EDAs to solve a real-valued optimization problem has been developed that is based on copula theory. In [5], it is possible to find a review of this research and to find an R package for working with copula-based EDAs.

In copula-based EDAs, the step for estimating the probabilistic model is divided into two parts: i) estimating the marginal distributions, F_i, for $i = 1, ..., n$, and ii) estimating the dependency structure. Typically, a specific distribution is assumed to be the correct distribution for each marginal distribution, and its parameters are estimated based on maximum likelihood. In other cases, kernel density estimators (KDE), marginal empirical distributions or estimators based on Kendall's inversion were used.

After the marginal distributions are estimated, the selected population is transformed into uniform variables in the interval [0, 1] by evaluating each cumulative marginal distribution. This transformed population is then used to estimate a copula-based model C that describes the dependency among the variables. The sampling step creates a population by sampling the dependent distributions $\left(u_1^{(k)}, u_2^{(k)}, ..., u_n^{(k)}\right)$ for $k = 1, ..., s$ samples that were created by the copula. New individuals are calculated by $x_i^{(k)} = F_i^{-1}(u_i)$, where F_i^{-1} is the inverse function of F_i. According to (2) the new individual $(x_1, x_2, ..., x_n)$ is a sample that obeys the joint distribution combined by the copula and the marginal distributions. This entire process is repeated until a stop condition is attained, and the best individual is found, as seen in figure 1.

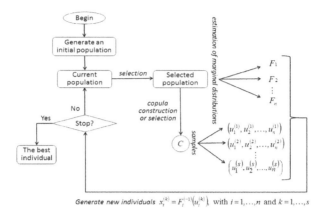

Fig. 1. Flowchart of a general copula-based EDA. Adapted from [11].

3.2   Proposed Model

The main reason to propose a copula-based EDA is the copulas' flexibility to build effective joint distributions of the search space at the expense of the common and convenient assumption asserted in most EDAs, i.e., that the variables follow a Gaussian distribution. This conjecture often generates inconsistent results and, therefore, a poor performance of the algorithm, particularly in problems of numerical optimization in multivariate spaces. Thus, the use of copulas in EDAs can simplify their operations, after the construction of the probability models is performed in less time and with greater accuracy compared to other estimation techniques.

The first step of MEC-EDA initializes a random population within the search space for the problem. This population is then evaluated, and the correlation matrix for all variables is calculated using the current population as the data inputs. Subsequently, it is possible to calculate Kendall's tau and to obtain the parameter θ for Clayton's copula using equation (6). Then, the algorithm simulate pseudo-random uniform variables from a Clayton's copula using the conditional approach described in [4]. This technique is useful for recursively generating observations of random variables uniformly distributed over [0, 1] whose joint distribution function is a chosen copula. Thus, using the definition (3), the Clayton's bivariate copula and its generator function and the parameter θ, the algorithm can generate samples of a n-variate distribution.

The next step is estimating the marginal distributions. Our model employed empirical marginal distributions. The use of empirical marginals CDF rather than raw data made the experiments comparable, and the rank statistics tend to cope better with real-world systematic biases and errors.

Finally, a new population is obtained through the linear interpolation of the sample u in the selected population:

$$x = F_i^{-1}(u) = interpolate(x_i^{<j>}, x_i^{<j+1>}) \tag{6}$$

where $interpolate\left(x_i^{<j>}, x_i^{<j+1>}\right)$ denote an interpolated number between two consecutive samples of the sorted current population.

This new population is merged to a mutated version of the older. A new evaluation is performed, and the worst individuals are eliminated. Subsequently, a percentage of individuals obtained from a variation of the learned CDF and with a lower probability of emerging in the evolutionary process (rebel individuals) [12], is added to the final population. The algorithm proceeds with the joint distribution estimation and sampling until a stop condition is satisfied. A pseudocode of this algorithm is given in the following figure.

Whenever the perfect dependency among the random variables occurs ($\theta \to \infty$), a population's percentage is restarted uniformly in the real interval of search space and the remaining population is restarted uniformly within the limits of the best individual variables at that generation. Another heuristic used is a periodic reset of the limits of the best individual variables according to the limits of search space. All these approaches help to preserve diversity.

```
1.   begin
2.     t → 0;
3.     Generate an initial population P(t) randomly;
4.     while (termination criteria are not satisfied) do
5.         Evaluate Φ(P(t));
6.         Compute the matrix of Kendall's tau: (n, m);
7.         if (any of n variables generate the same value in all individuals) do
8.             Restart population P(t), retaining only the best individual;
9.         end
10.        Compute Clayton's copula parameter θ using Kendall's tau average;
11.        Select the best k individuals from P(t): S(P(t));
12.        Generate samples u from a multivariate extension of Clayton's bivariate;
13.        Determine the marginal empirical cumulative distributions Fᵢ;
14.        Generate k new individuals: x = Fᵢ⁻¹(u) = interpolate(xᵢ^<j>, xᵢ^<j+1>);
15.        Apply a mutation operator in certain individuals from the older population;
16.        Merge the populations (lines 14, 15), evaluate and eliminate the worst individuals;
17.        Add rebel individuals to the final population;
18.        t → t + 1;
19.    end
20.  end
```

Fig. 2. MEC-EDA pseudocode

4   Experimental Results and Discussion

4.1   Benchmark Functions

The performance of the proposed model was evaluated using the traditional benchmark functions: Sphere, Ackley, Rastrigin, Griewank, Rosenbrock and the Summation Cancellation[1]. Sphere, Griewank, Ackley and Rastrigin have their global optimum at $x=(0, \cdots, 0)$, with evaluation zero; Summation Cancellation has its global optimum at $x=(0, \cdots, 0)$, with evaluation -10^5, and Rosenbrock has its global optimum at $x=(1, \cdots, 1)$, with evaluation zero. The search space was: [−600, 600] in Sphere and

---

[1] Modified to a minimization problem.

Griewank, [−30, 30] in Ackley, [−0.16, 0.16] in Summation Cancellation, [−5.12, 5.12] in Rastrigin, and [−9, 11] in Rosenbrock. All these benchmark functions are noncon-vex, except the Sphere and the Summation Cancellation.

All tests were performed with n =10 variables to be optimized. For each function, 30 independent experiments were performed using two stop criteria: i) reaching the global optimum with a precision greater than 1e-6 or ii) reaching 300,000 evaluations. In all experiments, 55% of individuals were restarted uniformly in the real interval of the search space, and 45% of individuals were restarted uniformly in the real interval of the population's maximum and minimum in each respective generation, retaining always the best individual. These limits were reset to the search space in each of the 10 consecutive restarts. The mutation operator was applied in the first 5 individuals (step 15 of pseudocode), with a rate of 70% for Rosenbrock and 100% to all the other functions. Moreover, 5% of rebel individuals were added to the final population. The results are shown in the table 2. For each function, we present results for our model (MEC-EDA) and for the models described in [5]. Each column shows the population size, the best fitness and the worst fitness found, the average final fitness and the number of evaluations that were performed. Different population sizes were evaluated (5, 10, 20, 30, 50, 100 and 200). We selected the minimum population size required by the algorithm to find the global optimum of each function with a high success rate and the minimum number of function evaluations.

Table 2. Performance comparison for 10 variables

Sphere Problem

Algorithm	Pop. Size	Success Rate	Best Fitness	Worst Fitness	Average Fitness	N° Eval.
MEC-EDA	5	30/30	2.9e-7	9.9e-7	7.9e-7 ± 1.9e-7	3,102 ± 357
UMDA	81	30/30	–	–	6.9e-7 ± 2.3e-7	3,823 ± 128.3
GCEDA	310	30/30	–	–	6.5e-7 ± 2.0e-7	13,082 ± 221.4
CVEDA	104	30/30	–	–	6.7e-7 ± 1.8e-7	4,777 ± 118.8
DVEDA	104	30/30	–	–	6.7e-7 ± 2.0e-7	4,787 ± 100.2
Cópula MIMIC	150	30/30	–	–	6.9e-7 ± 1.7e-7	6,495 ± 209.0

Ackley Problem

Algorithm	Pop. Size	Success Rate	Best Fitness	Worst Fitness	Average Fitness	N° Eval.
MEC-EDA	5	30/30	7.2e-7	9.9e-7	9.1e-7 ± 8.2e-8	4,900 ± 1,041
UMDA	81	30/30	–	–	8.1e-7 ± 1.1e-7	4,998 ± 88.0
GCEDA	310	30/30	–	–	6.5e-7 ± 2.0e-7	13,082 ± 221.4
CVEDA	104	30/30	–	–	8.1e-7 ± 1.0e-7	6,330 ± 163.2
DVEDA	104	30/30	–	–	7.9e-7 ± 1.4e-7	6,678 ± 133.8
Cópula MIMIC	150	30/30	–	–	8.0e-7 ± 1.2e-7	10,784 ± 143.7

Rastrigin Problem

Algorithm	Pop. Size	Success Rate	Best Fitness	Worst Fitness	Average Fitness	N° Eval.
MEC-EDA	7	30/30	1.9e-7	9.9e-7	7.9e-7 ± 2.0e-7	6,667 ± 1,789
UMDA	447	30/30	–	–	6.7e-7 ± 2.3e-7	33,614 ± 2,452
GCEDA	721	30/30	–	–	6.8e-7 ± 1.8e-7	46,095 ± 2,158
CVEDA	447	30/30	–	–	6.6e-7 ± 1.7e-7	32,914 ± 2,011
DVEDA	325	30/30	–	–	7.3e-7 ± 1.7e-7	24,710 ± 1,754
Cópula MIMIC	386	30/30	–	–	6.4e-7 ± 2.1e-7	27,315 ± 1,673

Griewank Problem

Algorithm	Pop. Size	Success Rate	Best Fitness	Worst Fitness	Average Fitness	N° Eval.
MEC-EDA	5	30/30	1.4e-7	9.9e-7	8.2e-7 ± 1.6e-7	5,919 ± 5,265
UMDA	111	30/30	–	–	6.6e-7 ± 1.9e-7	5,224 ± 231.2
GCEDA	355	30/30	–	–	6.9e-7 ± 1.8e-7	15,099 ± 414.1
CVEDA	142	30/30	–	–	7.0e-7 ± 1.8e-7	6,579 ± 389.9
DVEDA	150	30/30	–	–	6.5e-7 ± 2.4e-7	6,785 ± 338.1
Cópula MIMIC	188	30/30	–	–	6.6e-7 ± 1.8e-7	8,221 ± 220.4

Rosenbrock Problem

Algorithm	Pop. Size	Success Rate	Best Fitness	Worst Fitness	Average Fitness	N° Eval.
MEC-EDA	30	20/30	5.5e-7	6.5e+0	1.2e+0 ± 2.5e+0	180,805 ± 105,783
UMDA	2000	0/30	–	–	8.0e+0 ± 2.6e-2	300,000 ± 0.0
GCEDA	2000	0/30	–	–	7.5e+0 ± 1.9e-1	300,000 ± 0.0
CVEDA	2000	0/30	–	–	7.5e+0 ± 1.1e-1	193,867 ± 48,243
DVEDA	2000	0/30	–	–	7.5e+0 ± 1.5e-1	172,200 ± 35,184
Cópula MIMIC	2000	0/30	–	–	7.6e+0 ± 1.3e-1	139,000 ± 5,139

Summation Cancellation Problem

Algorithm	Pop. Size	Success Rate	Best Fitness	Worst Fitness	Average Fitness	N° Eval.
MEC-EDA	10	0/30	-99,999.95	-99,996.86	-99,996.12 ± 7.8e-1	300,011 ± 4.0
UMDA	2000	0/30	–	–	-5.7e+2 ± 3.4e+2	300,000 ± 0.0
GCEDA	355	30/30	–	–	-1.0e+5 ± 1.3e-7	42,434 ± 305.4
CVEDA	325	30/30	–	–	-1.0e+5 ± 1.3e-7	44,622 ± 858.3
DVEDA	965	30/30	–	–	-1.0e+5 ± 9.3e-8	117,408 ± 959.4
Cópula MIMIC	2000	0/30	–	–	-2.3e+4 ± 2.7e+4	300,000 ± 0.0

Only linear and independence relationships are considered in the models of [5], and all algorithms use normal marginal distributions. Because these models have not restarted the population, they use a third stop criterion: namely, whenever the stan-dard deviation of the evaluation of the solutions in the population is less than 1e-8.

It is possible to verify that the proposed algorithm provides superior results for most functions, thus yielding a smaller number of evaluations to achieve the optimization target. The proposed model failed to optimize Rosenbrock, even though it generated 20 successful experiments and the best average fitness compared to other copula-based EDAs. MEC-EDA also failed to optimize the Summation Cancellation. The algorithms with the best average fitness and number of evaluations are highlighted in the table 2.

4.2   Discussion

The results show that, despite the algorithm's simplicity, small improvements can improve the performance of a copula-based EDA. Our approach uses an alternative method for the parameter estimation of the Archimedean copula, as well as strategies related to restarting the population, a traditional mutation operator and rebel individuals to avoid premature convergence and stagnation in local optima.

The preliminary results are good and suggest that with some adjustment, it might be possible to improve the results, particularly for the Rosenbrock function, for which the algorithm was incapable of achieving the global minimum in all experiments and for the Summation Cancellation, for which the average fitness nearly reached the global optimum. A learning strategy to define the proportion of individual restarts and to minimize the number of individuals restarting should solve these failed cases and improve the evaluations number because MEC-EDA works well with a considerably smaller population compared to other CEDAs. Nevertheless, an additional number of benchmark functions must be used to verify the proposed algorithm's performance.

5   Conclusion

This paper presented a new model (MEC-EDA) based on Estimation of Distribution Algorithms and copulas to optimize numerical functions. The presented results are promising, although further adjustments and tests must be performed. Specifically, the impact of estimating Kendall's tau on each generation of the EDA must be verified. Additional tests could be performed with different Archimedean copulas to determine whether the type of copula used is important and the results compared to other efficient heuristics, such as Covariance Matrix Adaptation Evolution Strategy (CMA-ES).

Acknowledgments. This work has been financially supported by CNPq and FAPERJ.

References

1. Hauschild, M., Pelikan, M.: An introduction and survey of estimation of distribution algorithms. Swarm and Evolutionary Computation 3(1), 111–128 (2011)
2. Larrañaga, P., Lozano, J.A. (eds.): Estimation of distribution algorithms. A new tool for evolutionary computation. Kluwer Academic Publishers, London (2002)

3. Nelsen, R.B.: An introduction to copula. Springer, New York (1998)
4. Cherubini, U., Luciano, E., Vecchiato, W.: Copula Methods in Finance. Wiley (2004)
5. González-Fernández, Y., Soto, M.: copulaedas: An R Package for Estimation of Distribution Algorithms Based on Copulas (2013), Preprint arXiv:
   http://arxiv.org/abs/1209.5429
6. Póczos, B., Ghahramani, Z., Schneider, J.: Copula-based Kernel dependency measures. In: Proceedings of the International Conference on Machine Learning, ICML 2012 (2012)
7. Joe, H.: Families of m-variate distributions with given margins and m(m-1)/2 bivariate dependence parameters. In: Rüschendorf, L., Schweizer, B., Taylor, M.D. (eds.) Distributions with Fixed Marginals and Related Topics, pp. 120–141 (1996)
8. Cuesta-Infante, A., Santana, R., Hidalgo, J.I., Bielza, C., Larrañaga, P.: Bivariate empirical and n-variate archimedean copulas in estimation of distribution algorithms. In: Proceedings of the IEEE Congress on Evolutionary Computation, CEC 2010, pp. 1355–1362 (2010)
9. Joe, H.: Multivariate Models and Dependence Concepts. Chapman & Hall (1997)
10. Genest, C., Rivest, L.-P.: Statistical inference procedures for bivariate Archimedean copulas. J. Am. Stat. Assoc. 88(423), 1034–1043 (1993)
11. Guo, X., Wang, L., Zeng, J., Zhang, X.: VQ Codebook Design Algorithm Based on Copula Estimation of Distribution Algorithm. In: First International Conference on Robot, Vision and Signal Processing, pp. 178–181 (2011)
12. DelaOssa, L., Gámez, J.A., Mateo, J.L., Puerta, J.M.: Avoiding premature convergence in estimation of distribution algorithms. In: Proceedings of the IEEE Congress on Evolutionary Computation, CEC 2009, pp. 455–462 (2009)

News Mining Using Evolving Fuzzy Systems

José Antonio Iglesias, Alexandra Tiemblo,
Agapito Ismael Ledezma, and Araceli Sanchis

Carlos III University of Madrid,
Avda. de la Universidad, 30, 28911 Leganés (Madrid), Spain
jiglesia@inf.uc3m.es, 100073062@alumnos.uc3m.es,
{ledezma,masm}@inf.uc3m.es

Abstract. Online news has become one of the major channels for Internet users to get news. Modern society generates huge amounts of online newspapers every day. Thus, the processing and analysis of this information is an important challenge. In this paper, we present an approach for classifying different news articles into various topic areas based on the text content of the articles. In order to achieve this task, we need to take into account that there are thousands of new articles each day and also that articles of the same topic can vary according to the present time. For this reason, the presented approach is based on Evolving Fuzzy Systems (EFS) and the model that describes a topic area changes according to the change in the text content of the articles. This approach has been successfully tested using real on-line news.

Keywords: News Mining, Evolving Intelligent Systems, Fuzzy Rules.

1   Introduction

Modern society generates huge amounts of information every day, specially in digital format, which obstruct the storage and further processing and analysis. In particular, an important part of this information is generated by the on-line newspapers. News websites are daily overwhelmed with plenty of news articles. According to the Statistical Report on Internet Development released by China Internet Network Information Center (CNNIC) in July 2013 [1], the number of online news users had reached 461 million by the end of June 2013 (a growth of 68.60 millions from June 2012), and the utilization ratio of online news was 78.0%. Thus, online news has become one of the major channels for Internet users to get news and its utilization ratio has remaining high due to the following reasons: 1) in the era of mobile Internet, it is one of major activities of Internet users to read news in their fragmented time; 2) Internet users can get news through more channels 3) all news media vied with each other to make inroads into the mobile Internet.

Because of this explosion of information from news paper, data mining is an essential issue. How to extract knowledge from on-line news is currently an important research topic and challenge [2] [3]. For this reason, news mining tools,

E. Corchado et al. (Eds.): IDEAL 2014, LNCS 8669, pp. 327–335, 2014.

techniques, and algorithms are emerging strong during these times. These techniques help to analyze the overflow of information and extract value knowledge from on-line news sources.

During the last years, there have been many approaches related with web news mining and news exploration systems [4] [5] [6]. In this sense, there are many systems which classify news into predefined categories. However, most of these systems use a predefined and statistic classifier over time. However, the news articles of the different categories change constantly and these changes should be considered in the classifier. For this reason, we propose an approach which not only collects, analyzes and extracts relevant terms from different web news, but also classifies web news in an evolving manner. Thus, the evolving classifier of web news that we propose is updated according to the changes in the web news. This classifier is based on Evolving Fuzzy Systems [7] which allows not only update the classifier but also cope with huge amounts of web news and process data in on-line and real time - which is essential in this (web) environment.

The remainder of the paper is organized as follows: Section 2 describes existing researches and approaches related with the area of web news mining. Section 3 describes our proposed (evolving) approach for the classification of web news. Section 4 presents the results and analysis of the evaluation of our approach. Finally, section 5 presents the conclusions and future work guidelines.

2   Background and Related Work

Web news mining is one application of the text mining. The term *text mining* or *Knowledge Discovery from Text* (KDT) was mentioned for the rst time in 1995 by Feldman et al. [2]. They propose to structure the text documents by means of information extraction, text categorization, or applying NLP techniques as pre-processing step before performing any kind of KDTS.

Text mining can be defined as the analysis of semi-structured or unstructured text data. As the text is in unstructured form, it is quite difficult to deal with it. Thus, the goal of the text mining is to turn text information into numbers so that data mining algorithms can be applied. It arose from the related fields of data mining, artificial intelligence, statistics, databases, library science, and linguistics. As it is detailed in [8], the term text mining has been used to describe different applications such as text categorization [6], text clustering [9] and finding patterns in text databases [4].

The term *web news mining* describes the analysis of web news. During the last years, there has been many approaches related with web news mining and news exploration systems. In [3], the authors describe the use of data mining techniques to analyze web news collected from published on the Web news. In a different research [5] the authors propose that in order to facilitate an in-depth analysis of the news it is necessary to extract structured information (ideally, identifying *who, what, whom, when, where and why* [10]). In [11] a quantitative method that identifies weak signal topics by exploiting keyword-based text mining is presented.

3   Our Approach

The goal of the approach presented in this research is to classify different news articles (from the web) into various topic areas (categories) based on the text content of the articles. This approach consists of two well differentiated phases (or modules): *Term extraction* and *Evolving Classification*. These phases are represented in the figure 1 and the following subsections will explain them in detail.

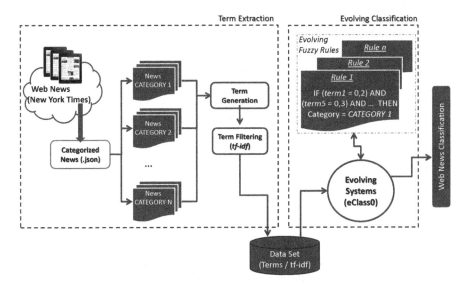

Fig. 1. Our approach: Web News Classification

3.1   Term Extraction

This module is responsible for extracting documents of different topic areas from the web and label each document with a set of terms. Each term (extracted from the document) has its corresponding relevance value which allows us to score the terms.

The first part of this module is to collect web news from online newspapers. Although we can use any source for this process, we are collecting the news articles from the New York Times (NYT) online newspaper[1]. In this research, we use the NYT API[2] called *Article Search API v2* with which we can search NYT articles from September 18, 1851 to today, retrieving headlines, abstracts, lead paragraphs, links to associated multimedia and other article metadata. Using this API, we can obtain a JSON response with several parameters. In this research, we will need only the content of the parameter *lead_paragraph* which contains the first paragraph of the web news as it was published in the web.

---

[1] http://www.nytimes.com/

[2] http://api.nytimes.com/svc/search/v2/articlesearch

This step was done by using an automated process which allows to obtain as many web news articles as we need. The news articles obtained (which are already categorized) are converted into a string. For the purpose of this paper, we collect hundreds of web news articles which are already categorized in 7 different topics: *Art, Bussines, Health, Science, Sports, Technology* and *Travel*.

As we can see in figure 1, once the categorized web news articles (strings) have been collected, they need to be analyzed by applying the following 2 steps:

Term Generation: In this step we will obtain the relevant terms of the articles. This task will be done by using the open-source tool *RapidMiner* [12]. *RapidMiner* is one of the most popular tools for data mining and predictive analysis. In this research, we will use this tool for executing the following steps:

1) Tokenization: This step breaks a stream of text up into phrases, works, symbols, or other meaningful elements called tokens.

2) Stopword elimination: The most common words that unlikely to help text mining such as prepositions, articles, and pro-nouns are considered as stopwords. This step eliminates these words from the text document because they are not useful for the text mining applications.

3) Stemming or lemmatization: This step reduces the words into their stems, base or root. In this case, we have used the Snowball stemmer [13].

Term Filtering: The *term generation* module produces a set of terms associated with each document. However, the relevance of these terms should also take into account. In order to assign a *relevance value* to each term, we propose an information retrieval (IR) approach. In this case, the term relevance with respect to a news article collection is obtained by using one of the most successful and well-tested techniques in IR: *tf-idf* [14] (Term Frequency - Inverse Document Frequency). *tfidf* is a metric that determines the relative frequency of words in a specific news article compared to the inverse proportion of that word over the entire news article corpus. This calculation provides how relevant a given word is in a document. Those words which appear in a small group of articles will have higher *tf-idf* value than those words which are very common such as prepositions or pronouns. Thus, in this process what we get is a set of words (strings) per web news article, and its corresponding *tf-idf* value which determines how relevance is that word in the set of articles.

3.2 Evolving Classification

This module is responsible for not only classifying a specific news article (set of scored terms) into a topic area, but also updating the structure of the classifier according to the changes in the articles. The structure consists of a set of (evolving) fuzzy rules which can be interpreted by humans.

eClass is a fuzzy rule-based classifier which uses (fuzzy) rules that evolve from streaming data. In particular, *eClass0* possesses a zero-order Takagi-Sugeno consequent, so a fuzzy rule in the *eClass0* model has the following structure:

$$Rule_i = IF(X_1 \ is \ Prot_1) \ AND \ldots AND(X_n \ is \ Prot_n)$$
$$THEN \ \ Category \ = \ Category_i$$

where i represents the number of rule; n is the number of input variables (corpus); the X_i stores the *tf-idf* of the set of words (article) to classify, and the $Prot_i$ stores the *tf-idf* of the words of one of the prototypes (cluster center) of the corresponding class (category). $Category \in \{\text{set of different categories}\}$.

The *eClass0* model is composed of one or several fuzzy rules per category (the number of rules depends on the heterogeneity of the news articles in the same category). During the training process, a set of rules is formed "from scratch" using an evolving clustering approach to decide when to create new rules. The inference in *eClass0* is produced using the "winner takes all" rule and the membership functions that describe the degree of association with a specific prototype are of Gaussian form. The *potential* (equation 1) is a Cauchy function of the sum of distances between a certain data sample and *all* other data samples in the feature space, and it is used in the partitioning algorithm.

$$P(x_k) = \frac{1}{1 + \frac{\sum_{i=1}^{k-1} distance(x_k, x_i)}{k-1}} \tag{1}$$

where *distance* represents the distance between two samples in the data space.

However, in these classifiers, the potential (P) is calculated recursively (which makes the algorithm faster and more efficient)[15]. The result of this function represents the *density* of the data that surrounds a certain data sample.

In this case, the distance (similarity) between two samples is measured by the cosine distance(*cosDist*). However, this expression requires all the accumulated data sample available to be calculated, which contradicts to the requirement for real-time and on-line application proposed in this research. For this reason, in [15] it is developed a recursive expression cosine distance. All details about the *eClass0* model and the learning algorithm can be found in [7].

The procedure of this classifier for creating and updating the fuzzy rules is:

1. Calculate the potential of the *new* news article (set of words/strings) to be a prototype. This calculation is done by using a function of the accumulated distance between a sample and all the other set of strings in the data space [15]. The result represents the *density* of the data that surrounds a certain data sample.

2. Update all the prototypes considering the new article. The *density* of the data space surrounding certain prototype changes with the insertion of each new article and the existing prototypes need to be updated.

3. Insert the new article as a new prototype if needed. The potential of the new article is calculated recursively and the potential of the other prototypes is updated.

4. Remove existing prototypes if needed. After adding a new prototype, we check whether any of the already existing prototypes are described *well* by the newly added prototype.

As we can see in Figure 1, this process updates the Fuzzy Rules that define the different categories. More details about this procedure and the learning algorithm can be found in [7] (and other application of these algorithm in [16]).

4   Experimental Design and Results

In order to evaluate the presented approach, we have collected 200 news articles for each of the 7 categories (in total, 1400 categorized web news articles). Although this approach is designed to be used in real-time, we have used this dataset in order to have comparable results with other techniques. The main terms of these 1400 articles were extracted as it was proposed in the *Term Extraction* module. However, since the number of different terms is very high (more than 5000 different terms), we have applied a terms reduction technique. This reduction has been done taking into account that the *tf-idf* of several terms is very low. For this reason, we have removed those terms with a low tf-idf value (or 0) in all the documents. Specifically, we sum the tf-idf values of a particular term in all the documents. If this value is lower than a threshold, it is removed from the dataset. In this research, we have used several thresholds from 0.3 to 3. As higher the threshold is, as smaller the dataset is. We should consider that the number of terms removed using the threshold 3 is very high. In this case, if there is no reduction, the number of different terms is 5315; however, using a threshold of 3, this number is drastically reduced to 26. In addition, after applying this terms reduction, there were several articles which were represented only by terms with value 0. These articles were removed from the dataset.

Once this data reduction was applied, we created several sets of data which combine two or more categories. The datasets that we created were: 1) *Health vs. Science*, 2) *Science vs. Techonology*, 3) *Health vs. Science vs. Sports*, 4) *Business vs. Health vs. Science vs. Sports*, 5) *Arts vs. Business vs. Health vs. Science vs. Sports vs. Travel*.

The results are shown in the figure 2, where the different lines (different colors) represent the percentage of articles correctly classified in the corresponding categories. Also, the x-asis represent the threshold that has been used to reduce the original dataset.

Figure 2 shows how if we classify a news into only two categories (Science or Technology), the percentage of correctly classified is around 90%. Also, we can observe that the best classification is done using a reduction tf-idf threshold of 2 (59 terms). It is important to highlight that using this classification, the results are better when the terms reduction is high (using a tf-idf threshold higher than 1). This aspect is important since we do not need all the terms of the news in order to be able to classify them with a high percentage. Also, as we already supposed, if the number of categories (classes) is more than two, the percentage of news correctly classified decrease. However, for example, using 6 different categories, we obtain a percentage of news correctly classified of 38,85 - what is a good results if we take into account the reduced number of terms that we have used for this task.

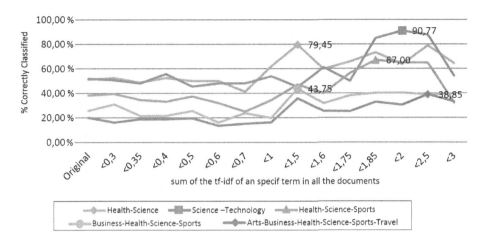

Fig. 2. Results taking into account the data reduction using tf-idf

Finally, in order to evaluate the performance of these results, the proposed classifier is compared with 3 different types of classifiers: 1) The C4.5 algorithm (decision-tree-based classifier), 2) The Naive Bayes Classifier, which is a simple probabilistic classifier based on applying Bayes theorem with independence assumptions and 3) The K Nearest Neighbor Classifier (kNN) which is an instance-based learning technique based on closest training examples in the feature space. This classification is done taking into account 3 datasets with a data reduction using the sum of the tf-idf as 1,5. In table 1 we observe that the proposed evolving classifiers achieves the better results. However, what it more important, the proposed classifier is the only classifier which is able to process streaming data in online and in real time.

Table 1. Comparison of the Different Classifiers

DataSets (Categories in the classification)	eClass0 (evolving classifier)	Decision Tree (C4.5)	Naive Bayes	K-NN
Health-Science	79,45	49,32	57,53	67,12
Science-Technology	90,77	49,23	66,15	73,85
Health-Science-Sports	67,00	34,00	54,00	53,00

5   Conclusions and Future Work

An evolving approach for classifying different web news articles into various topic areas based on the text content of the articles has been proposed in this paper. In order to get the relevance of the different terms of the web news, we have seen that *tf-idf* is a simple but powerful and efficient numeric measure for this task. Related to the proposed (evolving) classifier, it is important to highlight that it is very simple and it works very fast. Also, the proposed classifier is one pass, non-iterative, recursive and it can be used in an interactive mode. Since the number of web news articles is very high, this method can also cope with huge amounts of news and process them quickly. Although the amount of terms from the articles is huge, we can extract the most important terms with no need to store all the news in memory. The approach has been successfully tested using real on-line news, and according to the results we can see that it works quite well even using a (very) small number of terms.

As future work we propose to increase the number of web news and categories and employ other different measures to score the different terms of an article. Also, since a web news article is represented as a set of terms, this approach could be used in other areas such as the classification of tweets or short messages.

Acknowledgments. This work has been supported by the Spanish Government under i-Support (Intelligent Agent Based Driver Decision Support) Project (TRA2011-29454-C03-03).

References

1. Statistical report on internet development in china, Tech. rep., China Internet Network Information Center (July 2013)
2. Feldman, R., Dagan, I.: Kdt - knowledge discovery in texts. In: International Conference on Knowledge Discovery (KDD), pp. 112–117 (1995)
3. Hsu, L.-F.: Mining on terms extraction from web news. In: Pan, J.-S., Chen, S.-M., Nguyen, N.T. (eds.) ICCCI 2010, Part I. LNCS, vol. 6421, pp. 188–194. Springer, Heidelberg (2010)
4. Lent, B., Agrawal, R., Srikant, R.: Discovering trends in text databases. In: Proceedings of KDD, pp. 227–230 (1997)
5. Li, J., Li, J., Tang, J.: A flexible topic-driven framework for news exploration. In: Proceedings of KDD (2009)
6. Sadiq, A.T., Abdullah, S.M.: Hybrid intelligent techniques for text categorization. International Journal of Advanced Computer Science and Information Technology (IJACSIT) 2, 23–40 (2014)
7. Angelov, P., Filev, D.: An approach to online identification of takagi-sugeno fuzzy models. IEEE Trans. Systems, Man, and Cybernetics, Part B: Cybernetics 34(1), 484–498 (2004)
8. Kosala, R., Blockeel, H.: Web mining research: A survey. SIGKDD Explor. Newsl. 2(1), 1–15 (2000)
9. Lu, Y., Zhang, P., Liu, J., Li, J., Deng, S.: Health-related hot topic detection in online communities using text clustering. PloS One 8(2), e56221 (2013)
10. Appelt, D.E.: Introduction to information extraction. AI Commun. 12(3), 161–172 (1999)
11. Yoon, J.: Detecting weak signals for long-term business opportunities using text mining of web news. Expert Systems with Applicat. 39(16), 12543–12550 (2012)

12. Mierswa, I., Scholz, M., Klinkenberg, R., Wurst, M., Euler, T.: Yale: Rapid proto-typing for complex data mining tasks. In: KDDM Conference, pp. 935–940. ACM (2006)
13. Snowball web site (2014), `http://snowball.tartarus.org/`
14. Salton, G., Buckley, C.: Term-weighting approaches in automatic text retrieval. Information Processing and Management 24(5), 513–523 (1988)
15. Angelov, P.P., Zhou, X.: Evolving fuzzy-rule-based classifiers from data streams. IEEE Transactions on Fuzzy Systems 16(6), 1462–1475 (2008)
16. Iglesias, J.A., Angelov, P., Ledezma, A., Sanchis, A.: Creating evolving user behavior profiles automatically. IEEE Trans. Knowledge and Data Engineering 24(5), 854–867 (2012)

Predicting Students' Results Using Rough Sets Theory

Anca Udristoiu, Stefan Udristoiu, and Elvira Popescu

University of Craiova, Faculty of Automation, Computers and Electronics
Bvd. Decebal, Nr. 107, 200440, Craiova, Dolj, Romania
{anca_soimu,s_udristoiu,elvira_popescu}@yahoo.com

Abstract. This paper proposes the utilization of rough set theory for predicting student scholar performance. The rough set theory is a powerful approach that permits the searching for patterns in e-learning database using the minimal length principles. Searching for models with small size is performed by means of many different kinds of reducts that generate the decision rules capable for identifying the final student grade.

Keywords: education data mining, rough set theory, decision rules, learning management system.

1 Introduction

Predicting students' performance is one of the most important and useful applications of educational data mining and its goal is to score or mark from student course behavior and activity [14]. In this study, the rough set technique is used to offer methods for understanding, processing and modelling data, resolving the limitations of the e-learning systems.

The rough set theory was discovered by Zdzislaw Pawlak [1] and is a powerful mathematical tool for modeling the imperfect and incomplete knowledge [2]. Rough set theory has also excellent results in approximate reasoning [5], mathematical logic analysis and reduct [6, 7, 8], building of predictive models [9], and decision support system [10, 11, 12]. Many studies have shown that the use of rough set theory formulate a clear decision-making projects and enhance the effectiveness of the research while doing optimization [7]. The research related to education of Qu and Wang [11] provided a basis of personalized teaching strategies in distance learning website by analysis of reduct and attribute significance. In [13] the study analyzed students' misconception based on rough set theory. Although the rough set theory is rarely used in education, in this study we use its characteristics, which are very suitable for discovering rules useful in educational process.

2 Student Representation and Discretization

We have collected data from on-line course activity provided by Moodle [15] that is one of the most widely used open source learning management system. In fact, we have used the following data based on student 'Database' course activity [14]:

E. Corchado et al. (Eds.): IDEAL 2014, LNCS 8669, pp. 336–343, 2014.

Nassignment – number of assignments taken; *Nquiz* - number of quiz taken; *Nquiz_p* - number of quiz passed; *Nquiz_f* - number of quiz failed; *Nmessages* - number of messages sent to the chat; *Nmessages_ap* - number of messages sent to the teacher; *Nposts* - number of messages sent to the forum; *Nread* - number or forum messages read; *Total_time_assignment*- total time spent on assignment; *Total_time_quiz* – total time used in quizzes; *Total_time_forum*- total time used in forum; *Mark*- final mark the student obtained in the course.

Since the data provided by Moodle are structured, they didn't necessitate preparation [14]. So, we directly discretise them, transforming numerical values into categorical ones for a good interpretation and understanding. We have used the manual method for discretising all attributes, so the teacher has to specify the cut off points. The mark descriptor has four values: *insufficient*, if value < 5; *average*, if value > 5 and < 7; *good* if value >7 and < 9; *excellent* if value > 9. The other attributes have the values: LOW, MEDIUM and HIGH   [14].

A student is represented in Prolog by means of a term:

```
student(ListofDescriptors)
```

where the argument is a list of terms used to specify the student attributes.
The term used to specify the student attributes is of the form:

```
descriptor(DescriptorName,DescriptorValue)
```

The model representation of students is in the following example:

```
student([
descriptor(Nassignment,medium),descriptor(Nquiz,low),
descriptor(Nquiz_p,low),descriptor(Nquiz_f,high),
descriptor(Nmessages, medium), descriptor(Nmessages_ap,
medium),descriptor(Nposts,low),descriptor(Nread,low),
descriptor(Total_time_assignment,low),
descriptor(Total_time_quiz,low),
descriptor(Total_time_forum,low)]).
```

3   Modelling of Student Information Using Rough Sets

3.1   Rough Sets Foundations

Rough sets theory is an intelligent mathematical tool and it is based on the concept of approximation space [1], [3]. In rough sets theory, the notion of information system determines the knowledge representation system. In this section, we recall some basic definitions from literature [1, 2, 3].

Let U denote a finite non-empty set of objects (students) called the universe. Further, let A denote a finite non-empty set of attributes. Every attribute $a \in A$, there is a function $a: U \rightarrow V_a^c$ where V_a is the set of all possible values of a, to be called the domain of $a \triangleright$ A pair IS = (U, A) is an information system. Usually, the specification of an information system can be presented in tabular form. Each subset of attributes

$B \subseteq A$ determines a binary *B–indiscernibility* relation *IND(B)* consisting of pairs of objects indiscernible with respect to attributes from B like in (1):

$$IND(B) = \{(x, y) \in U \times U : \forall a \in B, a(x) = a(y)\} \tag{1}$$

IND(B) is an equivalence relation and determines a partition of U which is denoted by $U/IND(B)$. The set of objects indiscernible with an object $x \in U$ with respect to the attribute set, B, is denoted by $I_B(x)$ and is called *B–indiscernibility* class:

$$I_B(x) = \{y \in U : (x, y) \in IND(B)\} \tag{2}$$

$$U/IND(B) = \{I_B(x) : x \in U\} \tag{3}$$

Table 1. Student Information System

U	Nmessages	Nmessages_ap	Mark
R_1	medium	low	average
R_2	medium	low	average
R_3	medium	low	average
R_4	medium	high	good
R_5	high	high	good
R_6	medium	medium	average
R_7	medium	medium	average
R_8	medium	high	good
R_9	medium	high	good
R_{10}	high	low	good
R_{11}	high	low	average

It is said that a pair ASB = (U, IND(B)) is an approximation space for the information system IS=(U, A), where $B \subseteq A$. The information system from Table 1 represents the students enrolled into a course represented in terms of descriptors values, as described in Section 2. For simplicity we consider only two descriptors as attributes, namely the Nmessages and Nmessages_ap. So, our information systems *is* IS = (U, B), where U = {R1, R2, R3, R4, R5, R6, R7, R8, R9, R10, R11} and B={Nmessages, Nmessages_ap}.

Let $W = \{w_1, ..., w_n\}$ be the elements of the approximation space $AS_B=(U, IND(B))$. We want to represent X, a subset of U, using attribute subset B. In general, X cannot be expressed exactly, because the set may include and exclude objects which are indistinguishable on the basis of attributes B, so we could define X using the lower and upper approximation.

The *B-lower approximation X*, $\underline{B}X$, is the union of all equivalence classes in *IND(B)* which are contained by the target set X. The lower approximation of X is called the positive region of X and is noted *POS(X)*.

$$\underline{B}X = \bigcup \{w_i \mid w_i \subseteq X\} \tag{4}$$

The *B-upper approximation* $\overline{B}X$ is the union of all equivalence classes in *IND(B)* which have non-empty intersection with the target set *X*.

$$\overline{B}X = \bigcup \{w_i \mid w_i \cap X \neq \varnothing\} \tag{5}$$

Example: Let $X = \{R_1, R_2, R_3, R_4, R_5, R_6, R_7, R_8\}$ be the subset of *U* that we wish to be represented by the attributes set $B = \{Nmessages, Nmessages\_ap\}$. We can approximate *X*, by computing its *B-lower approximation*, $\underline{B}X$ and *B-upper approximation*, $\overline{B}X$. So, $\underline{B}X = \{\{R1, R2, R3\}, \{R5\}, \{R6, R7\}\}$ and $\overline{B}X = \{\{R1, R2, R3\}, \{R5\}, \{R6, R7\}, \{R4, R8, R9\}\}$. The tuple ($\underline{B}X$, $\overline{B}X$) composed of the lower and upper approximation is called a rough set; thus, a rough set is composed of two crisp sets, one representing a *lower boundary* of the target set *X*, and the other representing an *upper boundary* of the target set *X*. The accuracy of a rough set is defined as: cardinality($\underline{B}X$)/cardinality($\overline{B}X$). If the accuracy is equal to 1, then the approximation is perfect.

3.2   Dispensable Features, Reducts and Core

An important notion used in rough set theory is the decision table. Pawlak [1] gives also a formal definition of a decision table: an information system with distinguished conditional attributes and decision attribute is called a decision table. So, a tuple DT = (U, C, D), is a decision table. The attributes C = {Nmessages, Nmessages_ap} *are* called conditional attributes, instead D = {Mark} is called decision attribute. The classes *U/IND(C)* and *U/IND(D)* are called condition and decision classes, respectively. The *C-Positive* region of *D* is given by:

$$POS_C(D) = \bigcup_{X \in IND(D)} \underline{C}X \tag{6}$$

Let $c \in C$ a feature. It is said that *c* is dispensable in the decision table *DT*, if $POS_{C-\{c\}}(D) = POS_C(D)$; otherwise the feature *c* is called indispensable in *DT*. If *c* is an indispensable feature, deleting it from *DT* makes it to be inconsistent.

A set of features *R* in *C* is called a reduct, if $DT' = (U, R, D)$ is independent and $POS_R(D) = POS_C(D)$. In other words, a reduct is the minimal feature subset preserving the above condition.

3.3   Producing Rules by Discernibility Matrix

We transform the decision table into discernibility matrix to compute the reducts. Let $DT = (U, C, D)$ be the decision table, with $U = \{R_1, R_2, R_3, R_4, R_5, R_6, R_7, R_8, R_9, R_{10}, R_{11}\}$. By a discernibility matrix of *DT*, denoted *DM(T)*, we will mean *nxn* matrix defined as:

$$m_{ij} \overset{a(R_i)}{=} \{(a \in C : a(R_i) \neq a(R_j)) \text{ and } (d(R_i) \neq d(R_j))\} \tag{7}$$

where $i, j = 1, 2, \ldots, 11$.

The items within each cell of the discernibility matrix, *DM(DT)* are aggregated disjunctively, and the individual cells are then aggregated conjunctively. To compute the reducts of the discernibility matrix we use the following theorems that demonstrate equivalence between reducts and prime implicants of suitable Boolean functions [2], [12]. For every object Ri $\in$ U, the following Boolean function is defined:

$$g_{R_i}(\text{Nmessages, Nmessages\_ap}) = \underset{R_j \in U}{\wedge} \left(\underset{a \in m_{ij}}{\vee a} \right) \tag{8}$$

The following conditions are equivalent [3]:

1. $\{a_{i1}, \ldots, a_{in}\}$ is a reduct for the object Ri, i = 1..n. and
2. $a_{i1} \wedge a_{i2} \wedge .. \wedge a_{in}$ is a prime implicant of the Boolean function g_{Ri}

On Boolean expression the absorption Boolean algebra rule is applied. The absorption law is an identity linking a pair of binary operations. For example: $a \vee (a \wedge b) = a \wedge (a \vee b) = a$

From the decision matrix we form a set of Boolean expressions, one expression for each row of the matrix.

For the *average* mark, we obtain the following rules based on the table reducts:

- (Nmessages=mediu $\vee$ Nmessages_ap=low) $\wedge$ (Nmessages_ap =low) $\wedge$ (Nmessages=high)
- (Nmessages_ap =medium) $\wedge$ (Nmessages=medium $\vee$ Nmessages_ap =medium)
- (Nmessages=high $\vee$ Nmessages_ap =low) $\wedge$ (Nmessages_ap =low)

For the *good* mark we obtain the following rules based on the table reducts:

- Nmessages_ap = high
- Nmessages=high $\vee$ Nmessages_ap =high
- Nmessages_ap = high $\wedge$ (Nmessages=high $\vee$ Nmessages_ap =low)

By applying the absorption rule on the prime implicants, the following rules are generated:

- Rule 1: Nmessages_ap =low $\wedge$ Nmessages=high →average
- Rule 2: Nmessages_ap =medium→average
- Rule 3: Nmessages_ap =low→average
- Rule 4: Nmessages_ap = high→good
- Rule 5: Nmessages=high $\vee$ Nmessages_ap =high→good

3.4 Evaluation of Decision Rules

Decision rules can be evaluated along at least two dimensions: performance (prediction) and explanatory features (description). The performance estimates how well the rules classify unevaluated students. The explanatory feature estimates how interpretable the rules are [2]. Let be our decision table $DT = (U, C, D)$. We use the set-theoretical interpretation of rules. It links a rule to data sets from which the rule is discovered [2]. Using the cardinalities of sets, we obtain the 2×2 contingency table representing the quantitative information about the rule *if descriptors then mark*. Using the elements of the contingency table, we may define the support (s) and accuracy (a) of a decision rule by:

$$s(rule) = cardinality(descriptorSet \cap markSet) \tag{9}$$

$$a(rule) = \frac{cardinality(descriptorSet \cap markSet)}{cardinality(descriptorSet)} \tag{10}$$

where the set *descriptorSet ∩ markSet* is composed from student descriptors which have a certain *descriptorSet* and a certain *mark*.

The coverage(c) of a rule is defined by:

$$c(rule) = \frac{cardinality(descriptorSet \cap markSet)}{cardinality(markSet)} \tag{11}$$

For the generated Rule 2, the contingency Table 2 is obtained, where the descriptor *Nmessages_ap* is denoted by D. For the Rule 2, the support is 2, accuracy is 2/2 and coverage is 2/6. In [4], the study suggests that high accuracy and coverage are requirements of decision rules.

Table 2. Contingency Table Representing the Quantitative Information about the Rule 2

	mark = average	not(mark = average)	
D =medium	cardinality(D=medium and mark=average) = 2	cardinality(D=medium and not(mark=average)) = 0	cardinality(D =medium)=2
not(D=medium)	cardinality(not(D =medium) and mark=average)=4	cardinality(not(D =medium) and not(mark=average))=5	cardinality(not(D =medium))=9
	cardinality(mark=average) =6	cardinality (not(mark=average))= 5	cardinali- ty(U)=11

4   Decision Rule Extraction Using Rough Sets Models and Experiments

In this paper we present the application of rough set to discover student rules between students' descriptors and mark categories. A rule is represented using a Prolog fact:

```
rule(Mark, Accuracy, Coverage, ListofStudentDescriptors)
```

where *Mark*, the head of the rule, is the mark category, *Accuracy* is the rule accuracy computed as in (10), *Coverage* is the rule coverage computed as in (11) the body of the rule, is composed by conjunctions of student descriptors, while *Mark*, the head of the rule, is the mark category.

Decision rules are generated from reducts as described in Section 3. The student classification algorithm based on the discovered rules can be resumed as:

- collect all the decision rules in a classifier,
- compute for each rule the support, accuracy and coverage,
- eliminate the rules with the support less than the minimum defined support,
- order the rules by accuracy, than by coverage,
- if a student matches more rules select the first one: a student matches a rule, if all the descriptors, which appear in the body of the rule, are included in the descriptors of the student.

In the experiments realized through this study, two databases are used for the learning and testing process. The database used to learn the correlations between student behaviour and marks, contains information about 40 students, each described by 11 descriptors. For each mark class, the following metrics: accuracy, specificity, and sensitivity. The counted results are presented in Table 3.

Table 3. Results recorded for different marks

Mark	Accuracy(%)	Sensitivity(%)	Specificity(%)
Good	98.3	97	73.1
Average	97.7	96.1	72.8
Excellent	97.9	96	72.8
Insufficient	96.3	95.2	71.7

5   Conclusion

In this study, a method based on rough set theory is proposed and developed to assist the teacher by doing the pre-evaluation of students during a course study. For establishing correlations with the mark, we experimented and selected some descriptors of the student activity in the Moodle system for a "Database" course. The results of experiments are very promising and show that the methods based on rough set theory are very useful for predicting the results of the student during a course activity. The Prolog language used for representation of students' descriptors and rules makes a simple and flexible integration of our methods with other learning management systems.

In future work, it would be interesting to repeat the analysis   using more data from different types of courses and also to select   other student descriptors. It would be also very useful to do experiments using more experts in order to analyse the obtained rules for discovering interesting relashionships.

Acknowledgment. This work was partially supported by the grant number 16C/2014, awarded in the internal grant competition of the University of Craiova.

References

1. Pawlak, Z., Skowron, A.: Rough Membership Functions. In: Advances in the Dempster-Shafer Theory of Evidence, pp. 251–271. John Wiley and Sons, New York (1994)
2. Stepaniuk, J.: Rough – Granular Computing in Knowledge Discovery and Data Mining. SCI, vol. 152. Springer, Heidelberg (2009)
3. Hassanien, A.E., Abraham, A., Peters, J.F., Kacprzyk, J.: Rough Sets in Medical Imaging: Foundations and Trends. In: Computational Intelligence in Medical Imaging: Techniques and Applications, pp. 47–87. CRC Press, USA (2008)
4. Michalski, R.: A Theory and Methodology of Inductive Learning. Artificial Intelligence 20(2), 111–161 (1983)
5. Wang, H., Zhou, M., William, Z.: A New Approach to Establish Variable Consistency Dominance—Based Rough Sets Based on Dominance Matrices. In: International Conference on Intelligent System Design and Engineering Application, Sanya, Hainan, pp. 48–51 (2012)
6. Ren, Y., Xing, T., Quan, Q., Chen, X.: Attributes Knowledge Reduction and Evaluation Decision of Logistics Centre Location Based on Rough Sets. In: 4th International Conference on Intelligent Computation Technology and Automation, Shenzhen, pp. 67–70 (2011)
7. Zaras, K., Marin, J.C., Boudreau-Trude, B.: Dominance-Based Rough Set Approach in Selection of Portfolio of Sustainable Development Projects. American Journal of Operations Research 2(4), 502–508 (2012)
8. Ke, G., Mingwu, L., Yong, F., Xia, Z.: A Hybrid Model of Rough Sets and Shannon Entropy for Building a Foreign Trade Forecasting System. In: 4th International Joint Conference on Computational Sciences and Optimization, Yunnan, pp. 7–11 (2011)
9. Lai, C.J., Wen, K.L.: Application of Rough Set Approach to Credit Screening Evaluation. Journal of Quantitative Management 12(1), 69–78 (2005)
10. Chao, D., Sulin, P.: The BSC Alarm Management System Based on Rough Set Theory in Mobile Communication. In: 7th International Conference on Computational Intelligence and Security, Hainan, pp. 1557–1561 (2011)
11. Hossam, A.N.: A Probabilistic Rough Set Approach to Rule Discovery. International Journal of Advanced Science and Technology 30, 25–34 (2011)
12. Qu, Z., Wang, X.: Application of Clustering Algorithm and Rough Set in Distance Education. In: 1st International Workshop on Education Technology and Computer Science, Wuhan, pp. 489–493 (2009)
13. Sheu, T., Chen, T., Tsai, C., Tzeng, J., Deng, C., Nagai, M.: Analysis of Students' Misconception Based on Rough Set Theory. Journal of Intelligent Learning Systems and Applications 5(2), 67–83 (2013)
14. Romero, C., Zafra, A., Luna, J.M., Ventura, S.: Association rule mining using genetic programming to provide feedback to instructors from multiple-choice quiz data. Expert Systems 30(2), 162–172 (2013)
15. Cole, J.: Using Moodle. O'Reilly (2005)

Graph-Based Object Class Discovery from Images with Multiple Objects

Takayuki Nanbu and Hisashi Koga

Graduate Schools of Information Systems, University of Electro-Communications,
Chofugaoka 1-5-1, Chofu-si, Tokyo 182-8585, Japan
koga@is.uec.ac.jp

Abstract. Discovering objects models from image database has gained much attention. Although the BoVW (Bag-of-Visual-Words) approach has succeeded for this research topic, Xia and Hancock pointed out the two drawbacks of the BoVW: (1) it does not represent the spatial co-occurrence between local features and (2) it is difficult to select proper vocabulary size in advance. To overcome these drawbacks, they propose a novel unsupervised graph-based object discovery algorithm. However, this algorithm assumes that one image contains only one object. This paper develops a new unsupervised graph-based object discovery algorithm that treats images with multiple objects. By clustering the local features without specifying the number of clusters, our algorithm does not have to decide the vocabulary size in advance. Next, it acquires object models as frequent subgraph structures defined by a set of co-occurring edges which describe the spatial relation between local features.

1 Introduction

As the volume of digital images has increased recently, automatic object recognition without human intervention has become more important. However, to recognize various objects, a lot of object models must be laboriously prepared in advance. For this reason, unsupervised discovery of object models from an image database has gained much attention. In literatures, several researches [1] [2] [3] have devised algorithms to discover object models based on the Bag-of-Visual-Words (BoVW) scheme [4] which describes an image as a set of visual words, i.e., representative local features. To acquire object models, these previous works apply the latent topic model like pLSA (Probabilistic Latent Semantic Analysis) or LDA (Latent Dirichlet Allocation) to image processing. Then, object models are learned as visual common topics. Some recent approaches like the geometric LDA (gLDA) [3] have tried to learn visual topics by jointly modeling the appearance of local patches and their spatial arrangement.

Although the BoVW approach has succeeded in the context of object discovery from images, Xia and Hancock [5] pointed out the two drawbacks of the BoVW: (1) An individual visual word is not discriminative enough. Hence, the object models should be more directly related to the co-occurrence of local features with a specific spatial arrangement. (2) it is hard to determine properly the number of

E. Corchado et al. (Eds.): IDEAL 2014, LNCS 8669, pp. 344–353, 2014.

representative visual words in advance, because the optimal value shall depend on the diversity of the image database. The number of representative visual words is called the *vocabulary size*.

To solve these drawbacks, Xia and Hancock proposed a novel graph-based object discovery algorithm called SPGC (Similarity Propagation based Graph Clustering)[5]. SPGC represents an image as a graph whose vertices are the local features. Then, after clustering the graphs, each graph cluster becomes one object model. In the graph clustering, the similarity between graphs are measured based on the SIFT matching without clustering the local features. Hence, SPGC does not specify the vocabulary size in advance. In addition, SPGC encodes the spatial relation between local features into the object models with the graph edges. Despite the above merits, SPGC has the constraint that one image contains only one object.

Inspired by SPGC, this paper develops a new unsupervised graph-based object discovery algorithm that handles images with multiple objects, while keeping the merits of SPGC. First, our algorithm clusters the local features without specifying the number of clusters and, thus, avoids configuring the proper vocabulary size in advance. Then, object models are discovered as frequent subgraph structures expressed by co-occurring edges, where each edge captures the spatial relation between the local features.

This paper is organized as follows. Sect. 2 introduces the previous work SPGC. Sect. 3 explains our object discovery method. Here, we also discuss the related works. Sect. 4 reports the experimental results. Sect. 5 is the conclusion.

2   SPGC

SPGC [5] is stated here so that the readers may contrast it with our algorithm described later. Let S be the image database.

SPGC first extracts the stable SIFT features in each image $I \in S$. Let these stable features by V_I. A point in V_I is described with (1) the pixel location and (2) the 128-dimensional SIFT feature vector. Then, I is represented as the Delaunay graph G_I for V_I.

Next, SPGC clusters all the graphs, where the similarity between the graphs G_{I_1} and G_{I_2} denoted by $\mathrm{sim}(G_{I_1}, G_{I_2})$ is proportional to the size of the maximum common subgraph (MCS). In the computation of the MCS, the SIFT matching is used to judge the equality of two vertices. When $\mathrm{sim}(G_{I_1}, G_{I_2}) \geq R$ where R is a thresholding parameter, G_{I_1} and G_{I_2} are unified into the same cluster. The merge operations are performed iteratively such that if either $\mathrm{sim}(G_{I_1}, G_{I_3}) \geq R$ or $\mathrm{sim}(G_{I_2}, G_{I_3}) \geq R$ provided that $\mathrm{sim}(G_{I_1}, G_{I_2}) \geq R$, G_3 belongs to the same cluster as G_{I_1} and G_{I_2}. Finally, one graph cluster represents one object class. Thus, SPGC acquires an object model by seeking frequent graphs.

SPGC smartly avoids the difficulty in choosing the proper vocabulary size without clustering the local features, where the graph similarity is measured based on the SIFT matching. In SPGC, an object model can model the spatial relation between local features with graph edges.

However, SPGC assumes that one image contains only one object. Hence it cannot treat images which contain multiple objects. For example, suppose that we have two graph clusters C_1 and C_2 such that C_1 is associated with the car object and C_2 corresponds to the dog object. Then, if an image I which contains both a car and a dog appears, I may unify C_1 and C_2 and make an inappropriate larger graph cluster, because G_I is similar to both C_1 and C_2.

The above restriction of SPGC motivates us to develop a new unsupervised graph-based object discovery algorithm that treats images with multiple objects.

3   Our Object Discovery Algorithm

Our discovery algorithm is named as the DCG (Discovery of Co-occurring Graph structure). DCG aims to keep the merits of SPGC and has the next properties.

- DCG clusters the local features without knowing the number of clusters in advance unlike the k-means used in the BoVW scheme.
- DCG expresses each image as a graph whose edge represents the spatial relation between local features and acquires an objects model as a set of edges co-occurring frequently.

Unlike SPGC, DCG clusters the local features. However, the clustering algorithm in DCG is able to adapt to the diversity of given images, since it does not require the number of clusters beforehand. While the entire graph from an image forms an object instance in SPGC, DCG permits a subgraph structure to become an object instance, which enables DCG to process images with multiple objects.

DCG executes the next 3 steps in order, so as to find the object models. Hereafter, let $S = \{I_1, I_2, \ldots, I_n\}$ be the image database consisting of n images.
Step 1: The local features extracted from S are clustered.
Step 2: Each image $I \in S$ is represented as a graph G_I. Now, we have n graphs.
Step 3: Object models are discovered by mining co-occurring edges from them.

Remarkably, DCG uses the graph structure not only to express the images, but also to perform Step 1 and Step 3 both of which are reduced to some graph decomposition problems. From now on, we explain each step.

3.1   Clustering Local Features

Step 1 clusters the local features. First, we extract the SIFT features from every image I. Then, in the same way as SPGC, they are ranked with the ranking algorithm [6]: 20 artificial images are constructed by applying 20 kinds of transformation operations to I. A feature v in I is ranked according to the frequency that v is matched to the right features in the 20 artificial images. Then, the top T features in the ranking are saved as stable feature points and denoted by V_I

Next, the SIFT matching is performed between all the image pairs in S. Namely, V_{I_i} are (SIFT-)matched to V_{I_j} for $1 \leq i \leq n$, $1 \leq j \leq n$, $i \neq j$. Then, we can build a SIFT-Matching Graph $SMG = (V, E)$ such that $V = \cup_{i=1}^{n} V_{I_i}$ and the matched feature pairs are connected by the edges in E.

Then, we cluster the features in V by separating SMG into subgraphs. Let L be a list of the vertices in V in which they are sorted in the decreasing order of their degrees. That is, a feature v is prioritized more, as v is matched to more other features. With L and SMG, the clustering proceeds as follows.

Step (a): Let v be the first vertex in L. We classify v and v's two-hop neighbor vertices in SMG into the identical cluster C_v.
Step (b): After subtracting the members of C_v from L and SMG, Step (a) is recursively executed until all the vertices in V belong to some cluster.

Supported by the SIFT matching, this clustering does not need the cluster number beforehand. As for Step (a), the rationale to choose the two-hop neighbor is explained as follows: Despite the robustness of the SIFT matching, two features v and v' which correspond to the same object component might not be matched accidentally. Against such a failure, we expect that v' is matched to at least one of the features matched to v, so that v and v' may belong to the same cluster.

Execution Speed Acceleration
As n increases, the SIFT matching between all the image pairs will consume much time. Since most image pairs are usually very dissimilar, we introduce a pre-filtering technique to skip the SIFT matching between such dissimilar image pairs so as to shrink the execution time.

Our pre-filtering technique first gives some coarse labels to all the features with a light-weighted clustering algorithm named ROUGH. Then, for an image pair I_1 and I_2, we investigate how many features of the same label they share. I_2 is actually SIFT-matched to I_1, if I_2 belongs to the the top 20% images in S for I_1 with respect to this quantity.

ROUGH scans V only twice. Let v_i be the i-th local feature in V. At the first scan, an incremental clustering algorithm runs only to determine the cluster centroids. Suppose that v_i is inspected at time t_i and that $F = \{f_1, f_2, \cdots, f_m\}$ is the set of m clusters which have been already generated just before t_i. $F = \phi$ before t_1. If v_i is far from all of the m cluster centroids by more than a distance threshold D, a new cluster f_{m+1} which consists of v_i only is constructed, and v_i becomes the centroid of f_{m+1}. Otherwise, v_i joins the existing cluster, say f, whose centroid is the nearest to v_i. Then, we move the centroid of f to the average vector over all the members in f. After the first scan fixes the cluster centroids, the second scan assigns the final label to each feature v by relating v to the cluster centroid nearest to v. Despite its simpleness, ROUGH decides the cluster centroids adaptively to the point distribution for V in the 128-dimensional feature space. Furthermore, we need not teach ROUGH the number of clusters.

Graph Representation of Images
In Step 2, DCG transforms each image I into a graph G_I by constructing the Delaunay graph for V_I. The vertices have the labels computed as above. Now, we have the set of n graphs $\{G_{I_1}, G_{I_2} \cdots, G_{I_n}\}$.

3.2 Discovering Object Models by Mining Co-occurring Edges

First, we explain our idea to discover object models. If the images are represented as graphs and multiple instances of the same object O exist in different images, almost the same graph structure will appear at the locations of O's instances. Therefore, the edges composing the graph structure corresponding to O are likely to co-occur one another within each instance of O. Hence, we consider that the set of edges that co-occur in many images originate from the same object class. Thus, DCG obtains the object models by gathering the set of edges co-occurring in many graphs (i.e., images). Especially, DCG collects them via the next 3 steps.

(Step 1) Extraction of frequent edges
(Step 2) Discovering co-occurring frequent edge pairs
(Step 3) Agglomerative clustering of co-occurring frequent edges

(Step 1) counts the frequency of an edge e and regards e as frequent, if e appears more than ϵ times, where ϵ is a parameter. Since ϵ works as a so-called minimum support parameter of frequent pattern mining, user are responsible for configuring ϵ adequately. When the proper value is unknown, we recommend to set ϵ to a rather large value at the beginning and to reduce it gradually until the satisfying result is derived. This is because, even if ϵ is too large, it will only consume a small amount of execution time by accompanying few frequent edges.

(Step 2) examines the degree of co-occurrence between all the frequent edge pairs. For instance, let e_α and e_β be a frequent edge pair. Let S_α (S_β respectively) are the set of images which contain e_α (respectively e_β). As e_α and e_β co-occur more frequently, S_α becomes more similar to S_β. Hence, we measure the degree of co-occurrence with the Jaccard coefficient $\text{sim}(S_\alpha, S_\beta) = \frac{|S_\alpha \cap S_\beta|}{|S_\alpha \cup S_\beta|}$. When $\text{sim}(S_\alpha, S_\beta) \geq \theta$, we regard e_α and e_β as a co-occurring frequent edge pair. In the implementation, $\theta = 0.5$. Hereafter, the term "Co-occurring Frequent Edge Pair" is abbreviated as CFEP.

(Step 3) clusters co-occurring frequent edges agglomeratively. We repeat merging two CFEPs into the same cluster, if they share the same frequent edge. For instance, if e_α and e_β are a CFEP and e_β and e_γ are another CFEP, e_α, e_β and e_γ will belong to the same cluster, since e_β are common. The merge operation finishes, when there remains no CFEPs to be merged. After this clustering, one cluster becomes one object model. Thus, an object model is derived as a set of co-occurring frequent edges. (Step 3) is reduced to a graph decomposition problem as follows. We construct a graph named *frequent-edge graph* (FEG) whose vertices are the frequent edges. An edge of the FEG connects one CFEP. Then, (Step 3) is reduced to the decomposition of FEG into connected components, s.t., one connected component grows one cluster.

Finally, let us illustrate the algorithm to discover object models with the toy example in Fig. 1. Fig. 1 consists of 4 images with two kinds of fruit. The local features have been already clustered, so that every vertex has a label. Because DCG aims to discover frequent objects, DCG is expected to obtain the apple object model only, as other fruit appear only once. Note that the graph structures covering the apple instances are slightly different one another. When $\epsilon = 2$, the 4 edges

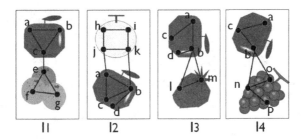

Fig. 1. Images with Apple Object Instances

(a-b), (b-c), (a-c) and (b-d) become frequent edges, where $S_{(a\text{-}b)} = \{I_1, I_2, I_3\}$, $S_{(a\text{-}c)} = \{I_1, I_2, I_3, I_4\}$, $S_{(b\text{-}c)} = \{I_1, I_2, I_4\}$ and $S_{(b\text{-}d)} = \{I_2, I_3\}$. Since all the other edges occur at most once, they cannot become frequent edges. Thus, sim(a-b,a-c) = 0.75, sim(a-b,b-c) = 0.5, sim(a-b,b-d) = 0.67, sim(a-c,b-c) = 0.75, sim(a-c,b-d) = 0.5 and sim(b-c,b-d) = 0.25. Hence, if $\theta = 0.6$, the 3 edge pairs ((a-b) and (a-c)), ((a-b) and (b-d)) and ((a-c) and (b-c)) become the CFEPs. By merging them at (Step 3), the 4 edges (a-b), (a-c), (b-d) and (b-c) form the same cluster and are identified as the object model.

3.3 Related Works

Our method in Sect. 3.2 is viewed as an application of frequent pattern mining to image processing which takes a spatial configuration into account. Hence, we mention several related works from this viewpoint. [7] [8] and [9] extend the frequent itemset mining to image processing such that a transaction consists of spatially close local features. These works expected a frequent itemset to be more discriminating than a single local feature in the context of image classification rather than object discovery. [10] discovered meaningful visual patterns by merging the spatial frequent itemsets. Gao *et al.* [11] also studied object discovery. Whereas [11] searched maximal frequent subgraphs directly, we cluster frequent edges agglomeratively to reduce the computational overhead. Recently, [12] utilizes the spatial relation among objects to grasp a group of objects for better scene understanding. Compared with these related works, DCG is unique in that it is easily accelerated with Min Hash [13], a hashing technique for the similarity search with the Jaccard coefficient. DCG extends the previous work [2] supported by Min Hash, so that the spatial relation between near feature points may be considered.

4   Experimental Results

This section evaluates DCG with the experiment using the the COIL100 image database which was also used to evaluate SPGC in [5]. The COIL100 consists of 100 classes of objects. Since every image in the COIL100 includes exactly one

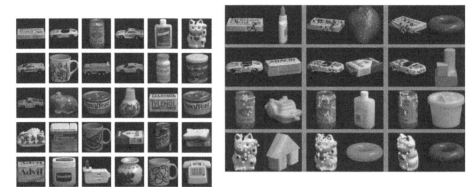

Fig. 2. 30 frequent object classes   **Fig. 3.** Examples of Discovered Objects

object, we synthesize an image with two objects by merging two images. Each synthesized image consists of one frequent object and an object randomly chosen from the whole COIL100 database. As the frequent objects, we choose 30 object classes shown in Fig. 2.

4.1 Object Discovery from Images with Multiple Objects

We demonstrate that DCG successfully discovers objects from multiple images. We prepare 300 images by selecting 10 instances for the 30 frequent object classes. Here, the orientations of the instances range from $0°$ to $45°$ at the interval of $5°$.

As for the parameters in DCG, T, the number of stable feature points per image is set to 80. This value is just twice as big as that used in SPGC [5], because SPGC deals with images with only one object, while each image consists of two objects in this experiment. ϵ is set to 3.

Fig. 3 shows several discovered object examples. In this figure, the color of an edge shows the edge cluster that it belongs to. When a image I contains many edges of the same color, it implies that an instance of the corresponding object exists in I. Thus, Fig. 3 shows that DCG discovers the frequent objects well. We also evaluate DCG quantitatively with *precision* and *recall*. The precision is the percentage that the discovered frequent object instances conform to the ground-truth. The recall evaluates the ratio of the ground-truth frequent object instances which are not missed by DCG. To compute the precision and recall, we must judge if I has an instance of the object model O. We judged that I contains an instance of O, when at least 3 edges in I belong to the object model (i.e. edge cluster) for O.

Table 1 summarizes the evaluation result. The left-most 5 columns displays the result for the 5 object IDs used in the COIL100. The right-most column termed "Total" presents the overall performance over the 30 frequent objects. Regarding to the rows, N_i is the number of the ground-truth object instances; N_d is the number of object instances discovered by DCG. N_d^+ presents the number of

Table 1. Quantitative Evaluation

ID	obj7	obj23	obj27	obj28	obj53	25 other objects	Total
N_i	10	10	10	10	10	10	300
N_d	10	8	9	7	9	10	293
N_d^+	10	8	9	7	9	10	293
precision	1	1	1	1	1	1	1
recall	1	0.8	0.9	0.7	0.9	1	0.97
N_c	2	2	1	1	1	1	32

the correct object instances out of the N_d discovered object instances. Therefore, the precision is equal to $\frac{N_d^+}{N_d}$, whereas the recall equals $\frac{N_d^+}{N_i}$. Finally, N_c displays the number of object models (that is, clusters of co-occurring edges) associated with each object class.

The total precision becomes 1 and the total recall reaches 0.97. For 25 classes, the precision and recall ideally reaches 1.0 both. For the two object classes obj7 and obj23, 2 object models are obtained. However, in the research of *unsupervised* object discovery, it is allowable that the number of the discovered object models goes beyond that of the ground truth object classes, since the object class is defined from the semantic viewpoint without concerning the feature variety among different instances of the same object class. Though the COIL100 database are sampled under controlled imaging conditions, this experimental result supports that DCG is of much promise and effective.

As for the execution speed, it takes 7.2 minutes on the Intel Single PC platform (Memory: 8GB, OS: Ubuntu Linux). Without this acceleration method, the execution time is prolonged up to 18.3 minutes. Hence, the acceleration method shrinks the execution time up to 40%. Interestingly, without the acceleration method, the total precision and recall decrease to 0.96 and 0.91 respectively. This phenomenon is interpreted as follows. Without the acceleration method, the obviously dissimilar image pairs are to be SIFT-matched, which only yields undesirable matched feature pairs corresponding to different object components. Such feature pairs affects the clustering of the local features in Sect. 3.1 harmfully. Thus, the acceleration method not only shrinks the execution time, but also improves the accuracy of DCG.

4.2   Vocabulary Size

As mentioned in Sect. 3.1, DCG can decide the vocabulary size adaptively to the diversity of the image database. To demonstrate this feature, we alter the diversity of the image database by changing the number of frequent object classes from 10 to 30 and examine the vocabulary size. Since each frequent object class has 10 instances, the number of images in the database varies from 100 to 300.

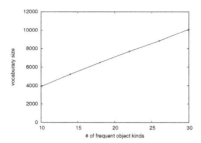

Fig. 4. Vocabulary Size and # of Frequent Object Classes

Fig. 4 shows the experimental result. The vocabulary size increases linearly to the number of frequent object classes. Therefore, DCG adjusts the vocabulary size adaptively to the diversity of the image database.

5   Conclusion

Motivated by the fact that the previous graph-based object discovery algorithm SPGC [5] can only deal with images including one object only, this paper proposes a new unsupervised graph-based object discovery algorithm DCG which treats images with multiple objects. DCG preserves the merits of SPGC as much as possible. Different from the well-known BoVW in which the k-means clustering fixes the vocabulary size, DCG adjusts the vocabulary size adaptively to the diversity of the image database. After representing each image as a graph, DCG acquires the object models as a set of frequent co-occurring edges. In this way, DCG associates a subgraph structure with one object class, which makes it possible to cope with images with multiple objects.

Acknowledgments. This work is supported by the Ministry of Education, Culture, Sports, Science and Technology, Grant-in-Aid for Scientific Research (C) 24500111, 2014.

References

1. Sivic, J., Russell, B., Efros, A., Zisserman, A., Freeman, W.: Discovering Objects and their Localization in Images. In: Proc. ICCV 2005, pp. 370–377 (2005)
2. Pineda, G.F., Koga, H., Watanabe, T.: Scalable Object Discovery: A Hash-Based Approach to Clustering Co-occurring Visual Words. IEICE Transactions on Information and Systems E94-D(10), 2024–2035 (2011)
3. Philbin, J., Sivic, J., Zisserman, A.: Geometric Latent Dirichlet Allocation on a Matching Graph for Large-scale Image Datasets. IJCV 95(2), 138–153 (2011)
4. Sivic, J., Zisserman, A.: Video Google: A Text Retrieval Approach to Object Matching in Videos. In: ICCV 2003, pp. 1470–1477 (2003)

5. Xia, S., Hancock, E.R.: Incrementally Discovering Object Classes Using Similarity Propagation and Graph Clustering. In: Zha, H., Taniguchi, R.I., Maybank, S. (eds.) ACCV 2009, Part III. LNCS, vol. 5996, pp. 373–383. Springer, Heidelberg (2010)
6. Xia, S., Ren, P., Hancock, E.R.: Ranking the Local Invariant Features for the Robust Visual Saliencies. In: Proc. ICPR 2008(2008)
7. Nowozin, S., Tsuda, K., Uno, T., Kudo, T., Bakir, G.: Weighted Substructure Mining for Image Analysis. In: Proc. CVPR 2007, pp. 1–8 (2007)
8. Quack, T., Ferrari, V., Leibe, B., Van Gool, L.: Efficient Mining of Frequent and Distinctive Feature Configurations. In: Proc. ICCV 2007 (2007)
9. Fernando, B., Fromont, E., Tuytelaars, T.: Effective Use of Frequent Itemset Mining for Image Classification. In: Fitzgibbon, A., Lazebnik, S., Perona, P., Sato, Y., Schmid, C. (eds.) ECCV 2012, Part I. LNCS, vol. 7572, pp. 214–227. Springer, Heidelberg (2012)
10. Yuan, J., Wu, Y., Yang, M.: From Frequent Itemsets to Semantically Meaningful Visual Patterns. In: ACM SIGKDD 2007, pp. 864–873 (2007)
11. Gao, J., Hu, Y., Liu, J., Yang, R.: Unsupervised Learning of High-order Structural Semantics from Images. In: Proc. ICCV 2009, pp. 2122–2129 (2009)
12. Li, C., Parikh, D., Chen, T.: Automatic Discovery of Groups of Objects for Scene Understanding. In: Proc. CVPR 2012, pp. 2735–2742 (2012)
13. Cohen, E., Datar, M., Fujiwara, S., Gionis, A., Indyk, P., Motwani, R., Ullman, J.D., Yang, C.: Finding Interesting Associations without Support Pruning. IEEE TKDE 13(1), 64–78 (2001)

A Proposed Extreme Learning Machine Pruning Based on the Linear Combination of the Input Data and the Output Layer Weights

Leonardo D. Tavares[1], Rodney R. Saldanha[1],
Douglas A.G. Vieira[2], and Adriano C. Lisboa[2]

[1] Graduate Program in Electrical Engineering — Federal University of Minas Gerais
Av. Antônio Carlos 6627, 31270-901 — Belo Horizonte, MG, Brazil
`tavares@dcc.ufmg.br, rodney@cpdee.ufmg.br`
[2] ENACOM - Handcrafted Technologies — Rua Professor José Vieira de Mendonça,
770, offices 406 and 407 — BH-TEC — Belo Horizonte, MG, Brazil
`{douglas.vieira,adriano.lisboa}@enacom.com.br`

Abstract. Extreme Learning Machines (ELMs) are gaining fairly popularity in training neural networks, since they are quite simple and have good performance. However, an open problem is the number of neurons in the hidden layer. This paper proposes a method for pruning the hidden layer neurons based on the linear combination of the hidden layer weights and the input data.

Keywords: Extreme Learning Machines, Pruning, Hidden layer, Linear dependency.

1 Introduction

The Extreme Learning Machine (ELM) is a training method for Single Layer Feedforward Neural Network (SLFNs) that gained popularity due to its simplicity and speed. The hidden layer weights are not adjusted, while the output layer weights are adjusted by using the least squares method. It presupposes a hidden space of high dimensionality, which is a feasible linearization of the regressors [6] [7] [5].

In this sense, a question naturally comes: is it possible to remove some hidden nodes without affecting significantly the machine capability ? This work objective is to propose a method to remove (prune) hidden layer nodes based on the linear combination between the hidden layer weights and the input data. For this purpose the method uses a QR decomposition. Additionally, it is expected with this pruning that the machine has faster learning and becomes less complex.

The paper is organized as follows: a brief explanation about the ELM training method is present in Section 2, then, in Section 3, the neural network pruning method used in the experiments is described. The experiments and main result are shown in the Section 4. Finally the conclusion and future works are discussed in Section 5.

E. Corchado et al. (Eds.): IDEAL 2014, LNCS 8669, pp. 354–361, 2014.

2   Extreme Learning Machine

The ELM is a training method for a SLFN, which does not require the adjustment of the hidden layer weights. Consider a set of N distinct samples in the form $(\mathbf{x}_i, \mathbf{t}_i)$ where $\mathbf{x}_i = [x_{i1} \; x_{i2} \; \ldots \; x_{in}]' \in \mathbb{R}^n$ are the inputs of real systems, $i = 1, 2, \ldots N$, apostrophe symbol $(')$ denotes the transpose of the vector or matrix, and $\mathbf{t}_i = [t_{i1} \; t_{i2} \; \ldots \; t_{im}]' \in \mathbb{R}^m$ are the real (or desired) system response. Consider also h as the number of hidden nodes and $f(\cdot)$ an activation function, mathematically, a SFLN can be modeled as [7]:

$$\mathbf{o}_i = \sum_{j=1}^{h} \beta_j f(\mathbf{x}_i' \mathbf{w}_j + b_j), \; i = 1, 2, \ldots, N \tag{1}$$

where $\mathbf{o}_i = [o_{i1} \; o_{i2} \; \ldots \; o_{im}]' \in \mathbb{R}^m$ are the responses found by SLFNs, $\mathbf{w}_i = [w_{i1} w_{i2} \ldots w_{ih}]' \in \mathbb{R}^h$ is the weight vector connecting the input and hidden neurons, $\beta_i = [\beta_{i1} \beta_{i2} \ldots \beta_{im}]' \in \mathbb{R}^m$ connecting hidden and output layer, b_j is the threshold of the jth hidden neuron.

In a matrix compact form we get:

$$\mathbf{H}\beta = \mathbf{O} \tag{2}$$

where:

$$\mathbf{H} = \begin{bmatrix} f(\mathbf{x}_1' \mathbf{w}_1 + b_1) \ldots f(\mathbf{x}_N' \mathbf{w}_h + b_h) \\ \vdots \quad \ldots \quad \vdots \\ f(\mathbf{x}_1' \mathbf{w}_1 + b_1) \ldots f(\mathbf{x}_N' \mathbf{w}_h + b_h) \end{bmatrix} \in \mathbb{R}^{N \times h} \tag{3}$$

$$\beta = [\beta_1 \; \ldots \; \beta_h]' \text{ and } \mathbf{O} = [\mathbf{o}_1 \; \ldots \; \mathbf{o}_N]' \tag{4}$$

As aforementioned, the ELM adjusts randomly the values of w and b. Considering $\mathbf{T} = [\mathbf{t}_1 \; \mathbf{t}_2 \; \ldots \; \mathbf{t}_N]'$, now the objective is to compute the value of β such that:

$$\min_{\beta} \|\mathbf{T} - \mathbf{H}\beta\|_2 \tag{5}$$

Using linear least squares, we can calculate the parameter β as:

$$\beta = (\mathbf{H}'\mathbf{H})^{-1}\mathbf{H}'\mathbf{T} \tag{6}$$

In this context, it is important to observe that the number of hidden nodes may have a strong influence on the β output weight, after the least square solution application and in its norm.

Moreover, the matrix $\mathbf{H}$ may acquire some good or bad properties depending on its hidden weights, input values and activation function. For example, the matrix may become singular if its conditioning number is high, or, if the matrix composed by the result of $\mathbf{X}'\mathbf{W}$ contains linear dependent columns. In this case, that column may be discarded.

3   Proposed Method

The proposed method uses the QR Decomposition, in order to remove useless nodes of the hidden layer, based on its inputs [8]. It is expected to build a faster and more compact machine with a smaller conditioning number. It is important to remark that what is sought by the proposed method is to remove only the unnecessary nodes and retain those that contribute to the input-output mapping, based on the training input data and the weights of hidden nodes. Didactically speaking, suppose a linear system of the type $\mathbf{Ax} = \mathbf{b}$, as follows:

$$\mathbf{A} = \begin{bmatrix} 2\ 3\ 1\ 1 \\ 4\ 5\ 1\ 2 \\ 6\ 8\ 1\ 3 \\ 8\ 9\ 1\ 4 \end{bmatrix} \quad \mathbf{b} = \begin{bmatrix} 15 \\ 25 \\ 38 \\ 45 \end{bmatrix}$$

We can calculate the value of x using least squares as: $\mathbf{x} = (\mathbf{A'A})^{-1}\mathbf{A'b}$. Nevertheless, the $(\mathbf{A'A})$ conditioning number is, approximately, 1.9×10^{16}, which can be considered a very high conditioning number, and clearly the matrix has one linearly dependent column. Furthermore, it's not possible to apply the linear least squares to the matrix $\mathbf{A}$, once it has not inverse (the $(\mathbf{A'A})$ determinant is equal to zero).

Let $\mathbf{QR} = \mathbf{A}$ be the QR decomposition of matrix $\mathbf{A}$, where $\mathbf{Q}$ is an orthogonal matrix and $\mathbf{R}$ is an upper triangular matrix. Now, we will pay attention to the $\mathbf{R}$ matrix, that have the following values:

$$\mathbf{R} = \begin{bmatrix} -13.3 & -10.9 & -1.8 & -5.4 \\ 0.0 & -0.9 & 0.4 & -0.4 \\ 0.0 & 0.0 & 0.5 & 0.0 \\ 0.0 & 0.0 & 0.0 & 0.0 \end{bmatrix}$$

where the absolute values of its diagonal are: $d = [13.3, 0.9, 0.5, 0.0]'$. The result of diagonal indicates the first three columns are linearly independent and the 4-th isn't, once its value is equal to 0, evidencing the need of removing it. In the case of this didactic example, by selecting just the first three columns we get the reduced matrix $\mathbf{C}$, as follows:

$$\mathbf{C} = \begin{bmatrix} 2\ 3\ 1 \\ 4\ 5\ 1 \\ 6\ 8\ 1 \\ 8\ 9\ 1 \end{bmatrix}$$

Now, the conditioning number of $(\mathbf{C'C})$ is 1.9×10^3, which means 13 orders of magnitude smaller. Applying the linear least square using the new $\mathbf{C}$ matrix we get:

$$\mathbf{x} = (\mathbf{C'C})^{-1}\mathbf{C'b} \tag{7a}$$
$$\mathbf{x} = [2\ 3\ 2]' \tag{7b}$$

3.1   Method Development

The QR factorization is a decomposition method of a matrix $\mathbf{A} \in \mathbb{R}^{m \times n}$ (with $m > n$) into orthogonal matrix $\mathbf{Q} \in \mathbb{R}^{m \times n}$ and an upper triangular matrix $\mathbf{R} \in \mathbb{R}^{n \times n}$ [3] [2] [4]. The $\mathbf{R}$ diagonal has some interesting properties, such as, to indicate if some column is or not linearly depend in relation to others. Consider $\mathbf{A} = \mathbf{X}'\mathbf{W}$ as the result of the product of the input data $\mathbf{X}$ and the hidden node weights $\mathbf{W}$, and $\mathbf{QR} = \mathbf{A}$ as its QR decomposition.

Using the properties of the diagonal of $\mathbf{R}$, it is possible to reduce the $\mathbf{A}$ dimension, such that just the linearly independent columns, it also means, just the nodes whose the relationship between inputs and weights contribute to the input-output mapping.

It is possible just selecting that value of diagonal of $\mathbf{R}$ which are greater than some user defined threshold ε. In the case of the present paper, the threshold is automatically defined as the greatest value of diagonal of R scaled by the machine floating-point relative accuracy (e.g. 2^{-52}). The algorithm 1 presents the basic steps of the proposed method.

Algorithm 1. Algorithm to produce a pruned ELM based on QR factorization

---

Require: $\mathbf{X} \in \mathbb{R}^{mn}$ {The elm inputs}
Require: $\mathbf{W} \in \mathbb{R}^{nh}$ {The elm hidden weights}
Require: $\mathbf{T} \in \mathbb{R}^{m}$ {The elm hidden weights}
Require: ε {Threshold}
  $\mathbf{A} = \mathbf{X}' \times \mathbf{W} \in \mathbb{R}^{mh}$
  $[\mathbf{Q}, \mathbf{R}] = qr(\mathbf{A})$ {QR factorization}
  $k = 1$
  while $k \leq h$ **do**
    if $R(k,k) > \varepsilon$ **then**
      $\mathbf{C} = [\mathbf{C}\ \mathbf{A}(:,k)]$ {Concatenate the k-th column of A in B}
    end if
    $k = k + 1$
  end while
  $\mathbf{H} = f(\mathbf{C})$ {$f(\cdot)$ is the activation function}
  $\beta = \mathbf{H}^{+}\mathbf{T}$

---

4   Experiments and Results

This section evaluates the performance of the proposed method. All simulations were carried out in MATLAB®. All ELMs use the simple sigmoidal activation function $f(u) = 1/(1 + \exp(-u))$ in the hidden layer. One synthetic case and five well known benchmark problems were chosen for the experiments. The benchmark data set collected from UCI Machine Learning Repository [1], all of them for regression purpose. In all benchmark experiments the inputs have been normalized into range the $[0, 1]$ (even for nominal features), and the targets in range the $[-1, 1]$.

Three measures were used to compare the proposed pruning methods, to assessment the aspects: accuracy, complexity and efficiency. The first is the empirical error (training and testing), the second is the conditioning number of the hidden space (as a measure of complexity of the machine), the third is the time elapsed for the training phase (as a measure of efficiency). The Root-Mean-Square-Error (RMSE) was chosen as the empirical error measure.

An one-dimensional synthetic data set was used, only for didactic purposes, to demonstrate that the proposed pruning method is capable of generating results better than the original ELM. The didactic function used is calculated by:

$$y = \frac{(x-2)(2x+1)}{1+x^2} + \xi \tag{8}$$

where $x \in [-10, 10]$ and ξ is a normal noise $N(\mu, \sigma)$ with $\mu = 0$ and $\sigma = 0.2$. For this experiment 500 samples were used and an initial ELM with $h = 40$.

The Table 1 shows the comparison between original and pruned results and the Figure 1.

Table 1. Comparison between the results found by the random initialization method and the proposed method

Criteria	Initial	Pruned
Number of hidden nodes	40	18
RMSE	0.0667	0.0674
Conditioning number	$10^{17.29}$	$10^{13.74}$
Time elapsed for training phase (in seconds)	0.0133	0.0018

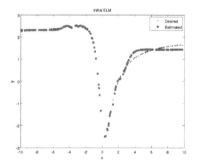 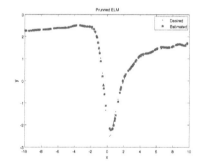

(a) Result of synthetic case using a not pruned ELM in training phase.

(b) Result of synthetic case using a pruned ELM in training phase.

Fig. 1. Comparison between initial and pruned ELM. The black dots are the original function and the red dots are the estimated by ELMs.

It is possible to observe that the empirical error RMSE for training phase was not significantly affected (only 1% higher). On the other hand, the other criteria have been improved: the conditioning number of the hidden layer reduced 4 orders of magnitude and the time elapsed became one tenth of the initial. This means that the proposed method is able to generate a faster and less complex learning machine without, thereby, its effectiveness is affected.

In order to verify those properties, the method was also evaluated using five real-world data set. The characteristic of each data set, number of data used for training and testing are presented in Table 2.

Table 2. Specification of real-world benchmark data sets

Data sets	# Observations		# Attributes	
	Training	Testing	Continuous	Nominal
Abalone	2, 000	2, 177	7	1
Bank	4, 500	3, 692	8	0
California housing	8, 000	12, 460	8	0
Census (house 8L)	10, 000	12, 784	8	0
Delta ailerons	3, 000	4, 19	6	0

For each data set the ELM was arbitrarily defined with a distinct initial number of hidden nodes. The number of hidden nodes and the number of hidden nodes after applying the proposed method are presented in the Table 3.

Table 3. Specification of initial ELM hidden nodes and after applying the proposed method, for each data set

Data sets	# initial hidden nodes	# hidden nodes after proposed method
Abalone	25	9
Bank	190	9
California housing	80	9
Census (house 8L)	160	9
Delta ailerons	45	6

Each experiment was run 50 times. The average and standard deviation are presented. The first comparison is related to the empirical risk. The results of the training and testing phase, are presented in Table 4. The best results in both, i.e. the lowest results for RMSE, are marked with the symbol *.

Table 4. Comparison of training and testing RMSE of ELM and Pruned ELM. Between parentheses are the standard deviation. The values smaller than 0.0001 for were ignored.

Data sets	ELM		Pruned ELM	
	Training	Testing	Training	Testing
Abalone	*0.0851(0.0016)	0.0891(0.0023)	*0.0870(0.0021)	0.0894(0.0029)
Bank	*0.0441(−)	0.0592(0.0331)	*0.0474(0.0012)	0.0595(0.0336)
California housing	*0.1218(0.0025)	0.1436(0.0026)	0.1242(0.0227)	*0.0885(0.0029)
Census (house 8L)	*0.0731(0.0094)	0.0862(0.0024)	0.0621(0.0079)	*0.0443(0.0020)
Delta ailerons	*0.0533(−)	0.0566(−)	*0.0554(−)	0.0571(−)

Based on this result, it can be observed that the RMSE values were not significantly affected and, in some scenarios in the testing phase, where, on average, were more successful. It must be emphasized that the initial objective of the proposed method is not reducing the RMSE. However, it is desired that the RMSE is not degraded.

The second measure used to compare the proposed method is the conditioning number of **H**. Table 5 presents the result of the comparison.

Table 5. Comparison between the Conditioning number of **H** of initial ELM and Pruned ELM. The values are presented in log_{10} base. The best results are marked with the symbol *.

Data sets	Initial ELM	Pruned ELM
Abalone	11.26	* 7.84
Bank	16.80	* 7.84
California housing	17.03	* 9.41
Census (house 8L)	19.17	*16.92
Delta ailerons	17.74	* 9.77

As can be observed, in all cases pruning nodes of the hidden layer generated a smaller conditioning number. In the most extreme case, the base "California housing" the difference has reached 8 orders of magnitude. Thus, the proposed method has demonstrated effective.

The third and last criterion is the time spent in training. The idea is to verify how much is the gain in training time in case of reduction of nodes in the hidden layer. Table 6 shows the comparison of the average time spent in training.

Table 6. Comparison between the time spent for training of initial ELM and Pruned ELM. The values are presented in seconds. The best results are marked with the symbol *.

Data sets	Initial ELM	Pruned ELM
Abalone	0.0039	* 0.0022
Bank	0.0690	* 0.0044
California housing	0.0460	* 0.0092
Census (house 8L)	0.1028	* 0.0104
Delta ailerons	0.0107	* 0.0025

Again it is possible to verify that for criterion of time, in all scenarios the pruning of neurons in the hidden layer led to a training time much lower than when compared to the initial ELM. With this, is possible to check that the method leads to ELM a more efficient ELM.

5   Conclusion and Future Works

The experiments demonstrated that the proposed pruning method is able to remove neurons, where the relationship between the input data and hidden layer weights are linearly dependent with other neurons. In addition, pruning of neurons did not significantly affect the empirical error, which may be considered that it has remained stable. As can be appreciated, in some cases the validation error was improved after the application of the method.

The method was well suited reducing the complexity of the machine and the significant improvement of the training time. The conditioning number of the hidden layer reached 8 orders of magnitude lower. Likewise, there was a reduction in training time, which came to be one tenth of the original time.

As future work we recommend: (i) use of other forms of decomposition, (ii) mathematical proof that the QR decomposition always be able to generate the minimal machine complexity and (iii) studies of multi objective aspects of the proposed method.

Acknowledgment. The authors would like to thank CNPq, FAPEMIG and CAPES for the financial support.

References

1. Bache, K., Lichman, M.: UCI Machine Learning Repository (2013),
   http://archive.ics.uci.edu/ml
2. Boyd, S., Vandenberghe, L.: Convex Optimization. Cambridge University Press, New York (2004)
3. Golub, G., Van Loan, C.: Matrix Computations. Johns Hopkins Studies in the Mathematical Sciences. Johns Hopkins University Press (1996)
4. Hastie, T., Tibshirani, R., Friedman, J.: The Elements of Statistical Learning. Springer Series in Statistics. Springer New York Inc., New York (2001)
5. Huang, G.B., Wang, D., Lan, Y.: Extreme learning machines: a survey. International Journal of Machine Learning and Cybernetics 2(2), 107–122 (2011)
6. Huang, G.B., Zhu, Q.Y., Siew, C.K.: Extreme learning machine: a new learning scheme of feedforward neural networks. In: 2004 IEEE International Joint Conference on Proceedings of the Neural Networks, vol. 2, pp. 985–990 (2004)
7. Huang, G.B., Zhu, Q.Y., Siew, C.K.: Extreme learning machine: Theory and applications. Neurocomputing 70(1-3), 489–501 (2006)
8. Strang, G.: Linear Algebra and Its Applications. Brooks Cole (1988)

A Drowsy Driver Detection System Based on a New Method of Head Posture Estimation

Ines Teyeb, Olfa Jemai, Mourad Zaied, and Chokri Ben Amar

REsearch Groups in Intelligent Machines (REGIM-Lab), University of Sfax,
National Engineering School of Sfax BP 1173, 3038 Sfax, Tunisia
ines.teyeb@gmail.com

Abstract. A drowsy driver detection system based on a new method
for head posture estimation is proposed. In the first part, we introduced
six possible models of head positions that can be detected by our algorithm which is explained in the second part. Indeed, there are three key
stages characterizing our method: First of all, we proceed with driver's
face detection by Viola and Jones algorithm. Then, we extract the image
reference and the non image reference coordinates from the face bounding's box. Finally, based on measuring both the head inclination's angle
and distances between the extracted coordinates, we classify the head
state (normal or inclined). Test results demonstrate that the proposed
system can efficiently measure the aforementioned parameters and detect
the head state as a sign of driver's drowsiness....

Keywords: Drowsiness detection, head posture analysis, face detection.

1 Introduction

Driving while drowsy is a major cause behind road accidents each year, and
exposes the driver to a much higher crash risk compared to driving while alert.
Therefore, the use of assisting systems that monitor a driver's level of vigilance
and alert the driver in case of drowsiness can be significant in the prevention of
accidents.

In This paper, we introduce a new approach towards detection of driver's
drowsiness based on video approach to analyse head movement. Thus, fatigue
is estimated through head posture analysis. Then, when head's inclination exceed a certain time T and a predefined angle X, that means, the driver is non
concentrated and his gaze direction is detached from the wheel.

Recalling the history, head movement related to sleepiness and fatigue had
been studied by many researchers. Ji and Yang[1] note that head position can
reflect a person's level of fatigue. If a driver frequently looks in other directions
for an extended time, he is either fatigued or inattentive. Studies that examine
sleep and sleepiness during long haul flights have used inclinometers to measure
head movement[2]. During wakefulness it showed activity, whereas during sleep
it showed no activity and the authors considered head movement to be more of
a general activity rather than an indicator of sleepiness. Kito and all [3] used

E. Corchado et al. (Eds.): IDEAL 2014, LNCS 8669, pp. 362–369, 2014.

a gyroscope while driving to measuring head movement. But they observed the driver's visual behaviour at an intersection rather than the effect of fatigue on head movements. In 2008, Wuand [4]applied the method of geometric analysis of five facial points for estimating the posture of the head. But its limit is that the location of these points is difficult and vulnerable to changes in facial expressions. Another method, nonlinear nuclear transformation that overcame shortcomings of linear estimation had been introduced by Lu [5]. However, it requires a large number of training examples. Chen [6] has used adaboost algorithm for face and eyes detection and analysed head posture by the characteristics of the triangle of attributes. But he considered only two simple postures.

The remainder of the paper is organized as follows: Section 2 explains our proposed algorithm of head posture estimation. Experimental results showing the robustness of our system is the topic of the third section and section 4 ends with a conclusion and discussion on possible enhancements.

2   Head Posture Analysis

2.1   Head Inclination Possible States

Our proposed method allows the detection of six possible states of head inclination, which are: right, left, backward and forward inclination besides the right and left rotation. Any head posture can be divided into these basic postures as Fig.1 indicates.

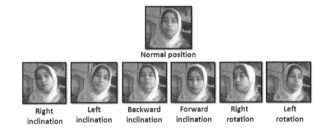

Fig. 1. Different states of head position

2.2   Our Proposed Method of Head Posture Estimation

In video analysis, the captured video stream is segmented into frames $F_1, \ldots,$ $F_i, \ldots, F_n$. We supposed that the first frame presents a normal state of alertness (start of driving), this involves the face detection image, presented in the frame, as our referencial (normal posture without inclination).
Either:
- Img_r : the reference head image detected in the first frame (F_1)
- Img_o : another detected head image in the sequence of frames (head's image in any position, not necessarily coinciding with the reference's image).

The main steps of our proposed method are:

1. The extraction of the reference image coordinates (face's image)(F_1):
As mentioned before, the first step towards head posture estimation is the detection of the driver's face. However, we have assumed that the monitoring camera is installed inside the vehicle under the front mirror facing the driver at a fixed angle. For face detection itself, several approaches have been used in the related literature, we have used in this works, the most famous and commonly used face detection scheme which is the viola-Jones face detection algorithm [7]. It is able to efficiently detect faces.
Once the face regions are detected based on viola-Jones algorithm, coordinates of the face bounding rectangle corners were extracted.
2. The same process is repeated in the sequence of frames ($F_i, \ldots, F_n$).
3. Comparison of the extracted coordinates in both images (Img_r and Img_o).

At this level, our technique can check the position of the head (normal or inclined). But to judge the state of alertness of the driver, we need other parameters such as inclination's angle and duration. In other words, if the head inclination angle exceeds a certain value X and time T, we can say that the level of alertness of the driver is down. The posture estimation algorithm is summarized in Fig.2.

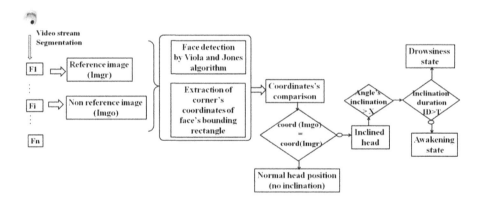

Fig. 2. Algorithm for head posture estimation

Step1: Reference Image Coordinates's Extraction
Once the face regions are detected based on viola-Jones algorithm, coordinates of the face bounding rectangle corners were extracted in order to compute two distances Xir and Yir:

- Xir is the distance between the two corners of the face bounding's rectangle(A_3 and A_2) on the X-axis (Xir=$|x_3 - x_2|$).
- Yir is the distance between the two corners of the face bounding's rectangle(A_3 and A_4) on the Y-axis (Yir=$|y_3 - y_4|$).

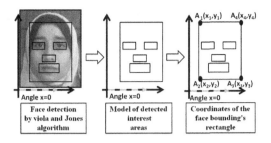

Fig. 3. Reference image coordinates's extraction process

Step2: Non Reference Image Coordinates's Extraction As indicated in Fig.4, when there is an inclination, we noted a gap on the position of inclined face's bounding rectangle (discontinuous rectangle) relative to the reference face's bounding rectangle (continuous rectangle).

Case (a and b)
In a right or left inclination, the size of the flunked face's bounding rectangle (discontinuous rectangle) increases in both height and width dimensions.
D_H ($D_H=|H_o - H_r|$) and D_W ($D_W=|W_o - W_r|$) represent the height and the width of the difference between the two face boundings rectangles respectively. With:

H_r: height of the reference face's bounding rectangle (continuous rectangle).
H_o: height of the flunked face's bounding rectangle (discontinuous rectangle).
W_r: width of the reference face's bounding rectangle (continuous rectangle).
W_o: width of the flunked face's bounding rectangle (discontinuous rectangle).

Case (c and d)
In both cases, just the height of the flunked face's bounding rectangle (discontinuous rectangle) varies.
    Either:

- Xio is the distance between the two corners of the face's bounding rectangle, A'_3 and A'_2, on the X-axis (Xio=$|x'_3 - x'_2|$).
- Yio is the distance between the two corners of the face's bounding rectangle, A'_3 and A'_4, on the Y-axis (Yio=$|y'_3 - y'_4|$).

Case (e and f):
These cases have the same characteristics as the left and right inclinations. However the difference lies in the relationship of belonging of both face's bounding rectangles (reference image and inclined one). Indeed, for the cases (a) and (b), we note an inclusion relationship, however for these cases (e and f), it was an intersection.

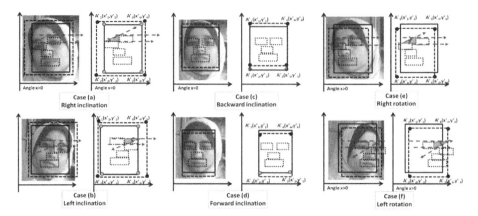

Fig. 4. Extraction of non reference image coordinates in different positions

Step3: Both Images (Img_r and Img_o) Coordinates's Comparison

When the head position changes, the face bounding's rectangle will change. Therefore, we can analyze head posture with rectangle attributes.

If $(Xir = Xio)$ and $(Yir = Yio)$ **Then** head in normal state

If $(Xir \neq Xio)$ or $(Yir \neq Yio)$ **Then** head in inclined state

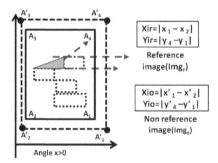

Fig. 5. Estimating head posture based on extracted coordinates

2.3 Inclination's Degree

For the left and right, in both cases inclination or rotation, according to geometric relationships, we can deduce the angle $\hat{P_3}$ between the eyes to compute inclination's degree:

$$cos(\hat{P_3}) = \frac{P_3 P_2}{P_3 P_1} \tag{1}$$

For the others cases (forward and backward inclination), as we have already said only the value of height varies, so we can know the inclination's degree (D) by computing the difference between H_r and H_o (D=$|H_o - H_r|$).

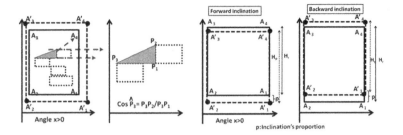

Fig. 6. How to estimate inclination degrees

3   Experimental Results

3.1   Head Posture Estimation

The experiment is based on face database which we have built. Table 1 summarizes part of experimental results. We note these characteristics of the reference image: $Xir = 288, Yir = 288, H_r = 288, W_r = 288$.

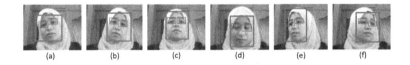

Fig. 7. Estimating head position

Table 1. *Results of head posture estimation*

image	Xio	Yio	D_H	D_W	result
(a)	329	329	41	41	right inclination
(b)	344	344	56	56	left inclination
(c)	283	283	5	0	backward inclination
(d)	291	283	3	0	forward inclination
(e)	393	393	105	105	right rotation
(f)	301	301	13	13	left rotation

Our algorithm can't estimate head posture correctly when the head rotation angle is too big, because Viola and Jones algorithm is not suitable for a large rotation angle[7]. To overcome this limit, we can improve the learning phase of Viola and Jones algorithm by extending the feature set of pseudo-Haar characteristics[8]. Besides, Viola and Jones proposed an improvement that can correct this defect, which involves learning a dedicated cascade for each orientation or view[9]. Another solution is applying the rotational invariant Viola-Jones detector [10] which allows the conversion of a trained classifier for any angle.

3.2 Drowsiness Detection System

Our system has been tested on many subjects (about 60 driver, man and women) and in various situations (with and without hats). To evaluate its performance, we compared it to another drowsiness detection system previously developed[11][12]. It was based on eyes closure duration. Figure 8 presents a case of experimental result of both systems:

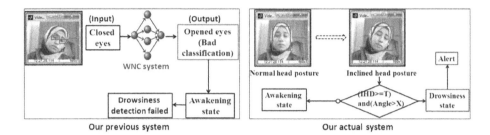

Fig. 8. Drowsiness detection systems

We have used the Correct Drowsiness Detection Rate (CDDR)to evaluate our actual system and the previous one for the different tests:

$$CDDR = \frac{Number\ of\ correct\ drowsiness\ detection\ cases}{Real\ number\ of\ drowsiness\ candidates}. \tag{2}$$

The proposed algorithm based only on eyes blinking can't estimate drowsiness correctly. Here the classification system may be affected by the lighting condition. As a result, we have a false drowsiness detection. We obtained 80% as CDDR. This problem is rectified by the head movement analysis presented in our actual system. However, our novel drowsiness detection system, in which we proceed with comparing the extracted coordinates of both key images (reference and non reference one) and also with head inclination degree, successfully trigger an alert to indicate a state of drowsiness.We obtained an accurate result of correct drowsiness detection equal to 88.33% as CDDR.

4 Conclusion

We have proposed a new system of drowsiness detection based on head position estimation. By this work we have contributed to establishing a novel approach of head movement estimation based on the extraction of reference and non reference images coordinates from the face bounding's box. This involves computing angle of head inclination and distances between the extracted coordinates in order to estimate the head state. Based on the experiment results, the proposed approach exhibits high performances, which involves the robustness of our new method.

Acknowledgments. The authors would like to acknowledge the financial support of this work by grants from General Direction of Scientific Research (DGRST), Tunisia, under the ARUB program.

References

1. Ji, Q., Yang, X.: Real-time eye, gaze, and face pose tracking for monitoring driver vigilance. Real-Time Imaging 8(5), 357–377 (2002)
2. Wright, N., McGown, A.: Vigilance on the civil flight deck: incidence of sleepiness and sleep during longhaul flights and associated changes in physiological parameters. Ergonomics 44, 82–106 (2001)
3. Kito, T., Haraguchi, M., Funatsu, T., Sato, M., Kondo, M.: Measurements of gaze movements while driving. Percept. Mot. Skills 68, 19–25 (1989)
4. Wu, J., Trivedi, M.M.: A two-state head pose estimation framework and evaluation. Pattern Recognition 41(3), 1138–1158 (2008)
5. Lu, Y., Wang, Z., Li, X.: Head Pose Estimation Based on Kernel Principal Component Analysis. Opto-Electronic Engineering 35(8), 63–66 (2008)
6. Chen, Z., Chang, F., Liu, C.: Pose parameters estimate based on adaboost algorithm and facial feature triangle. Geomatics and Information Science of Wuhan University 36(10), 1164–1168 (2011)
7. Viola, P., Jones, M.: Robust Real-time Object Detection. Int. J. Computer Vision 57(2), 137–154 (2001)
8. Lienhart, R., Maydt, J.: An Extended Set of Haar-like Features for Rapid Object Detection. In: IEEE ICIP (2002)
9. Viola, P., Jones, M.: Fast Multi-view Face Detection. In: IEEE CVPR (2003)
10. Barczak, A.L.C.: Toward an efficient implementation of a rotation invariant detector using haar-like features. In: IVCNZ, Dunedin, New Zealand, pp. 31–36 (2005)
11. Ines, T., Olfa, J., Tahani, B., Chokri, B.A.: Detecting Driver Drowsiness Using Eyes Recognition System Based on Wavelet Network. In: Proceedings of the 5th International Conference on Web and Information Technologies (ICWIT 2013), May 09-12, Hammamet, Tunisia, pp. 2456–2254 (2013)
12. Olfa, J., Ines, T., Tahani, B., Chokri, B.A.: A Novel Approach for Drowsy Driver Detection Using Eyes Recognition System Based on Wavelet Network. IJES: International Journal of Recent Contributions from Engineering, Science & IT 1(1), 46–52 (2013)

A CBR-Based Game Recommender
for Rehabilitation Videogames in Social Networks

Laura Catalá, Vicente Julián, and José-Antonio Gil-Gómez

Dept. Sistemas Informáticos y Computación,
Universidad Politécnica de Valencia Valencia, Spain
laucaad@epsg.upv.es, {vinglada,jgil}@dsic.upv.es

Abstract. Health care can be greatly improved through social activities. Present day technology can help through social networks and free internet games. A system can be built, combining present day technology with recommender systems to ensure supervision for the elderly and disabled. Using the behavior studied on social networking sites a system was created to match games to particular users. Common associations between a user's personality and a game's genre were considered in the process and used to create a formula for how appropriate a game suggestion is. We found that the games receiving the best results from the users were those games that trained those certain users' disabilities not others.

Keywords: Case Base Reasoning, Rehabilitation, Social Networks, Videogames, Recommendation Systems.

1   Introduction

Advances in technology have led to an increase in access to and usage of the Internet. Videogames and social networks once limited to PC are now available through mobile devices. Usage has gradually increased in elderly and disabled populations (as identified by the World Health Organization 1), groups generally lacking technological literacy. Scientists and engineers are aware these groups have trouble and have made great improvements in making the Internet easily accessible. They have accomplished this by adapting interfaces to fit these groups of people's needs, through methods such as modifying the graphics interface and creating ergonomic keyboards 2.

Studies have shown that people who socialize every day live longer and happier lives 3,4,5,6,7. This proves it necessary for elderly and disabled people to avoid solitude. Likewise, statistics show that everyone, including the elderly and disabled, are increasing their social network usage 8.

Medical supervision and services are sometimes insufficient 9. If a patient is in need of continuous therapy, for conditions like hemiplegia or aphasia, and left untreated, the outcome could be devastating. For these reasons and more, it is necessary to create more accessible, user-friendly health services.

E. Corchado et al. (Eds.): IDEAL 2014, LNCS 8669, pp. 370–377, 2014.
© Springer International Publishing Switzerland 2014

This recommender system (based in social networks) recommends videogames and activities to assist disabled individuals in utilizing personalized mental and physical exercises. A novel use of social networks allows ready access to personal and social information to help create more personalized, user-friendly recommendation software for rehabilitation. It is possible to maintain a simple yet intelligent graphic interface environment so that it can be adapted to the visually impaired. Furthermore, the user's feedback can be retrieved from the videogame's results once the user has finished playing and not necessarily asked to the user. In that way, user feedback could be voluntary to help to collect greater detail.

The article is structured as follows: Section 2 describes the background to our system, section 3 describes the system we have developed, section 4 validates the functions required of our system and section 5 reveals our conclusions.

2   Background

Recommendation systems [10][11], systems that provide effective recommendations about what action users can take or what information they can consume, can be effective tools for performing decision-support tasks. Traditional recommender systems base their recommendations on quantitative measures of similarity between the user's preferences and the current items to recommend (i.e. content-based recommenders [12]), between the user's profile and the profile of other users with similar preferences (i.e. collaborative filtering recommenders [13]) and on combinations of both (i.e. hybrid recommenders [14]).

In addition, online recommender systems suffer from problems inherent to complex social networks: the number of users and/or items to recommend can be very high. This can be seen in collaborative filtering, the process of comparing two users with the aim of extracting a similarity which requires that they have qualified the same objects. This can be unrealistic in large social networks. Another major weakness of online recommender systems is their trustworthiness. In an open network with a large number of users it is impossible to ensure that all views expressed are true opinions of users and there is no tampering with the resulting recommendations. In order to overcome these problems, it is necessary to embed a social layer in current recommender approaches, taking into account aspects such as reputation and trust among users. Therefore, there are a number of open challenges for the development of a new generation of recommender systems [15], such as exposing underlying assumptions behind recommendations, approaching trust and trustworthiness from the perspective of backing recommendations and providing rationally compelling arguments for recommendations.

The recommender system incorporates problem solving to the already discovered issues. It relies on social network for information but prevents tampering or faulty results. It would recommend videogames and activities that will help all people in need improve their disabilities, stay involved with social activities and utilize mental and physical exercises.

3   Elderly and Disabled Impairment Training – EDIT

Elderly & Disabled Impairment Training (EDIT) is a case-based recommender system that matches patients with condition-specific rehabilitation games.

In designing EDIT, a list of disabilities recognized by the World Health Organization (WHO) is used 1. In this list, the WHO grouped all notable disabilities and tagged them with a special code. This information was used to collect and combine with a list of games that help train each disability. It was then made into an ontology (see Figure 1), using both lists, with leaves representing games and branches representing codes describing disabilities.

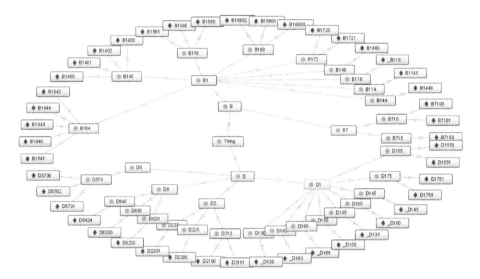

Fig. 1. Ontology tree for WHO disabilities classification

3.1   Case Structure

Each case has a game description and a solution, as in any normal case, but the novel query procedure takes a user description (social network information and questionnaire 16 results) and uses it within the game description (case description). Then the game description is used as a query to get a case solution.

Game descriptions are based on:

Table 1. Structure of Case Description

Disability and sub-disability	Based on WHO categorization
Level of difficulty	How hard the game is
User's personality	Matches genre to the user's taste
Any necessary extra details	Determined by whether or not the game trains any special areas of the body; defaults to none

An algorithm helps find the user's personality using social network information 17. User personality is used because their personality has to do with their tastes in Genres. Papers have documented a correlation between people with a specific personality and their taste in videogames18. For this reason, priority is given to the case solution correlated with the games a user might find appealing. Conversely, unappealing game suggestions are reduced.

The cases' solution includes information contained in a list of free internet games that are suitable for disability training:

Table 2. Structure of Case Solution

Max score	Maximum score reachable in the game
Number of levels	Highest level reachable in the game
Game's name	The name of the game
Game's URL	A web link to the game
Time limit	Maximum time that one can be playing
Number of times played	The sum of times users have played the game through the recommender
List of scores	Array of scores achieved by the users
List of times	Array of the length of time the users have played the game per session
Number of likes	Positive feedback and reviews
Number of dislikes	Negative feedback and reviews
Average level reached	Mean level played until by users
Ideal personality	Personality traits associated with the game

Static information (name, max level, etc.) and variable information (number of times played, number of likes, etc.) are both included. Also, games designed for a specific personality could be assumed to work with other personalities. The system will learn a new case when a person plays a game with a different personality so that this game can later relate with that kind of personality depending on the users' experiences.

3.2   CBR Phases

The user has to be logged into the social network in order to use EDIT. The network currently is operable solely on Facebook but plans are to expand. The system asks for permissions in order to gather profile and social information about the user. Afterwards, if needed, the user is prompted with a short questionnaire to identify all possible disabilities. Once all necessary fields are filled in the user description, the recommendation cycle will begin. Regular CBR recommendation process will be followed as proposed by Aamodt and Plaza 19.

3.2.1.   Retrieval and Reuse Phases

Using the user's description, a videogame description is derived. For example the user's disability helps determine the disability in the game description. Certain qualities like "level of difficulty" are not directly listed. An algorithm consisting of the age of the user, the level of disability, and their educational level, is used to imply the level of difficulty for the game description.

A user with more than one disability will have a set of games for each disability gathered. After the list of games to recommend is completed, the top k-Recommendations are reorganized with a custom function. Only the game with the highest score is recommended. Items considered for calculation:

- The user's votes on the game:
  ```
  votes=(NumLikes-NumDislikes)/(NumLikes+NumDislikes);
  ```
- The total amount of times played compared to other retrieved games:
  ```
  numPlayed = Num/totalTimesPlayedRetrievedCases;
  ```
- The average time users have played the game over the maximum time users can play the game. The same is done for average and maximum scores, and average and maximum levels:
  ```
  times = getAverageTimePlayed/getTimeLimit;
  scores = getAverageScores/getMaxScore;
  levelsPlayed = getLevel/getMaxLevel;
  ```
- The similarity level (sim). sim is set to 1 when the game's disability and the user's disability are on the same leaf of the ontology tree, set to 0.5 if they are not, but share the parent branch. If there is no correlation, sim is set to 0. This measures how suitable a retrieved game is for the user's disability.

This information is calculated as a value between zero and one. To do so, different weights for each of the above items are used as such:

```
weight = w1*votes + w2*numPlayed + w3*times + w4*scores + w5*levelsPlayed
                                + w6*sim;
```

Thus, various configurations can be used to study their effects on the results.

3.2.2.   Revision and Retention Phases

After the user finishes playing the game, the session information is added to the case. Each time a game is played, its case will be modified following the rules in Table 3:

Every time a new recommendation cycle is started, the above functions will depend on the feedback the users give to the game in order to reorganize the top k-Recommendations. The more the system is used, the better recommendations it will make. If the disability of the user matches the disability described by the game, and there isn't a case in which the user and the game's personality coincide, a new case is created where the game's ideal personality is set to the user's personality (as also shown in Table 3).

Table 3. Rules for Case Modifications

Number of times played	⇒ +1
List of scores	⇒ Add new score
List of times	⇒ Add new time
Number of likes	⇒ If the user liked the game: +1
	⇒ If the user didn't like it: +0
Number of dislikes	⇒ If the user liked the game: +0
	⇒ If the user didn't like it: +1
Number of levels reached	⇒ If the level reached is past the highest level previously reached: Change this to the last level reached
Ideal personality	⇒ If the ideal personality for the game is the same as the user's personality: No change necessary
	⇒ If the ideal personality for the game is different than the user's personality: a new case is created with the current session information

4   Validation

EDIT is implemented with *jColibry* 20. All users' information is retrieved from an experimental program for Augmented Reality rehabilitation. Also, users' information is only retrieved after agreement with the privacy policy.

Once the databases are created, they are used in the following steps:

1) EDIT retrieves a list of the most suitable games for the user by checking their disabilities and finding the best games that would improve the faster way the user's abilities.

2) Simultaneously, EDIT learns by performing recommendations to a set of users and retrieving their experiences.

3) For evaluating EDIT's functioning, more recommendations are sent. The system is a success when these recommendations coincide with *1)*.

Figure 2 shows how the system works as the cases database is expanded. 10% of the total amount of cases is added each time until the 100% is reached. In this prototype, 100% is equal to 124 cases. Each variable within the EDIT function is weighted the same except for the following. Each line of Figure 2 represents the behavioral change in the system:

- sim is emphasized when increasing w6.
- votes is emphasized when increasing w1.
- numPlayed is emphasized when increasing w2.

As seen in Figure 2, the best results are produced when the database is at 80%-90% completion. When `sim` is the weighted variable, the most accurate precision is produced (close to 20% higher than the least precise weighted variable).

Similarly to Figure 2, the system has been tested to evaluate the learning mechanism. The system was executed with iterations in which the system returned a set of recommendations and calculated the average precision. As illustrated in Figure 3, EDIT learns faster after 5-10 iterations are performed. Once again, the variable most closely related to a higher precision is `sim`.

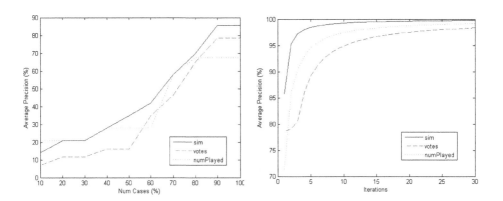

Fig. 2. Database size influence on precision   **Fig. 3.** Iterations influence on precision

5   Conclusions

EDIT is a social recommendation system for activities/games to help elderly or disabled people. The system has a virtual environment via the web. Instructions are simple and easy to understand and follow by any user with any disabilities, including those with cognitive impairment, because feedback is automatically retrieved from the videogame's results. The scope of the paper does not include the virtual environment. The recommendation process is becoming increasingly faster and more accurate and new functionalities are being added to the system.

Every recommendation was completed using *jColibry* recommender. *Lenskit* 21 for collaborative recommendation has also been tested and the results were satisfactory. Due the small user's database in this prototype, results with *Lenskit* were excluded.

Acknowledgments   This work was partially supported by MINECO/FEDER TIN2012-36586-C03-01 of the Spanish government. Special thanks to Edward Benjamin Lichtman for proofreading.

References

1. Classification, Assessment, Surveys and Terminology Team, International Classification of Functioning, Disability and Health. World Health Organization (2001)
2. Emiliani, P.L.: Assistive Technology (AT) versus Mainstream Technology (MST): The research perspective. Institute of applied Physics, National Research Council, Firenze, Italy (2006)
3. Shye, D., Mullooly, J.P., Freeborn, D.K., Pope, C.R.: Gender Differences in the relationship between social networks support and mortality: A longitudinal study of an elderly cohort. Elsevier Science Ltd. (1995)
4. Holmén, K., Furukawa, H.: Loneliness, health and social network among elderly people – a follow-up study. Archives of Gerontology and Geriatrics 35 (2002)
5. de Belvis, A.G., Avolio, M., Spagnolo, A., Damiani, G., Sicuro, L., Cicchetti, A., Ricciardi, W., Rosano, A.: Factors associated with health-related quality of life: the role of social relationships among the elderly in an Italian region. Public Health 122 (2007)
6. Thomas, P.A.: Gender, social engagement, and limitations in late life. University of Texas at Austin, Texas (2011)
7. Fiorillo, D., Sabatini, F.: Quality and quantity: The role of social interactions in self-reported individual health. Social Science & Medicine 73 (2011)
8. Fritsch, T., Steinke, F., Brem, D.: Analysis of Elderly Persons' Social Network: Need for an Appropriate Online Platform. Association for the Advancement of Artificial Intelligence (2012)
9. Kickbusch, I., Brindley, C.: Health in the post-2015 development agenda: an analysis of the UN-led thematic consultations, High-Level Panel report and sustainable development debate in the context of health. World Health Organization (2013)
10. Adomavicius, G., Tuzhilin, A.: Toward the Next Generation of Recommender Systems: A Survey of the State-of-the-Art and Possible Extensions. IEEE Transactions on Knowledge and Data Engineering 17(6), 734–749 (2005)
11. Zhou, X., Xu, Y., Li, Y., Josang, A., Cox, C.: The state-of-the-art in personalized recommender systems for social networking. Artificial Intelligence Review 37(2), 119–132 (2012)
12. Pazzani, M.J., Billsus, D.: Content-Based Recommendation Systems. In: Brusilovsky, P., Kobsa, A., Nejdl, W. (eds.) Adaptive Web 2007. LNCS, vol. 4321, pp. 325–341. Springer, Heidelberg (2007)
13. Schafer, J.B., Frankowski, D., Herlocker, J., Sen, S.: Collaborative Filtering Recommender Systems. In: Brusilovsky, P., Kobsa, A., Nejdl, W. (eds.) Adaptive Web 2007. LNCS, vol. 4321, pp. 291–324. Springer, Heidelberg (2007)
14. Burke, R.: Hybrid Recommender Systems: Survey and Experiments. User Modeling and User-Adapted Interaction 12(4), 331–370 (2002)
15. Chesñevar, C., Maguitman, A., González, M.: Empowering Recommendation Technologies Through Argumentation, pp. 403–422. Springer (2009)
16. Jitapunkul, S., Pillay, I., Ebrahim, S.: The abbreviated mental test: its use and validity. Age Ageing (1991)
17. Adali, S., Golbeck, J.: Predicting Personality with Social Behavior
18. Zammitto, V.L.: Gamers' Personality and their Gaming Preferences. University of Belgrano (2001)
19. Aamodt, A., Plaza, E.: Case-based reasoning: Foundational issues, methodological variations and system approaches. AI Communications 7(1), 39–59 (1994)
20. JColibri, http://gaia.fdi.ucm.es/research/colibri/jcolibri
21. LensKit, http://lenskit.grouplens.org/

A New Semantic Approach for CBIR
Based on Beta Wavelet Network Modeling Shape Refined
by Texture and Color Features

Asma ElAdel[1,3], Ridha Ejbali[2,3], Mourad Zaied[1,3], Chokri Ben Amar[4,3]

[1] National Engineering School, Gabes, Tunisia
[2] Faculty of sciences, Gabes, Tunisia
[3] REsearch Groups in Intelligent Machines, University of Sfax, ENIS,
B.P 1173, Sfax, 3038, Tunisia
[4] National Engineering School, Sfax, Tunisia

Abstract. Nowadays, large collections of digital images are being created. Many of these collections are the product of digitizing existing collections of analogue photographs, diagrams, drawings, paintings, and prints. Content-Based Image retrieval is a solution for information management. Image retrieval combining low level perception (color, texture and shape) and high level one is an emerging wide area of research scope. In this paper, we presented a new semantic approach based on extraction of shape refined with texture and color features extraction, using 2D Beta Wavelet Network (2D BWN) modeling. The shape descriptor is based on Best Detail Coefficients (BDC), the texture descriptor is based on Best Approximation Coefficients (BAC) and the one for color is calculated on the approximated image by applying the first two moments.

Experimental results for Wang database showed the effectiveness of the proposed method.

Keywords: Beta wavelet network modeling, color, texture, shape, features extraction.

1   Introduction

Historically, for humans, the image is an indispensable means of communication and backup that express the past, present and future times. Image search was, indeed, an essential ongoing need. Nowadays, with the evolution of the digital world, the creation and image management have become progressively easy and accessible. Therefore, the search for images is becoming more difficult and complicated. As a solution, many researchers have focused on the domain of Content-Based Image Retrieval (CBIR). CBIR aims to retrieve similar images to a query one, from large databases using visual contents. The problem is to how bridge the lack of coincidence between the information extracted from those data, and the interpretation that the same data have for a user in a given situation. This is called semantic CBIR. Many techniques

E. Corchado et al. (Eds.): IDEAL 2014, LNCS 8669, pp. 378–385, 2014.

have evolved for this problem, such as Machine Learning Techniques [1], Relevance Feedback (RF) [2]. Other approaches [3, 4, 5] are based on low features (shape/texture/color) combination. Also, we can cite the work of Prabhu [6] which present a new approach based on Haar wavelet in order to extract color histogram and texture features. Neural Network (NN) and Wavelet Neural Network (WNN), particularly Beta Wavelet Network (BWN) are computational models known by their capability of learning and pattern recognition, and used to solve a wide variety of complex tasks. Since Beta wavelet [7] was an effective tool in various fields such as image compression [8], face recognition [9], 3D face recognition [10], image classification [11, 12], phoneme recognition [13], speech recognition [14] and in particular Arabic word recognition [15] and hand tracking and recognition [16]; this study used the BWN modeling to propose a new semantic approach for CBIR.

Human eye is sensitive to the shape of the object since it represents the first characteristic defining the object [17]. So, a new 2D BWN modeling-based shape descriptor was presented in this study. Then, the output was refined by extracting texture and color features, based on 2D BWN modeling respectively.

The paper is organized as follows: Section 2 describes the main idea of proposed approach. Section 3 describes the methodology used to extract the three descriptors shape, texture and color. Section 4 is devoted to present the obtained results.

2   Proposed Approach

Any CBIR system is based on two stages: indexation stage and retrieval stage. The principle of our approach can be explained by the two following algorithms:

2.1   Indexing Algorithm (off-line stage)

1. Each reference image (RI) will be normalized into a same size 256*256 and converted to hsv color space.
2. Then, it will be analyzed with 2D BWN modeling.
3. Features extraction (shape, texture and color) based on 2D BWN modeling.
4. Features indexing in three vectors for shape texture and color respectively.

2.2   Retrieval Algorithm (on-line stage)

1. Each query image (QI) will be normalized into a same size 256*256 and converted to hsv color space.
2. Then, it will be analyzed with 2D BWN modeling.
3. BDC-based shape extraction.
4. Measuring the similarity between this QI and all of those RI.
5. The output of step4 will be refined by comparing their textures with those of QI: first, we extract BAC-based texture extraction of QI. Then, we select the texture descriptors only of the top 50 retrieved RI (based on the shape content) corresponding to the step4 output. Finally, we measure the similarity between them.

6. Refinement of step5 output, by extracting color features: first, we extract color descriptor of the approximated images (approximated QI and all of approximated RI corresponding to the output of step5). Then, we measure the similarity between them.

7. The system will retrieve the most relevant and similar images to the query one, based on visual contents (shape refined by texture and color).

The overview of the proposed approach is illustrated in Fig.1.

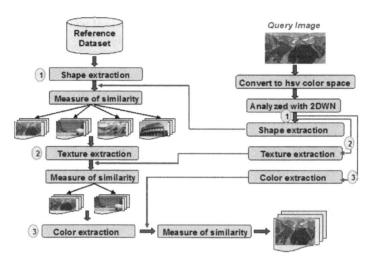

Fig. 1. Overview of proposed approach

3   Features Extraction

In this work, we propose a new method of descriptors extraction based on 2D BWN modeling (see Fig.2). This choice was selected for many reasons:

- A Wavelet Network is a Neural Network who's the hidden layer is composed with wavelet and scaling functions (more information about our Beta wavelet can be found in [7, 8].

- The specificity of our 2D BWN architecture is that the hidden layer is composed only by wavelet and scaling ones contributing more in the reconstruction of the signal (image); which reduces enormously and compresses the network size [10,11].

- With the learning aspect, wavelet network can model and edit unseen visual content (shape, texture and color).

- It's very important to determine features invariant to translation, scale and rotation and with low complexity and wavelet network can be an effective solution for such problem.

Given that an image can be characterized with three vision content (shape, texture and color), it was ensured, by scientists and psychologists, that the shape is the most important characteristic for human vision [17]. For this reason, in this paper we were beginning by comparing images based on their shape features.

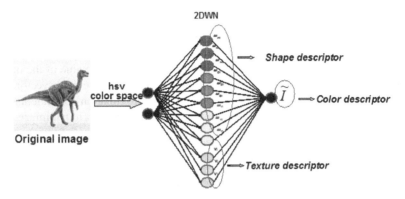

Fig. 2. 2D BWN modeling-based features extraction

3.1   Shape Feature Extraction

Shape detection, characterizing the content of an image, is obtained after the validation of the following steps:

- **Step1 (Image filtering):** This step consists on image denoising, which count the number of different colors containing in the matrix of the indexed image L. If this number is greater than a predetermined threshold (presence of noise), the colors of the entire image will be eliminated.

- **Step2 (Shape detection with 2D BWN modeling):** The image will be projected on the *2D B*WN. The coefficients of detail wavelets having best contributions in the reconstruction of image, called Best Detail Coefficients (BDC), will be summed in order to obtain the shape.
- **Step3 (Calculation of Hue moments):** Once we detected the shape with 2D BWN modeling, we will calculate the seven geometric moment's of Hue blocks by dividing the found binary image into four sub-equal parts. The moments are calculated by applying the following equation (1):

$$M_{p,q} = \sum_x \sum_y x^p y^q I(x,y) \tag{1}$$

Where: (p, q) correspond to the moment order and I(x, y) represents the pixel value at position (x, y).

3.2   Texture Feature Extraction

The proposed texture descriptor algorithm can be detailed in the following steps:

- Step1: The image will be projected on 2D-BWN.

- Step2: The approximation coefficients of scaling functions having optimal contributions in the reconstruction of image, called Best Approximation Coefficient (BAC) will be extracted.
- Step3: Energy computing of BAC for each level of performed wavelet decomposition (2).

$$E = \frac{1}{\frac{m}{k}\frac{n}{k}} \sum_{i=1}^{\frac{m}{k}} \sum_{j=1}^{\frac{n}{k}} |x(i,j)| \tag{2}$$

m and n are the dimensions of the image and X (i, j) is the pixel value of the image X, of i and j. K is the number of decomposition level. In total, we get a vector texture with five components corresponding to energies of approximation coefficients at each decomposition level and a sixth component is the energy of the original image.

3.3   Color Feature Extraction

An efficient color descriptor must characterize two important things: indication of the color content of an image and information about the spatial distribution of the colors. Color moments can be an effective descriptor.

In this work, all images will be converted into hsv color space for better correspondence with human perceptions of color similarities than other color spaces [18, 19].

So, each color channel (h, s and v) will be approximated with a 2D BWN. Then the first and second moments, of the color distribution given by Stricker and Orengo [20], will be computed.

The first moment is Mean (3), which can be considered as the average color value in the image:

$$E_i = \sum_{j=1}^{N} \frac{1}{N} p_{ij} \tag{3}$$

The second moment is Standard deviation, i.e. the square root of the variance of the distribution as shown in (4).

$$\sigma_i = \sqrt{(\frac{1}{N} \sum_{j=1}^{N} (p_{ij} - E_i)^2)} \tag{4}$$

Therefore, each image is characterized by six moments (two moments for each color channel).

3.4   Measure of Similarity

The similarity distances (shape, texture and color moments) of the image i are calculated by applying the Euclidean distance. If we note, for example, Minds and Maxds the minimum and maximum values of the similarity distances shape of n images of the dataset, the normalized value dsni is calculated by (5):

$$dsni = \frac{d_{si} - Minds}{Maxds - Minds} \tag{5}$$

4   Results

In our experiments, we have used Wang dataset in order to test the performance of our approach for CBIR. This base contains 1000 images. The images are of the size 256 x 384 and classified into ten clusters (100 images per cluster) as shown in Fig.3. From each cluster, 50 images are randomly selected as query images.

Fig. 3. Example of Wang clusters

The performance of retrieval of the system can be measured in terms of its precision.

Precision measures the ability of the system to retrieve only the models that are relevant. It is defined as:

$$Precision = \frac{Number\ of\ relevent\ images\ retrieved}{Total\ number\ of\ images\ retrieved} \tag{6}$$

Table 1 compares our approach with the approach combining color histogram and texture features based haar wavelet [6]. The results clearly show the robustness of the Wavelet Network, because of its capability of learning, instead of the use of simple wavelet. Furthermore, the combination of Hue moments and statistic approaches like energy computing with wavelet network increases the efficiency of the shape and texture descriptors respectively.

Table 1. Comparison between average precision of our approach with other approach

Categories	Precision	
	WBCH[6]	**Our approach**
Dinosaurs	97%	100%
Food	77%	82%
African people	65%	80%
Building	71%	78%
Buses	92%	100%
Elephants	86%	92%
Horses	87%	100%
Flowers	76%	100%
Beaches	62%	94%
Mountains	49%	80%
Average Precision	**76.2%**	**90.6%**

Concerning the execution time, the proposed approach was able to definitely decrease the execution time of our application, as shown in Table 2.

As we can see, the average execution time of a QI, of the proposed approach, is less about three times.

Table 2. Average execution time of proposed approach for a query image

	Use of three descriptors of all reference images	Proposed approach
Average execution time	15.73s	5.02s

5   Conclusion

In this paper, we presented a new local and semantic approach for CBIR. First, we extracted the 2D BWN modeling-based shape feature from a query image. Then, the results of the latter were refined by extracting 2D BWN-based texture and color features respectively, in order to strengthen our system efficiency. Results reported seem very promising and our approach provides better performance than using simple wavelet decomposition.

Acknowledgement. The authors would like to acknowledge the financial support of this work by grants from General Direction of Scientific Research (DGRST), Tunisia, under the ARUB program.

References

1. Jain, S.: A Machine Learning Approach: SVM for Image Classification in CBIR. Internationa Journal of Aplication or Annovation in Engineering & Management (IJAIEM) 2(4) (April 2013)
2. Shanmugapriya, N., Nallusamy, R.: A New Content Based Image Retrieval System Using GMM and Relevance Feedback. Journal of Computer Science 10(2), 330–340 (2014)
3. Singha, M., Hemachandran, K.: Content Based Image Retrieval using Color and Texture. Signal & Image Processing: An International Journal (SIPIJ) 3(1) (February 2012)
4. Mahantesh, K., Anusha, M., Manasa, K.R.: A Novel Aproach for Image Retrieval System Combining Color, Shape & Texture Features. International Journal Technology and Advanced Engineering (IJETAE) 3(3) (2013)
5. Jain, N., Sharma, S., Sairam, R.M.: Content Base Image Retrieval using Combination of Color, Shape and Texture Features. International Journal of Advanced Computer Research 3(1(8)) (2013)
6. Prabhu, J., Kumar, J.S.: Wavelet Based Content Based Image Retrieval Using Color and Texture Feature Extraction bY Gray Level Coocurence Matrix and Color Coocurence Matrix. Journal of Computer Science 10(1), 15–22 (2014)
7. Zaied, M., Ben Amar, C., Alimi, M.A.: Award a new wavelet based beta function. In: International Conference on Signal, System and Design, SSD 2003, Tunisia, pp. 185–191 (January 2003)
8. Ben Amar, C., Zaied, M., Alimi, M.A.: Beta wavelets synthesis and application to lossy image compression. J. Adv. Eng. Software 36, 459–474 (2005)
9. Zaied, M., Ben Amar, C., Alimi, M.A.: Beta wavelet networks for face recognition. Journal of Decision Systems 14, 109–122 (2005)

10. Zaied, M., Salwa, S., Jemai, O., Amar, C.: A novel approach for face recognition based on fast learning algorithm and wavelet network theory. International Journal of Wavelets, Multiresolution and Information Processing 19, 923–945 (2011)
11. Jemai, O., Zaied, M., Ben Amar, C., Alimi, M.A.: FBWN: an architecture of Fast Beta Wavelet Networks for Image Classification. In: IEEE World Congress on Computational Intelligence, Barcelona, Spain, pp. 18–23 (July 2010)
12. Jemai, O., Zaied, M., Ben Amar, C., Alimi, M.A.: Pyramidal hybrid approach: Wavelet network with OLS algorithm-based image classiffication. International Journal of Wavelets, Multiresolution and Information Processing 9, 111–130 (2011)
13. Ejbali, R., Benayed, Y., Zaied, M., Alimi, M.A.: Wavelet networks for phonemes recognition. In: International Conference on Systems and Information Processing (2009)
14. Ejbali, R., Zaied, M., Ben Amar, C.: Multi-input Multi-output Beta Wavelet Network: Modeling of Acoustic Units for Speech Recognition. International Journal of Advanced Computer Science and Applications (IJACSA) 3 (2012)
15. Ejbali, R., Zaied, M., Ben Amar, C.: Wavelet network for recognition system of Arabic word. International Journal of Speech Technology 13, 163–174 (2010)
16. Bouchrika, T., Zaied, M., Jemai, O., Ben Amar, C.: Ordering computers by hand gestures recognition based on wavelet networks. In: International Conference on Communications, Computing and Control Applications (CCCA), Marseilles, France, pp. 36–41 (2012)
17. Dickinson, S.J., Pizlo, Z.: Shape Perception in Human and Computer Vision. Springer (2013) ISBN: 978-1-4471-5195-1
18. Ma, J.Q.: 'Content-Based Image Retrieval with HSV Color Space and Texture Features. In: WISM 2009, pp. 61–63 (2009)
19. Sural, S., Qian, G., Pramanik, S.: Segmentation and Histogram Generation Using the HSV Color Space for Image Retrieval. In: International Conference on Image Processing (ICIP), VIIth Digital Image Computing: Techniques and Applications, pp. 589–592 (2002)
20. Stricker, M., Orengo, M.: Similarity of color images. In: Proc. SPIE Conference on Storage and Retrieval for Image and Video for Image and Video Databases, vol. 24(20), pp. 381–392 (1995)

Tackling Ant Colony Optimization Meta-Heuristic as Search Method in Feature Subset Selection Based on Correlation or Consistency Measures

Antonio J. Tallón-Ballesteros and José C. Riquelme

Department of Languages and Computer Systems,
University of Seville, Spain
atallon@us.es

Abstract. This paper introduces the use of an ant colony optimization (ACO) algorithm, called Ant System, as a search method in two well-known feature subset selection methods based on correlation or consistency measures such as CFS (Correlation-based Feature Selection) and CNS (Consistency-based Feature Selection). ACO guides the search using a heuristic evaluator. Empirical results on twelve real-world classification problems are reported. Statistical tests have revealed that InfoGain is a very suitable heuristic for CFS or CNS feature subset selection methods with ACO acting as search method. The use of InfoGain is shown to be the significantly better heuristic over a range of classifiers. The results achieved by means of ACO-based feature subset selection with the suitable heuristic evaluator are better for most of the problems comparing with those obtained with CFS or CNS combined with Best First search.

Keywords: Feature selection, classification, ant colony optimization, heuristic evaluator, filter, feature subset selection.

1 Introduction

Ant colony optimization (ACO) meta-heuristic [6] was proposed by Dorigo et al. and is inspired in the behaviour of real ant colonies. Depending on the amount of pheromone deposited by the ant in their walk there would be some points that are more likely to be visited by the next ants [4]. The (artificial) ants in ACO define a randomized construction heuristic which makes probabilistic decisions depending on the strength of artificial pheromone trails and available heuristic information. As such, ACO can be interpreted as an extension of traditional construction heuristics, which are readily available for many combinatorial optimization problems. Yet, an important difference with construction heuristics is the adaptation of the pheromone trails during algorithm execution to take into account the cumulated search experience. As construction algorithms work on partial solutions trying to extend these in the best possible way to complete problem solutions.

E. Corchado et al. (Eds.): IDEAL 2014, LNCS 8669, pp. 386–393, 2014.

In essence, feature selection (FS) is a NP-hard combinatorial optimization problem. Often, FS is tackled with classical search methods for avoiding the prohibitive exhaustive search. Instead of looking for the optimal solution, obtaining a good solution in a reasonable time might be preferable for certain problems. On one hand, the typical NP-hard Travelling Salesman Problem has been treated successfully with ACO. On the other hand, ACO was applied for feature selection in [9] in the context of rough set reducts with very promising results. Here, we apply ACO as a stochastic procedure for quickly finding high quality solutions, in the scope of ordinary or crisp sets for FS in classification tasks. In this context, feature subset selection (FSS) problem is formulated using a graph with the purpose of getting a subset of attributes that is relevant for the problem at hand. FSS needs a search method, that usually is any kind of artificial intelligence heuristic technique. The current proposal shifts the search from an heuristic non-stochastic perspective to a stochastic angle.

This paper goals to address the suitability of using the ant colony optimization meta-heuristic as a search method built-in in the CFS and CNS feature subset selectors for classification problems.

The rest of this article is organized as follows: Sect. 2 describes some concepts about ACO meta-heuristic and feature selection; Sect. 3 presents our proposal; Sect. 4 details the experimentation; then Sect. 5 shows and analyzes statistically the results obtained; finally, Sect. 6 states the concluding remarks.

2   Background

2.1   Ant Colony Optimization Meta-Heuristic

Artificial ants used in ACO are stochastic solution construction procedures that probabilistically build a solution by iteratively adding solution components to partial solutions by taking into account a) heuristic information about the problem instance being solved, if available, and b) (artificial) pheromone trails which change dynamically at runtime to reflect the agents' acquired search experience.

The problem representation is a graph where the nodes represent the different points that the ants can visit and the edges are the link between points. Links are unidirectional and there are no cycles, so it is not possible to go back to a point previously visited. At the beginning of the algorithm every ant is located in a point and will construct a solution taking several decisions until the stop condition is met. At the end each ant has found a candidate solution to the problem at hand.

The main steps of the algorithm are the following:

1. *Initialisation.* The algorithm starts and all the pheromone variables are initialized to a value τ_0 which is a key parameter.
2. *Construct Ant Solutions.* This action starts the algorithm loop and is related with sending ants around the construction graph. An ant in the node i chooses the j one according to a probabilistic decision rule, which is a function of the pheromone τ and the heuristic η. A set of n_a ants constructs solutions to the concrete problem being tackled.

3. *Update Pheromones.* The purpose of this part is to change the values of the pheromones, by both depositing and evaporating.

Among the several variants of ACO, hereinafter we focus on Ant System (AS) that was introduced in 1991 by Dorigo and published a few years later by Dorigo et al. [5].

2.2   Feature Selection

Feature selection methods try to pick a subset of features that are relevant to the target concept [2]. According to Langley [10], the different approaches for feature selection can be divided into two broad categories (i.e., filter and wrapper) based on their dependence on the inductive algorithm that will finally use the selected subset. Filter methods are independent of the inductive algorithm, whereas wrapper methods use the inductive algorithm as the evaluation function.

FS involves two stages: a) to obtain a list of attributes according to an attribute evaluator and b) to perform a search on the initial list. All candidate lists would be assessed using a measure evaluation and the best one will be returned.

Two of the most widespread feature subset selectors are Correlation-based Feature Selection (CFS) [8] and Consistency-based feature selection (CNS) [3] that work in combination with a search method such as Greedy Search, Best First (BF) or Exhaustive Search. Generally speaking, BF is a powerful search method [7] which is the reason to be used very frequently by the machine learning community nowadays. We have chosen CFS and CNS as representative FSS methods, because they are based on different kind of measures, have few parameters and have provided a good performance inside the supervised machine learning area. Often, BF search is the preferable option by the researchers for both FSS algorithms. CFS is probably the most used FSS in data mining. CNS is also powerful, however the amount of published works is more reduced.

3   Proposal

Firstly, the graph meaning may be reformulated in order to deal with a feature selection problem by means of ACO meta-heuristic. Here, the nodes represent features and edges the link between nodes and the possibility to add another feature to the current solution. The search for a candidate solution is a walk through the graph. Once an ant visits an edge it contains a weight indicating the strength of this solution component.

In the current work, ACO, implemented following the AS model, is considered as search strategy in the context of CFS and CNS methods after the attribute evaluation phase. ACO guides the search by means of a heuristic evaluator. As heuristic evaluators, on one hand we have considered for CFS and CNS approaches the own attribute evaluator, obtaining the pure versions for CFS-AS and CNS-AS. On the other hand, we have tried Information Gain (InfoGain as abbreviation) [1] as evaluator resulting an hybrid approach. Moreover, CFS,

CNS and InfoGain compute different kinds of measure to evaluate the relevance, such as correlation, consistency and information, respectively.

The probabilistic transition rule is defined in the same way as in [9] and is the most widely used in AS [5]:

$$p_{ij} = \frac{\tau_{ij}^{\alpha} \cdot [\eta_{ij}]^{\beta}}{\sum_{il \in \mathcal{N}(x)} \tau_{il}^{\alpha} \cdot [\eta_{il}]^{\beta}}, \quad \forall ij \in \mathcal{N}(x). \tag{1}$$

where p_{ij} represents the probability that current ant at feature i would travel to feature j, τ_{ij} is the amount of pheromone on the ij edge, η_{ij} is the heuristic desirability of the ij transition and $\mathcal{N}(x)$ the set of current feasible components. Lastly, α and β are parameters that may take real positive values –according to the recommendations on parameter setting in [5]– and are associated with heuristic information and pheromone trails, respectively.

All ants update pheromone level with an increase of small quantities, depending directly on the heuristic desirability of the ij transition given by the measure (merit) of the subset attribute evaluator used as heuristic evaluator and inversely proportional to the subset size.

4   Experimentation

Table 1 depicts the feature subset selection methods applied in the experimental process. We have grouped them according to the attribute evaluator and search method.

Table 2 summarizes the main parameters along with their symbols and numerical or conceptual values for all the feature subset selection methods used for the experiments. On one hand, in relation to ACO-based feature subset selection, the n_a and gen parameters have been set to fix values to our choice. For τ_0 parameter, in [5] there is a suggestion to assign a small positive constant and hence we have defined a value of 0.5. The trade-off between α and β parameters may influence in the behaviour of the algorithm thus for their determination a preliminary experimental design by means of a five-fold cross validation on the training set has been carried out with a couple of values for each one parameter (1 and 2). On the other hand, for BF-based search method in the context of CFS and CNS, we have followed the recommendations of the authors ([8] and [3] for the number of expanded nodes; the search direction has been fix according to our previous experiences).

Table 3 represents the data sets employed throughout the experimentation. They come mostly from binary and multi-class classification real-world problems (*Cl.* column specifies the number of classes) taken from the public UCI repository. The number of instances ranges from more than one hundred to approximately fourteen thousands, thus problems with a medium size, and the dimensionality varies between nineteen and one hundred and twenty nine. Also, we have included the number of selected features for every feature subset selector obtained in the training set. Last row shows the dimensionality reduction (higher is better) in mean over the original data sets for each filter.

Table 1. List of feature subset selectors for the experimentation

Attribute evaluator	Search method	Heuristic evaluator	*Abb. Name*
CFS	*AntSearch*	*CFS*	*CFS − AS h*1
		InfoGain	*CFS − AS h*2
	BestFirst	−	*CFS − BF*
CNS	*AntSearch*	*CNS*	*CNS − AS h*1
		InfoGain	*CNS − AS h*2
	BestFirst	−	*CNS − BF*

Table 2. Parameter values for ACO-based feature subset selection approaches (CFS-AS and CNS-AS) and BestFirst-based ones (CFS-BF and CNS-BF)

Search method	Parameter	Symbol	Value
Ant search	*Number of ants*	n_a	10
	Number of generations	gen	10
	Pheromone trail influence	α	1
	Heuristic informacion value	β	2
	Pheromone initial value	τ_0	0.5
Best First	*Consecutive expanded nodes without improving*		5
	Search direction		*Forward*

Table 3. Summary of the data sets used and selected features for each feature subset selector

Data set	*Size*	*Train*	*Test*	*Feat.*	*Cl.*	*Selected features*				
						CFS			*CNS*	
						AS		*BF*	*AS*	*BF*
						h1	h2		h1	h2
batch(gas)	13910	10432	3478	129	6	7	13	20	7	10	6
cardiotoc.	2126	1595	531	22	10	12	12	12	15	15	13
hepatitis	155	117	38	19	2	10	9	10	15	11	11
ionosphere	351	263	88	33	2	9	10	11	9	9	9
libras	360	270	90	90	15	8	34	23	47	55	20
lymph.	148	111	37	38	4	11	8	12	19	13	10
promoter	106	80	26	58	2	10	10	10	15	8	8
satimage	6435	4435	2000	36	6	20	21	23	13	13	12
sonar	208	104	104	60	2	4	7	8	14	9	9
soybean	683	511	172	82	19	22	22	25	39	58	16
SPECTF	267	80	187	44	2	10	8	12	12	10	8
waveform	5000	3750	1250	40	3	12	14	14	14	13	12
Averages	2479.08	1812.33	666.75	54.25	6.08	11.25	14.00	15.00	18.25	18.67	11.17
Dim. reduction						79.26	74.19	72.35	66.36	65.59	79.42

The experimental design follows a stratified hold-out cross validation with three and one quarters for the training and test sets, respectively. Sometimes, these proportions do not match since the original data are prearranged. For the statistical analysis between two feature subset selection methods we have carried a Wilcoxon signed-ranks test with the test accuracy results obtained.

5   Results

Tables 4 and 5 report the accuracy test results for the FSS based on ACO with CFS and CNS -that is, CFS-AS and CNS-AS, respectively- with two different heuristic evaluators, their difference ($Diff.$) and its ranking ($R.$). Ten executions with different seeds were run and the most frequent solution was considered for the assessment. We have carried out experiments with three kind of deterministic classifiers such: a) C4.5, based on decision trees, b) SVM, founded in support vectors, and c) PART, a rule-based approach. The reason for the choice of these classifiers is motivated by the fact that their overall performance is good in the feature selection scope [11]. The best results, excluding ties, for each pair (classifier, FSS method) have been highlighted with boldface. According to Wilcoxon signed-ranks test, since there are 12 data sets, the T value at $\alpha = 0.05$ should be less or equal than 14 (the critical value) to reject the null hypothesis. On one hand, in relation to CFS-AS, for classifiers C4.5 and SVM the h2 heuristic evaluator is significantly better than h1. On the other hand, for CNS-AS the performance of h2 with PART is the significant best option.

Table 4. CFS-AS: Accuracy test results and statistical tests

Data set	C4.5				SVM				PART			
	$CFS-AS$				$CFS-AS$				$CFS-AS$			
	h1	h2	$Diff.$	R.	h1	h2	$Diff.$	R.	h1	h2	$Diff.$	R.
$batch(gas)$	93.99	**96.87**	2.88	8	72.28	**78.21**	5.92	9	93.73	**96.35**	2.62	7
$cardiotoc.$	61.58	61.58	0.00	2	**67.98**	67.80	−0.19	2	60.45	**62.52**	2.07	5
$hepatitis$	84.21	**89.47**	5.26	10	86.84	**89.47**	2.63	6	84.21	**86.84**	2.63	8
$ionosphere$	**90.91**	87.50	−3.41	9	82.95	**89.77**	6.82	10	90.91	**93.18**	2.27	6
$libras$	52.22	**65.56**	13.33	12	50.00	**63.33**	13.33	12	54.44	**65.56**	11.11	12
$lymph.$	81.08	81.08	0.00	2	81.08	**83.78**	2.70	8	**72.97**	70.27	−2.70	9
$promoter$	73.08	73.08	0.00	2	73.08	73.08	0.00	1	80.77	80.77	0.00	1
$satimage$	86.05	**86.25**	0.20	5	83.80	**84.50**	0.70	3	**85.00**	83.55	−1.45	2
$sonar$	67.31	**74.04**	6.73	11	67.31	**75.00**	7.69	11	68.27	**75.96**	7.69	11
$soybean$	88.95	**91.28**	2.33	6	94.19	**95.35**	1.16	5	90.70	**92.44**	1.74	3
$SPECTF$	66.84	**69.52**	2.67	7	**66.31**	63.64	−2.67	7	70.59	**76.47**	5.88	10
$waveform$	74.32	**74.40**	0.08	4	85.92	**86.88**	0.96	4	**78.88**	77.04	−1.84	4
	$T = min\{66, 12\} = 12$				$T = min\{68.5, 9.5\} = 9.5$				$T = min\{62.5, 15.5\} = 15.5$			

Table 6 outlines the global accuracy test results of the best CFS-AS and CNS-AS approaches versus based-BF CFS or CNS. The best results, in each data set, are marked in bold. We can assert the following statements in relation to the

Table 5. CNS-AS: Accuracy test results and statistical tests

Data set	C4.5 CNS − AS				SVM CNS − AS				PART CNS − AS			
	h1	h2	Diff.	R.	h1	h2	Diff.	R.	h1	h2	Diff.	R.
batch(gas)	**96.00**	95.54	−0.46	3	64.49	**69.15**	4.66	10	95.46	95.46	0.00	1.5
cardiotoc.	63.84	**66.48**	2.64	6	61.39	**64.41**	3.01	9	58.76	**64.22**	5.46	8
hepatitis	84.21	**86.84**	2.63	5	**86.84**	84.21	−2.63	8	76.32	**84.21**	7.89	10
ionosphere	88.64	**93.18**	4.55	12	81.82	**84.09**	2.27	7	89.77	**95.45**	5.68	9
libras	56.67	**61.11**	4.44	11	67.78	**70.00**	2.22	6	55.56	**64.44**	8.89	11
lymph.	78.38	**81.08**	2.70	8	**91.89**	83.78	−8.11	11	**78.38**	64.86	−13.51	12
promoter	**84.62**	80.77	−3.85	10	73.08	**84.62**	11.54	12	76.92	**80.77**	3.85	5
satimage	84.70	**84.75**	0.05	2	**84.50**	83.70	−0.80	4	85.50	**86.25**	0.75	4
sonar	**75.96**	73.08	−2.88	9	**75.96**	75.00	−0.96	5	72.12	**75.96**	3.85	6
soybean	91.86	91.86	0.00	1	**95.35**	94.77	−0.58	3	92.44	92.44	0.00	1.5
SPECTF	**69.52**	66.84	−2.67	7	64.71	64.71	0.00	1	65.24	**69.52**	4.28	7
waveform	**76.00**	74.88	−1.12	4	85.12	**85.36**	0.24	2	77.36	**77.84**	0.48	3
	$T = min\{44.5, 33.5\} = 33.5$				$T = min\{51.5, 26.5\} = 26.5$				$T = min\{64.5, 13.5\} = 13.5$			

Table 6. Global accuracy test results of CFS-BF and CNS-BF versus CFS-AS and CNS-AS

Data set	C4.5 CFS − BF	CFS − AS	SVM CFS − BF	CFS − AS	PART CNS − BF	CNS − AS
		h2		h2		h2
batch(gas sensor)	95.92	96.87	83.04	78.21	**98.30**	95.46
cardiotoc.	61.58	61.58	**67.80**	**67.80**	61.02	64.22
hepatitis	84.21	**89.47**	86.84	**89.47**	84.21	84.21
ionosphere	92.05	87.50	88.64	89.77	88.64	**95.45**
libras	61.11	**65.56**	57.78	63.33	55.56	64.44
lymph.	81.08	81.08	81.08	**83.78**	67.57	64.86
promoter	73.08	73.08	73.08	73.08	69.23	**80.77**
satimage	85.60	**86.25**	83.85	84.50	85.45	**86.25**
sonar	73.08	74.04	75.00	75.00	**75.96**	**75.96**
soybean	93.02	91.28	94.77	**95.35**	93.02	92.44
SPECTF	66.84	69.52	**73.26**	63.64	66.31	69.52
waveform	74.40	74.40	**86.88**	**86.88**	76.16	77.84
Wins by pairs	2	6	2	6	3	7
Global wins	0	3	3	5	2	4
Averages						
Accuracy	78.50	79.22	79.33	79.23	76.79	79.29
Selected feat.(%)	27.65	25.81	27.65	25.81	20.58	34.41

achieved results. First, the comparison between pairs of FSS methods for each classifier and data sets points out that: a) C4.5 with CFS-AS gets better results 6 times, b) SVM with CFS-AS wins in 6 problems, and c) PART classifier with CNS-AS 7 times. Second, a global analysis from a qualitative point of view means that (SVM, CFS-AS) pair reaches the best results 5 times, followed by (PART, CNS-AS) with 4 wins. Third, the percentage of selected attributes in (SVM, CFS-AS) pair is close to 25, while with (PART, CNS-AS) is slightly greater and takes a value near 35.

6 Conclusions

CFS-AS and CNS-AS were presented. Experiments revealed that ACO-based search via AS in feature subset selection with InfoGain heuristic is better, and in some cases with significant differences, than the pure versions of ACO-based filters (that is, a concrete subset attribute evaluator with the homonymous heuristic evaluator, e.g. CFS-AS with h1). It is very important to stress that ACO-based feature subset selector with the proper heuristic evaluator is better in more than the half of the problems that the traditional Best First search in CFS or CNS. The two preferred classifier-FSS pairs, bearing in mind the performance regarding the accuracy and number of selected attributes are, in this order, (SVM, CFS-AS) and (PART, CNS-AS).

Acknowledgments. This work has been partially subsidized by TIN2007-68084-C02-02 and TIN2011-28956-C02-02 projects of the Spanish Inter-Ministerial Commission of Science and Technology (MICYT), FEDER funds and P11-TIC-7528 project of the "Junta de Andalucía" (Spain).

References

1. Cover, T.M., Thomas, J.A.: Elements of Information Theory. Wiley (1991)
2. Dash, M., Liu, H.: Feature selection for classification. Intelligent Data Analysis 1(3), 131–156 (1997)
3. Dash, M., Liu, H.: Consistency-based search in feature selection. Artificial Intelligence 151(1), 155–176 (2003)
4. Dorigo, M., Di Caro, G., Gambardella, L.M.: Ant algorithms for discrete optimization. Artificial Life 5(2), 137–172 (1999)
5. Dorigo, M., Maniezzo, V., Colorni, A.: Ant system: optimization by a colony of cooperating agents. IEEE Transactions on Systems, Man, and Cybernetics, Part B: Cybernetics 26(1), 29–41 (1996)
6. Dorigo, M., Stützle, T.: Ant colony optimization: overview and recent advances. In: Handbook of Metaheuristics, pp. 227–263. Springer (2010)
7. Gaschig, J.: Performance measurement and analysis of certain search algorithms. Technical report, DTIC Document (1979)
8. Hall, M.A.: Correlation-based feature selection for machine learning. PhD thesis, The University of Waikato (1999)
9. Jensen, R., Shen, Q.: Finding rough set reducts with ant colony optimization. In: Proceedings of the 2003 UK Workshop on Computational Intelligence, vol. 1 (2003)
10. Langley, P.: Selection of relevant features in machine learning. Defense Technical Information Center (1994)
11. Tallón-Ballesteros, A.J., Hervás-Martínez, C., Riquelme, J.C., Ruiz, R.: Feature selection to enhance a two-stage evolutionary algorithm in product unit neural networks for complex classification problems. Neurocomputing 114, 107–117 (2013)

EventStory: Event Detection Using Twitter Stream Based on Locality

Sorab Bisht and Durga Toshniwal

Department of Computer Science and Engineering,
Indian Institute of Technology Roorkee, India
{bisht.saurabh57,durgatoshniwal}@gmail.com

Abstract. Increased popularity of social media sites such as Twitter, Facebook, Flickr, etc. have produced an enormous amount of spatio-temporal data. One of the application of this type of data is event detection. Most of event detection techniques have focused on temporal feature of data for detecting an event. However, location associated with data has to be taken into consideration to detect locality based event (local event) such as local festival, sporting event or emergency situations. Users in proximity of the location of an event are more likely to post messages about an event compared to users distant from the location of that event. In this paper, we are proposing a framework, called EventStory. Our framework first identifies locally significant key-words (LSK) by monitoring changes in the bursty nature of keywords in both local and global regions. Candidate event clusters are created based on co-occurrence of locally significant keywords (LSK) in the each keyword cluster. A cluster scoring scheme is used which uses the features of cluster to filter irrelevant clusters. A case study is presented to show effectiveness of our approach.

Keywords: Event Detection, Keyword Clustering, Local Event Detection, Social Media Analytics.

1 Introduction

Today various social media sites such as Facebook, Twitter, Google+, Flickr, etc., provide a medium to share something interesting with other users. Twitter is one of the social media sites which has shown an exceptional growth in the amount of data being generated. Currently, twitter have 640 million active users generating 9,100 tweets per second[1]. Twitter allows its user to share their thoughts, an activity or information about an event by using short text messages called tweets with a restriction of 140 characters per tweet. Along with text, tweets have features associated with it such as timestamp of tweet, user information, and geo-location. But there are following issues with tweet. First, amount of data is posted with rapid rate which requires scalable and efficient techniques. Secondly, not all tweets are significant compared to news articles

---

[1] http://www.statisticbrain.com/twitter-statistics/

E. Corchado et al. (Eds.): IDEAL 2014, LNCS 8669, pp. 394–403, 2014.

where all documents are significant. Lastly, due to limits in number of characters, tweet contain noisy data in the form of misspellings and abbreviations such as "2mmrw, tomorow" for word "tomorrow".

Most of the event detection techniques does not exploit the location associated with tweets. Hence, these techniques are not helpful in detecting locality based events, i.e. events specific to small geographic regions such as local festival, local sporting event, emergency situations.

In this paper, we propose a framework EventStory, which takes temporal occurrence of a keyword in both local region and region global region in consideration to identify locally significant keywords (LSK). The granularity of local region ranges from a sub-area in a city to an entire state while the granularity of global region ranges from a country to entire world. The reason behind comparing temporal occurrence is that, a keyword related to a local event will have more significance in local region compared to global region. A LSK vector for each keyword cluster is created based on occurence of LSKs in associated tweets in cluster. This vector is used to find similarity between keywords for clustering related keywords. Relevance of an event is computed using temporal feature, locality feature and textual feature of event clusters. Finally, event representative keywords are generated to describe the event.

2   Related Work

Event detection can be defined as identifying unusual occurrence of an activity compared to the previous occurrence of that activity. Research has been primarily focused on detecting, summarizing and tracking of events based on temporal aspect of messages[1,2,7].

Most of the techniques do not take geographic property of tweets into consideration. Hence, these approaches have issues detecting event happening in a local region only. There are some existing approaches for local event detection but they come with certain disadvantages. [3] used spatio-temporal information present in Flickr photos for detecting events. However, this approach is suitable for static data only, which is not useful for continuous stream of twitter data. [4] exploited mentions of popular places in tweets to detect an event. The drawback of this approach is that the tweet related an event might not have mention of a place in it. In [5], their method uses log-likelihood ratio for geographic and time dimension to find significant terms while our approach uses the global burst score and local burst score to find locally significant terms. [6] proposed an approach in which each bursty keyword is assigned a spatial signature and keywords are clustered using that spatial signature. Relevance of events are determined using the score for each keyword. The problem with their approach is in detecting two different events happening simultaneously at nearby location. Since keywords related to these events have same spatial signature, a single cluster is created rather than different clusters.

Related to clustering related terms and finding relevant clusters, [2] clustered event segments using the frequency distribution similarity and content similarity in sub-windows. They used metadata of Wikipedia to find newsworthiness of an event segment. But, due to unavailability of metadata related to a new type of event, the event segments will be considered less newsworthy. We are using only the features associated with cluster to find relevant clusters. In [7] author used co-occurrence between frequent bigrams in multiple time windows to cluster related bigrams. Co-occurrence doesnt work properly for tweets because they are shorter than normal text. To overcome this limitation, we are using content similarity in the form of vectors of occurrence of LSK.

3 Proposed Work

In this paper, we propose a framework EventStory which mainly focusses on finding locally significant keywords (LSK) and clustering related keywords. The flow diagram of framework is given in fig. 1. The framework can be divided into following blocks:

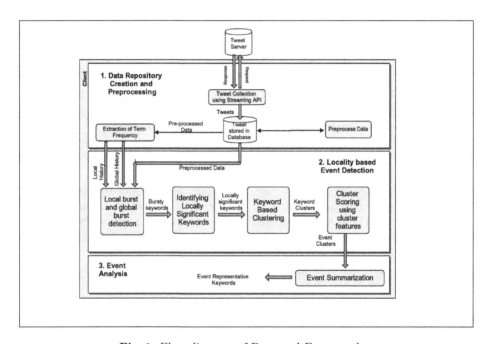

Fig. 1. Flow diagram of Proposed Framework

- Data Repository Creation and Pre-processing
- Locality based Event Detection
- Event Analysis

These blocks can be further divide into sub-blocks. These sub-blocks are described below -

3.1 Preprocessing

Twitter data is noisy in nature. Pre-processing steps are taken to handle issues such as misspellings, non-event words, abbreviations and stop words. Fig 2 shows the flow of preprocessing. Blocks in preprocesing are described below

- **Remove multiple ocurrences of word** because we are interested in only document (tweet) frequency of words, not the term frequency.
- **Remove non-ascii words** such as punctuations, emoticons, etc. because they doesnot provide any information about an event.
- **Resolve Abbreviations and Misspellings**: To handle misspellings and abbreviations, a dictionary is used which contains misspellings, abbreviations and their corresponding correct word(for e.g. "2day", "2dae" for "today" or "kno", "knw" for "know").
- **Remove Stop words**: Stop words are removed with the help of list containing english stop words and social media specific stop words such as lol, hmm, etc.

In our approach, retweets are removed because they does not provide new information about an event.

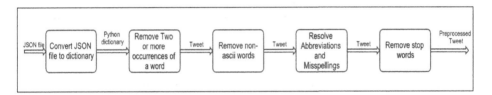

Fig. 2. Preprocessing Flow diagram

3.2 Term Frequency Extraction

The function of this block is to divide the tweets in multiple buckets such that each bucket contains tweets posted in a particular time window. All the time windows have equal time duration and are non-overlapping (w_1, w_2,..., w_{c-1}, w_c) where w_c represents the current time window. For a keyword k, u_i^k (where i <c) denotes the number of users mentioning keyword k in a time window i. u_i^k is normalized by dividing by total number of users in window w_i.

   As described in [2] we are using user frequency in place tweet frequency because it might be possible that a same set of user is mentioning same term multiple times in a time window. Each u_i^k is taken as a data point and value of u_i^k is independent of previous values, so according to central limit theorem the distribution of u_i^k can be considered normal distribution.

3.3 Burst Detection

Burst detection is based on the discrepancy paradigm [8] to measure the deviation of actual frequency from expected frequency. Given a set of keywords K_c in time window (w_c), we want to extract a subset of keywords (bursty keywords) , $L_c \subset K_c$ which might rpresent an event. To find these bursty keywords, normalized frequency of users mentioning keyword k i.e. u_c^k in current window w_c is compared with the expected frequency (μ_k) calculated using historic data. Expected value (μ_k) and deviation (σ_k) are estimated using data from fixed number of previous time windows by applying maximum likelihood estimation considering distribution as normal distribution.

We are using the mechanism used in [9] to calculate burst-score, but for calculating expected frequency we have taken all the time windows of previous week, compared to taking only selected time window from previous time windows. It is not necessary to have historic time window of 1 week only but it needs to be of fixed size. Keywords showing high deviation from expected value have higher burst score. Burst-score is defined by z-score value of u_c^k as,

$$Burst - score = \frac{u_c^k - \mu_k}{\sigma_k} \qquad (1)$$

Keywords with burst score two standard deviation more than mean are considered as bursty keywords. Bursy keywords are identified for local region and in next block it is determined whether a bursty keyword is locally significant keyword (LSK) or not.

3.4 Identification of Locally Significant Keywords

Burst score is calculated for keyword with respect to both local and global region. Burst score calculated using local region data is defined as bur_{local} and burst score calculated using global region data is defined as bur_{global}.

Burst-scores are z-scores of keyword frequency and z-scores can be directly compared[10]. A significance value is calculated by equation 2 where relative difference between bur_{local} and bur_{global} with respect to bur_{local}. Value of $signifcane$ ranges between 0 to 1 (1 when bur_{global} is negative or zero, 0 when bur_{local} is less than equal to bur_{global}). If significance is above a threshold (α) then keyword is added to locally significant keywords (LSK) set for current window w_c. Threshold value is established using empirical analysis of previous data related to same local and global region.

$$significance = \frac{bur_{local} - bur_{global}}{bur_{local}} \qquad (2)$$

The reason behind using significance is that the frequency distribution of keywords related to a global event will closer in both local and global region, compared to a local event. Significance is important for two reasons. First, it helps in filtering global event keywords. Secondly, virtual events keywords such as memes gets less significance score.

3.5   Keyword Based Clustering

Similarity Measure. Co-occurrence [7] has been used as similarity measure for finding similarity between keywords but it doesnt work properly in case of tweets because of short nature of tweets. To find the relatedness between two keywords, we need to find whether the tweets belonging to both keywords contain the similar pattern of locally significant keywords (LSK). Keywords having higher similarity are more likely to represent same event.

A set of LSK, $K = \{K_1, K, \ldots, K_l\}$ is obtained from previous block, where l is the number of LSK identified. For each K_i, a cluster (C_i) is formed containing associated tweets (T_i), keyword list (L_i) which initially contains K_i only.

Based on occurrence of LSK in set T_i, a LSK Vector V_i is created using eq. 3

$$V_i = K_1.o_1 + K_2.o_2 + \ldots + K_j.o_j + \ldots + K_l.o_l \tag{3}$$

where o_j is the occurrence value of K_j in cluster C_i and is defined by

$$o_j = \sum_{i=1}^{|T_i|} I_{t_k} \tag{4}$$

,I is the indicator function of presence of K_j in tweet t_k.

Keywords belonging to same event share similarpattern of keywords in their cluster. Similarity between LSK vectors will be more for the keywords belonging to same event. Finding similarity between two LSK vectors can be reduced to finding similarity between two documents where LSK vectors represent the document vectors. To find the correlation between two LSK vectors we are using cosine similarity.

$$cos(C_m, C_n) = \frac{V_m.V_n}{\sqrt{\sum_{i=1}^{l} V_{m_i}^2} \cdot \sqrt{\sum_{i=1}^{l} V_{n_i}^2}} \tag{5}$$

where $V_m.V_n$ is the dot product between LSK vectors of clusters C_m and C_n

Clustering Keywords. We are using agglomerative hierarchical clustering with modified similarity metric to cluster related keywords that might represent same event as shown in algorithm 1. The algorithm 1 returns a list of candidate event clusters $E = \{E_1, E_2, \ldots, E_z\}$. For an event, at least two keywords are required to represent the event. Only those clusters are considered as candidate event clusters which contain two of more keywords in its keyword list.

3.6   Cluster Scoring

For each candidate event E_i present in candidate event list E, cluster score is computed to determine the relevance of event E_i. E_i is represented by tuple $(L_i, B_i, Sig_i, sim_i, U_j)$ where B_i is the set of burst score (b_j) of keywords in keyword list L_i, Sig_i is the set of significance score of keywords in keyword list

---

Algorithm 1. Clustering Related Keywords Using LSK Vector

Input: A finite set of $C = \{C_1, C_2, \ldots, C_l\}$ individual LSK clusters
Output: A finite set of $E = \{E_1, E_2, \ldots, E_z\}$ candidate event clusters
1 $max\_sim = cos(C_1, C_2)$
2 **repeat**
3   **for** $each\,pair\,of\,clusters\,(C_i, C_j)\,in\,C$ **do**
4        $calculate\,cos(C_i, C_j)$
5      $max\_sim = max\{cos(C_i, C_j)\}, 1 \leq i \leq |C|$
                                                $, 1 \leq j \leq |C|, i \neq j$

    $max\_pair = m, n$ //**cluster pair with max similarity if** $max\_pair \geq \beta$
    then
      /***Merging***/
      $Create\,Cluster$,
      $C_{mn} = \{L_{mn}, T_{mn}, G_{mn}, V_{mn}\}$
      $L_{mn} = L_m \cup L_n$
      $T_{mn} = T_m \cup T_n$
      $G_{mn} = G_m \cup G_n$
      $V_{mn}\,is\,calculated\,from\,T_{mn}\,and\,K\,using\,eq.\,3$
      $remove\,C_m, C_n\,from\,C$
      $add\,C_{mn}\,to\,C$
6 **until** $max\_sim > \beta$
7 **return** C

L_i, sim_i is the cosine similarity obtained from merging clusters, U_j is the set of number of users mentioning keywords in keyword list L_i. Cluster score is computed as,

$$Score(E_i) = (\sum_{j=1}^{m} b_j.log(u_j).sig_j).sim_i \qquad (6)$$

where m is the number of keywords in keyword list L_i. Following features are used in computing cluster score :

- *Temporal Feature*, b_j : As described in [2], logarithm of number of users (u_j) mentioning keyword (l_j) can be used to give more weightage to bursty keyword with higher frequency, compared to bursty keyword having same burst score with lower frequency.
- *Locality Feature*, sig_j : It is used to give more weightage to keyword having higher level of significance in the locality.
- *Content Feature*, sim_i : It is used to define the content similarity between clusters which formed the current cluster C_i

Based on score computed, it is determined whether the candidate event cluster is relevant or not (i.e the intensity of cluster is determined).

3.7   Event Summarization

Event summarization is done by generating event representative keywords from tweets present in event cluster. If LSK K_j is present in equal to or more than half of the tweets in T_i, then K_j is added to event representative terms set. Final result generated contains event clusters alongwith their corresponding event representative keywords and cluster score calculated using equation 6.

4   Experiments

4.1   Dataset

Data is collected from Twitter using their public streaming API[2]. It provide only 1% sample of public live tweet stream. Details of the datset is given in table 1. Data from 1/3/2014 to 8/3/2014 has been used to calculate the expected frequency (μ) and standard deviation (σ) for all the keywords occuring during this period. Events were detected on 9/3/2014. The parameters are dependent on the regions chosen for local and global regions. A suggested guessing and result comparison on already existing database is done to determine these parameters. We set α=0.45 and β=0.3 to find a reasonable balance between results.

Table 1. Details of dataset collected during 1 Mar 2014 to 9 Mar 2014

Region	Bounding Boxes	No. of Users	No. of Tweets
London (*Local Region*)	[-0.563, 51.2613, 0.2804, 51.6860]	95105	871776
United Kingdom (*Global Region*)	[-13.4139, 49.1621, 1.5690, 60.8547]	405260	3553428

4.2   Case Study of Detected Events

We have done a case study of events happened in London on 9/3/2014 during the duration 11.30 AM to 8.30 PM, GMT Timezone (12.30 PM to 9.30 PM Local Time, BST) as shown in Table 2. We have chosen 9/3/2014 for detection purpose because many such as England vs Wale Rugby, FA Cup draw, Ellie Goulding concert were happening on same day.

As seen from Table 2, related to *England vs Wales rugby* match, before start of the match event representative keywords were about match("engvwal") and place ("twickerham") where match was scheduled. During first time window the intensity of keywords is high even though game started at 3.00 pm because people were posting about their activity of going to the stadium. Around the game time (3.00 pm) , the events representative keywords included were people supporting Teams ("comeonengland") and name of players ("owen farell", "lawes courtney"," leigh halfpenny"). At the end of the game people are talking about

---

[2] https://dev.twitter.com

Table 2. Represents events detected 11.30 am to 8.30 pm (in GMT) on 9 March 2014 in the London region, Timezone (12.30 pm to 9.30 pm Local Time, BST). Each column under event representative keywords represent a one hour time window (Here last character in each cell represents the intensity (score) of event cluster. H represents high cluster intensity, M represents moderate cluster intensity and L represents low cluster intensity).

Event	Event Representative Keywords								
England vs Wales Rugby	engvwal, twickenham H	twickenham, engvwal	twickenham, engvwal	davies, curtis, wales, eng, rugby, comeonengland H	engvwal, carrythemhome, wales, ht, jenkins, gethin, leigh, halfpenny, farrell, owen, lawes, courtney H	lawes,courtney, chariot, swing H			
FA Cup					fa, draw, hull, sheff, utd, semi, finals M	wafc,facup, fa, fixtures, semis, wigan M	semi, finals, fa M	finals, semi	
Ellie Goulding Concert								greater, o2	ellie, goulding M
Westwood Vivienne Interview	westwood, vivienne, shami L								
Dancing on ice						doi,bolero M	doi,christine M		
Other Events	burnley, blackburn M hydepark, hyde M	burnley, blackburn M	hampstead, heath suspected, vietnam	hampstead,heath	hampstead, heath	lescott, interfering	hydepark, hyde M		

lawes courtney who was the best player in match and people singing Sweet Low, Sweet Chariot song. Other event detected was *FA Cup Semi-Final Draw*. Event representative keywords realted to this event were semi-final and name of teams playing in semi-final (for e.g. Hull City vs Sheffield United).

One of the event deteted was *Ellie Goulding concert*. Event representative keywords obtained were singer's name ("Ellie Goulding") and place of concert ("O2 Arena"). Our framework is able to detect low key event such as *Vivionne Westwood Interview* taken by Shami. Other events detected were *Dancing on ice* with performers name ("bolero", "christine") as representative keywords, mentions of popular places ("hyde park" and "hamstead heath") in london on that day.

Events occurring outside London such as match between Burnley and Blackburn occurring in Lancashire ("blackburn","burnley"), for suspicion regrading missing malaysian plane MH370 near vietnam ("suspected","vietnam") and a player activity in a game happening outside london("lescott","interfering") were also detected. The possible reason behind occurrence of these terms is that the frequency of people tweeting about these event were higher in london, causing higher local burst than global burst.

5   Conclusion and Future Work

In this paper, we have proposed a framework called EventStory to detect locality based real world event using twitter stream. Events detection has been done by detecting locally significant keywords (LSK) using comparison of burst score in local and global region. Repeated terms such as "Happy", "Morning","Birthday" are eliminated by the means of calculating expected frequency and deviation of terms using historic data. Our framework have shown a novel approach to cluster related keywords based on LSK vector which is created based on occurence of LSK in a cluster and finding relevant clusters using cluster features which also include locality feature. We were able to detect many local events and find the relevance of detected events. Locality feature is used to give more weightage to locally significant keywords. Our case study has shown the evolution of events representative terms during different phases of the event.

One of the issue with our framework is the value of parameter α depends on the combination of local and global region. This framework is not specific to twitter data. Other social media data can be used by making changes in parameters. In future work, we would like classify events based on tweet content. Also analysis of sentiments of people related to event in local and global region can be done to analyse change in reactions related to event in different regions.

References

1. Becker, H., Naaman, M., Gravano, L.: Beyond Trending Topics: Real-World Event Identification on Twitter. In: International AAAI Conference on Weblogs and Social Media, Barcelona, Spain (July 2011)
2. Li, C., Sun, A., Datta, A.: Twevent: segment-based event detection from tweets. In: Proceedings of the 21st ACM International Conference on Information and Knowledge Management (CIKM 2012), pp. 155–164. ACM, New York (2012)
3. Chen, L., Roy, A.: Event detection from Flickr data through wavelet-based spatial analysis. In: Proceedings of the 18th ACM International Conference on Information and Knowledge Management, pp. 523–532 (2009)
4. Watanabe, K., Ochi, M., Okabe, M., Onai, R.: Jasmine: a real-time local-event detection system based on geolocation information propagated to microblogs. In: Proceedings of the 20th ACM International Conference on Information and Knowledge Management, CIKM 2011, pp. 2541–2544. ACM, New York (2011)
5. Weiler, A., Scholl, M.H., Wanner, F., Rohrdantz, C.: Event identification for local areas using social media streaming data. In: Proceedings of the ACM SIGMOD Workshop on Databases and Social Networks (DBSocial 2013), pp. 1–6. ACM, New York (2013)
6. Abdelhaq, H., Sengstock, C., Gertz, M.: EvenTweet: online localized event detection from twitter. Proceedings of the VLDB Endowment 6(12), 1326–1329 (2013)
7. Parikh, R., Karlapalem, K.: Et: events from tweets. In: Proceedings of the 22nd International Conference on World Wide Web Companion, pp. 613–620. International World Wide Web Conferences Steering Committee (2013)
8. Lappas, T., Vieira, M.R., Gunopulos, D., Tsortas, V.J.: On the spatiotemporal burstiness of terms. Proceedings of the VLDB Endowment, 826–847 (2012)
9. Abdelhaq, H., Gertz, M., Sengstock, C.: Spatio-temporal characteristics of bursty words in Twitter streams. In: Proceedings of the 21st ACM SIGSPATIAL International Conference on Advances in Geographic Information Systems, pp. 194–203. ACM (2013)
10. Abdi, H.: Z-scores. Encyclopedia of measurement and statistics. Sage, Thousand Oaks (2007)

Univariate Marginal Distribution Algorithm with Markov Chain Predictor in Continuous Dynamic Environments

Patryk Filipiak and Piotr Lipinski

Computational Intelligence Research Group
Institute of Computer Science, University of Wroclaw, Poland
{patryk.filipiak,lipinski}@ii.uni.wroc.pl

Abstract. This paper presents an extension of the continuous Univariate Marginal Distribution Algorithm with the prediction mechanism based on a Markov chain model in order to improve the reactivity of the algorithm in continuous dynamic optimization problems. Also a population diversification into exploring, exploiting and anticipating fractions is proposed with the auto-adaptation mechanism for updating dynamically the sizes of these fractions. The proposed approach is tested on the popular benchmark functions with the recurring type of changes.

1 Introduction

A great deal of Dynamic Optimization Problems (DOPs) at some point becomes *predictable*, i.e. after observing the history trail of the environmental changes for some time, one can often anticipate the future location of an optimum using e.g. extrapolation, linear regression or pattern matching techniques. Particularly, the class of DOPs where optima tend to recur in the same (or nearly the same) points in the search space seem most likely to become predictable in that sense.

Van Hemert et al. [12] suggested a "futurist" approach to DOPs that empowered Simple Genetic Algorithm (SGA) by directing the population into the future promising regions thus helping it track moving optima. Numerous other anticipation mechanisms based on Kalman filter [7,10], auto-regressive models [1] or time series identification [13] were proposed. An interesting approach using an auto-regressive model for predicting when the next change will occur and a Markov chain predictor [6] for anticipating the future landscape was recently presented in [9]. The latter approach was tested on the popular bit-string problems where it proved its rapidness in reacting to the recurring changes.

In this paper the applicability of the above Markov chain predictor is extended to the continuous domain. Note that, unlike bit-string problems where identifying the states of the environment is rather straightforward, in the case of continuous domain some parametrization is required for storing information about the landscape. For this purpose Evolution Strategies [8] or Evolutionary Programming [11] could be used. However, a stochastic parametrization utilized

E. Corchado et al. (Eds.): IDEAL 2014, LNCS 8669, pp. 404–411, 2014.

in Estimation of Distribution Algorithms (EDAs) [2] seems most adequate since it models the exact spatial locations of the best candidate solutions.

EDAs are the class of Evolutionary Algorithms (EAs) that search for the optima by building up a probabilistic model of the most promising regions in the vector space. They typically begin with a zero knowledge or a very rough estimation concerning the expected location of the optimum. Later on, the more exact model of the landscape is acquired iteratively by performing a random sampling in the area of interest and then narrowing down this area by cutting off the regions containing "the worst samples". Then, sampling it again and so on until convergence.

Univariate Marginal Distribution Algorithm (UMDA) [5] falls under the umbrella of EDAs. It estimates the distribution of the best individuals in the d-dimensional search space $(d > 1)$ however the mutual independence of all d coefficients is assumed in order to simplify the computation. Although UMDA was originally designed for discrete optimization problems, a modification of this algorithm that is capable of exploring the continuous search spaces (by replacing the original binomial distribution model with Gaussian) was introduced [4].

The main contribution of this paper is in applying a Markov chain predictor to (the continuous) UMDA[1] in order to improve the reactivity of the algorithm to the environmental changes in continuous DOPs. Also a population diversification into *exploring*, *exploiting* and *anticipating* fractions is proposed with the auto-adaptation mechanism for updating dynamically the sizes of these fractions.

In the next section the proposed prediction model is presented in details. Later on, the suggested modification of UMDA is introduced. In Section 4 the experimental results are discussed. Section 5 concludes the paper.

2   Prediction Model

The proposed prediction mechanism is based on a Markov chain model [6] — a tool that is frequently applied in statistics and data analysis. In the most general form a *Markov chain* can be formulated as a sequence of random variables $(X_n)_{n\in\mathbb{N}}$, i.e. $X_1, X_2, X_3, \ldots$, such that for all $n \in \mathbb{N}$

$$\mathbb{P}(X_{n+1} = x \mid X_n = x_n, \ldots, X_1 = x_1) = \mathbb{P}(X_{n+1} = x \mid X_n = x_n). \quad (1)$$

It means that the instance of X_{n+1} depends only on its predecessor X_n, regardless the remaining of the history trail.

A *discrete chain Markov model* can be defined as a tuple (S, T, λ), where $S = \{S_1, S_2, \ldots, S_m\} \subset \mathbb{R}^d$ (for $d \geq 1$) is a finite set of possible states (i.e. a domain of X_n), $T = [t_{ij}]$ is a transition matrix containing probabilities of switching from state S_i to S_j (for $1 \leq i, j \leq m$), and $\lambda = (\lambda_1, \lambda_2, \ldots, \lambda_m)$ is the initial probability distribution of all the states in S.

Starting from the prior knowledge, which in practice is often a *zero knowledge*, i.e. $\lambda = (1/m, 1/m, \ldots, 1/m)$, and by assuming that the variables X_n are sampled from some unknown stochastic process, the characterization of this process is built iteratively by observing instances of X_n and updating the transition

---

[1] For the remainder of this paper the continuous UMDA will be referred to as UMDA.

matrix T as follows. Let $C = [c_{ij}]$ be an intermediary matrix for counting the past transitions. For all $1 \leq i, j \leq m$ the coefficients c_{ij} hold the number of transitions from state S_i to S_j that occurred so far. After each such transition the value of c_{ij} is incremented. The matrix T is then built up with the row-vectors $T_i = [c_{i1}/c, c_{i2}/c, \ldots, c_{im}/c]$ for $i = 1, \ldots, m$, where $c = \sum_{j=1}^{m} c_{ij}$.

The assumption stated in Equation 1 implies that any row T_i of matrix T indeed estimate the probability distribution of switching from the state S_i to any of the possible states $S_j \in S$. As a consequence, the model of changes encoded in T can be used further for predicting the future value of X_n for at least one step ahead by simply picking up the most probable transition originating at the present state. This mechanism is typically referred to as a *Markov chain predictor*.

In this paper a Markov chain predictor is used for anticipating the most probable future landscapes in the changing environment. It is assumed that the environmental changes follow some unknown yet repeatable pattern, i.e. the same sequences of transitions recur many times, which is substantial for training the transition matrix correctly.

The fundamental aspect in applying a Markov chain predictor is the definition of S. Ideally, the states $S_1, S_2, \ldots, S_m$ should play the role of *snapshots*, i.e. the exact models of the landscape at the given time. However, when dealing with optimization problems the main impact is on localizing the optimum, thus the quality of the model for the remaining part of the search space is negligible.

In the proposed approach, the states of the environment are characterized with Gaussian distributions that model the spatial locations of best candidate solutions in the search space. For all $i = 1, \ldots, m$ holds $S_i \equiv (\mu_i, \Sigma_i)$ where $\mu_i \in \mathbb{R}^d$ is a mean vector and $\Sigma_i = [\sigma_{kl}]_{1 \leq k, l \leq d}$ is a covariance matrix. These parameters are estimated and delivered by UMDA as described further in Section 3. Later on, the outputs of a Markov chain predictor are utilized by the same UMDA to direct the optimization towards the future promising regions.

3   Algorithm

The proposed modification of UMDA, named UMDA-MI[2], maintains the three fractions of candidate solutions — *exploring*, *exploiting* and *anticipating*. The first one is filled up with random immigrants sampled with the uniform distribution across the search space. It is aimed at identifying the newly appearing optima located presumably beyond the present area of interest.

Let Φ and $\tilde{\Phi}$ be Gaussian distributions. The remaining two fractions are formed with the individuals sampled as follows. The exploiting fraction uses the present distribution model Φ that is updated iteratively during the run of UMDA-MI whereas the anticipating fraction uses the foreseen distribution model $\tilde{\Phi}$ obtained from the Markov chain predictor.

One of the key aspects of UMDA-MI is a selection of the proper sizes of the three fractions such that all of them could participate in tracking the moving

---

[2] The letter 'M' stands for a Markov chain predictor while 'I' for random immigrants.

Algorithm 1. Univariate Marginal Distribution Algorithm with the Markov Chain predictor (N_{gen} – number of generations, N_{sub} – number of subiterations)

Initialize estimation of distribution Φ_1 randomly
Initialize Markov Chain predictor by setting: $(S, T, \lambda) = (\emptyset, [], 1)$ and $\widetilde{\Phi}_1 = \Phi_1$
for $t = 1 \to N_{gen}$ **do**
    $\Phi_{t_1} = \Phi_t$
    Initialize fraction sizes: $size_{explore}$, $size_{exploit}$ and $size_{anticip}$
    for $k = 1 \to N_{sub}$ **do**
        $P_{explore} = \text{RandomImmigrants}(size_{explore})$
        $P_{exploit} = \text{GenerateFraction}(\Phi_{t_k}, size_{exploit})$
        $P_{anticip} = \text{GenerateFraction}(\widetilde{\Phi}_t, size_{anticip})$
        $P_{t_k} = P_{explore} \cup P_{exploit} \cup P_{anticip}$
        $\text{Evaluate}(P_{t_k})$
        $\text{UpdateEstimation}(\Phi_{t_k}, P_{t_k})$
        $\text{UpdateFractionSizes}(P_{explore}, P_{exploit}, P_{anticip})$
    end for
    $S = S \cup \{\Phi_{t_k}\}$
    if $\text{size}(S) = \text{max-size}(S)$ **then**
        Find a pair $\{S_i, S_j\}$ of the most similar states in the set S
        $\overline{S_{ij}} = \text{AverageState}(\{S_i, S_j\})$
        Replace $\{S_i, S_j\}$ with $\overline{S_{ij}}$ in the set S
    end if
    Update transition matrix T
    Assign Φ_{t+1} to the most similar state to Φ_{t_k} in the set S
    $\widetilde{\Phi}_{t+1} = \text{PredictNextState}(\Phi_{t+1})$
end for

---

optima without dominating the whole process. Note that too strong exploring fraction (parameter $size_{explore}$) may lead to the excessive diversification of candidate solutions. On the other hand, too many exploiting individuals (parameter $size_{exploit}$) would significantly slow down the convergence while the surplus of anticipating ones (parameter $size_{anticip}$) may result in loosing the track of moving optima in case of misleading forecasts. The exact values of $size_{explore}$, $size_{exploit}$ and $size_{anticip}$ may either be the predefined constants or the self-adapted variables, yet they are required to sum up to the population size $M > 0$.

Algorithm 1 presents the main loop of UMDA-MI. It begins with a random selection of the distribution $\Phi_1 = (\mu_1, \Sigma_1)$ and defining the empty chain predictor $(S, T, \lambda) = (\emptyset, [], 1)$. The anticipated distribution $\widetilde{\Phi}_1$ is set to Φ_1 as for UMDA-MI it is established that in case of no clear forecast about the next environmental change, the best prediction available is in fact the present distribution Φ_t.

The algorithm runs for $N_{gen} > 0$ generations, each of which is split into $N_{sub} > 0$ intermediary steps called *subiterations*. Every k-th subiteration (for $k = 1, \ldots, N_{sub}$) includes generating the three fractions of candidate solutions as follows. The exploring fraction is filled with $size_{explore}$ random individuals, while the exploiting and anticipating fractions are formed with $size_{exploit}$ and $size_{anticip}$ candidate solutions sampled with the distribution models Φ_{t_k} and $\widetilde{\Phi}_{t_k}$

Algorithm 2. UpdateFractionSizes($P_{explore}, P_{exploit}, P_{anticip}$) step ($M$ – population size)

$(best, medium, worst) = \text{RankFractions}(P_{explore}, P_{exploit}, P_{anticip})$

$dist_{medium} = |\text{BestEvaluation}(P_{best}) - \text{BestEvaluation}(P_{medium})|$

$dist_{worst} = |\text{BestEvaluation}(P_{best}) - \text{BestEvaluation}(P_{worst})|$

$size_{best} = \min \{size_{max}, size_{best} + \varepsilon\}$

$size_{medium} = \max \left\{ size_{min}, (M - size_{best}) \cdot \frac{dist_{worst}}{dist_{worst} + dist_{medium}} \right\}$

$size_{worst} = M - size_{best} - size_{medium}$

---

respectively. Next, all the fractions are grouped together and evaluated. Among them the top $0 < M_{best} < M$ individuals $p_1, p_2, \ldots, p_{M_{best}} \in \mathbb{R}^d$ are filtered out and utilized for updating the estimation of distribution $\Phi_{t_k} = (\mu_{t_k}, \Sigma_{t_k})$ using the following formula

$$\mu_{t_k} = \sum_{i=1}^{M_{best}} \frac{p_i}{M_{best}}, \qquad \sigma_{t_k} = \frac{\sqrt{\sum_{i=1}^{M_{best}} (p_i - \mu_{t_k})^2}}{M_{best}}.$$

Unless the fraction sizes are fixed constants, UMDA-MI sets all of them to $M/3$ at the beginning of each generation and then updates them automatically N_{sub} times, i.e. once per subiteration. The updating rule is presented in Algorithm 2. It begins with finding a single best individual per fraction as its representative. Then, all the three fractions are given labels adequate to the fitness of their respective representatives. The fraction containing the best representative is labeled *best*, the second best is labeled *medium* and the last one — *worst*. Next, the size of the *best* fraction is increased by the small constant $\varepsilon > 0$. The remaining $M - size_{best}$ "vacant slots" are disposed between the *medium* and *worst* fractions proportionally to the differences in fitness of their representatives and the representative of the *best* fraction. As it was mentioned, all the three sizes must sum up to M. They are also restricted to the range $[size_{min}, size_{max}]$ where $0 < size_{min} < size_{max} < M$ in order to prevent from the excessive domination of a certain fraction causing the exclusion of the others.

After completing all of the N_{sub} subiterations, the Markov chain predictor is launched (once per generation). Firstly, it adds the obtained distribution Φ_{t_k} to the set of possible states S. Later, if the number of elements of S (after extending) reaches the predefined maximum size, the following replacement procedure is executed. Among all the elements in S a single pair of the two most similar states $\{S_i, S_j\} \subset S$ is identified and replaced with their average $\overline{S_{ij}}$. In order to find the most similar state to S_i, for all $i = 1, \ldots, \text{size}(S)$ a number of random samples is generated according to S_i and compared with $S \setminus \{S_i\}$. A Gaussian with the highest response to these samples is selected as S_j. In the case when S_i or S_j is the present state of environment, the state $\overline{S_{ij}}$ takes over its role.

The transition matrix T is built and maintained using the intermediary counting matrix C as described in Section 2. However, each time the two states S_i and S_j are unified, the corresponding j-th row and j-th column of C are removed while their values are added to the respective elements of i-th row and i-th column.

Finally, the most probable future state according to T is picked and used as the distribution model for the next anticipating fraction $\widehat{\Phi}_{t+1}$.

4   Experimental Results

Experiments were performed on the DOPs generated with the Dynamic Composition Benchmark Generator (DCBG) [3] which rotates and aggregates the functions given as input in a time-dependent manner following the formula

$$F^{(t)}(x) = \sum_{j=1}^{m} w_j \left[f_j \left(\frac{x - O_j(t) + O_j^{old}}{\lambda_j M_j} \right) + H_j(t) \right],$$

where $x = (x_1, \ldots, x_d)$, M_j are the orthogonal rotation matrices for the mixture components f_j, $O_j(t)$ are the optima of the rotated f_j functions at the time t, O_j^{old} are the optima of the original f_j functions, w_j and λ_j are the weighting and stretching factors (respectively) and $H_j(t)$ are the peak heights. The detailed description of the above parameters can be found in [3].

During the experiments, the following component functions [3] were used

(1) Sphere function: $f(x) = \sum_{i=1}^{d} x_i^2$

(2) Rastrigin function: $f(x) = \sum_{i=1}^{d} (x_i^2 - 10 \cos(2\pi x_i) + 10)$

(3) Griewank function: $f(x) = \frac{1}{4000} \sum_{i=1}^{d} x_i^2 - \prod_{i=1}^{d} \cos(\frac{x_i}{\sqrt{i}}) + 1$

(4) Ackley function: $f(x) = -20 \exp\left[-0.2 \sqrt{\frac{1}{d} \sum_{i=1}^{d} x_i^2} \right]$
$$- \exp\left[\frac{1}{d} \sum_{i=1}^{d} \cos(2\pi x_i) \right] + 20 + e$$

All the examined algorithms were run for $N_{gen} = 100$ generations, each of which contained $N_{sub} = 2, 5, 10$ or 20 subiterations. For every benchmark problem the environment was changing recurrently (which means that the same landscape recurred from time to time) in a synchronous manner — one change per generation. The population size (or its equivalent in EDA) M was set to 100, thus the changes took place every $N_{sub} \cdot 100$ evaluations. The two variants of dimensionality of each problem were considered: $d = 5$ and $d = 10$.

Regarding the Markov chain predictor: the size for the set of states S was limited to 10 and (for the auto-adaptation mechanism) the updating step $\varepsilon = 10$ was used with the lower and upper bounds: $size_{min} = 10$ and $size_{max} = 80$.

The proposed algorithm UMDA-MI was compared with the Simple Genetic Algorithm with a re-initialization after each environmental change (SGA-R) and continuous UMDA. The results presented as UMDA-M are the best scores reached by the separate runs of UMDA-MI with the fixed fraction sizes set to $size_{explore} = 0$ and each of the variants among $size_{anticip} = 0, 10, 20, \ldots, 90$. Analogously, UMDA-I are the best scores among $size_{explore} = 0, 10, 20, \ldots, 90$ and $size_{anticip} = 0$. The algorithms were compared with the best-of-generation measure that takes into account the best solution found just before the next environmental change.

Table 1 summarizes the results averaged over 50 independent runs. Note, that although UMDA-I always outperformed UMDA-M, however it was UMDA-MI that scored the best results in nearly all the cases due to the adaptive balancing of the three fraction sizes.

The evaluation of best individuals during the sample runs of SGA-R, UMDA and UMDA-MI for the composition of Sphere's functions is illustrated on Figure 1. It is evident that after a short period (≈ 12 generations) the Markov chain

(a) SGA-R

(b) UMDA

(c) UMDA-MI

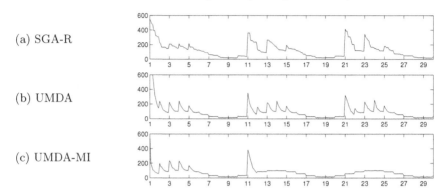

Fig. 1. The evaluation of best individuals during the sample runs of SGA-R, UMDA and UMDA-MI for the composition of Sphere's functions ($d = 10, N_{sub} = 10$)

Table 1. The best-of-generation results averaged over 50 independent runs for the compositions of Sphere, Rastrigin, Griewank and Ackley functions

Dimensions		5				10			
Function	N_{sub}	2	5	10	20	2	5	10	20
	SGA-R	216.0	149.7	90.4	59.7	441.0	350.2	246.5	129.0
	UMDA	129.6	92.8	78.1	71.5	151.5	107.7	89.2	77.5
Sphere	UMDA-M	120.3	79.4	63.5	62.4	143.7	97.4	75.4	69.1
	UMDA-I	102.9	78.7	67.1	61.2	120.9	97.4	81.9	69.9
	UMDA-MI	**94.4**	**62.1**	**57.7**	**56.5**	**109.8**	**77.4**	**62.2**	**59.0**
	SGA-R	586.3	438.1	350.0	312.1	845.6	717.1	621.4	585.8
	UMDA	560.3	429.6	438.8	354.8	713.1	608.6	533.9	462.0
Rastrigin	UMDA-M	533.1	418.1	390.7	354.8	713.1	608.6	533.9	462.0
	UMDA-I	**335.3**	**290.7**	**256.4**	**220.4**	**517.3**	**467.6**	**429.9**	**389.5**
	UMDA-MI	370.9	312.1	263.7	229.5	532.9	488.0	441.7	409.2
	SGA-R	350.9	272.7	202.3	159.9	522.8	437.7	335.7	224.3
	UMDA	224.5	188.4	153.5	132.2	234.6	201.7	176.7	158.2
Griewank	UMDA-M	222.9	177.7	144.2	120.6	231.1	187.4	164.6	150.7
	UMDA-I	**203.5**	172.9	144.9	122.7	214.8	189.4	170.9	153.2
	UMDA-MI	203.9	**160.7**	**129.9**	**113.2**	**205.4**	**172.0**	**148.9**	**140.2**
	SGA-R	1687.4	1508.6	1176.4	665.5	1886.9	1831.2	1740.7	1507.2
	UMDA	1657.3	1613.6	1476.2	1364.6	1597.2	1587.4	1441.6	1343.2
Ackley	UMDA-M	1578.7	1254.0	1128.7	1178.3	1597.2	1452.1	1259.0	1248.3
	UMDA-I	1015.6	835.3	624.7	495.2	1119.5	979.1	837.9	692.5
	UMDA-MI	**965.9**	**742.3**	**485.6**	**403.9**	**1093.4**	**858.9**	**645.7**	**516.0**

predictor began recognizing correctly the environmental changes thus the locations where anticipating individuals were being injected became very accurate. That is why the plot in Figure 1(c) is rather flat compared to the remaining two.

5 Conclusions

The preliminary experiments revealed that the application of a Markov chain predictor together with the introduction of random immigrants into UMDA significantly improved the algorithm's reactivity to the recurrently changing environments. The comparison to SGA-R proved the superiority of a predictive approach over a re-initialization of the population after each environmental change. It is probable that the accuracy of a Markov chain predictor may deteriorate in other than recurrent types of changes. The future work should cover that as well as dealing with the presence of constraints.

References

1. Hatzakis, I., Wallace, D.: Dynamic multi-objective optimization with evolutionary algorithms: A forward-looking approach. In: Proceedings of the 8th Annual Conference on Genetic and Evolutionary Computation, pp. 1201–1208 (2006)
2. Larrañaga, P., Lozano, J.A.: Estimation of Distribution Algorithms: A New Tool for Evolutionary Computation. Kluwer, Norwel (2001)
3. Li, C., Yang, S., Nguyen, T.T., Yu, E.L., Yao, X., Jin, Y., Beyer, H.-G., Suganthan, P.N.: Benchmark generator for CEC 2009 competition on dynamic optimization, University of Leicester, University of Birmingham, Nanyang Technological University, Tech. Rep. (2008)
4. Liu, X., Wu, Y., Ye, J.: An Improved Estimation of Distribution Algorithm in Dynamic Environments. In: Proceedings of the IEEE Fourth International Conference on Natural Computation (ICNC 2008), vol. 6, pp. 269–272 (2008)
5. Mühlenbein, H., Paaß, G.: From recombination of genes to the estimation of distributions I. Binary parameters. In: Ebeling, W., Rechenberg, I., Voigt, H.-M., Schwefel, H.-P. (eds.) PPSN 1996. LNCS, vol. 1141, pp. 178–187. Springer, Heidelberg (1996)
6. Norris, J.R.: Markov chains, No. 2008. Cambridge University Press (1998)
7. Rossi, C., Abderrahim, M., Díaz, J.C.: Tracking moving optima using Kalman-based predictions. Evolutionary Computation 16(1), 1–30 (2008)
8. Schönemann, L.: Evolution strategies in dynamic environments. In: Yang, S., Ong, Y.-S., Jin, Y. (eds.) Evolutionary Computation in Dynamic and Uncertain Environments. SCI, vol. 51, pp. 51–77. Springer, Heidelberg (2007)
9. Simões, A., Costa, E.: Prediction in evolutionary algorithms for dynamic environments. Soft Computing, 1–27 (2013)
10. Stroud, P.D.: Kalman-extended genetic algorithm for search in nonstationary environments with noisy fitness evaluations. IEEE Transactions on Evolutionary Computation 5(1), 66–77 (2001)
11. Tinós, R., Yang, S.: Evolutionary programming with q-Gaussian mutation for dynamic optimization problems. In: Proceedings of the IEEE Congress on Evolutionary Computation (CEC 2008), pp. 1823–1830 (2008)
12. van Hemert, J., van Hoyweghen, C., Lukschandl, E., Verbeeck, K.: A "futurist" approach to dynamic environments. In: GECCO EvoDOP Workshop, pp. 35–38 (2001)
13. Zhou, A., Jin, Y., Zhang, Q., Sendhoff, B., Tsang, E.: Prediction-Based Population Re-initialization for Evolutionary Dynamic Multi-objective Optimization. In: Obayashi, S., Deb, K., Poloni, C., Hiroyasu, T., Murata, T. (eds.) EMO 2007. LNCS, vol. 4403, pp. 832–846. Springer, Heidelberg (2007)

A Diversity-Adaptive Hybrid Evolutionary Algorithm to Solve a Project Scheduling Problem

Virginia Yannibelli[1,2] and Analía Amandi[1,2]

[1] ISISTAN Research Institute, UNCPBA University
Campus Universitario, Paraje Arroyo Seco, Tandil (7000), Argentina
[2] CONICET, National Council of Scientific and Technological Research, Argentina
{vyannibe,amandi}@exa.unicen.edu.ar

Abstract. In this paper, we address a project scheduling problem. This problem considers a priority optimization objective for project managers. This objective implies assigning the most effective set of human resources to each project activity. To solve the problem, we propose a hybrid evolutionary algorithm. This algorithm incorporates a diversity-adaptive simulated annealing algorithm into the framework of an evolutionary algorithm with the aim of improving the performance of the evolutionary search. The simulated annealing algorithm adapts its behavior according to the fluctuation of diversity of evolutionary algorithm population. The performance of the hybrid evolutionary algorithm on six different instance sets is compared with those of the algorithms previously proposed in the literature for solving the addressed problem. The obtained results show that the hybrid evolutionary algorithm significantly outperforms the previous algorithms.

Keywords: project scheduling, human resource assignment, multi-skilled resources, hybrid evolutionary algorithms, evolutionary algorithms, simulated annealing algorithms.

1   Introduction

A project scheduling problem implies defining feasible start times and feasible human resource assignments for project activities in such a way that the optimization objective, defined as part of the problem, is reached. Besides, to define human resource assignments, it is necessary to have knowledge about the effectiveness of the available human resources in relation to different project activities. This is mainly because the development and the results of an activity depend on the effectiveness of the human resources assigned to it [1, 2].

Until now, many different kinds of project scheduling problems have been formally described and addressed in the literature. However, to the best of our knowledge, only few project scheduling problems have considered human resources with different levels of effectiveness [3, 4, 5, 6, 7, 10], a fundamental aspect in real project scheduling problems. These problems state different assumptions about the effectiveness of the human resources.

E. Corchado et al. (Eds.): IDEAL 2014, LNCS 8669, pp. 412–423, 2014.

The project scheduling problem described in [6, 7] considers that the effectiveness of a human resource depends on various factors inherent to its work context (i.e., the activity to which the resource is assigned, the skill to which the resource is assigned within the activity, the set of human resources that has been assigned to the activity, and the attributes of the resource). This is a really significant aspect of the project scheduling problem described in [6, 7]. This is mainly because, in real project scheduling problems, the human resources usually have different effectiveness levels in relation to different work contexts [8, 1, 2] and, therefore, the effectiveness of a human resource needs to be considered in relation to its work context. To the best of our knowledge, the influence of the work context on the effectiveness of the human resources has not been considered in other project scheduling problems. Because of this, we consider that the project scheduling problem described in [6, 7] states valuable and novel assumptions about the effectiveness of the human resources in the context of project scheduling problems. In addition, this problem considers a priority optimization objective for managers at the early stage of scheduling. This objective involves assigning the most effective set of human resources to each project activity.

The project scheduling problem described in [6, 7] can be seen as a special case of the RCPSP (Resource Constrained Project Scheduling Problem) [9] and, therefore, is a NP-Hard problem. For this reason, exhaustive search algorithms only can solve very small instances of the problem in a reasonable period of time. Thus, heuristic search algorithms have been proposed in the literature to solve the problem: an evolutionary algorithm was proposed in [6], and a memetic algorithm was proposed in [7] that incorporates a hill-climbing algorithm into the framework of an evolutionary algorithm. The memetic algorithm is the best of both algorithms, as reported in [7].

In this paper, we address the project scheduling problem described in [6, 7] with the aim of proposing a better heuristic search algorithm to solve it. In this respect, we propose a hybrid evolutionary algorithm. This algorithm integrates a diversity-adaptive simulated annealing algorithm within the framework of an evolutionary algorithm. The behavior of the simulated annealing algorithm is adaptive based on the fluctuation of diversity of evolutionary algorithm population. The integration of a diversity-adaptive simulated annealing algorithm is meant to improve the performance of the evolutionary search [18, 19].

We propose the above-mentioned hybrid evolutionary algorithm based on the following reasons. The hybridization of evolutionary algorithms with other search and optimization techniques has been proven to be more effective than the classical evolutionary algorithms in the resolution of a wide variety of NP-Hard problems [18, 19, 20, 21] and, in particular, in the resolution of scheduling problems [22, 7, 21, 20, 19]. Besides, the hybridization of evolutionary algorithms with simulated annealing algorithms has been shown to be more effective than the hybridization of evolutionary algorithms with hill-climbing algorithms in the resolution of different NP-Hard problems [18, 19]. Thus, we consider that the proposed hybrid evolutionary algorithm could outperform the heuristic algorithms previously proposed to solve the problem.

The remainder of the paper is organized as follows. In Section 2, we give a brief review of published works that consider the effectiveness of human resources in the context of project scheduling problems. In Section 3, we describe the problem addressed in this paper. In Section 4, we present the proposed hybrid algorithm. In Section 5, we present the computational experiments carried out to evaluate the performance of the hybrid algorithm and an analysis of the results obtained. Finally, in Section 6 we present the conclusions of the present work.

2 Related Works

Different project scheduling problems described in the literature have considered the effectiveness of human resources. However, these project scheduling problems state different assumptions about the effectiveness of human resources. In this respect, only few project scheduling problems have considered human resources with different levels of effectiveness [3, 4, 5, 6, 7, 10], a fundamental aspect in real project scheduling problems. In this section, we focus the attention on analyzing the way in which the effectiveness of human resources is considered in project scheduling problems reported in the literature.

In [12, 11, 13, 14, 15, 16, 17], the authors describe multi-skill project scheduling problems. In these problems, each project activity requires specific skills and a given number of human resources (employees) for each required skill. Each available employee masters one or several skills, and all the employees that master a given skill have the same effectiveness level in relation to the skill (homogeneous levels of effectiveness in relation to each skill).

In [3], the authors describe a multi-skill project scheduling problem with hierarchical levels of skills. In this problem, given a skill, for each employee that masters the skill, an effectiveness level is defined in relation to the skill. Thus, the employees that master a given skill have different levels of effectiveness in relation to the skill. Then, each project activity requires one or several skills, a minimum effectiveness level for each skill, and a number of resources for each pair skill-level. This work considers that all sets of employees that can be assigned to a given activity have the same effectiveness on the development of the activity. Specifically, with respect to effectiveness, such sets are merely treated as unary resources with homogeneous levels of effectiveness.

In [4, 5], the authors describe multi-skill project scheduling problems. In these problems, most activities require only one employee with a particular skill, and each available employee masters different skills. Besides, the employees that master a given skill have different levels of effectiveness in relation to the skill. Then, the effectiveness of an employee in a given activity is defined by considering only the effectiveness level of the employee in relation to the skill required for the activity.

Unlike the above-mentioned problems, the project scheduling problem described in [6, 7] considers that the effectiveness of a human resource depends on various factors inherent to its work context. Then, for each human resource, it is possible to define different effectiveness levels in relation to different work contexts. This is a very important aspect of the project scheduling problem described in [6, 7]. This is because, in real project scheduling problems, the human resources have different effectiveness levels in relation to different work contexts [8, 1, 2] and, therefore, the effectiveness of a human resource needs to be considered in relation to its work context. Based on the above-mentioned, we consider that the project scheduling problem described in [6, 7] states valuable assumptions about the effectiveness of the human resources in the context of project scheduling problems.

3   Problem Description

In this paper, we address the project scheduling problem presented in [6, 7]. A description of this problem is presented below.

Suppose that a project contains a set A of N activities, $A = \{1, ..., N\}$, that has to be scheduled (i.e., the starting time and the human resources of each activity have to be defined). The duration, precedence relations and resource requirements of each activity are known.

The duration of each activity j is notated as d_j. Moreover, it is considered that pre-emption of activities is not allowed (i.e., the d_j periods of time must be consecutive).

Among some project activities, there are precedence relations. The precedence relations establish that each activity j cannot start until all its immediate predecessors, given by the set P_j, have completely finished.

Project activities require human resources – employees – skilled in different knowledge areas. Specifically, each activity requires one or several skills as well as a given number of employees for each skill.

It is considered that organizations and companies have a qualified workforce to develop their projects. This workforce is made up of a number of employees, and each employee masters one or several skills.

Considering a given project, set SK represents the K skills required to develop the project, $SK = \{1,..., K\}$, and set AR_k represents the available employees with skill k. Then, the term $r_{j,k}$ represents the number of employees with skill k required for activity j of the project. The values of the terms $r_{j,k}$ are known for each project activity.

It is considered that an employee cannot take over more than one skill within a given activity. In addition, an employee cannot be assigned more than one activity at the same time.

Based on the previous assumptions, an employee can be assigned different activities but not at the same time, can take over different skills required for an activity but not simultaneously, and can belong to different possible sets of employees for each activity.

As a result, it is possible to define different work contexts for each available employee. It is considered that the work context of an employee r, denoted as $C_{r,j,k,g}$, is made up of four main components. The first component refers to the activity j which r is assigned (i.e., the complexity of j, its domain, etc.). The second component refers to the skill k which r is assigned within activity j (i.e., the tasks associated to k within j). The third component is the set of employees g that has been assigned j and that includes r (i.e., r must work in collaboration with the other employees assigned to j). The fourth component refers to the attributes of r (i.e., his or her experience level in relation to different tasks and domains, the kind of labor relation between r and the other employees of g, his or her educational level in relation to different knowledge areas, his or her level with respect to different skills, etc.). It is considered that the attributes of r could be quantified from available information about r (e.g., curriculum vitae of r, results of evaluations made to r, information about the participation of r in already executed projects, etc.).

The four components described above are considered the main factors that determine the effectiveness level of an employee. For this reason, it is assumed that the effectiveness of an employee depends on all the components of his or her work context. Then, for each employee, it is possible to consider different effectiveness levels in relation to different work contexts.

The effectiveness level of an employee r, in relation to a possible context $C_{r,j,k,g}$ for r, is notated as $e_{rCr,j,k,g}$. The term $e_{rCr,j,k,g}$ represents how well r can handle, within activity j, the tasks associated to skill k, considering that r must work in collaboration with the other employees of set g. The mentioned term $e_{rCr,j,k,g}$ takes a real value over the range [0, 1]. The values of the terms $e_{rCr,j,k,g}$ inherent to each employee available for the project are known. It is considered that these values could be obtained from available information about the participation of the employees in already executed projects.

The problem of scheduling a project entails defining feasible start times (i.e., the precedence relations between the activities must not be violated) and feasible human resource assignments (i.e., the human resource requirements must be met) for project activities in such a way that the optimization objective is reached. In this sense, a priority objective is considered for project managers at the early stage of the project schedule design. The objective is that the most effective set of employees be assigned each project activity. This objective is modeled by Formulas (1) and (2).

Formula (1) maximizes the effectiveness of the sets of employees assigned to the N activities of a given project. In this formula, set S contains all the feasible schedules for the project in question. The term $e(s)$ represents the effectiveness level of the sets of employees assigned to project activities by schedule s. Then, $R(j,s)$ is the set of employees assigned to activity j by schedule s, and the term $e_{R(j,s)}$ represents the effectiveness level corresponding to $R(j,s)$.

Formula (2) estimates the effectiveness level of the set of employees $R(j,s)$. This effectiveness level is estimated calculating the mean effectiveness level of the employees belonging to $R(j,s)$.

For a more detailed discussion of Formulas (1) and (2), we refer to [6].

$$\max_{\forall s \in S} \left(e(s) = \sum_{j=1}^{N} e_{R(j,s)} \right) \tag{1}$$

$$e_{R(j,s)} = \frac{\sum_{r=1}^{|R(j,s)|} e_{rCr,j,k(r,j,s),R(j,s)}}{|R(j,s)|} \tag{2}$$

4   Hybrid Evolutionary Algorithm

To solve the problem, we propose a hybrid evolutionary algorithm. This algorithm incorporates a diversity-adaptive simulated annealing stage into the framework of an evolutionary algorithm. The behavior of the simulated annealing stage is adaptive based on the diversity within the underlying evolutionary algorithm population. The incorporation of a diversity-adaptive simulated annealing stage pursues two aims. When the evolutionary algorithm population is diverse, the simulated annealing stage behaves like an exploitation process to fine-tune the solutions in the population. When the evolutionary algorithm population starts to converge, the simulated annealing stage changes its behavior from exploitation to exploration in order to introduce diversity in the population and thus to prevent the premature convergence of the evolutionary search. Thus, the evolutionary search is augmented by the addition of one stage of diversity-adaptive simulated annealing [18, 19].

The general behavior of the hybrid evolutionary algorithm is described as follows. Given a project to be scheduled, the algorithm starts the evolution from a random initial population of feasible solutions. Each of these solutions codifies a feasible project schedule. Then, each solution of the population is decoded (i.e., the related schedule is built), and evaluated according to the optimization objective of the problem by a fitness function. As explained in Section 3, the objective is to maximize the effectiveness of the sets of employees assigned to project activities. In relation to this objective, the fitness function evaluates the assignments of each solution based on knowledge about the effectiveness of the employees involved in the solution.

After the solutions of the population are evaluated, a parent selection process is used to determine which solutions of the population will compose the mating pool. The solutions with the greatest fitness values will have more chances of being selected. Once the mating pool is composed, the solutions in the mating pool are paired, and a crossover process is applied to each pair of solutions with a probability P_c to generate new feasible ones. Then, a mutation process is applied to each solution generated by the crossover process, with a probability P_m. The mutation process is applied with the aim of introducing diversity in the new solutions generated by the crossover process. Then, a survival selection strategy is used to determine which solutions from the solutions in the population and the solutions generated from the mating pool will compose the new population. Finally, a diversity-adaptive simulated annealing algorithm is applied to the solutions of the new population. The simulated annealing algorithm behaves like either an exploitation process or an exploration process depending on the diversity of the population. Thus, the simulated annealing algorithm modifies the solutions of the population.

This process is repeated until a predetermined number of iterations is reached.

4.1   Encoding of Solutions and Fitness Function

In relation to the encoding of solutions, we used a representation proposed in [6]. Each solution is represented by two lists having as many positions as activities in the project. The first list is a standard activity list. This list is a feasible precedence list of the activities involved in the project (i.e., each activity j can appear on the list in any position higher than the positions of all its predecessors). The activity list describes the order in which activities shall be added to the schedule.

The second list is an assigned resources list. This list contains information about the employees assigned to each activity of the project. Specifically, position j on this list details the employees of every skill k assigned to activity j.

In order to build the schedule related to the representation, we used the serial schedule generation method proposed in [6]. In this method, each activity j is scheduled at the earliest possible time.

To evaluate a given encoded solution, a fitness function is used. This function decodes the schedule s related to the solution by using the serial method above-mentioned. Then, the function calculates the value of the term $e(s)$ corresponding to s (Formulas (1) and (2)). This value determines the fitness level of the solution. The term $e(s)$ takes a real value over $[0, ..., N]$.

To calculate the term $e(s)$, the function utilizes the values of the terms $e_{rCr,j,k,g}$ inherent to s (Formula 2). As was mentioned in Section 3, the values of the terms $e_{rCr,j,k,g}$ inherent to each available employee r are known.

4.2   Parent Selection, Crossover, Mutation and Survival Selection

To develop the parent selection, we used the process called deterministic tournament selection with replacement [18], where the parameter t defines the tournament size. This process is a variant of the traditional tournament selection process [18].

To develop the crossover and the mutation, we considered feasible processes for the representation used for the solutions. Thus, the crossover operator contains a feasible crossover operation for activity lists and a feasible crossover operation for assigned resources lists. For activity lists, we considered the one-point crossover proposed by Hartmann [22]. For assigned resources lists, we considered the traditional uniform crossover [18]. The crossover operator is applied with a probability of P_c.

The mutation operator contains a feasible mutation operation for activity lists and a feasible mutation operation for assigned resources lists. For activity lists, we considered the simple shift operator described in [22]. For assigned resources lists, we considered the traditional random resetting [18].

To develop the survival selection, we applied the process called deterministic crowding [18]. This process preserves the best solutions found by the hybrid evolutionary algorithm and preserves the diversity of the population. For a detailed description of the deterministic crowding process, we refer to [18].

4.3   Diversity-Adaptive Simulated Annealing Algorithm

A diversity-adaptive simulated annealing algorithm is applied to each solution of the population obtained by the survival selection process, except to the best solution of this population. The best solution of the population is maintained into this population.

The simulated annealing algorithm applied here is an adaptation of the one proposed in [23], and is described as follows.

The simulated annealing algorithm is an iterative process that starts considering a given encoded solution s for the problem and a given initial value T_0 for a parameter called temperature. In each iteration, a new solution s' is generated from the current solution s by a move operator. If the new solution s' is better than the current solution s (i.e., the fitness value of s' is higher than the fitness value of s), the current solution s is replaced by s'. Otherwise, if the new solution s' is worse than the current solution s, the current solution s is replaced by s' with a probability equal to $exp(-delta / T)$, where T is the current temperature value and $delta$ is the difference between the fitness value of s and the fitness value of s'. Thus, the probability of accepting a new solution that is worse than the current solution mainly depends on the temperature value. If the temperature is high, the acceptance probability is also high, and vice versa. The temperature value is decreased by a cooling factor at the end of each iteration. The described process is repeated until a predetermined number of iterations is reached.

Before applying the simulated annealing algorithm to the solutions of the population, the initial temperature T_0 is defined. In this case, the initial temperature T_0 is inversely proportional to the diversity of the population, and this diversity is represented by the spread of fitnesses within the population. Specifically, T_0 is calculated as detailed in Formula (3), where f_{max} is the maximal fitness into the population and f_{min} is the minimal fitness into the population. Therefore, when the population is very diverse, the value of T_0 is very low, and thus, the simulated annealing algorithm only accepts movements that improve the solutions to which it is applied, behaving like an exploitation process. When the population converges, the value of T_0 rises, and thus, the simulated annealing algorithm increases the probability of accepting worsening movements. A consequence of this is that the simulated annealing algorithm will try to move away from the solutions to which it is applied, exploring the search space. Eventually, the diversity of the population will increase, and thus, the temperature T_0 will decrease. Based on the above-mentioned, the self-adaptation of the simulated annealing algorithm to either an exploitation or exploration behavior is governed by the diversity of the population.

$$T_0 = \frac{1}{\left| f_{max} - f_{min} \right|} \tag{3}$$

In relation to the move operator used by the simulated annealing to generate a new solution from the current solution, we considered a feasible move operator for the representation used for the solutions. Thus, the move operator contains a feasible move operation for activity lists and a feasible move operation for assigned resources lists. For activity lists, we considered a move operator called adjacent pairwise interchange [22]. For assigned resources lists, we considered a move operator that is a variant of the traditional random resetting [18]. In this variant, only one randomly selected position of the list is modified.

5   Computational Experiments

To develop the computational experiments, we used the six instance sets presented in [7]. Table 1 shows the main characteristics of these instance sets. Each instance of these six instance sets contains information about a number of activities to be planned and information about a number of employees available to develop the activities. For a detailed description of these instance sets, we refer to [7].

Each instance of the six instance sets has a known optimal solution with a fitness level $e(s)$ equal to N (N is the number of activities of the instance). These optimal solutions are considered here as references.

The hybrid evolutionary algorithm was run 20 times on each of the instances of the six instance sets. After each one of the 20 runs, the algorithm provided the best solution of the last population. To perform these runs, the algorithm parameters were set with the values shown in Table 2. The parameters were fixed thanks to preliminary experiments that showed that these values led to the best and most stable results.

Table 1. Characteristics of instance sets

Instance set	Activities per instance	Possible sets of employees per activity	Instances
j30_5	30	1 to 5	40
j30_10	30	1 to 10	40
j60_5	60	1 to 5	40
j60_10	60	1 to 10	40
j120_5	120	1 to 5	40
j120_10	120	1 to 10	40

Table 2. Parameter values of the hybrid evolutionary algorithm

Parameter	Value
Population size	90
Number of generations	300
Parent Selection process	
t (tournament size)	5
Crossover process	
Crossover probability P_c	0.8
Mutation process	
Mutation Probability P_m	0.05
Simulated annealing algorithm	
Number of iterations	25
Cooling factor	0.9

Table 3 reports the results obtained by the experiments. The second column reports the average percentage deviation from the optimal solution (Av. Dev. (%)) for each instance set. The third column reports the percentage of instances for which the value of the optimal solution is achieved at least once among the 20 generated solutions (Optimal (%)).

Table 3. Results obtained by the computational experiments

Instance set	Av. Dev. (%)	Optimal (%)
j30_5	0	100
j30_10	0	100
j60_5	0	100
j60_10	0	100
j120_5	0.64	100
j120_10	0.8	100

The results obtained by the algorithm for j30_5, j30_10, j60_5 and j60_10 indicate that the algorithm has reached an optimal solution in each of the 20 runs carried out on each instance of the sets.

The Av. Dev. (%) obtained by the algorithm for j120_5 and j120_10 is greater than 0%. Taking into account that the instances of j120_5 and j120_10 have known optimal solutions with a fitness level $e(s)$ equal to 120, we analyzed the meaning of the average deviation obtained for each one of these sets. In the case of j120_5 and j120_10, average deviations equal to 0.64% and 0.8% indicate that the average value of the solutions reached by the algorithm is 119.232 and 119.04 respectively. Therefore, we may state that the algorithm has obtained very high quality solutions for the instances of j120_5 and j120_10.

In addition, the Optimal (%) obtained by the algorithm for j120_5 and j120_10 is 100%. These results indicate that the algorithm has reached an optimal solution in at least one of the 20 runs carried out on each instance of the sets.

5.1 Comparison with a Competing Algorithm

To the best of our knowledge, only two algorithms have been previously proposed for solving the addressed problem: a classical evolutionary algorithm [6], and a classical memetic algorithm [7] that incorporates a hill-climbing algorithm into the framework of an evolutionary algorithm. According to the experiments reported in [7], both algorithms have been evaluated on the six instance sets presented in Table 1 and have obtained the results that are shown in Table 4. Based on the results in Table 4, the memetic algorithm is the best of both algorithms. Below, we compare the performance of the memetic algorithm with that of the hybrid evolutionary algorithm.

The results in Table 3 and Table 4 indicate that the hybrid evolutionary algorithm and the memetic algorithm have reached the same effectiveness level (i.e., an optimal effectiveness level) on the first three instance sets (i.e., the less complex sets). However, the effectiveness level reached by the hybrid evolutionary algorithm on the last three instance sets (i.e., the more complex sets) is higher than the effectiveness level reached by the memetic algorithm on these sets. Thus, the performance of the hybrid evolutionary algorithm on the three more complex instance sets is better than that of the memetic algorithm. The main reason for this is that the simulated annealing algorithm in the hybrid evolutionary algorithm adapts its behavior according to the fluctuation of population diversity and thus prevents the premature convergence of the evolutionary search, whereas the hill-climbing algorithm in the memetic algorithm usually leads to a premature convergence of the evolutionary search. Thus, the hybrid evolutionary algorithm can reach better solutions than the memetic algorithm on the more complex instance sets.

Table 4. Results obtained by the algorithms previously proposed for the addressed problem

Instance set	Evolutionary algorithm [6]		Memetic algorithm [7]	
	Av. Dev. (%)	Optimal (%)	Av. Dev. (%)	Optimal (%)
j30_5	0	100	0	100
j30_10	0	100	0	100
j60_5	0.42	100	0	100
j60_10	0.59	100	0.1	100
j120_5	1.1	100	0.75	100
j120_10	1.29	100	0.91	100

6   Conclusions

We proposed a hybrid evolutionary algorithm to solve the project scheduling problem described in [6, 7]. This algorithm incorporates a diversity-adaptive simulated annealing algorithm into the framework of an evolutionary algorithm to improve the performance of the evolutionary search. The behavior of the simulated annealing algorithm is adaptive to either an exploitation or exploration behavior based on the fluctuation of population diversity. The presented computational experiments show that the hybrid evolutionary algorithm significantly outperforms the algorithms previously proposed for solving the addressed problem.

In future works, we will evaluate other parent selection, crossover, mutation and survival selection processes. Besides, we will investigate the incorporation of other search and optimization techniques into the framework of the evolutionary algorithm.

References

1. Heerkens, G.R.: Project Management. McGraw-Hill (2002)
2. Wysocki, R.K.: Effective Project Management, 3rd edn. Wiley Publishing (2003)
3. Bellenguez, O., Néron, E.: Lower Bounds for the Multi-skill Project Scheduling Problem with Hierarchical Levels of Skills. In: Burke, E., Trick, M. (eds.) PATAT 2004. LNCS, vol. 3616, pp. 229–243. Springer, Heidelberg (2005)
4. Hanne, T., Nickel, S.: A multiobjective evolutionary algorithm for scheduling and inspection planning in software development projects. European Journal of Operational Research 167, 663–678 (2005)
5. Gutjahr, W.J., Katzensteiner, S., Reiter, P., Stummer, C., Denk, M.: Competence-driven project portfolio selection, scheduling and staff assignment. Central European Journal of Operations Research 16(3), 281–306 (2008)
6. Yannibelli, V., Amandi, A.: A knowledge-based evolutionary assistant to software development project scheduling. Expert Systems with Applications 38(7), 8403–8413 (2011)

7. Yannibelli, V., Amandi, A.: A Memetic Approach to Project Scheduling that Maximizes the Effectiveness of the Human Resources Assigned to Project Activities. In: Corchado, E., Snášel, V., Abraham, A., Woźniak, M., Graña, M., Cho, S.-B. (eds.) HAIS 2012, Part I. LNCS, vol. 7208, pp. 159–173. Springer, Heidelberg (2012)
8. Barrick, M.R., Stewart, G.L., Neubert, M.J., Mount, M.K.: Relating member ability and personality to work-team processes and team effectiveness. Journal of Applied Psychology 83, 377–391 (1998)
9. Blazewicz, J., Lenstra, J., Rinnooy Kan, A.: Scheduling Subject to Resource Constraints: Classification and Complexity. Discrete Applied Mathematics 5, 11–24 (1983)
10. Yannibelli, V., Amandi, A.: Project scheduling: A multi-objective evolutionary algorithm that optimizes the effectiveness of human resources and the project makespan. Engineering Optimization 45(1), 45–65 (2013)
11. Bellenguez, O.: A reactive approach for the multi-skill Project Scheduling Problem. In: 7th International Conference on the Practice and Theory of Automated Timetabling (PATAT 2008), pp. 1–4. Université de Montréal, Montréal (2008)
12. Bellenguez, O., Néron, E.: A branch-and-bound method for solving multi-skill project scheduling problem. RAIRO - Operations Research 41(2), 155–170 (2007)
13. Drezet, L.E., Billaut, J.C.: A project scheduling problem with labour constraints and time-dependent activities requirements. International Journal of Production Economics 112, 217–225 (2008)
14. Li, H., Womer, K.: Scheduling projects with multi-skilled personnel by a hybrid MILP/CP benders decomposition algorithm. Journal of Scheduling 12, 281–298 (2009)
15. Valls, V., Pérez, A., Quintanilla, S.: Skilled workforce scheduling in service centers. European Journal of Operational Research 193(3), 791–804 (2009)
16. Aickelin, U., Burke, E., Li, J.: An Evolutionary Squeaky Wheel Optimization Approach to Personnel Scheduling. IEEE Transactions on Evolutionary Computation 13(2), 433–443 (2009)
17. Heimerl, C., Kolisch, R.: Scheduling and staffing multiple projects with a multi-skilled workforce. OR Spectrum 32(4), 343–368 (2010)
18. Eiben, A.E., Smith, J.E.: Introduction to Evolutionary Computing, 2nd edn. Springer (2007)
19. Rodriguez, F.J., García-Martínez, C., Lozano, M.: Hybrid Metaheuristics Based on Evolutionary Algorithms and Simulated Annealing: Taxonomy, Comparison, and Synergy Test. IEEE Transactions on Evolutionary Computation 16(6), 787–800 (2012)
20. Talbi, E.-G. (ed.): Hybrid Metaheuristics. SCI, vol. 434. Springer, Heidelberg (2013)
21. Blum, C., Puchinger, J., Raidl, G.R., Roli, A.: Hybrid metaheuristics in combinatorial optimization: A survey. Applied Soft Computing 11(6), 4135–4151 (2011)
22. Kolisch, R., Hartmann, S.: Experimental Investigation of Heuristics for Resource-Constrained Project Scheduling: An Update. European Journal of Operational Research 174, 23–37 (2006)
23. Yannibelli, V., Amandi, A.: Hybridizing a multi-objective simulated annealing algorithm with a multi-objective evolutionary algorithm to solve a multi-objective project scheduling problem. Expert Systems with Applications 40(7), 2421–2434 (2013)

Computing Platforms for Large-Scale Multi-Agent Simulations: The Niche for Heterogeneous Systems

Worawan Marurngsith

Department of Computer Science, Faculty of Science and Technology,
Thammasat University 99 Phaholyothin Road, Pathum Thani, 12120, Thailand
wdc@cs.tu.ac.th

Abstract. A rapid shift of computing platforms for large-scale multi-agent simulation (MAS) towards higher parallelism using tools from simulation frameworks has made the impact of MAS logic on performance become transparent. This limits the perspective of developing MAS logic towards a sustained high performance direction. This paper presents a review of 62 works related to large-scale MASs published on Scopus from 2010 – April 2014. The review was compiled in three aspects (a) the recent direction of computing platforms, (b) the state of the art in simulation frameworks, and (c) the synergy between MAS logic and scalable performance achieved. The results confirm that the nature of dynamic interactions of autonomous agents among themselves, groups, and environments has most impact on performance of computing platforms. The analysis of the results shows the correspondence between the nature of MAS logic and the execution model of heterogeneous systems. This features heterogeneous systems as a promising platform for the even larger-scale MASs in the future.

Keywords: agent-based simulation, multi-agent simulation, platform, simulation framework, review.

1   Introduction

Multi-agent simulation (MAS) is a powerful technique for predicting consequences or finding out patterns, emerged from dynamic interactions of multiple autonomous entities, of specific complex phenomena. The technique is pure bottom-up. That is, the overall behaviour of a target system is derived from unique microscopic actions of autonomous entities, so called *agents*. In MAS, agents can represent life entities, vehicles, digital devices, wearable systems, equipment, buildings *etc*. Conceptually most agents have five features [1-3]. First, they are *self-contained*, individual, heterogeneous, and can be uniquely identifiable. Second, agents are *autonomous*. They can make their own decisions without a central controller. Third, agents are *active*. They can be both 'reactive', *i.e.* having a sense of their surrounding or 'proactive', *i.e.* having goals to achieve and their behaviour will correspond to the goals they are set for. Forth, agents have a state that varies over time. Last, agents are social. They can have dynamic interactions with other agents. Consequently, their behaviour can be adapted in response to the interactions.

E. Corchado et al. (Eds.): IDEAL 2014, LNCS 8669, pp. 424–432, 2014.

When the more realistic and complex phenomena are of interest, MAS models are normally extended to *large-scale*[1]. The term large-scale indicates the degree of complexity of a model in many ways [1, 4] *i.e.*, (a) comprising millions of agents; (b) having complex rules and parameters; (c) covering the impact or complexity of target system to a large extent; or (d) covering a large spatial dimension. Large-scale MAS models have been applied as a research tool in many fields such as traffic[5], social science [6], emergency response [7], ambient intelligence [8], geography [9], and environmental study [10, 11]. To deal with model complexity, large-scale MASs requires a high demand on computing resources. Thus, model restructuring or exploiting the scalable performance of parallel computing resources have been popular solutions to keep up the model to the expected complexity [1].

Since modellers of large-scale MAS are professionals in various fields, the characteristics of models are diverse. Many legacy MAS simulation frameworks have been proposed to support MAS implementation on various parallel computing platforms *e.g.*, FLAME [12], JADE [13], Repast HPC[14], P-HASE [15]. These frameworks have successfully provided modellers the high performance, ease of use features for executing large-scale MASs. However as the number of frameworks, and simulation models has growing rapidly, the perspective of developing MAS logic towards a sustained high performance direction is still an open issue.

This paper presents a review of recent literatures related to large-scale MASs available on the Scopus bibliographic database from 2010 – April 2014. The aim of this review is to summary the current direction of MAS computing platforms, and to highlight the links between algorithms, logics and strategies used in large-scale MAS to the key characteristics of those platforms. Thus, the paper makes three contributions as follows:

(1)   The recent direction of computing platform for large-scale MAS and the state of the art in simulation frameworks are depicted (Section 2-3),

(2)   The synergy between MAS algorithms, logics, strategies and scalable performance achieved in various computing platforms is outlined (Section 4),

(3)   The perspective of developing MAS logic towards a sustained high performance direction is discussed (Section 5).

In this review, a list of literatures related to MASs published on the Scopus bibliographic database has been obtained. The search criteria was set to find the words 'multi-agent' or 'agent-based' with 'system' or 'simulation' in the article title, abstract, and keywords. Four document types were selected including, conference paper, article, review and article in press. The search results showed 605 literatures, which were then stored in a saved list[2]. The literatures in the saved list were screened using three steps. First, 164 papers having irrelevant titles were discards. Second, the abstracts of the remaining literatures were reviewed. The total of 341 papers addressed

---

[1] This class of MAS is also called massively MAS.
[2] The list of literatures saved contains the search results obtained on April 20, 2014.

irrelevant topics[3] was discarded. At the final step of screening, ten papers were discarded as their introduction and conclusion confirmed that they addressed irrelevant topics. The total of 90 literatures[4] remained was grouped into three categories *i.e.*, simulation techniques and platforms (21 papers), review papers (7 papers), and model implementations (62 papers). The key characteristics of the model implementation presented in 62 papers were captured. This is (a) the computing platform and simulation framework used, (b) the MAS strategies presented and (c) the scalable performance achieved.

From 62 papers presenting large-scale MAS models from 15 different fields, the results depict that, in average, 77 percent of large-scale MAS studies were based on parallel computing platforms (*i.e.*, high performance computers (HPC), multicores, distributed systems, and clusters). This follows by 15 percent of models which used graphical processing units (GPU) as their platforms. C++ and Java were dominant languages used. Three simulation frameworks were most popular. These are Repast, JADE and FLAME. The nature of dynamic interactions of autonomous agents among themselves, groups, and environments has most impact on performance of computing platforms. The analysis of the results shows that the nature of logic and strategies commonly used in large-scale MAS corresponds to the execution of model of heterogeneous systems (*i.e.* having CPUs work together with GPUs or any acceleration processors). This findings features heterogeneous systems as a promising platform for the even larger-scale MASs in the future.

2   Computing Platforms

Since the emergence of multicore era in 2001, the development of computing platforms both hardware and software has shifted towards maximising parallelism at every level. This impacts the rapid development of parallel programming library and system software to support higher level of parallelism. The computing platforms used in large-scale MAS clearly support this trend. From hardware perspective, the computing platforms used in literatures were grouped into four categories, *i.e.* (a) parallel computing platforms which comprises HPC/multicores or distributed systems/clusters, (b) Grid and Cloud computing, (c) GPUs or heterogeneous systems, (d) others. The Fig. 1 depicts the numbers of literatures which used each type of platforms. The information related to programming languages and implementation techniques of the reviewed MAS models is shown in Table 1. The literatures were grouped into 15 different fields *i.e.*, (1) traffic and transportation, (2) behavioural and social science, (3) disaster/climate/emergency/epidemic management, (4) computer science/digital forensic and software engineering, (5) ecology, (6) geography, (7) energy and power management, (8) engineering, (9) business/finance/management, (10) biology, (11) environmental science, (12) agriculture, (13) military, (14) geology, and (15) immunology.

---

[3] Irrelevant works are the models that are not large-scale MASs or those related to *multi-agent control system*, *mobile agent computing* or *multi-agent technology* used in federated control, distributed computing or business process, and *agents from medical subjects*.

[4] The complete list of literatures can be found at http://parlab.cs.tu.ac.th/MAS.

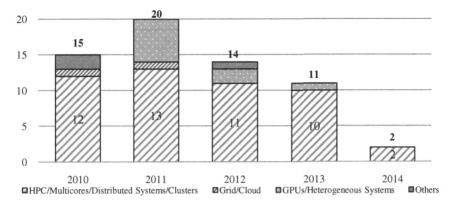

Fig. 1. Computing platforms used in large-scale MAS researches from 2010 – April 2014

The *HPC/multicores or distributed systems/clusters* group of platforms are dominant. From 2010 - 2013, 65 - 91 percent of the reviewed models used this type of hardware platforms for large-scale MASs experiments. The ratio of usage for HPC, multicores, distributed systems and clusters are quite equal. The platforms are used in both custom implementation and the model developed by using an MAS simulation framework. As show in Table 1, Java, C and C++ are dominant languages used in the MAS model development. These languages are combined with a parallel library such as Message Passing Interface (MPI). A few works have been implemented by using X10 [5], a partitioned global address space language developed at IBM. From all types of literatures reviewed, none of them have used OpenMP, the Intel Threading Building Block (TBB) library, or the Microsoft .NET Concurrency Runtime.

A few works in 2010 – 2011 mentioned the use of *Grid and Cloud* computing platform to support large-scale MASs. Wittek and Rubio-Campillo [6] showed that using Amazon EC2 Cloud platform for social-based research is viable. Despite the lower performance than HPC was observed, the lower cost per hour made it competitive.

The use of *GPUs or heterogeneous systems* for large-scale MASs is still a challenge. Since 2000, the proof-of-concept programs demonstrated that the use of GPUs could give dramatic speedups over central processing units (CPUs) for certain algorithms [16]. Maturity of general purpose languages and tools has made GPU platforms viable. Among several GPU languages (*e.g,* Brook, CUDA, OpenCL, C++ AMP, stream), CUDA offers the most advanced development tools. As shown in Table 1, many models used CUDA for performance improvement. Moser et al. [4] compared the performance of large-scale MASs implemented by CUDA and Cilk, the shared-memory parallel extension of C++ implemented in Intel C/C++ compilers. The research found that, using an artificial workload, more than 500 folds of speedup observed from the CUDA implementation, up to 36 times speedup on the Cilk. To accelerate the MAS performance, many simulation frameworks also offer features to support GPU execution (see Section 3).

Table 1. Information of literatures reviewed in this article

Fields	No. of Papers	Use for[1]		Implementation		Programming Languages			
		R	M	Cus-tom	Frame-works	Java	C/C++	Others	N/A
Traffic and Transportation	11	10	1	4	Aglobe, IBM, OMNET+, MATSim, FLAME	3	3	X10, C#, CUDA, MPI	2
Behavioural and Social Science	11	11	-	10	Jason	7	4	Map-reduce	-
Disaster, Climate, Emergency, Epidemic Management	11	11	-	8	Repast, RRAS, JADE, pDYN, GSAM	3	5	CUDA, MPI	-
Computer Science & SW Engineering	9	7	2	7	THEATRE, CDS	-	9	-	-
Ecology	4	4	-	4	Virtual Tawaf	2	2	CUDA	-
Geography	4	3	1	1	Reprint Of, MAG, MAGS	1	3	CUDA, MPI	-
Energy and Power Management	3	2	1	2	JADE	1	2	CUDA	-
Engineering	2	-	2	1	SBAP	-	2	MPI	-
Others	7	5	2	2	JADE,TIM-SIRS, I-ABM, CMSS, EvoDOE	2	5	JADE-Android	-
Total	**62**	**53**	**9**	**39**		**19**	**35**		

Note[1]: MAS is used as a research tool (R) or as a management tool (M).

A few literatures fall into the *others* group. Cui et al. [17] showed the effectiveness of using field-programmable gated arrays (FPGAs) to run a cellular model Conway's Game of Life. Interestingly, two works have effectively incorporated some part of their MAS models to run on Android *mobile devices* [10, 13]. The models have been implemented on JADE 4.1.1 platform which supports the execution of agents on Android. Despite it is not so popular to accelerate models on mobile devices due to power limitation; such devices can be competitive as they offer mobility, new parallel tools and rapid performance improvement. On Android, applications can be parallelised by interfacing with native codes implemented in OpenMP, OpenCL and MARE, a library for parallel computing on mobile from Qualcomm. Thus, mobile devices can be an attractive candidate platform for large-scale MASs in the future.

3   State of the Art in MAS Simulation Frameworks

Table 2 summarised the key features of four legacy MAS frameworks in terms of implementation approaches. The majority of the framework was implemented in C++.

The reason can be that there are many C++ libraries which support data and task parallelisation, and communication among parallel tasks. The implementation of all frameworks is based on object-oriented design. Consequently, MAS models developed on the frameworks are highly modular and extensible. All frameworks listed here provide support for both homogeneous and heterogeneous agents.

Techniques used for parallelising simulation models are grouped into two approaches as presented in [1] *i.e.*, an 'agent-parallel' and an 'environment-parallel' approach. Every framework implemented agents and environment in a parallel execution context. Thus they can offer high level of parallelism in the model. Almost all frameworks target for executing MAS models on heterogeneous platform (*i.e.*, the cooperation on model execution on HPC and another platform) with GPU being a strong candidate.

Table 2. Legacy simulation frameworks supporting large-scale MASs

Simulation Framework	Language used for implementation	Approach		Target Exe. Platform	Key Features
		Agent-parallel	Environmental-parallel		
JADE [18]	Java	Java Threads	Java Threads	Hetero-genous	Android enabled, plenty extensions
FLAME-GPU [12]	Java, C++, CUDA	GPU Threads	Thread Blocks	Hetero-geneous	XML to X-machine formal
RePast (HPC) [14]	C++, MPI	within &between processes	inter-process comm.	HPC	Logo language, support MMPs
P-HASE [15]	C++, OpenCL	GPU work items	CPU Threads	Hetero-geneous	using GPU with CPUs

JADE has many strong features *e.g.*, (a) it can be used as mobile agent system, not only for simulation; (b) it supports Android devices which extends the scope of target systems for modellers; (c) it has plenty of extensions available for various types of MAS models. JADE implementation complies with the FIPA specifications. It also offers a set of tools that supports the debugging and deployment phase.

FLAME-GPU is a variant of the FLAME platforms. It translates modeller's XML models into the internal X-machine formal. The formal is then used to generate the corresponding C code for the simulation executable on CPU and Nvidia GPUs. Using X-machine formal, FLAME-based models can be verified by using a state machine. Thus modellers exploit the GPU performance without requiring any specific knowledge of the underlining platforms.

Repast HPC is designed especially to support large-scale, distributed platforms. The framework can support the execution of models on massively parallel computers (MPP) having more than 300,000 nodes. It has a compilation toolchain to translate the model definition, in Logo, into the corresponding C++ codes. Thus, it offers the flexibility for modellers to develop MAS models with an easy to use language.

P-HASE is a framework designed for developing cellular models. It offers the feature to add a GPU tag inside a model behaviour script written in a C++-like language. The code inside the GPU tags is translated into OpenCL kernel functions. Using this technique, the framework can exploit the performance of CPUs with any OpenCL-enabled GPUs.

4   Synergy between MAS Logic and Scalable Performance

The meta-analysis method [19] was used to extract, from unrelated literatures, the synergy between the MAS logics and the scalable performance of the underlining platforms. Unfortunately, most of the papers only present the results related to their proposed models; only 14 papers [7, 11, 17, 20-30] (out of 62 papers) present the usable performance results. These results were collected, and organised as pairs of model logic and the speedup obtained (*e.g.*, the number of agents and obtained speedup). The Pearson's correlation was calculated to quantify the impact of the MAS logics to the speedup, in both positive and negative ways. When many papers have the same types of results, the median correlation value was used as the representative.

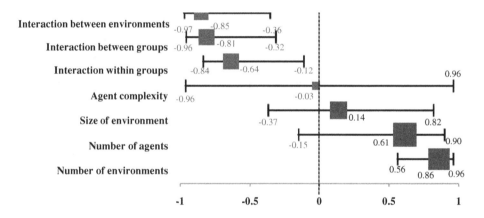

Fig. 2. Standard effects of MAS logic to speedup (correlation with 95% confidence interval)

Fig. 2 depicts the forest plot showing the effects of seven logics extracted from the meta-analysis and their impact on speedup (the performance of the platforms). The size of the square markers represents relative weight calculated by the number of data pairs collected from the literatures. Interestingly, by looking at the standard effects, two observations can be drawn *i.e.*, (a) all types of interactions have negative impact on speedup and (b) increasing number of agents or environments can strongly benefit speedup. The reason for the latter case can be that isolated agents are embarrassingly parallel (*i.e.* having intrinsic parallelism without any communication overhead). Increasing the size of environment seemed to help improving speedup despite to a lower degree. This suggested that when scheduling well, large size of environments can be handled in most computing platforms.

Interactions among agents at every level caused the platform to suffer from speedup loss. The interactions among environments and groups were observed as the key factors that cause performance loss in large-scale MAS models. Thus, to develop MAS logics towards a sustained high performance direction attention should be paid on optimising these factors.

The impact of agent complexity on speedup is still unclear. The standard effect shows slightly negative impact (-0.03). However as the number of data pairs is very small (4), the 95% confidence interval of the standard effect is nearly ±1. Thus more detail studies are required, and no conclusion could yet be drawn on this factor.

5   Conclusion and Perspectives

In this paper, the review of literatures featuring large-scale MAS models from 2010-April 2014 are presented. The review highlights parallel computing platforms (such as multicores, distributed systems, and clusters) as the dominant platform used in large-scale MAS studies, followed by GPUs. All models implemented by using object-oriented languages, mostly C++ and Java. Three popular simulation frameworks, Repast, JADE and FLAME, show the direction to support model execution on heterogeneous platform. The reviewed results show that the nature of logic and strategies commonly used in large-scale MAS corresponds to the execution of model of heterogeneous system. This makes the heterogeneous systems as a promising platform for the even larger-scale MASs in the future.

References

1. Parry, H.R., Bithell, M.: Large Scale Agent-Based Modelling: A Review and Guidelines for Model Scaling. In: ABMs of Geographical Systems (2012)
2. Hawe, G.I., et al.: Agent-based simulation for large-scale emergency response: A survey of usage and implementation. ACM Comput. Surv. 45(1), 1–51 (2012)
3. Crooks, A., Heppenstall, A.: Introduction to Agent-Based Modelling. In: Agent-Based Models of Geographical Systems, pp. 85–105. Springer, Netherlands (2012)
4. Moser, D., et al.: Comparing Parallel Simulation of Social Agents Using Cilk and OpenCL. In: IEEE/ACM DS-RT 2011, pp. 88–97. IEEE Computer Society (2011)
5. Osogami, T., et al.: Toward simulating entire cities with behavioral models of traffic. IBM Journal of Research and Development 57(5) (2013)
6. Wittek, P., Rubio-Campillo, X.: Scalable agent-based modelling with cloud HPC resources for social simulations (2012)
7. Parker, J., Epstein, J.M.: A distributed platform for global-scale agent-based models of disease transmission. ACM T. Model Comput. S. 22(1) (2011)
8. Serrano, E., Botia, J.: Validating ambient intelligence based ubiquitous computing systems by means of artificial societies. Information Sciences 222, 3–24 (2013)
9. Tang, W., Bennett, D.A.: Reprint of: Parallel agent-based modeling of spatial opinion diffusion accelerated using GPU. Ecological Modelling 229 (2012)
10. Mocanu, A., Ilie, S., Badica, C.: Ubiquitous multi-agent environmental hazard management (2012)

11. Cicirelli, F., et al.: Efficient environment management for distributed simulation of large-scale situated multi-agent systems. Concurr. Comp.-Pract. E (2014)
12. Richmond, P., et al.: High performance cellular level agent-based simulation with FLAME for the GPU. Briefings in Bioinformatics 11(3), 334–347 (2010)
13. Su, C.-J., Wu, C.-Y.: JADE implemented mobile multi-agent based, distributed information platform for pervasive health care monitoring. Applied Soft Computing 11(1), 315–325 (2011)
14. Collier, N., North, M.: Parallel agent-based simulation with Repast for High Performance Computing. Simulation 89(10), 1215–1235 (2013)
15. Marurngsith, W., Mongkolsin, Y.: Creating GPU-Enabled Agent-Based Simulations Using a PDES Tool. In: Omatu, S., Neves, J., Rodriguez, J.M.C., Paz Santana, J.F., Gonzalez, S.R. (eds.) Distrib. Computing & Artificial Intelligence. AISC, vol. 217, pp. 227–234. Springer, Heidelberg (2013)
16. Brodtkorb, A.R., et al.: Graphics processing unit (GPU) programming strategies and trends in GPU computing. J. Parallel Distrib. Comput. 73(1), 4–13 (2013)
17. Cui, L., et al.: Acceleration of multi-agent simulation on FPGAs (2011)
18. Telecom Italia SpA. Jade: Java Agent DEvelopment Framework an Open Source platform for peer-to-peer agent based appications (2014), http://jade.tilab.com/
19. Cumming, G.: Understanding The New Statistics: Effect Sizes, Confidence Intervals, and Meta-Analysis. Multivariate Applications Series. Routledge (2011)
20. Alberts, S., et al.: Data-parallel techniques for simulating a mega-scale agent-based model of systemic inflammatory response syndrome on gpu. Simulation (2012)
21. Dimakis, N., Filippoupolitis, A., Gelenbe, E.: Distributed building evacuation simulator for smart emergency management. Computer Journal 53(9) (2010)
22. Fernández, V., et al.: Evaluating Jason for distributed crowd simulations (2010)
23. Šišlák, D., et al.: Agentfly: NAS-wide simulation framework integrating algorithms for automated collision avoidance (2011)
24. Lämmel, G., Grether, D., Nagel, K.: The representation and implementation of time-dependent inundation in large-scale microscopic evacuation simulations. Transportation Research Part C: Emerging Technologies 18(1), 84–98 (2010)
25. Long, Q., et al.: Agent scheduling model for adaptive dynamic load balancing in agent-based distributed simulations. Simul. Model. Pract. Th. 19(4) (2011)
26. Mao, T., et al.: Parallelizing continuum crowds (2010)
27. Nouman, A., Anagnostou, A., Taylor, S.J.E.: Developing a distributed agent-based and DES simulation using poRTIco and repast (2013)
28. Razavi, S.N., et al.: Multi-agent based simulations using fast multipole method: Application to large scale simulations of flocking dynamical systems. Artificial Intelligence Review 35(1), 53–72 (2011)
29. Tang, W., Bennett, D.A.: Parallel agent-based modeling of spatial opinion diffusion accelerated using graphics processing units. Ecological Modelling (2011)
30. Tang, W., Bennett, D.A., Wang, S.: A parallel agent-based model of land use opinions. Journal of Land Use Science 6(2-3), 121–135 (2011)

Fuzzy Tool for Proposal of Suitable Products in Online Store and CRM System

Bogdan Walek, Jiří Bartoš, and Radim Farana

Department of Informatics and Computers, University of Ostrava, 30. dubna 22,
701 03 Ostrava, Czech Republic
{bogdan.walek,jiri.bartos,radim.farana}@osu.cz

Abstract. Nowadays, people often encounter a wide range of products that are suitable for them. Offers of suitable products are created based on people's buying preferences and previous purchases. In the area of CRM systems it is possible, and usually necessary, to design suitable products for customers. In this paper, a fuzzy tool connected with an expert system to design suitable products for customers within CRM system is proposed. All the main steps of the proposed procedure are described in this article.

Keywords: fuzzy, fuzzy tool, expert system, CRM, product, customer relationship.

1 Introduction

Nowadays, products are more and more tailored to customers. For example, repeated purchases in an online store are leading the information system of the online store to suggest probable product interests next visit. These offers are supported by the targeted advertising of the products in the form of short promo videos or newsletters. Information about suitable products are obtained on the grounds of buying preferences, the type of goods that were purchased and other factors. A similar principle can be used in the CRM system. Customer Relationship Management is a comprehensive approach for creating, managing and expanding customer's relationships and it is the basis of marketing and sales branches [1] [2] [3]. The concept of CRM describes a model for managing relationships between a company and a customer [4]. CRM systems organize and automate business processes for marketing and sales activities. They also provide customer service and support, and other activities that help companies to increase sales and profit [5]. The main goals of CRM are:

- to care about current customers,
- to find and win new customers,
- to regain former clients,
- to reduce the cost of marketing,
- to improve efficiency of business processes in company,
- to track product sales,
- to track projects and their activities.

E. Corchado et al. (Eds.): IDEAL 2014, LNCS 8669, pp. 433–440, 2014.

In case of a CRM system, the basics are to design and propose appropriate products that can be offered to individual customers. Products that are being offered to customers should reflect the buying habits, products previously purchased and other factors linked to it.

Nowadays, the primary challenge is to offer customer exactly such products which he needs and is interested in in a way leading to immediate purchase [10] [11]. If an online shop is able to offer such products within a reasonable time scale, the chances of purchase get higher. Therefore, the problem domain, which this paper will focus on, is offering suitable products to customers.

In this paper, a fuzzy tool connected to an e-shop for the selection of appropriate products within the CRM system is proposed. An expert system will suggest suitable products for specific customers based on various input information. This article is a continuation of paper [7].

2   Problem Formulation

Currently there are several approaches to offering suitable products for customers, which can be used both in e-shops and CRM systems.

The first of them is described in [8]. This approach is based on creating so called recommendation agent, which offers customers a wide range of alternative products on the ground of a given type of product and customer's preferences. Furthermore, the approach proposes so called comparison matrix, which makes alternative products as well as their attributes accessible to the customer in a clearly arranged matrix. The customer is then able to evaluate which product can be the most suitable for him. In reality, this approach is used in several e-shops (namely amazon.com or shopping.com). However, it is determined mainly for the selection of the most suitable product within one purchase, other factors, e. g. customer's online purchase history or his long-term preferences based on previous purchases data-mining, are not taken into consideration.

The other approach is described in [9]. This one is based on the idea of the extraction of semantic qualities of from that a gradual buildup of the knowledge base. The suggested recommender system then analyses the descriptions of the products which the customer checks or has purchased and creates his model. Subsequently, the system offers not only products within the same category (product alternatives), but also is able to offer suitable products across various product categories. This approach is certainly suitable for some types of e-shops, however, the customer's personal profile and the customer's qualities are not taken into consideration.

On the basis of previous purchases, similar products are chosen with the consideration that the customer does not own them yet. Due to this, there are chances that the customer will want to make a purchase and therefore it is appropriate to form an attractive offer, using various newsletters, time- limited discounts or offers which look like tailor-made for the customer. The disadvantage of this procedure is that the design of suitable products is based solely on information about previous products purchases or choices. In the CRM system, however, there is not only an overview about the products purchased by the customer but all the information about communication with the customer, about his whole history – the detailed information about the customer and his satisfaction with the products, detailed information about all purchases, including the underlying frequency of purchases, the total amount of purchases and other details of them. This information should be comprehensively utilized.

3   Problem Solution

For these reasons, a fuzzy tool for designing appropriate products based on various information obtained about customers and their purchases is proposed.

Choosing a fuzzy tool enables creating a knowledge base and describing rules of the future system by means of linguistic variables. Another reason is a possible extension of the tool by further expert knowledge easily and without any intervention into the system logic, which can be more difficult while working with other types of tools (agent systems, semantic systems).

The fuzzy tool will consist of several steps and it will include an expert system knowledge base. The proposal of appropriate products will be carried out with regard to the real probability of buying these products. The design of the fuzzy tool is visually displayed in the following figure:

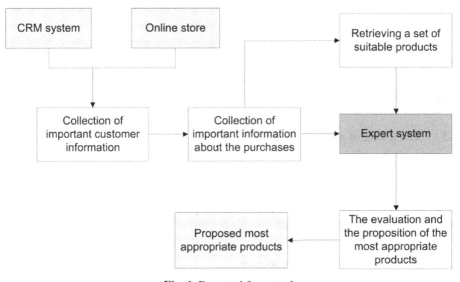

Fig. 1. Proposed fuzzy tool

The following sections will describe all the individual steps.

3.1   Collection of Important Customer Information

In the first step it is very important to retrieve relevant information about customers and their shopping preferences. This information can be obtained from the customer profile in the CRM system and also the summary information about his purchases. All such information should be stored in the CRM system anyway and therefore there should not be a problem to simply obtain them.

Important customer information includes:

- areas of interest
- age
- the average amount of purchases

All retrieved information then forms one of the inputs of the proposed expert system.

3.2   Collection of Important Information about the Purchases

The next step is to load and analyze in detail the information on the purchases, especially on the products. Each product can be classified into a certain category of products and with price as an important factor. Based on the history of purchased products, it is thus possible to determine other important inputs to the proposed expert system. Important information about products includes:

- type of product
- the price of the product
- the number of similar products purchases

Based on this information, it is possible to retrieve a set of products that may be suitable for the customer. The information retrieved in these steps will be saved in a relational database.

3.3   Retrieving of a Set of Suitable Products

In this step the tool retrieves a set of suitable products, from which it will choose the most suitable one. A set of suitable products can be retrieved mainly due to differentiation of the type and price of the product. The result of this step will therefore complete the set of suitable products, which can be processed by the expert system that will make the selection of the most suitable products.

3.4   The Design of the Expert System

This step designs the expert system composed from a knowledge base comprised of the IF-THEN rules. IF-THEN rules are established on the basis of the input information obtained in the first two steps. The output is a linguistic variable. Every IF-THEN rule generates a linguistic variable and it represents the level of the suitability of the product. The set of IF-THEN rules is therefore gradually evaluating the appropriate values of the products that were retrieved in the previous step. Five main input parameters were defined by the expert. The expert also defined more than hundred of IF-THEN rules.

The input linguistic variables are:

- the age of the customer
- the average amount of purchases
- the level of similarity of the purchased products
- the level of price similarities of the purchased products
- number of similar purchases

Here, similarity is defined as the degree of correspondence between a chosen product and its parameters (price, type, qualities) and products previously purchased by the customer.

The output is represented by one linguistic variable, which describes the level of the suitability of the product.

The examples of the IF-THEN rules are as follows:

```
IF (USER_AGE IS VERY LOW) AND
(AMOUNT_OF_PURCHASES IS VERY HIGH) AND
(PU_PR_TYPE_SIMILARITY IS HIGH) AND
(PU_PR_PRICE_SIMILARITY IS HIGH) AND
(NUM_OF_SIMILAR_PR IS MANY) THEN
(PRODUCT_SUITABILITY IS VERY HIGH)

IF (USER IS LOW) AND
(AMOUNT_OF_PURCHASES IS HIGH) AND
(PU_PR_TYPE_SIMILARITY IS MEDIUM) AND
(PU_PR_PRICE_SIMILARITY IS HIGH) AND
(NUM_OF_SIMILAR_PR IS MANY) THEN
(PRODUCT_SUITABILITY IS HIGH)

IF (USER IS HIGH) AND
(AMOUNT_OF_PURCHASES IS LOW) AND
(PU_PR_TYPE_SIMILARITY IS MEDIUM) AND
(PU_PR_PRICE_SIMILARITY IS LOW) AND
(NUM_OF_SIMILAR_PR IS FEW) THEN
(PRODUCT_SUITABILITY IS LOW)
```

For creating the knowledge base of the expert system, the LFLC tool can be used. This tool is able to define input and output linguistic variables and IF-THEN rules. LFLC tool also has inference mechanisms and implemented defuzzification procedures, so a complete expert system can be created with this tool. LFLC tool is more described in [6].

The fuzzy sets have both triangular and trapezoid forms, these forms and kernels and supreme were defined by experts in a problem domain and reviewed during testing and prototyping. For the defuzzification method, after extensive testing, the COG (Simple Center Of Gravity) was selected.

The creating of the knowledge base of the expert system in the LFLC tool is shown in Fig. 2:

General | Input variables | Output variables Rules | Input / Output |

user_age & amount_of_purchases & pu_pr_type_similarity & pu_pr_price_similarity & num_of_similar_pr —:

	user_age	amount_of_purchases	pu_pr_type_similarity	pu_pr_price_similarity	num_of_similar_pr	product_suitability
1.☑	very low	very high	high	high	many	very high
2.☑	low	high	med	high	many	high
3.☑	high	low	med	low	few	low

Fig. 2. Creating knowledge base in the LFLC tool

3.5   The Evaluation and the Proposition of the Most Appropriate Products

In this step, the expert system conducts evaluation of the complete set of suitable products. As already mentioned in the previous step, each product is assigned with some level of suitability. Evaluated products are then sorted in descending order according to the level of suitability. This will allow presenting the most suitable products to the users of the system.

To verify the proposed tool, an online shop selling clothes was used. The chosen customer is a young man who has been purchasing middle-expensive everyday use T-shirts in this shop. After evaluating, the fuzzy tool offers this customer three T-shirts which match his preferences and previous purchases, then one T-shirt which matches the type but is rather expensive. As for a sporty T-shirt, the evaluation is low as the customer has bought a T-shirt of this kind before, but its price and type do not meet the requirements. A golf T-shirt has been purchased only once and the price and the type do not meet the requirements, consequently, the evaluation is very low.

The visualization of evaluated products is depicted in the following picture:

Proposal of suitable products

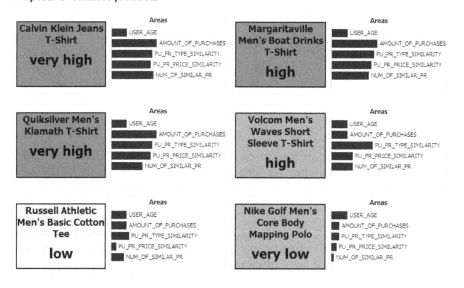

Fig. 3. Visualization of evaluated products

3.6   The Selection of the Most Suitable Product

In the last step, the user of the CRM can choose the most appropriate products and therefore to present appropriate purchase offers. Due to the large amount of information that is entering the expert system, the assessment of the most suitable products is comprehensive. The proposed products have therefore a good chance that the customer will be interested in the purchases. When the purchase of one or more products is conducted, the goal of the expert system is met.

4   Conclusion

In this article the selection of the most appropriate products for customers within the CRM system was discussed. First, the area of interest was identified and the problem domain was defined. Furthermore, a fuzzy tool connected with the expert system was proposed. The objective of this expert system is based on the information about customers and their purchases and purchased products and afterwards suggests the best products that the customer may also be interested in. The main steps of the fuzzy tool were introduced in this article. Along with the individual steps of the proposed tool, the way of creating a knowledge base of an expert system, as well as the principle of operation of the expert system on several examples were introduced. In conclusion, the visualization and evaluation of the selection of the appropriate products for a customer was presented.

Future work will be devoted to the verification of the proposed approach in other selected CRM systems connected to Internet shops. Also, the aim is to generalize the proposed procedure and clearly formalize the main steps of the proposed instrument.

The plans also involve the expansion of the knowledge base driven by the results gained from the practical verification of the proposed procedure.

Acknowledgment. This paper was supported by the European Regional Development Fund in the IT4Innovations Centre of Excellence project (CZ.1.05/1.1.00/02.0070) and by the internal grant SGS15/PřF/2014, called Fuzzy modeling tools for securing intelligent systems and adaptive search burdened with indeterminacy, at Department of Informatics and Computers, University of Ostrava.

References

1. Brown, S.A.: Customer Relationship Management: A Strategic Imperative in the World of E-Business. John Wiley, New York (2000)
2. Swift, R.S.: Accelerating Customer Relationships: Using CRM and Relationship Technologies. Prentice Hall, Upper Saddle River (2001)
3. Wong, K.W.: Data Mining Using Fuzzy Theory for Customer Relationship Management. In: 4th Western Australian Workshop on Information Systems Research (2001)
4. Jeong, Y.A.: On the design concepts for CRM system. Industrial Management & Data Systems 103(5), 324–331 (2003)

5. Morgan, R.M., Hunt, S.D.: The commitment – trust theory of relationship marketing. Journal of Marketing 58(3), 20–38 (1994)
6. Habiballa, H., Novák, V., Dvořák, A., Pavliska, V.: Using software package LFLC 2000. In: 2nd International Conference Aplimat 2003, Bratislava, pp. 355–358 (2003)
7. Walek, B.: Fuzzy tool for customer satisfaction analysis in CRM systems. In: 36th International Conference on Telecommunications and Signal Processing, TSP 2013, Rome, pp. 11–14 (2013)
8. Häubl, G., Trifts, V.: Consumer Decision Making in Online Shopping Environments: The Effects of Interactive Decision Aids. Marketing Science 19(1), 4–21 (2000)
9. Ghani, R., Fano, A.: Building Recommender Systems Using a Knowledge Base of Product Semantics. In: 2nd Int'l Conf. on Adaptive Hypermedia and Adaptive Web Based Systems, Malaga (2002)
10. Moorthy, S., Ratchford, B.T.: Consumer information search revisited: theory and empirical analysis. Journal of Consumer Research 23(4), 263–277 (1995)
11. Alba, J., Lynch, J.: Interactive home shopping: consumer retailer, and manufacturer incentives to participate in electronic marketplaces. Journal of Marketing 61(3), 38–53 (1997)

Giving Voice to the Internet
by Means of Conversational Agents*

David Griol, Araceli Sanchis de Miguel, and José Manuel Molina

Universidad Carlos III de Madrid
Avda. de la Universidad, 30
28911 - Leganés, Spain
{david.griol,araceli.sanchis,josemanuel.molina}@uc3m.es

Abstract. In this paper we present a proposal to develop conversational agents that avoids the effort of manually defining the dialog strategy for the agent and also takes into account the benefits of using current standards. In our proposal the dialog manager is trained by means of a POMDP-based methodology using a labeled dialog corpus automatically acquired using a user modeling technique. The statistical dialog model automatically selects the next system response. Thus, system developers only need to define a set of files, each including a system prompt and the associated grammar to recognize user responses. We have applied this technique to develop a conversational agent in VoiceXML that provides information for planning a trip.

Keywords: Conversational Agents, Spoken Interaction, POMDPs, Machine Learning, User Modeling, Neural Networks.

1 Introduction

A conversational agent can be defined as a software that accepts natural language as input and generates natural language as output, engaging in a conversation with the user [4]. Thus, these interfaces make technologies more usable, as they ease interaction, allow integration in different environments, and make technologies more accessible, especially for disabled people. Usually, these agents carry out five main tasks: Automatic Speech Recognition (ASR), Spoken Language Understanding (SLU), Dialog Management (DM), Natural Language Generation (NLG), and Text-To-Speech Synthesis (TTS). These tasks are typically implemented in different modules of the system's architecture.

When designing this kind of agents, developers need to specify the system actions in response to user utterances and environmental states that, for example, can be based on observed or inferred events or beliefs. This is the fundamental task of dialog management [4], as the performance of the system is highly

---

* This work has been supported in part by the Spanish Government under i-Support (Intelligent Agent Based Driver Decision Support) Project (TRA2011-29454-C03-03), and Projects MINECO TEC2012-37832-C02-01, CICYT TEC2011-28626-C02-02, and CAM CONTEXTS (S2009/TIC-1485).

E. Corchado et al. (Eds.): IDEAL 2014, LNCS 8669, pp. 441–448, 2014.

dependent on the quality of this strategy. Thus, a great effort is employed to empirically design dialog strategies for commercial systems. In fact, the design of a good strategy is far from being a trivial task since there is no clear definition of what constitutes a good dialog strategy [7].

Once the dialog strategy has been designed, the implementation of the system is leveraged by programming languages such as the standard VoiceXML [6], for which different programming environments and tools have been created to help developers. These programming standards allow the definition of a dialog strategy based on scripted Finite State Machines. With the aim of creating dynamic and adapted dialogs, the application of statistical approaches to dialog management makes it possible to consider a wider space of dialog strategies [7].

The most extended methodology for machine-learning of dialog strategies consists of modeling human-computer interaction as an optimization problem using Markov Decision Process (MDP) and reinforcement methods [2]. The main drawback of this approach is the large state space, whose representation is intractable if represented directly [9]. Partially Observable MDPs (POMDPs) outperform MDP-based dialog strategies since they provide an explicit representation of uncertainty [7]. Other interesting approaches for statistical dialog management are based on modeling the system by means of Hidden Markov Models, stochastic Finite-State Transducers, or using Bayesian Networks.

Additionally, the design of speech recognition grammars for the ASR and SLU tasks have been usually built on the basis of handcrafted rules that are tested recursively, which in complex applications is very costly [3]. However, as stated by [4], many sophisticated commercial systems already available receive a large volume of interactions. Therefore, industry is becoming more interested in substituting rule based grammars with other statistical techniques based on the large amounts of data available.

As an attempt to improve the current technology, we propose to combine the flexibility of statistical dialog management with the facilities that VoiceXML offers, which would help to introduce statistical methodologies for the development of commercial (and not strictly academic) dialog systems. To this end, our technique employs a POMDP-based dialog manager. Expert knowledge about deployment of VoiceXML applications, development environments and tools can still be exploited using our technique. The only change is that transitions between dialog states is carried out on a data-driven basis (i.e., it is not a deterministic process). In addition, the system prompts and the grammars for ASR are implemented in VoiceXML-compliant formats (e.g., JSGF or SRGS).

Pietquin and Dutoit [5] described a similar proposal based on a graphical interface dedicated to ease the development of VoiceXML-based dialog systems. The main aim is focused on enabling non-specialist designers to semi-automatically create their own systems. In this case, the results of a MDP-based strategy learning method are provided in order to facilitate the design of the dialog strategy for the VoiceXML system. Speech grammars are not automatized by the proposal. Our goal is to make developers' work even easier with a very simple design of each VoiceXML file (they are only reduced to a system prompt and an automatic generated speech grammar) and the complete automation of the dialog strategy (the next system prompt, i.e. the next VoiceXML file, is automatically selected by the statistical dialog model).

The remainder of the paper is as follows. Section 2 describes our proposal to integrate spoken interaction to web-based systems by means of the combination of a statistical dialog manager and a Voice-XML complaint platform. Section 3 presents the application of our proposal to develop a commercial system for planning a trip. This section also presents the results of its preliminary evaluation. Finally, Section 4 presents some conclusions and future research lines.

2   Our Proposal to Provide a Spoken Access to the Web

The application of POMDPs to model a spoken conversational agent is based on the classical architecture of these systems shown in Figure 1 [7]. As this figure shows, the user has an internal state S_u corresponding to their goal and the dialog state S_d represents the previous history of the dialog. Based on the user's goal prior to each turn, the user decides some communicative action (also called intention) A_u, expressed in terms of dialog acts and corresponding to an audio signal Y_u. Then, the speech recognition and language understanding modules take the audio signal Y_u and generate the pair $(\tilde{A}_u, C)$.

This pair consists of an estimate of the user's action A_u and a confidence score that provides an indication of the reliability of the recognition and semantic interpretation results. This pair is then passed to the dialog model, which is in an internal state S_m an decides what action A_m the conversational agent should take. This action is also passed back to the dialog manager so that S_m may track both user and machine actions. The language generator and the text-to-speech synthesizer take A_m and generate an audio response Y_m. The user listens to Y_m and attempts to recover A_m. As a result of this process, users update their goal state S_u and their interpretation of the dialog history S_d.

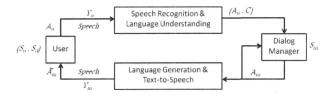

Fig. 1. Classical architecture of a conversational agent

One of the main reasons to explain the challenge of building conversational agents is that $\tilde{A}_u$ usually contains recognition errors (i.e., $\tilde{A}_u \neq A_u$). As a result, the user's action A_u, the user's state S_u, and the dialog history S_d are not directly observable and can never be known to the system with certainty. However, $\tilde{A}_u$

and C provide evidence from which A_u, S_u, and S_d can be inferred. Therefore, when using POMDPs to model a conversational agent, the POMDP state S_m expresses the unobserved state of the world and can naturally be factored into three distinct components: the user's goal S_u, the user's action A_u, and the dialog history S_d. Hence, the factored POMDP state S is defined as:

$$s_m = (s_u, a_u, s_d) \tag{1}$$

The belief state b is then a distribution over these three components:

$$s_m = b_s = b(s_u, a_u, s_d) \tag{2}$$

The observation o is the estimate of the user dialog act $\tilde{A}_u$. In the general case this will be a set of N-best hypothesized user acts, each with an associated probability

$$o = [(\tilde{a_u}^1, p_1), (\tilde{a_u}^2, p_2), \cdots, (\tilde{a_u}^N, p_N)] \tag{3}$$

where $p_n = P(\tilde{a_u}^N | o)$ for $n = 1 \cdots N$.

The transition function for an SDS-POMDP follows directly by substituting the factored state into the regular POMDP transition function and making independence assumptions:

$$P(s'_m | s_m, a_m) = P(s'_u, a'_u, s'_d | s_u, a_u, s_d, a_m) =$$
$$= P(s'_u | s_u, a_m) P(a'_u | s'_u, a_m) P(s'_d | s'_u, a'_u, s_d, a_m) \tag{4}$$

This is the transition model. Making similar reasonable independence assumptions regarding the observation function gives,

$$P(o' | s'_m, a_m) = P(o' | s'_u, a'_u, s'_d, a_m) = P(o' | a'_u) \tag{5}$$

This is the observation model. The above factoring simplifies the belief update equation since substituting (8) and (9) into (1) gives

$$b'(s'_u, a'_u, s'_d) = k \cdot \underbrace{P(o' | a'_u)}_{Observation\ model} \underbrace{P(a'_u | s'_u, a_m)}_{User\ action\ model}$$

$$\sum_{s_u} \underbrace{P(s'_u | s_u, a_m)}_{User\ goal\ model} \cdot \sum_{s_d} \underbrace{P(s'_d | s'_u, a'_u, s_d, a_m)}_{Dialog\ model} b(s_u, s_d) \tag{6}$$

As shown in Equation 6, the probability distribution for a'_u is called the user action model. It allows the observation probability to be scaled by the probability that the user would speak a'_u given the goal s'_u and the last system prompt a_m. The user goal model determines the probability of the user goal switching from s_u to s'_u following the system prompt a_m. Finally, the dialog model enables information relating to the dialog history to be maintained such as grounding and focus.

The optimization of the policy is usually carried out by using techniques like the Point-based Value Iteration or Q-learning, in combination with a user simulator. Q-learning is a technique for online learning where a sequence of sample dialogs are used to estimate the Q functions for each state and action. Given that a good estimate of the true Q-value is obtained if sufficient dialogs are done, user simulation has been introduced to reduce the too time-consuming and expensive task to obtain these dialogs with real users.

Simulation is usually done at a semantic dialog act level to avoid having to reproduce the variety of user utterances at the word or acoustic levels. At the semantic level, at any time t, the user is in a state s_u, takes action a_u, transitions into the intermediate state s'_u, receives machine action a_m, and transitions into the next state s''_u. To do this, we propose the use of a recently developed user simulation technique based on a classification process in which a neural network selects the next user response by considering the previous dialog history [1].

We also propose to merge statistical approaches with VoiceXML. To do this, a VoiceXML-compliant platform (such as Voxeo Evolution[1]) is used for the creation of Interactive Voice Response (IVR) applications and the provision of telephone access. Static VoiceXML files and grammars can be stored in the voice server. We propose to simplify these files by generating a VoiceXML file for each specific system prompt. Each file contains a reference to a grammar that defines the valid user's inputs for the corresponding system prompt.

The conversational agent selects the next system prompt (i.e. VoiceXML file) by consulting the probabilities assigned by the POMDP-based statistical dialog manager to each system prompt given the current state of the dialog. This module is stored in an external web server. The result generated by the statistical dialog manager informs the IVR platform about the most probable system prompt to be selected for the current dialog state. The platform just selects the corresponding VoiceXML file and reproduces it to the user.

3   Development of a Conversational Agent to Plan a Trip

We have applied our proposal to develop and evaluate the *Your Next Trip* system, which provides tourist information useful to plan a trip. The system was developed to provide telephonic access to the contents of a web portal that is updated dynamically from different web pages, databases, and the contribution of the users, who can add and edit the contents.

Figure 2 shows different snapshots of the portal, which contents include cities, places of interest, weather forecast, hotel booking, restaurants and bars, shopping, street guide, cultural activities (cinema, theater, music, exhibitions, literature and science), sport activities, festivities, and public transportation. In addition to provide specific information related to the previously described categories, the system also provides user-adapted recommendations based on the opinions and highest rated places in the application.

---

[1] http://evolution.voxeo.com/

Users can access these functionalities visually by means of the different web pages or orally by means of the application of our proposal with the combination of the POMDP-based dialog manager and the Voxeo Evolution Voice-XML complaint platform.

Fig. 2. Different snapshots of the *Your Next Trip* system

With regard the POMDP-based dialog manager, rewards in the conversational agent were given based on the task completion rate and the number of turns in the dialog. The user modeling module described in [1] was initially trained using the 100 dialogs acquired with a Wizard of Oz experiment in which an expert simulated the system operation. The dialog manager was implemented and trained via interactions with the simulated user model to iteratively learn a dialog policy. A total of 150,000 dialogs was simulated. Using the definitions described in [8] for the summary Q-learning algorithm, the POMDP system was given 20 points for a successful dialog and 0 for an unsuccessful one. One point was subtracted for each dialog turn.

To assess the benefits of our proposal, we have already completed a preliminary evaluation of the developed system with recruited users and a set of scenarios covering the different functionalities of the system. A total of 150 dialogs for each agent was recorded from the interactions of 25 recruited users. These users followed a set of scenarios that specify a set of objectives that must be fulfilled by the user at the end of the dialog and are designed to include the complete set of functionalities previously described for the system.

We asked the recruited users to complete a questionnaire to assess their opinion about the interaction. The questionnaire had seven questions: i) Q1: *How well did the system understand you?*; ii)Q2: *How well did you understand the system messages?*; iii) Q3: *Was it easy for you to get the requested information?*; iv) Q4: *Was the interaction with the system quick enough?*; v) Q5: *If there were system errors, was it easy for you to correct them?*; vi) Q6: *How did the system adapt to your preferences?*; vi) Q7: *In general, are you satisfied with the performance of the system?* The possible answers for each questions were the same: *Never/Not at all, Seldom/In some measure, Sometimes/Acceptably, Usually/Well*, and *Always/Very Well*. All the answers were assigned a numeric value between one and five (in the same order as they appear in the questionnaire).

Also, from the interactions of the users with the system we completed an objective evaluation of the application considering the following interaction parameters: i) question success rate (SR), percentage of successfully completed questions: system asks - user answers - system provides appropriate feedback about the answer; ii) confirmation rate (CR), computed as the ratio between the number of explicit confirmations turns and the total of turns; iii) error correction rate (ECR), percentage of corrected errors.

Table 1 shows the average results of the subjective evaluation using the described questionnaire. It can be observed that the users perceived that the system understood them correctly. Moreover, they expressed a similar opinion regarding the easiness to understand the system responses. In addition, they assessed that it was easier to obtain the information specified for the different objectives, and that the interaction with the system was adequate and adapted to their preferences. An important point remarked by the users was that it was difficult to correct the errors and misunderstandings generated by the ASR and NLU processes in some scenarios. Finally, the satisfaction level also shows the correct operation of the system.

Table 1. Results of the preliminary evaluation with recruited users (For the mean value M: 1=worst, 5=best evaluation)

Q1	M = 4.45, SD = 0.49
Q2	M = 4.37, SD = 0.47
Q3	M = 4.05, SD = 0.55
Q4	M = 3.66, SD = 0.53
Q5	M = 3.19, SD = 0.61
Q6	M = 3.89, SD = 0.46
Q7	M = 4.21, SD = 0.32

SR	CR	ECR
94.36%	19.00%	92.11%

The results of the objective evaluation for the described interactions show that the developed system could interact correctly with the users in most cases, achieving a success rate of 94.36%. The fact that the possible answers to the user's responses are restricted made it possible to have a very high success in speech recognition. Additionally, the approaches for error correction by means of confirming or re-asking for data were successful in 92.11% of the times when the speech recognizer did not provide the correct input.

4   Conclusions and Future Work

In this paper, we have described a proposal to provide spoken interaction to the web. Our proposal works on the benefits of the POMDP statistically method for dialog management and VoiceXML, respectively. The former provides an efficient means to explore a wider range of dialog strategies and also introduce user adaptation, whereas the latter makes it possible to benefit from the advantages of using the different tools and platforms that are already available to simplify system development.

We have applied our technique to develop a conversational agent that provides information to plan a trip, and have . The results of its evaluation show that the described technique can predict coherent system answers in most of the cases, also obtaining a high user's satisfaction level. As a future work, we plan to study ways for adapting the proposed statistical model to more complex domains.

References

1. Griol, D., Carbo, J., Molina, J.: A statistical simulation technique to develop and evaluate conversational agents. AI Communications Journal 26(4), 355–371 (2013)
2. Levin, E., Pieraccini, R., Eckert, W.: A stochastic model of human-machine interaction for learning dialog strategies. IEEE Transactions on Speech and Audio Processing 8(1), 11–23 (2000)
3. McTear, M.F.: Spoken Dialogue Technology: Towards the Conversational User Interface. Springer (2004)
4. Pieraccini, R.: The Voice in the Machine: Building Computers That Understand Speech. MIT Press (2012)
5. Pietquin, O., Dutoit, T.: Aided Design of Finite-State Dialogue Management Systems. In: Proc. of ICME 2003, vol. 3, pp. 545–548 (2003)
6. Rouillard, J.: Web services and speech-based applications around VoiceXML. Journal of Networks 2(1), 27–35 (2007)
7. Schatzmann, J., Weilhammer, K., Stuttle, M., Young, S.: A Survey of Statistical User Simulation Techniques for Reinforcement-Learning of Dialogue Management Strategies. Knowledge Engineering Review 21(2), 97–126 (2006)
8. Thomson, B., Schatzmann, J., Weilhammer, K., Ye, H., Young, S.: Training a real-world POMDP-based Dialog System. In: Proc. of HLT 2007, pp. 9–16 (2007)
9. Young, S., Schatzmann, J., Weilhammer, K., Ye, H.: The Hidden Information State Approach to Dialogue Management. In: Proc. of ICASSP 2007, vol. 4, pp. 149–152 (2007)

Multivariate Cauchy EDA Optimisation

Momodou L. Sanyang[1,2] and Ata Kaban[1]

[1] School of Computer Science, University of Birmingham, Edgbaston, UK, B15 2TT
{M.L.Sanyang,A.Kaban}@cs.bham.ac.uk
[2] School of Information Technolgy and Communication, University of the Gambia
Brikama Campus, P.O. Box 3530, Serekunda, The Gambia
MLSanyang@utg.edu.gm

Abstract. We consider Black-Box continuous optimization by Estimation of Distribution Algorithms (EDA). In continuous EDA, the multivariate Gaussian distribution is widely used as a search operator, and it has the well-known advantage of modelling the correlation structure of the search variables, which univariate EDA lacks. However, the Gaussian distribution as a search operator is prone to premature convergence when the population is far from the optimum. Recent work suggests that replacing the univariate Gaussian with a univariate Cauchy distribution in EDA holds promise in alleviating this problem because it is able to make larger jumps in the search space due to the Cauchy distribution's heavy tails. In this paper, we propose the use of a multivariate Cauchy distribution to blend together the advantages of multivariate modelling with the ability of escaping early convergence to efficiently explore the search space. Experiments on 16 benchmark functions demonstrate the superiority of multivariate Cauchy EDA against univariate Cauchy EDA, and its advantages against multivariate Gaussian EDA when the population lies far from the optimum.

Keywords: Multivariate Gaussian distribution, Multivariate Cauchy Distribution, Estimation of Distribution Algorithm, Black-box Optimization.

1   Introduction

Black-box global optimization is an important problem which has many applications in lots of disciplines. Optimization is at the core of many scientific and engineering problems. Mathematical optimization only deals with very specific problem types, while on the other hand the search heuristics like evolutionary computation work in a black box manner. They are not specialized on specific kinds of functions although they don't have the guarantees that the mathematical optimizations do. This paper presents a method which is classified as a search heuristic, and it is an extension of a recent version called Estimation of Distribution Algorithm (EDA).

EDA is a population based stochastic black-box optimization method that guides the search to the optimum by building and sampling explicit probability models of promising candidate solutions [2]. In EDA, the new population of individuals is generated without using neither crossover nor mutation operations, which is in contrast to

E. Corchado et al. (Eds.): IDEAL 2014, LNCS 8669, pp. 449–456, 2014.

other evolutionary algorithms. In classical EDA, Gaussian distribution is used as the search operator to build a probabilistic model to fit the fittest individuals and create new individuals by sampling from the created model. It has been established that Gaussian EDA is prone to premature convergence [1], [2], [6] when its parameters are estimated using the maximum likelihood estimation (MLE) method. It converges too fast and does not get to the global optimum.

The premature convergence of classical Gaussian EDA attracted many efforts geared towards solving this problem [6], [10]. This paper presents the usage of Multi-variate Cauchy distribution, an extension of [8] with a full matrix valued parameter that encodes dependencies between the search variable as an alternative search opera-tor in EDA. We utilize its capability of making long jumps so as to escape premature convergence, which is typical of Gaussian in order to enable EDA algorithms get to the global optimum. Although Cauchy has already been used as an alternative search distribution for EDA [4], [5], [6], [10], it was the univariate version of Cauchy that was utilized, discarding statistical dependences among the search variables. We com-pare the performance of Multivariate Gaussian EDA with Multivariate Cauchy EDA to establish whether and when the long jumps that Cauchy is able make will be advan-tageous. We also compare the performance to Univariate Cauchy EDA to establish the advantages of multivariate modelling.

2   Differences between Multivariate Gaussian and Multivariate Cauchy Distributions

The main advantage of EDAs is the explicit learning of the dependences among va-riables of the problem to be tackled and utilizing this information efficiently to gener-ate new individuals to drive the search to the global optimum [10]. Using univariate Cauchy will make it hard to achieve this goal since it does not take on dependences. Therefore, this paper to the best of our knowledge is the first to include the modelling of dependencies in a Cauchy search distribution based EDA algorithm for black-box global Optimization.

An important difference between Gaussian and Cauchy is that Cauchy is heavier-tailed. This means that it is more prone to producing values that fall far from its mean, thus, giving Cauchy more chance of sampling further down its tail than the Gaussian. This gives Cauchy a higher chance of escaping premature convergence than the Gaus-sian [6], [10].

For the same reason, the Gaussian search distribution is good when the individuals are close to the optimum while Cauchy is better when the individuals are far from the optimum. Both of these findings were previously made in the context of traditional evolutionary computation [8] where univariate version of these distributions were used to implement the mutation operator. Our hypothesis, which we test in this paper, is that these advantages are carried forward to EDA based optimization where in addi-tion, multivariate modelling enables a more directed search.

Figure 1 shows the probability density functions for both Gaussian and Cauchy dis-tributions in one and two dimensions.

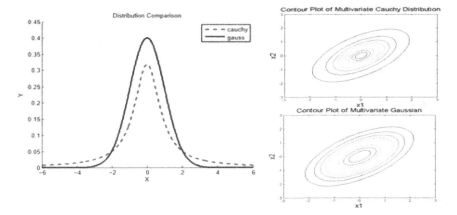

Fig. 1. Cauchy density (red dashed), along with standard normal density (blue) (Left), Multivariate Cauchy (Top Right) and Multivariate Gaussian (Bottom Right)

The leftmost plot in figure 1 shows the probability density function of a Cauchy versus a Gaussian in 1D. We see the heavy tail of Cauchy falling down slower than Gaussian. On the right, the plots depict contour plots of the 2D versions of these densities, with Cauchy on the top right and Gaussian at the bottom right. In the 2D versions, Cauchy has a flatter tail on the base as can be seen by the wider space between the second outermost and the outermost contour lines than those of the Gaussian. The parameter matrix $\Sigma = [1 \ .6; .6 \ 1]$ was used on both 2D version for plotting the contours.

3   Algorithm Presentation

The algorithms used in this paper for comparison are Multivariate Gaussian EDA (MGEDA), Multivariate Cauchy EDA (MCEDA) and Univariate Cauchy EDA (UCEDA). MGEDA takes on board the sample correlations between the variables of the selected individuals through a full covariance matrix, and MCEDA encodes pairwise dependencies among the search variables through its matrix valued parameter. UCEDA neglects dependences among the search variables. Algorithm 1 describes generic EDA algorithm.

1. Set t: =0. Generate M points randomly to give an initial population P.
   Do
2.   Evaluate fitness for all M points in P.
3.   Select some individuals P^{sel} from P.
4.   Estimate the statistics of P^{sel}
5.   Use statistics in step (4) to sample new population P^{new}.
6.   Set P to P^{new}

Until Termination criteria are met.

Algorithm 1. The Pseudocode of a simple EDA with Population size M.

This algorithm is a typical EDA, which proceeds by initially generating a population of individuals and then evaluates their fitness to select the fittest ones based on their fitness using the truncation selection. For the MGEDA, we compute the maximum likelihood estimates (MLE) of the mean (μ) and the covariance (Σ) of the fittest individuals and use these parameters to generate new ones by sampling from a multivariate Gaussian distribution with parameters μ and Σ. For MCEDA, we use the same estimates to sample from a Multivariate Cauchy distribution in step 5. In UCEDA, we use μ and the diagonal elements of Σ to sample each from Univariate Cauchy. The new population is formed by replacing the old individuals by the new ones.

3.1   A Note on Parameter Estimation

The philosophy in EDA is to estimate the density of the selected individuals so that when new individuals are sampled from the model, they will follow the same distribution. Fortunately, for Gaussian this works. Parameter estimation in Cauchy distributions was studied in statistics [3] where an Expectation and Maximization (EM) algorithm was developed to find the maximum likelihood estimate of a multivariate Cauchy distribution from a set of points, which we implemented for our study.

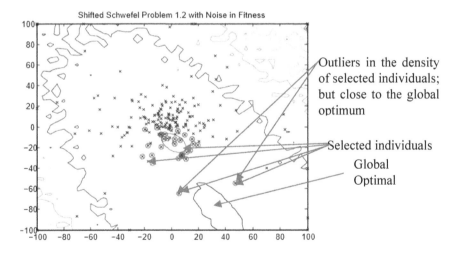

Fig. 2. A plot showing the behavior of EDA when the search distribution is a Cauchy distribution

However, we found that when we estimate the Cauchy's parameter (Using EM), then the obtained model of the selected individuals (a Cauchy density) will disregard any outliers. This is of course what a robust density estimator is meant to do- however for optimization those outliers may be some rare and very good solutions that got close to an optimum. Fig 2 illustrates such an example.

As you can see in fig. 2, which was a snap shot taken from an iteration of the experiments we conducted, two selected individuals are close to the optimum and as such are good individuals but they are outliers with respect to the density of the rest of the selected individuals. This is the reason why in algorithm 1, the Multivariate Cauchy distribution was used only in the sampling step.

4   Experiments

Our hypothesis is that MCEDA has better performance than both UCEDA and MGEDA when the initial population is far from the optimum and also when the population size is small. In turn Multivariate Gaussian should perform better when the population is close to the optimum. To test this hypothesis, we conducted an extensive experiment on 16 benchmark functions taken from [7]. In the following subsections, we will describe the functions, parameter settings, then we present results with analysis and conclude.

4.1   Benchmark Test Functions

The comparisons of the three EDA algorithms were carried out on the suite of benchmark functions from the CEC'2005 competition. 16 test functions were used in this set of experiments. Among the functions tested, 5 are unimodal and 11 multimodal. All the global optima are within the given box constrains. However, problem 7 was without a search range and with the global optimum outside of the specified initialization range.   All problems are minimization. Please see details of the functions in [7].

4.2   Parameter Setting

The dimensionality of all the problems is 2. We carried out three sets of experiments with the initial population size set to 20, 200 and 500 respectively. The percentage of individuals retained is 30%, which is a most widely used selection ratio. We did 25 independent runs for each problem on a fixed budget of 10,000 function evaluations in each case. The initialization was uniformly random within the search space. We also created harder versions of these problems by initializing far from the optimum to establish whether MCEDA can still perform in this situation.

4.3   Performance Criteria

The main performance criterion was the difference in fitness values (error) between the best individual found and the global optimum.

Table 1. Statistical Comparison of MCEDA and MGEDA on Problems 01-16 with initial Population far from the optimum and has size 200

	MCEDA		MGEDA		Rank sum Test	
	Mean	Std. dev.	Mean	Std. dev.	H	P
P01	**0**	**0**	1.6180e+03	508.7709	1	9.72e-011
P02	**0**	**0**	1.4116e+03	287.7206	1	9.72e-011
P03	**0**	**0**	3.8746e+08	1.2402e+08	1	9.72e-011
P04	**0**	**0**	953.7703	183.1595	1	9.72e-011
P05	**0**	**0**	7.0883e+03	411.0024	1	9.72e-011
P06	**0**	**0**	8.0103e+08	1.5906e+08	1	9.72e-011
P07	**0.0621**	**0.0464**	0.7204	0.7672	1	0.0010
P08	18.0903	3.9646	19.2004	4.0001	0	0.0625
P09	0.9025	0.6808	1.3412	0.9730	0	0.1057
P10	**0.6757**	**0.6134**	3.8126	2.4724	1	4.04e-007
P11	1.1429	0.4346	**0.0359**	**0.0993**	1	2.47e-010
P12	193.2284	184.7226	**0.4156**	**0.6739**	1	3.93e-009
P13	**0.0067**	**0.0147**	23.858	8.4094	1	6.57e-010
P14	**0.0220**	**0.0036**	0.7357	0.2264	1	1.41e-009
P15	**0.0649**	**0.0402**	52.3710	7.9035	1	1.41e-009
P16	**0**	**0**	1.7225	4.6068	1	0.0412

Table 2. Statistical Comparison of MCEDA and MGEDA on Problems 01-16 with uniform initialisation and small Population size 20

	MCEDA		MGEDA		Rank sum Test	
	Mean	Std. dev.	Mean	Std. dev.	H	P
P01	**0**	**0**	35.1160	163.9960	1	4.4787e-009
P02	**0**	**0**	116.3862	213.6704	1	1.3101e-009
P03	**0**	**0**	7.3128e+04	3.5850e+05	1	3.6574e-010
P04	**0**	**0**	132.4800	311.6559	1	1.3101e-009
P05	**0**	**0**	1.4107e+03	1.2712e+03	1	4.4787e-009
P06	**0**	**0**	14.0273	20.0025	1	3.6574e-010
P07	0.0858	0.1346	0.4516	1.1137	0	0.2288
P08	19.7487	4.1438	**17.6353**	**6.5513**	1	2.4248e-008
P09	3.5818	4.0211	0.6134	0.5887	0	0.2111
P10	**0.7562**	**0.9646**	1.3387	1.9365	1	0.0092
P11	0.5856	1.1481	**0.2505**	**0.3295**	1	0.0195
P12	2.0355e+03	1.1826e+03	**10.2325**	**20.2760**	1	9.2160e-010
P13	0.1538	0.2352	0.0507	0.0531	0	0.1633
P14	0.6806	0.2313	**0.1408**	**0.0194**	1	1.4634e-007
P15	0.1183	0.1278	1.1885	2.9844	0	0.8613
P16	**0.6865**	**2.9408**	2.2038	3.8294	1	1.1752e-007

Table 3. Statistical Comparison of UCEDA and MCEDA on Problems 01-16 with uniform initialisation and small Population size 20

	UCEDA		MCEDA		Rank sum Test	
	Mean	Std. dev.	Mean	Std. dev.	H	P
P01	0	0	0	0	0	N/A
P02	0	0	0	0	0	N/A
P03	5.2295e+03	6.0844e+03	**0**	**0**	1	9.7282e-011
P04	0	0	0	0	0	N/A
P05	0	0	0	0	0	N/A
P06	5.6064	14.1692	**0**	**0**	1	9.7282e-011
P07	0.2708	0.2092	**0.0858**	**0.1346**	1	1.4256e-004
P08	20.5476	0.5120	19.7487	4.1438	0	0.9690
P09	**3.5818**	**4.0211**	6.1954	4.4285	1	0.0289
P10	0.9154	0.9916	0.7562	0.9646	0	0.3855
P11	0.1991	0.5776	0.5856	1.1481	0	0.3731
P12	**1.0293e+03**	722.7733	2.0355e+03	1.1826e+03	1	0.0108
P13	0.3313	0.3508	**0.1538**	**0.2352**	1	1.7045e-004
P14	0.8139	0.0550	**0.6806**	**0.2313**	1	0.0085
P15	0.3948	0.1526	**0.1183**	**0.1278**	1	5.0089e-007
P16	0.4866	2.4328	0.6865	2.9408	0	0.5717

5   Results and Discussion

The results of our experiments are summarized in tables 1 to 3 and bold font indicates statistically significant out performance at 95% confidence level.

Tables 1 and 2 report results from experiments that compare MCEDA with MGEDA. MCEDA performed better than MGEDA in most of the 16 benchmark functions, see tables 1 and 2. From results omitted for space constraints, we also found when the population size was increased to 200, MGEDA outperformed the MCEDA, and this was also confirmed in the results of the experiments with population size 500. The reason for this is the MGEDA is better when the best individuals are close to the optimum, so when we initialized lots of them everywhere, there are chances that some of them will be close. To test this hypothesis, we devised two sets of experiments with initial population far from the optimum (Results shown in Table 1), and with uniform initialization everywhere in the search space, but a smaller population size of only 20 (Results shown in Table 2). Taken together, these results confirm our hypothesis discussed above. MCEDA has performed better than MGEDA, in both of these settings. This is because of the long jumps of Cauchy.

In table 3, we report a comparison between MCEDA and UCEDA. We can clearly see from table 3 that MCEDA performed better than UCEDA in most of the functions.

6   Conclusions and Future Work

In this paper, we studied the use of a multivariate Cauchy distribution in black-box continuous optimization by EDA. Our MCEDA blends together the advantages of multivariate modelling with the Cauchy sampling's ability of escaping early convergence and efficiently explore the search space. We conducted extensive experiments on 16 benchmark functions and found that MCEDA outperformed MGEDA when the population is far from the global optimum and is able to work even with small population sizes. We also demonstrated the superiority of multivariate Cauchy EDA against univariate Cauchy EDA.

Future work is to extend this study to high dimensional search spaces, possibly leveraging recent techniques of random projection for optimization [1].

References

1. Kaban, A., Bootkrajang, J., Durrant, R.: Towards large scale continuous EDA: A random matrix theory perspective. In: Proc. of Genetic and Evolutionary Computation Conference (GECCO), Amsterdam, The Netherlands, pp. 383–390 (2013)
2. Larranaga, P., Lozano, J.A.: Estimation of Distribution Algorithms: A new tool for Evolutionary Computation. Kluwer Academic Publishers (2001)
3. Peel, D., McLachlan, G.: Robust mixture modelling using the t distribution. Statist. Comput. 10, 339–348 (2000)
4. Posik, P.: BBOB-Benchmarking a simple Estimation of Distribution Algorithm with Cauchy Distribution. In: GECCO 2009, Montreal, Quebec, Canada (July 2009)
5. Posik, P.: Comparison of Cauchy EDA and BIPOP-CMA-ES Algorithms on the BBOB Noiseless Testbed. In: GECCO 2010, Portland, Oregon, USA (July 2010)
6. Pošík, P.: Preventing premature convergence in a simple EDA via global step size setting. In: Rudolph, G., Jansen, T., Lucas, S., Poloni, C., Beume, N. (eds.) PPSN 2008. LNCS, vol. 5199, pp. 549–558. Springer, Heidelberg (2008)
7. Suganthan, P.N., Hansen, N., Liang, J.J., Deb, K., Chen, Y.P., Auger, A., Tiwari, S.: Problem Definitions and Evaluation Criteria for the CEC 2005 Special Session on Real-Parameter Optimisation (2005)
8. Yao, X., Liu, Y., Lin, G.: Evolutionary Programming Made Faster. IEEE Transactions on Evolutionary Computations 3(2) (July 1999)
9. Yao, X., Liu, Y.: Fast evolution strategies. Control and Cybernetics 26(3), 467–496 (1997)
10. Yuan, B., Gallagher, M.: On the importance of diversity maintenance in estimation of distribution algorithms. In: Beyer, H.G., O'Reilly, U.M. (eds.) Proceedings of the Genetic and Evolutionary Computation Conference, GECCO 2005, vol. 1, pp. 719–729. ACM Press, New York (2005)

Continuous Population-Based Incremental Learning with Mixture Probability Modeling for Dynamic Optimization Problems

Adrian Lancucki[1], Jan Chorowski[1], Krzysztof Michalak[2],
Patryk Filipiak[1], and Piotr Lipinski[1]

[1] Computational Intelligence Research Group, Institute of Computer Science,
University of Wroclaw, Wroclaw, Poland
{adrian.lancucki,jan.chorowski,
patryk.filipiak,piotr.lipinski}@cs.uni.wroc.pl
[2] Institute of Business Informatics,
Wroclaw University of Economics, Wroclaw, Poland
krzysztof.michalak@ue.wroc.pl

Abstract. This paper proposes a multimodal extension of PBIL$_C$ based on Gaussian mixture models for solving dynamic optimization problems. By tracking multiple optima, the algorithm is able to follow the changes in objective functions more efficiently than in the unimodal case. The approach was validated on a set of synthetic benchmarks including Moving Peaks, dynamization of the Rosenbrock function and compositions of functions from the IEEE CEC'2009 competition. The results obtained in the experiments proved the efficiency of the approach in solving dynamic problems with a number of competing peaks.

Keywords: evolutionary algorithms, estimation of distribution algorithms, dynamic optimization problems, multimodal optimization.

1 Introduction

Many real-world optimization problems have a dynamic character, often defined as the tendency of the objective function or the constraints to change as time goes by. However, when virtually any parameter can be variable over time, reliable evaluation of algorithms becomes even harder than with static problems. It might be the case that, when solving real-world problems, one might benefit from multimodality. Tracking multiple optima and actively searching for emerging ones would enable reacting instantly to disruptive changes of objective function.

Estimation of distribution algorithms (EDAs) employ probabilistic models instead of traditional genetic operators. Typically, at each generation a new population is drawn from the distribution represented by the model, which is afterwards updated based on the fittest specimen. Thus, EDAs construct probability distributions describing good solutions, while carrying the search for optima. EDAs are characterized by the family of those distributions. The model can be calculated based on frequency of particular genes [4]. Often, estimation of

E. Corchado et al. (Eds.): IDEAL 2014, LNCS 8669, pp. 457–464, 2014.
© Springer International Publishing Switzerland 2014

one-dimensional distributions takes place, as in UMDA [9,13]. In [14] estimation
of two variables has been proposed. More advanced models include: decision
trees [18], one- and multidimensional Gaussian models [17,5], Boltzmann Machines [12], Bayesian Networks [8].

Applications of EDAs to DOPs have been little explored [13,17]. Yang et al. [15]
proposed maintaining diversity in the population by using two probabilistic models; the former optimizes solutions, while the latter is a model with high variance;
mechanisms similar to random immigrants were also investigated. In [17] an approach is mentioned that controls the pace of convergence of the population through
adjusting the Gaussian model proposed in [16], where the additional model is being
activated, when the main model is not able to generate good solutions.

Despite the benefits, EDAs are not yet effective at solving DOPs. Probabilistic modeling techniques are either too simple, ignoring intrinsic dependencies, or
too complex for successful estimation of density based on a fairly small population. Conversely, EDAs which take into account dependencies between random
variables are usually too computationally demanding to keep up with changes in
the objective function. Most of EDAs model unimodal probability distributions,
focusing on a particular area of the search space. However, broad monitoring of
optimization landscape changes is essential to dynamic optimization.

Recently, there were studies on applications of mixture models, which enable
modeling of multimodal distributions, and therefore facilitate solving optimization problems with many similar local optima. This paper proposes a novel
approach to solving DOPs with multimodal EDAs.

2   Algorithm

Multimodality in an EDA might be promoted by modeling the problem at hand
with a mixture model. This publication proposes to employ the Gaussian mixture
model; with M modes, pdf of distribution given by the mixture is

$$p(\boldsymbol{x}) = \frac{1}{M} \sum_{i=1}^{M} \frac{1}{\sqrt{(2\pi)^d |\boldsymbol{\Sigma}_i|}} \exp\left(-\frac{1}{2}(\boldsymbol{x} - \boldsymbol{\mu}_i)^\mathsf{T} \boldsymbol{\Sigma}_i^{-1}(\boldsymbol{x} - \boldsymbol{\mu}_i)\right), \qquad (1)$$

where $\boldsymbol{\Sigma}$ is diagonal (elements σ_{jj} are referred to simply as σ_j). Such distribution
is used at every generation to draw the entire population, the model is then
adapted to the population based on its fitness.

This paper proposes a new algorithm based on PBIL$_\mathrm{C}$ [11], Multimodal Continuous Population-Based Incremental Learning (MPBIL$_\mathrm{C}$), for approaching
DOPs. In the original work PBIL$_\mathrm{C}$ has been applied to continuous domains
by modeling gene distributions with normal distribution. Proposed extension
incorporates mixtures of Gaussians. M models of $\mathcal{N}(\boldsymbol{\mu}_i, \boldsymbol{\Sigma}_i)$ are maintained at
all times, allowing to model sub-populations focused on different local optima,
which might become global in subsequent timeslices. Update rules parametrized
with $(\alpha, \beta, \delta, \rho)$, similar to those in [11], govern updates of $\boldsymbol{\Sigma}_i$ and $\boldsymbol{\mu}_i$:

$$\boldsymbol{\mu}_i^{(t+1)} = (1 - \alpha)\boldsymbol{\mu}_i^{(t)} + \alpha(\boldsymbol{P}_i^1 - \boldsymbol{P}_i^N),$$

$$\boldsymbol{\Sigma}_i^{(t+1)} = \max\left(\delta, \; (1 - \beta)\boldsymbol{\Sigma}_i^{(t)} + \beta \cdot \mathrm{stdev}(\underbrace{\boldsymbol{P}_i^1, \ldots, \boldsymbol{P}_i^{\lceil\rho\cdot\mathrm{count}(i)\rceil}}_{\lceil\rho\cdot\mathrm{count}(i)\rceil})\right), \qquad (2)$$

where $\boldsymbol{P}_i^j$ is the jth fittest individual among the ones assigned to the ith Gaussian, and stdev($\cdot$) returns a diagonal matrix of standard deviations.

Pseudocode for MPBIL$_C$ is shown in Algorithm 1. Gaussians are initialized randomly in the search space with $\boldsymbol{\Sigma}_{init}$; with each timeslice, their $\boldsymbol{\mu}_i$ parameters are maintained while setting $\boldsymbol{\Sigma}_i$ back to $\boldsymbol{\Sigma}_{init}$. During sub-evolution, consecutive populations are drawn at each generation with the underlying Gaussian mixture. After evaluation, each individual is being assigned to the closest Gaussian, and their parameters are being updated. Gaussians are expected to iteratively converge towards Dirac delta function and halt with all σs equal to δ.

Algorithm 1. Pseudocode of the MPBIL$_C$ algorithm

1: $M \leftarrow InitializeModels(N, \boldsymbol{\Sigma}_{init})$
2: **while** not $TerminationCondition$ **do**   ▷ For each timeslice
3:      **for** $t \leftarrow 1$ **to** N_{gen} **do**   ▷ Begin sub-evolution
4:          $P \leftarrow DrawPopulation(M)$
5:          $EvaluatePopulation(P)$
6:          **if** $t \equiv 0 \pmod{t_r}$ **then**
7:              $EvaluateModels(M)$
8:              $m_w \leftarrow WorstModel(M)$
9:              $RandomlyInitializeModel(M, m_w, \boldsymbol{\Sigma}_r)$
10:         **end if**
11:         $AssignIndividualsToModels$
12:         $M \leftarrow AdaptModels(M, \alpha, \beta, \delta, \rho)$
13:     **end for**
14:     $ResetModelsStdev(\boldsymbol{\Sigma}_{init})$
15: **end while**

Depending on the objective, it is possible for many modes to converge to the same optima. A mechanism of randomly scattering the Gaussians, governed by parameters $(t_r, \boldsymbol{\Sigma}_r)$ prevents it. We measure the importance of each mixture's component by computing its overlap with other components:

$$\sum_{j=1}^{M} \exp\left(-\prod_{l=1}^{d}\left(\frac{\boldsymbol{\mu}_i^{(t)}[l] - \boldsymbol{\mu}_j^{(t)}[l]}{\boldsymbol{\sigma}_i^{(t)}[l]}\right)^2\right), \qquad (3)$$

where $\boldsymbol{\mu}[l]$ denotes lth element of $\boldsymbol{\mu}$. Every t_r iterations the Gaussian with the highest overlap with other mixture components is reinitialized in a random search space location with $\boldsymbol{\Sigma}_r$. Those parameters have to be chosen with care: t_r should be large enough to allow modes to converge, while Σ_r should reflect the size of the search space. Large values create an adverse effect of scattering individuals across all other Gaussians.

3   Validation of the Approach

3.1   Performance Measures

To assess performance of the proposed algorithm, different measures were employed during experiments. Nguyen et al. [7] summarizes various measures for DOPs; not all of them are feasible for EDAs, and some require full knowledge of the objective function (location of optima). The most general ones were chosen, with emphasis on measuring multimodality:

$$m_1: \quad E_{MO} = \tfrac{1}{n} \sum_{j=1}^{n} e_{MO}(j) \qquad m_6: \quad PC_{err}^{(t)} = \tfrac{1}{G} \sum_{i=1}^{G} \frac{\sum_{j=1}^{\#opt} err_{best(j)}^{(t)}}{\#opt}$$

$$m_2: \quad E_B = \tfrac{1}{m} \sum_{i=1}^{m} e_B(i) \qquad m_7: \quad err_{best}^{(t)} = \tfrac{1}{N} \sum_{i=1}^{N} F(best_{EA}^{(t)}) - Min_F^{(t)}$$

$$m_3: \quad I = \sum_{i=1}^{N} \sum_{j=1}^{P} (x_{ij} - c_i)^2 \qquad m_8: \quad \bar{F}_{BOG} = \tfrac{1}{G} \sum_{i=1}^{G} \left(\tfrac{1}{N} \sum_{j=1}^{N} F_{BOG_{ij}} \right)$$

$$m_4: explr^{(t)} = \frac{\sum_{i=1}^{N} dist(c_i, i) \times f(i)}{N} \qquad m_9: \quad stab_{F,EA}^{(t)} = \max\{0, acc_{F,EA}^{(t-1)} - acc_{F,EA}^{(t)}\}$$

$$m_5: acc_{F,EA}^{(t)} = \frac{F(best_{EA}^{(t)}) - Min_F^{(t)}}{Max_F^{(t)} - Min_F^{(t)}} \qquad m_{10}: ARR = \tfrac{1}{m} \sum_{i=1}^{m} \frac{\sum_{j=1}^{p(i)} [f_{best}(i,j) - f_{best}(i,1)]}{p(i)[f^*(i) - f_{best}(i,1)]}$$

along with *peak cover*, measuring the number of peaks found at moment t by having an individual within a peak's catchment area. In the experiments, catchment area has been simplified to a d-sphere. In addition, this paper introduces *peak cover error* (m_6) by analogy with *avg error of best individual* (m_7), and *exploration* (m_4) by analogy with *moment-of-inertia* (m_3) also considering quality of individuals.

3.2   Experiments

Variants of common benchmarks for DOPs were used in the experiments like the Moving Peaks Benchmark (MPB) [2], denoted by M_1. This paper also introduces dynamization of the Rosenbrock function's generalization, denoted by R_1:

$$f_{ros}(\boldsymbol{x}) = \sum_{i=1}^{d-1} \left((1 - x_i)^2 + 100(x_{i+1} - x_i^2)^2 \right). \tag{4}$$

For each pair of dimensions (i, j), where $1 \le i < j \le d$, rotation angle θ_{ij} is drawn from a normal distribution $\mathcal{N}(0, 0.5)$ to construct a rotation matrix $R_{ij}(\theta_{ij})$. Also, translation vector $(s_1, \ldots, s_d)^{\mathsf{T}}$ is drawn from $\mathcal{N}(0, 1)$ to construct a translation matrix. The final transformation matrix $M(t)$ is a product of all $\binom{d}{2}$ rotation matrices with translation as $M(t) = R_{12}R_{13} \cdot \ldots \cdot R_{(d-1)d}T$. Then, the objective function is $R_1(\boldsymbol{x}, \phi, t) = f_{ros}(\vec{M}(t) \cdot \boldsymbol{x})$.

To hinder solving problems M_1 and R_1, their variants have been introduced as M_2 and R_2 with applied cosine noise similar to that of the Griewank function:

$$M_2^{(i)}(\boldsymbol{x}, \phi, t) = M_1^{(i)}(\boldsymbol{x}, \phi, t) - w \prod_{i=1}^{n} \cos\left(\frac{c_i x_i}{\sqrt{i}}\right), \tag{5}$$

with R_2 constructed similarly. To compare the raw performance on compositions of popular functions (Sphere, Rastrigin, Weierstrass, Griewank, Ackley), MPBIL$_C$ has been also tested on IEEE CEC'2009 Competition on Dynamic Optimization [6] problems $F_2 - F_6$, dynamized using $T_1 - T_6$ change types. Revisions of the benchmark generator have also been present on other similar events.

On M_1, M_2, R_1 and R_2, MPBIL$_C$ was compared with a multimodal adaptation of Separable NES (SNES) [10], similar to Algorithm 1, but using SNES $\boldsymbol{\mu}$ and $\boldsymbol{\sigma}$ update rules. Tests on problems $F_2 - F_6$ with change types $T_1 - T_6$ were carried out using implementation available online [1]. The results allow for a direct comparison with a range of other contesting algorithms, using measures proposed in the competition. However, it should be noted that the benchmarks nor the employed measures promote multimodality. For this reason, the limitation on the number of function evaluations (FES) between timeslices was increased.

3.3   Results

Objective functions were parametrized with 5 random seeds chosen as numbers $1, \ldots, 5$, and results averaged over 30 runs for each seed. Thus, objective functions were always deterministic with respect to a random seed, while the algorithm behaved randomly with each subsequent run. F (its dynamization T) parametrized with random seed i is denoted by $F^{(i)}$ ($T^{(i)}$).

All benchmarks featured similar parameters concerning the amount of involved computations: population size $N = 300$, sub-evolution generations $N_{gen} = 800 - 1800$. In each case, MPBIL$_C$ featured $M = 10$ Gaussians, performing best in preliminary experiments for most of functions with considered population size. Every objective function underwent $T = 60$ timeslices (changes).

Results are presented in Table 1. In MPBs, MPBIL$_C$ was able to maintain and track a fair number of optima, resulting in lower errors, though exploration of search space suffers from fast convergence of Gaussians. Significant performance was achieved in R_1 and R_2.

Table 2 presents results of computations. Even though those objective functions might not necessarily promote multimodality, obtained results are in most cases comparable with best-performing algorithms of the competition, like jDE [3]. However, MPBIL$_C$ performed poorly on problem F_4.

Table 1. Performance of MPBIL$_C$ and multimodal modification of SNES on M_1, M_2, R_1 and R_2 under $m_1 - m_{10}$ measures. M_1, M_2 were run with $K = 10$ peaks, $d = 5$ dimensions, $(\alpha, \beta, \delta, \rho) = (0.4, 0.01, 0.0001, 0.15)$, Gaussians initialized with $\sigma_{init} = 5$ and restarted with $(t_r, \sigma_r) = (800, 0.5)$. R_1, R_2 were run with $d = 10$, domain restricted to $[-30, 30]^d$, adaptation parametrized with $(\alpha, \beta, \delta, \rho) = (0.4, 0.01, 0.00001, 0.15)$, Gaussians initialized with $\sigma_{init} = 0.6$ and restarted with $(t_r, \sigma_i) = (1000, 0.3)$. Cosine noise was parametrized with $(w, c_i) = (2.0, 1.0)$ in M_2 and $(w, c_i) = (20, 0.1)$ in R_2. In all cases positions of Gaussians were initialized with uniform distribution.

Measure	MPBIL$_C$					SNES					MPBIL$_C$					SNES				
	$M_1^{(1)}$	$M_1^{(2)}$	$M_1^{(3)}$	$M_1^{(4)}$	$M_1^{(5)}$	$M_1^{(1)}$	$M_1^{(2)}$	$M_1^{(3)}$	$M_1^{(4)}$	$M_1^{(5)}$	$R_1^{(1)}$	$R_1^{(2)}$	$R_1^{(3)}$	$R_1^{(4)}$	$R_1^{(5)}$	$R_1^{(1)}$	$R_1^{(2)}$	$R_1^{(3)}$	$R_1^{(4)}$	$R_1^{(5)}$
fbog	115	152.5	119.3	111.7	107.3	50.2	69.62	67.84	45.41	73.2	2.806	2.866	2.731	2.763	2.664	106.7	107.2	67.58	64.85	109.6
emo	84.4	45.51	4.587	41.86	48.5	89.53	56.32	17.16	64.65	145.5	10.44	11.76	122.4	118.3	131.3	442.2	214.1	173.1	178.5	231.1
eb	63.96	114.2	86.45	8.617	69.53	60.54	55.62	93.16	76.29	49.9	.0433	.043	.043	.0466	.0432	28.99	8.14	9.668	7.328	19.25
accur	.6603	.9213	.8609	.8585	.7415	.3225	.4175	.6713	.5716	.2091	1	1	1	1	1	.9989	.9989	.999	.9994	.9989
ln(inrt)	14.75	14.29	14.61	14.7	14.8	14.05	13	13.67	15.01	13.75	1.812	1.796	1.788	11.04	11.04	1.715	1.694	1.639	1.932	1.941
pk-cvr	2.5	4.146	4.121	2.92	2.285	.9988	.9987	.9998	.965	.9866	.6599	.6439	.6599	.6477	.6636	.0303	.0283	.0131	.0253	.0162
pk-err	52.74	33.16	18.89	25.54	42.6	51.87	39.8	48	54.11	85.2	7.778	9.518	1.275	8.723	9.932	107.6	108.7	68.12	65.51	11.21
stab	.6606	.9242	.8637	.8613	.7439	.3233	.4115	.6812	.568	.218	.855	.8553	.8597	.8573	.8566	.9989	.9989	.999	.9994	.9989
arr	.2978	.6736	.6404	.5729	.5199	.0422	.1242	.568	.2379	.1181	1	1	1	1	1	.898	.9005	.896	.8992	.8957
ln(explr)	13.47	14.46	14.24	14.37	14.25	13.62	13.12	13.8	14.7	13.18	11.76	11.77	11.61	11.91	11.79	1.685	1.712	1.618	1.802	1.822

Measure	MPBIL$_C$					SNES					MPBIL$_C$					SNES				
	$M_2^{(1)}$	$M_2^{(2)}$	$M_2^{(3)}$	$M_2^{(4)}$	$M_2^{(5)}$	$M_2^{(1)}$	$M_2^{(2)}$	$M_2^{(3)}$	$M_2^{(4)}$	$M_2^{(5)}$	$R_2^{(1)}$	$R_2^{(2)}$	$R_2^{(3)}$	$R_2^{(4)}$	$R_2^{(5)}$	$R_2^{(1)}$	$R_2^{(2)}$	$R_2^{(3)}$	$R_2^{(4)}$	$R_2^{(5)}$
fbog	117.5	89.05	85.12	10.42	104.8	4.049	4.088	4.033	4.017	4.11	1.089	1.087	1.087	1.086	1.089	56.76	85.47	77.34	57.65	126.3
emo	37.15	15.2	15.48	18.04	21.41	113.8	82.52	83.75	8.878	98.39	111.8	105.1	115.3	116.4	111.8	252.8	226.2	196.6	193.8	306.5
eb	86.23	7.181	75.97	73.2	8.903	3.668	3.69	3.65	3.638	3.696	.985	.9856	.9839	.9803	.9852	1.316	1.206	9.277	9.321	2.391
accur	.8291	.9091	.949	.9687	.7876	.0303	.0429	.0459	.0413	.033	1	1	1	1	1	.999	.9992	.9992	.999	.9987
ln(inrt)	18.01	17.95	18.38	18.36	17.81	8.798	9.447	8.718	8.549	8.356	1.815	1.813	1.81	11.05	11.05	11.5	11.37	11.6	11.44	11.45
pk-cvr	2.962	2.298	1.576	2.182	2.698	0	0	0	0	0	.0566	.0578	.063	.0576	.0561	.0318	.0226	.0259	.0292	.0251
pk-err	35.27	25.47	35.35	34.6	39.4	62	48.36	46.76	54.1	73.31	1	1	1	1	1	57.4	86.68	78.32	58.28	127.6
stab	.8292	.9091	.949	.9687	.7876	.0303	.0429	.0459	.0413	.033	.8541	.8518	.8565	.8522	.8538	.999	.9992	.9992	.999	.9987
arr	.597	.7165	.7963	.7675	.7368	0	0	0	0	0	1	1	1	1	1	.9237	.9257	.9243	.9262	.9234
ln(explr)	13.68	13.41	13.52	13.67	13.68	8.648	8.983	8.605	8.516	8.444	1.271	1.29	1.056	1.565	1.496	1.917	1.884	1.999	1.951	1.966

Table 2. Performance of MPBIL$_C$ on $F_2 - F_6$ objectives. $F_2 - F_6$ were run with $(\alpha, \beta, \delta, \rho) = (0.1, 0.05, 0.0001, 0.5)$ and $N_{gen} = 800$; problems F_2, F_5, F_6 with $(\alpha, \beta, \delta, \rho) = (0.4, 0.01, 0.0001, 0.15)$, $N_{gen} = 1800$ and $(t_r, \sigma_r) = (1000, 5)$. Results obtained for remaining random seeds $2, \ldots, 5$ were comparable with those presented.

Problem	Errors	$T_1^{(1)}$	$T_2^{(1)}$	$T_3^{(1)}$	$T_4^{(1)}$	$T_5^{(1)}$	$T_6^{(1)}$	$T_1^{(2)}$	$T_2^{(2)}$	$T_3^{(2)}$	$T_4^{(2)}$
F_2	Avg_best	.0305	.0606	.2396	.2517	.21	.2296	.0549	.0483	.1861	.2995
	Avg_worst	83.71	90.28	66.68	396.8	90.26	90.25	87.73	9.29	7.384	396.8
	Avg_mean	38.25	33.35	33.99	230.7	43.88	43.92	39.07	37.82	41.29	229.5
	STD	43.8	44.44	28.35	180	47.46	47.86	45.67	51.11	5.691	18.41
F_3	Avg_best	50.03	47.21	62.59	60.87	49.87	53.87	153.6	192.5	489	424.2
	Avg_worst	671	694.4	688.6	669.5	685.5	674.7	713.3	73.27	724.4	725.5
	Avg_mean	268.8	267.7	368.2	506.8	278.8	264.8	646.1	653.5	651.6	625.4
	STD	240.8	237.4	254	200.2	242.7	230.3	79.55	8.264	36.06	61.98
F_4	Avg_best	444.2	408.6	443.4	208.1	424.3	421.2	382.6	375.8	373.1	198.8
	Avg_worst	542.1	555.3	523.6	682.1	534.2	532.8	75.15	761.1	689.6	635.2
	Avg_mean	492.7	486.8	482.0	393.6	478.9	479.2	537.9	536.7	489.4	388.5
	STD	23.5	33.52	18.39	108.2	27.49	26.64	61.66	7.741	122.8	137.4
F_5	Avg_best	.1047	.1237	.1058	.3768	.2857	.3146	.1047	.1122	.1218	.4221
	Avg_worst	31.83	44.92	35.55	28.93	20.12	19.53	34.59	35.96	48.5	31.09
	Avg_mean	3.471	3.8	4.1493	4.195	2.471	1.999	2.682	2.075	1.027	4.735
	STD	5.971	35.5	8.7358	1.68	4.833	2.157	6.2	8.532	143.6	2.177
F_6	Avg_best	.1034	.1021	.1051	.1916	.192	.2021	.0992	.0977	.3466	.1913
	Avg_worst	42.09	8.692	4.355	424.6	67.56	69.22	64.09	87.78	73.47	361.8
	Avg_mean	14.21	17.02	6.943	24.23	21.45	21.62	2.008	2.088	29.15	23.81
	STD	152.4	42.16	23.58	73.3	42.07	41.73	44.27	45.8	41.43	57.88

4   Conclusions

This paper presents a multimodal estimation of distribution algorithm MPBIL$_C$, capable of solving dynamic optimization problems. The algorithm models problem at hand with a Gaussian mixture model and controls parameters of the mixture using PBIL$_C$-like rules. By iteratively adjusting parameters throughout sub-generations, modes of the mixture are expected to converge to optima, which would then be utilized at timeslice to track changes of the objective function.

To assess performance of the proposed algorithm, numerous benchmarks have been run on popular DOPs under different change functions, noise and random seeds. Multimodality has been verified through a set of measures promoting diversity and peak coverage. The algorithm has outperformed a multimodal modification of SNES on benchmarks M_1, M_2, R_1, R_2 and gave results comparable in most cases to those CEC'2009 winning algorithm jDE on problems $F_2 - F_6$. It should be noted, however, that as the wide range of possible DOPs is far from being covered by synthetic benchmarks, MPBIL$_C$ favors multimodality which is expected to be found in real-world problems, and not necessarily in benchmarks. Employed measures revealed that multimodality has been achieved, proving suitability of Gaussian mixtures for solving DOPs.

References

1. Special session & competition on "Evolutionary computation in dynamic and uncertain environments", http://www3.ntu.edu.sg/home/epnsugan/index_files/CEC-09-Dynamic-Opt/CEC09-Dyn-Opt.htm (accessed: May 1, 2014)
2. Branke, J.: Memory enhanced evolutionary algorithms for changing optimization problems. In: Congress on Evolutionary Computation, CEC 1999, pp. 1875–1882. IEEE (1999)
3. Brest, J., Zamuda, A., Boskovic, B., Maucec, M.S., Zumer, V.: Dynamic optimization using self-adaptive differential evolution. In: IEEE Congress on Evolutionary Computation, pp. 415–422. IEEE (2009)
4. Chen, S.H., Chen, M.C., Chang, P.C., Zhang, Q., Chen, Y.M.: Guidelines for developing effective estimation of distribution algorithms in solving single machine scheduling problems. Expert Syst. Appl. 37(9), 6441–6451 (2010)
5. Dong, W., Yao, X.: Unified eigen analysis on multivariate gaussian based estimation of distribution algorithms. Inf. Sci. 178(15), 3000–3023 (2008)
6. Li, C., Yang, S., Nguyen, T.T., Yu, E.L., Yao, X., Jin, Y., Beyer, H.G., Suganthan, P.N.: Benchmark generator for CEC 2009 competition on dynamic optimization. University of Leicester, University of Birmingham, Nanyang Technological University, Tech. Rep. (2008)
7. Nguyen, T.T., Yang, S., Branke, J.: Evolutionary dynamic optimization: A survey of the state of the art. Swarm and Evolutionary Computation 6, 1–24 (2012)
8. Pelikan, M., Goldberg, D.E.: Hierarchical problem solving by the bayesian optimization algorithm. In: Proceedings of GECCO 2000, pp. 267–274. Morgan Kaufmann (2000)
9. Santana, R., Larrañaga, P., Lozano, J.A.: Side chain placement using estimation of distribution algorithms. Artificial Intelligence in Medicine 39(1), 49–63 (2007)
10. Schaul, T., Glasmachers, T., Schmidhuber, J.: High dimensions and heavy tails for natural evolution strategies. In: GECCO, pp. 845–852. ACM (2011)
11. Sebag, M., Ducoulombier, A.: Extending population-based incremental learning to continuous search spaces. In: Eiben, A.E., Bäck, T., Schoenauer, M., Schwefel, H.-P. (eds.) PPSN 1998. LNCS, vol. 1498, pp. 418–427. Springer, Heidelberg (1998)
12. Shim, V.A., Tan, K.C., Cheong, C.Y., Chia, J.Y.: Enhancing the scalability of multi-objective optimization via restricted boltzmann machine-based estimation of distribution algorithm. Inf. Sci. 248, 191–213 (2013)
13. Yan, W., Xiaoxiong, L.: An improved univariate marginal distribution algorithm for dynamic optimization problem. AASRI Procedia 1, 166–170 (2012), AASRI Conference on Computational Intelligence and Bioinformatics
14. Yang, S.: Evolutionary computation for dynamic optimization problems. In: GECCO (Companion), pp. 667–682 (2013)
15. Yang, S., Yao, X.: Experimental study on population-based incremental learning algorithms for dynamic optimization problems. Soft Comput. 9(11), 815–834 (2005)
16. Yuan, B.: On the importance of diversity maintenance in estimation of distribution algorithms. In: Proceedings of GECCO 2005, pp. 719–726. ACM Press (2005)
17. Yuan, B., Orlowska, M.E., Sadiq, S.W.: Extending a class of continuous estimation of distribution algorithms to dynamic problems. Optimization Letters 2(3), 433–443 (2008)
18. Zhong, X., Li, W.: A decision-tree-based multi-objective estimation of distribution algorithm. In: 2013 Ninth International Conference on Computational Intelligence and Security, pp. 114–11/8 (2007)

Zero–Latency Data Warehouse System Based on Parallel Processing and Cache Module

Marcin Gorawski, Damian Lis*, and Anna Gorawska

Silesian University of Technology,
Institute of Computer Science,
ul. Akademicka 16, 44-100 Gliwice, Poland
{Marcin.Gorawski,Damian.Lis,Anna.Gorawska}@polsl.pl

Abstract. Zero–Latency Data Warehouse (ZLDW) cannot be developed and formed on the basis of a standard ETL process, where time frames are limiting access to current data and blocking the ability to take users needs into account. Therefore, after profound analysis of this issue and ones related to workload balancing, an innovative system based on a Workload Balancing Unit (WBU) was created. In this paper we present innovative workload balancing algorithm – CTBE (Choose Transaction By Election), which allows to analyze all incoming transactions and create a schema of dependencies between them. Also, cache in the created WBU ensures ability to store information on incoming transactions and exchange messages with systems transmitting updates and users' queries. By this work we intend to present an innovative system designed to support Zero–Latency Data Warehouse.

Keywords: Cache, CTBE algorithm, ETL, Workload Balancing, WBU, WINE–HYBRIS algorithm, Zero–Latency Data Warehouse.

1   Introduction

Continuous development in data analysis and processing domain has led to increase of data warehouses' significance in both science and industry. Moreover, growing need for processing the most current informations, without applying any time frames, became basis for the creation of new requirements, all fulfilled in the Zero–Latency Data Warehouse (ZLDW) model. Zero latency [1] in context of data warehousing means that each incoming data must be automatically processed and stored in the ZLDW without any additional delays. The concept of ZLDW [8,9] emphasizes user's preferences for analysis of updates and queries, while abandoning time frames. Thus, it is impossible to use classical ETL processes [13] to support a new type of data warehouse, since among variety of problems with adapting aforementioned process [2] to real–time processing model, its time constraints are unacceptable.

---

* Project co-financed by the European Union under the European Social Fund. Project no. UDA-POKL.04.01.01-00-106/09.

E. Corchado et al. (Eds.): IDEAL 2014, LNCS 8669, pp. 465–474, 2014.

Our first performed work on the ZLDW concerned creation of a suitable Work-load Balancing Unit [3,6,7,12,10] and modeling of a brand new ETL process, which would be able to meet imposed requirements. Conducted research resulted in a system equipped with a pioneering ETL process based on a WINE–HYBRIS algorithm. The process has been implemented as a part of the WBU, which alone is responsible for selection of next transaction, response time and data updates. For increasing computing power, required for analysis and additional calcula-tions, cloud computing with Windows Azure and Google App Engine were used. We have also carried out work on CUDA architecture [5] as one of many alter-natives to provide adequate computing power.

2   WBU–Based Zero–Latency Data Warehouse Architecture

The ZLDW environment (Figure 1) was created through integration of few com-ponents, i.e. source devices, Hostimp servers, web–services, cache memory mod-ule, WBU and finally ZLDW.

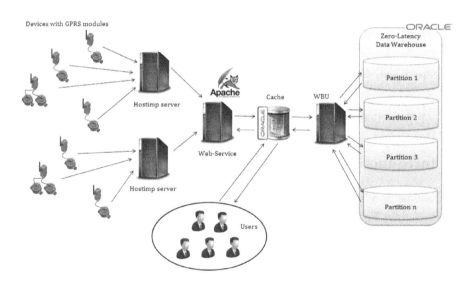

Fig. 1. Zero–Latency Data Warehouse system

Measuring devices are able to transfer data to any location through GPRS modules. Due to the nature of their output stream, in each and every case information must be decoded and then converted to correct format. This task involves hostimp servers, which store schemes of reading and processing them into a format suitable for the WBU. Hostimp servers, which cardinality is not limited, transmit data via a web–service to cache. Only then information is available to the WBU.

Workload Balancing Unit is responsible for global planning and execution of transactions as well as managing ZLDW's partitions. WBU was based on the workload balancing algorithm CTBE, which will be described later. Additionally, through use of cache, it is possible to accelerate information exchange with source systems. In systems based on the 'WINE'–based algorithms additional memory is not utilized, which results in transferring parts of system's resources for data sources monitoring.

The WBU was built on the basis of two types of transactions – read–only and write–only, which are executed in three independent processing pipelines. Created system is designed to ensure information freshness and minimization of response time. Data is transferred to cache through web–service so that all informations received from the hostimp servers are being passed directly into memory. Processing ends with download to the WBU.

Hostimp Servers are used to receive data from measuring devices and decode information. Web–service layer in turn mediates communication between servers and the WBU. With the development of the system it will be possible to use a greater number of servers, both Hostimp and web–service. Only WBU cannot be duplicated because of the direct access to the data warehouse. If it had a larger number of units, it would not be possible to maintain consistency in the ZLDW system.

3   Workload Balancing Algorithm CTBE

Choose–Transaction–By–Election algorithm was developed specifically for the ZLDW to manage data extraction. It is implemented in the WBU, where all transactions are analyzed before sending to the ZLDW and then converted in an appropriate manner.

The CTBE, through the sake of user's preferences, is able to correctly prioritize queries as well as updates, thus retaining appropriate balance between quality of data and service along with allowing to meet high demands of users. However, due to modifications of enqueueing method, changes in prioritizing and increasing number of processing queues from two to three, the algorithm is now able to minimize processing operations. The algorithm is constructed to support parallel processing, so that it is possible to reduce processing and analysis times to minimum.

In contrast to algorithms from the WINE group, the CTBE algorithm is supported by cache, through which all informations are being transferred to and from the WBU. Transactions entering processing pipeline are analyzed in terms of type (query or update) and assigned to two of three queues. The first queue, the queue of transactions T, buffers all queries and updates. The next two are divided into read–only (queries) and write–only transactions (updates). Queries are enqueued in two separated queues Q and T: $\{q_i : q_i \in Q \vee q_i \in T\}$. Updates as well as queries are written to the T queue, however they are linked to the update queue, which gives us third one: $\{u_j : u_j \in U \vee u_j \in T\}$.

All transactions brought to the T queue are marked with timestamp tq_i, which spells out time of arrival to cache memory. However, queries in both T and Q queues, in addition to the assigned timestamp tq_i, are redescribed further by QoS (Quality of Service) and QoD (Quality of Data) values. In turn, updates arriving to the U queue are defined only by timestamp tu_i in the initial phase, as it is for the queue of transactions.

Only in the analysis phase, each transaction will be examined and assigned additional parameters, allowing swift prioritizing.

Preliminary Analysis and Data Partitioning. Before beginning queries prioritization and determining degree of their freshness, each transaction must be examined with emphasis to identifying dependencies with ZLDW's partitions. The ZLDW system cannot be considered as a coherent whole, because then every single update would have to have a relationship with every query, which would lead to unworkable system policies. Therefore, system is divided into n separated partitions P, which can be represented as ZLDW $= \{P_i | 1 \leq i \geq n\}$. This allows more precise mapping of relationships between queries and updates. Moreover, it is possible to link query with more than one partition. Although the system provides possibility to create more threads, number of links between updates and partitions should not exceed value of 1.

3.1   Transactions Scheduling

After assigning transactions to appropriate partitions scheduling takes place. The very first operation is checking number of items in queries, updates and transactions queues. Number of transactions in the T queue should be as follows:

$$\sum_{a>0} T_a = \sum_{i \geq 0} Q_i + \sum_{j \geq 0} U_j. \tag{1}$$

Depending on the number of transactions in queues one of three different steps is executed:

1. when one of U–updates or Q–queries queues is empty,
2. when both U and Q queues are not empty and $\sum QoD > \sum QoS$,
3. when both U and Q queues are not empty and $\sum QoD < \sum QoS$.

First variant, when one of queues (U or Q) is empty, allows to reduce number of operations and is the simplest of all three options. In processing pipeline all operations are performed only on transactions buffered in T queue. As a result, transactions are sorted with ascending timestamp t_a order. The basis for this assumption lies in the absence of one of two, read or write, modes which makes impossible to link them and determine degree of queries freshness. As a result, transactions are executed with already established timestamp order, i.e. from the oldest one.

Two remaining steps are taken into account only when updates and queries queues have at least one element each. At given time for each query from the Q

queue values of QoS and QoD are added up. If $\sum QoS \geq \sum QoD$ only queue of queries will be selected to analysis and a request will be sent to the ZLDW. First value of QoS' is calculated, which for incoming queries is 0, in accordance to following formula:

$$QoS'(i) = \begin{cases} QoS(i) & \text{if } QoS'(i) = 0; \\ QoS'(i) + \left(\frac{1}{|T|*r_{max}}\right) & \text{if } QoS'(i) > 0. \end{cases} \tag{2}$$

Parameter r_{max} is an integral part of the system, because every modification of its value affects the rate of growth of QoS'. Another words, r_{max} enables management of processing speed by determining mentioned rate of QoS' growth.

Calculations for all queries residing in the Q–queries queue are followed by scheduling – descending in terms of QoS' values and ascending with timestamp. Only then query with the greatest value of QoS is executed on the data warehouse. With the sort based on the values of QoS', none of queries will be wrinkled. Therefore, waiting time may lengthen a bit, however due to low value of QoS' data availability and safety requirements will be fulfilled.

On the other hand, if a higher priority was granted to the update queue $(\sum QoD > \sum QoS)$, users are focused on retrieving the most current responses and it is required to analyze the update queue. Setting priorities for updates must be preceded by query's prioritization. It is made in the same manner as in the second stage. Without proper ordering the system is unable to determine which update must be executed first. Updates are scheduled upon values of (u) measure, also used in the algorithms from WINE group:

$$w(u) = \sum_{\forall q_i, |P_{qi} \cap P_u| = 1} \frac{qod_{qi}}{1 + pos_{qi}} \tag{3}$$

Upon this equation, updates are dependant of queries, more precisely of their place in the Q queue and original value of QoD. Calculating value of $w(u)$ for each of the updates is followed by scheduling – descending according to $w(u)$ and ascending with timestamp t_i. Thanks to scheduling of timestamps we are avoiding risk of overwriting data with obsolete information.

Example presented in Figure 1 shows how analysis is performed and illustrates character of input and output data sequences. Input sequence stored in cache contains only basic informations which are used later in the analysis of transactions. After downloading data from cache by the WBU, transactions are being allocated to appropriate queues. Each and every inquire is always enqueued in the T queue, meanwhile queries and updates are assigned the Q and U queues respectively. After queues' analysis are updated and filled with relevant data – such as information on partitions individual queries and updates belong to. After verification of aggregated values of QoS and QoD, it can be seen that QoS is grater, which means that query to execution will be chosen on the highest QoS' value and the smallest timestamp – for example, Q_1 query on the Figure 1. After successful transaction's commit it will be deleted from the T and Q queue as well. Then the whole process repeats with another transaction selected – query or update.

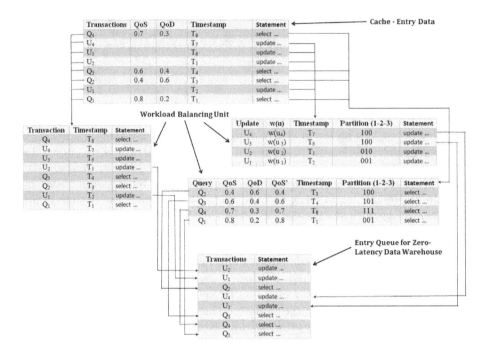

Fig. 2. Queues in transactions processing pipeline

4 Tests and Comparison

Entire data warehouse environment has been configured on two PCs. On a computer with a processor i5 and 8 GB of RAM, there have been Workload Balancing Unit and Web–Service launched. In turn, on the second computer based on the E8400 processor and 3GB of RAM, cache was configured and the ZLDW installed. During testing phase there has been about 500 measurements obtained, which allowed to set optimal dependencies between all consecutive elements of the system. The first test was drawn to the primary system, in which parallel processing was not taken into account in the WBU. In each test, trial and during each iteration number of updates and queries was equal.

To illustrate effectiveness of created solution we have performed several tests of the CTBE algorithm that measure processing times for inquires in update and queries queues. Moreover, experiments took into account two separate approaches to processing – single thread and parallel.

Figure 3 presents outcomes for short queues of queries and updates, i.e. less than 500 items in each queue. By analyzing results it stands out that even for a small number of queries and updates, processing time is significantly different for both queues. The reason is hidden in different approaches to processing. Processing tasks from the updates queue takes the most time – from 100 to 300 ms, as in the analysis and sorting stage, each individual update is analyzed in

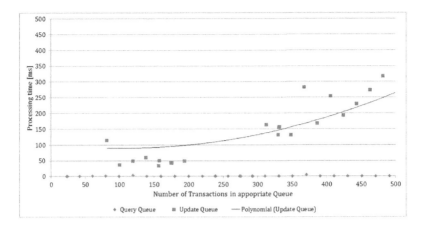

Fig. 3. Processing short queues in a single thread

accordance to each query stationed at that time in the system. In turn, queries are analyzed and prioritized in isolation from buffered updates.

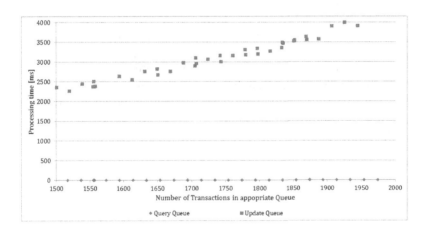

Fig. 4. Processing semi–long queues in a single thread

During another round of tests system without parallel processing unit was verified again, but this time at an angle of semi–long queues (Figure 4) processing. Results did not bring any surprise – it is ripening very large discrepancies in obtained processing times. While for short queues processing time was from 100 to 300 ms, results for slightly longer ones have increased to 2500 – 4000 ms. Due to such great differences the CTBE algorithm had to be adapted to parallel processing, so for huge inflows of transactions it would be possible to utilize results in a timely manner.

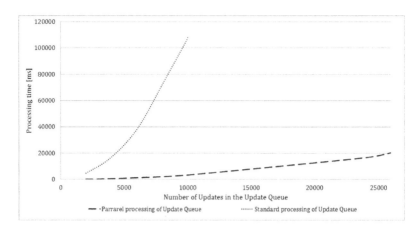

Fig. 5. Processing time of update queue's inquiries

Moreover, Figure 5 shows timing differences in processing queue of updates, for which time costs were too high. Chart illustrates another point of view – results obtained for the same data sets but with two different processing methods. First in which processing time is much higher, was launched in a single–thread mode, which results in faster increase in processing time. It is not acceptable for larger systems, like DWs, due to lengthening of delays. However, outcomes from the system based on parallel processing, highlights possibility of delay minimization while times did not exceed 20 000 ms for 25000 updates. With this example, it should also be mentioned that number of queries which were taken for testing, was equal to number of updates. This is very important because updates depend on queries and are analyzed with respect to them. This assumption implies that number of calculations grows polynomially.

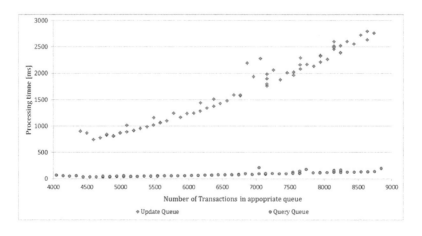

Fig. 6. Updates and queries processing time comparison – parallel model

The last of experiments, results of which are shown in Figure 6, was carried out for the final version of the system handling the ZLDW. Each of threads in the initial phase is processing not more than 1000 operations. With increase in number of operations, each thread (number of threads must not exceed number of updates) is supplied with additional operations, which is indirectly visible in values of processing time.

In the chart query queue processing time increases linearly without any radical strokes. Situation is different for updates where for 4500 – 9000 updates enqueues (with query queue size is 4500 – 9000 transaction), time begins to grow rapidly, due to quadratic increase in number of calculations to be performed. For example, with 100 objects in each queue, when analyzing query queue only 100 calculations are executed, meanwhile for 100 enqueued updates there will be 100 x 100, which gives us 10 000 main calculation.

With results presented above, it is possible to notice that the system is stable when processing 50 000 transactions at ones. Challenge of processing such number of transactions is not only deliverable but trivial, even though created system was configured for classic PC and not a specialized server or even in cloud computing environment.

5 Summary

Introduction of a new type of system based on the CTBE workload balancing algorithm has streamlined transaction processing in the ZLDW. Systems designed based on WINE and WINE–HYBRIS algorithms [4] need high computing power. With an increase in number of updates and queries such systems were not able to timely process transactions. With introduction of the CTBE algorithm, there is no need to use cloud computing or CUDA architecture, because power of a standard processor i5 is sufficient. In turn, cache enables effortless communication between modules and reduces Workload Balancing Unit's resource utilization that periodically retrieves data. Application of the algorithm makes it possible to locate the entire system inside corporate infrastructure limiting vulnerability to attacks and intrusion attempts [11] (compared to cloud computing, where whole system is placed on external servers). Moreover, such construction is compatible with different types of data sources. The WBU does not work directly on data contained in the transactions but on informations that describe them, which allows yielding responses in a short period. Therefore aforementioned specific requirements that emerged from the ZLDW model, has been successfully met.

References

1. Bruckner, R.M., Tjoa, A.M.: Capturing Delays and Valid Times in Data Warehouses - Towards Timely Consistent Analyses. J. Intell. Inf. Syst. 19(2), 169–190 (2002)
2. Gorawski, M., Gorawska, A.: Research on the Stream ETL Process. In: Kozielski, S., Mrozek, D., Kasprowski, P., Małysiak-Mrozek, B. z. (eds.) BDAS 2014. CCIS, vol. 424, pp. 61–71. Springer, Heidelberg (2014)

3. Gorawski, M., Gorawski, M., Dyduch, S.: Use of Grammars and Machine Learning in ETL Systems that Control Load Balancing Process. In: HPCC 2013. Fourth International Workshop on Frontiers of Heterogeneous Computing, FHC 2013, vol. 370, pp. 1709–1714. Institute of Electrical and Electronics Engineers, Piscataway (2013)
4. Gorawski, M., Lis, D., Gorawski, M.: The Use of a Cloud Computing and the CUDA Architecture in Zero-Latency Data Warehouses. In: Kwiecień, A., Gaj, P., Stera, P. (eds.) CN 2013. CCIS, vol. 370, pp. 312–322. Springer, Heidelberg (2013)
5. Gorawski, M., Lorek, M., Gorawska, A.: CUDA powered user-defined types and aggregates. In: 27th International Conference on Advanced Information Networking and Applications Workshops, WAINA 2013, pp. 1423–1428. IEEE Computer Society (2013)
6. Gorawski, M., Wardas, R.: The workload balancing ETL system basing on a learning machine. Studia Informatica 31(2A(89)), 517–530 (2010)
7. Karagiannis, A., Vassiliadis, P., Simitsis, A.: Scheduling strategies for efficient ETL execution. Inf. Syst. 38(6), 927–945 (2013)
8. Nguyen, T.M., Brezany, P., Tjoa, A.M., Weippl, E.: Toward a Grid-Based Zero-Latency Data Warehousing Implementation for Continuous Data Streams Processing. IJDWM 1(4), 22–55 (2005)
9. Nguyen, T.M., Tjoa, A.M.: Zero-latency data warehousing (ZLDWH): the state-of-the-art and experimental implementation approaches. In: 4th International Conference on Computer Sciences: Research, Innovation and Vision for the Future RIFV, pp. 167–176. IEEE (2006)
10. Piórkowski, A., Kempny, A., Hajduk, A., Strzelczyk, J.: Load balancing for heterogeneous web servers. In: Kwiecień, A., Gaj, P., Stera, P. (eds.) CN 2010. CCIS, vol. 79, pp. 189–198. Springer, Heidelberg (2010)
11. Skrzewski, M.: Monitoring malware activity on the LAN network. In: Kwiecień, A., Gaj, P., Stera, P. (eds.) CN 2010. CCIS, vol. 79, pp. 253–262. Springer, Heidelberg (2010)
12. Thiele, M., Fischer, U., Lehner, W.: Partition-based workload scheduling in living data warehouse environments. Inf. Syst. 34(4-5), 382–399 (2009)
13. Waas, F., Wrembel, R., Freudenreich, T., Thiele, M., Koncilia, C., Furtado, P.: On-Demand ELT Architecture for Right-Time BI: Extending the Vision. IJDWM 9(2), 21–38 (2013)

Distributed Multimedia Information System
for Traffic Monitoring and Managing

Aleksandar Stjepanovic and Milorad Banjanin

University of East Sarajevo, Faculty of Transport and Traffic Engineering, Doboj, RS, BIH
aco_stjepanovic@yahoo.com, banjaninmilorad@gmail.com

Abstract. In this paper, we created the multimedia web application with ITS (Intelligent Transportation Systems) implemented in DMIS (Distributed Multimedia Information System)(http://www.itsdoboj.hostoi.com/cns_admi- nistrator.php),for traffic monitoring and managing in Doboj- Prnjavor regional road section, the Republic of Srpska (RS), Bosnia and Hercegovina (BH). The focus of the research goals is oriented on the CCMS (Central Control Multimedia System) modul for experimental monitoring of public bus transport and the creation of opportunities for informing the users proactively about the current location of the buses in the Doboj-Prnjavor road section and the time of arrival at the destination stop, and monitoring the number of contextual parameters of this system.

Keywords: Intelligent Transportation System, Web application, Quality of Services , Quality of Experiences, traffic monitoring, page load time, Central Control Multimedia System, system performance.

1 Introduction

For the functions of the distributed multimedia information systems for transport and traffic management many multimedia applications were developed. Most of them have been developed for the traffic monitoring and measuring rating of the quality of services (QoS). The standard ITU-T-E.800 defines QoS as „a set of requirements for quality of service which describes the behavior of one or a group of media objects of a multimedia system". Seen from a broader point of view, QoS "is a set of quantitative and qualitative characteristics of a distributed multimedia system that are needed to realize the desired functionality of the application" [3]. According to this definition, QoS is on one hand user-oriented (includes subjective user satisfaction), and on the other hand reflects the specific obligation of the network to meet those needs. In a narrow sense, QoS identifies the factors that can be directly observed and measured at the point where the user accesses the service [3]. In this research, an application was developed to complement QoS with the some dimensions of Quallity of Experience (QoE) through the implementation of Intelligent Transportation System (ITS). Although QoE is much more difficult to determine. The ITU-T P.10/G.100 standard classifies this dimension of quality as a general acceptability of an application or service by the end user perception of the function. Thus, QoE includes the

E. Corchado et al. (Eds.): IDEAL 2014, LNCS 8669, pp. 475–483, 2014.

overall functioning of the system from end to end (user, terminal, network, service infrastructure), where as the general acceptability depends on several factors such as the content provider, the user expectations, the emotional factor [3]. The analysis of the service performance and contribution to the quality of customer satisfaction resulting from the implementation of ITS is possible using different methods [1],[5]. The information about the degree of satisfaction with the quality of the implemented ITS can be obtained by solving the Multi-faceted complex problems. Different user groups react differently to certain events beacuse of variety of parameters that have direct and indirect impact on that (the previous customer experiences, the level of education, emotional factors, etc.) [1]. Relevant information about the quality of service and the levels of customer satisfaction can be obtained from the established relationship between QoE and QoS [1]. To calculate QoS, the network parameters or the communication channels, such as the information flow, the number of packet loss, the packet delay, the speed of loading web pages is being observed [2]. For the assessment of the level of the user satisfaction with a service [4], the MOS (Mean Opinion Score) method is often used and it focuses on an "average" user of a product [3]. Another method for assessing the degree of the user satisfaction with the application or the service is the WQL hypothesis which shows the relation between QoS and QoE as a logarithmic function of Waiting time and QoE, on the linear ACR (Absolute Category Rating) scale. According to the WQL Hypothesis *"The relationship between Waiting time t and its QoE evalution on a linear ACR scale is Logarithmic"*) [3], the QoE function can be summarized with the form [3]:

$$QoE = k \cdot \ln(t) + c \qquad (1)$$

where k and c are the experimentally obtained constants, t is the load time of web pages. By the subjective measurement or estimation of MOS, the level of satisfaction with the average "type" users with CCMS service is determined. This approach has its strengths and weaknesses. In article [17] the authors made another approach to QoE measurements, without the use of MOS factors.

2   Basic Hypothesis

For the purposes of research and experiments in this paper, we selected an experimental section of road Doboj-Prnjavor regional road, the Republic of Srpska, BH, and proposed a model of Distributed Multimedia Information System (DMIS) for traffic monitoring and managing at University of East Sarajevo, Faculty of Transport and Traffic Engineering, Doboj, BH, which is a separate CCMS (Central Control Multimedia System) modul for monitoring and control of the intercity bus transport. The structure of DMIS is modular (Fig.1) with four main subsystems: the one for the traffic management, for the transport management, the contextual modul and the residual modul with implementing ITS and CCMS. The idea is to design the first implemented DMIS with ITS service in the territory of BH, because currently there is not any application for the traffic monitoring and managing. The concept of the proposed architecture is based on the recommendations of the European architecture for

ITS (FRAME European ITS Framework Architecture) and is in line with the international standard ISO 14813-1 according to which the service and the domain of ITS were identified as an eleven group system. In the contextual subsystem the CCMS functional modules are responsible for certain processes. The sensor module includes a GPS/GPRS (Global Positioning System/General Pocket Radio Service) device (m:smart phone based on Android OS), which is installed in the bus and is responsible for determining the geolocation of the bus in the experimental section of the road.

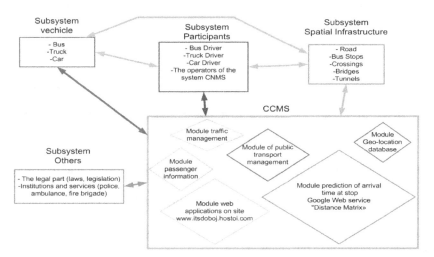

Fig. 1. Modular structure of the DMIS and CCMS for monitoring and managing of bus transport

The obtained data of the geolocation of the buses using the installed Android apps and GPRS technology are sent to the CCMS at the module responsible for the monitoring and managing of the public bus services. In order to detect the geolocation of the vehicles and send the data by the GPRS technology to the desired web page, the publicly available application "Self Hosted" [5], which is installed on the "smart phone", is used. This idea is based on the example of the use of "smart phones" as intelligent sensors, which is treated in this paper [6] with the aim of monitoring the degree of occupancy of a bus. For the purposes of the experiment the contextual functionality of the system is extended so the system is enriched with two cameras for monitoring the traffic at critical points and the software sensor based on a web camera and software SpeedCam2012 [7] for measuring the speed of the vehicles on the observed section. The graphical display of the geolocation of the elements of the system is shown in Figure 2. For the purposes of the experiment and in order to study the interaction between the users and the systems, a multimedia tablet device at a bus stop in a village of Stanari was built in. The Android applications for the interaction with the users and for the connection to CCMS were installed on this device. The flowchart of the data between the users, the tablet devices and CCMS is depicted in Figure 3. By using web applications on the bus stop location the users have three options for monitoring the system performance:

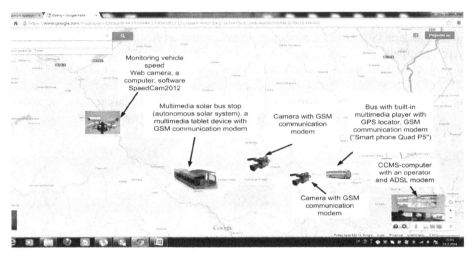

Fig. 2. Space map of layout elements on the road section Doboj-Prnjavor

(1) to identify the current location of the buses in the road section and the expected arrival time to the bus stop;
(2) to review the timetable and the service information related to the road conditions and weather parameters, and
(3) to receive information on possible causes of bus delays, failures, and other contextual information to the end users.

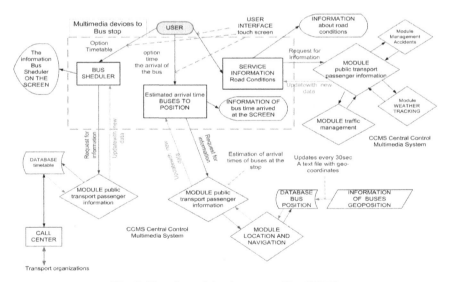

Fig. 3. Flowchart of the data-user-tablet-CCMS

Another form of interactive access to the system is via the multimedia CCMS website at: http://www.itsdoboj.hostoi.com. The user interaction with the individual modules of CCMS is designed as multi-modal, including text, audio, haptic and graphic visual channel with multimedia web applications. Figure 4 shows the homepage of the site. The algorithm of CCMS includes the following procedures and operations:

1)Vehicle location via the built-in GPS device (m:smart phone Allview P5 Quad);
2)The obtained data on the geolocation of the buses via HTTP protocol are sent every 15 seconds to the web site: http://www.itsdoboj.hostoi.com/cns_administrator.php;

Fig. 4. Home Page of CCMS Doboj

3) PHP (Hypertext Preprocessor) script on the server computer in the CCMS receives the data in a text file that contains information about the location of the bus and the time of sending data;
4) The application performs the processing of the data and on the basis of that a marker icon with a bus ticket is set on Google.Maps;
5) The current position of the buses and the distance to the bus stop on the route are the input to Google's web service "Distance Matrix API", which was used for the estimation of the required time of arrival of the buses and the calculation of the current distance;
6) The obtained data are returned to the user in the form of information of the HTTP protocol on the multimedia device about the position or on the website through which the users access the applications.

3   Results and Presentation of Methods

The research results of the level of the customer satisfaction are related to the registration of the speed of loading a Web application and the user's sense of applications quality expressed by a subjective evaluation of the tested users. In addition, important parameters are also the visual user experience, ease of access to the applications and navigation through the application [8]. As already mentioned, parameters such as

Bandwidth, Bit Rate, Delay, Jitter and Loss Rate [9], are usually associated with measuring QoS, while on the other hand Responsiveness (Promptness), Interactivity, Availability, Resilience, Task completion, Acceptability, Fatigue (Tiredness), Satisfaction, Delight (Annoyance), Joy [10] and other are usually associated with QoE. From the above mentioned factors, it is evident that the QoE evaluation becomes very complex and QoE can be expressed in terms of functions which includes several parameters [3]: f(System, User State, Content, Context). The methods of measuring QoE can be divided into two groups [3]:

1.Subjective assessment of QoE based on the measurements of: the customer reviews, the technology assessments; Objective measurement: performance of execution of tasks, behavior;
2. "Objective" QoE prediction based on: analytical/statistical models, translation of the input parameters in the estimated QoE.

3.1   Web Applications for Evaluation User Experience

The multimedia web applications used in the experiment are available at (www.itsdoboj.hostoi.com) and were developed for the experiment based on the technology with open source software such as HTML5, PHP, CSS3. For hosting the multimedia web application a free provider of hosting services was selected, which is located on the following site www.000webhost.com [11] where the registration is made and subdomain is selected (www.itsdoboj.hostoi.com). For Web applications testing, the tools freely available on the Internet were selected (XAMPP - Apache web server) [12]. The selected parameter for test was the speed of opening pages connected via a form (1) for the user QoE. Using the tools available on the web pages: http://tools.pindom.com/,[13]; http://www.webpagetest.org, [14]; http://gtmetrix.com, [15] the testing was carried out and the results obtained are shown in Table 1:

Table 1.

TOOLS	PERF/GR.	REQUS T.	LOAD TIME	PAGE SIZE
pingdom	80%	17	0.874s	244.1kB
webpa-getest	41%	19	4.096s	246kB
gtmetrix	46%	16	4.65s	238kB

Table 1. shows that the speed of loading the pages of the web site for the two test programs is fairly homogeneous, (4.096s and 4.65s), while the PLT (page load time) to "Pingdom" tool was 0.874s.

According to the analysis [3],[16] the users considered the page load time of 0.100s very fast, while the load over 10s usually gets negative ratings and the users generally give up on the access to such sites. Locations for "Pingdom" tool are in New York City-USA, for "webpagetest" in Vienen-Austria and for "gtmetrix" in Vancuver-Canada. On the other hand, the analysis of loading the entire page is done with regard to various parameters such as *the state of the load cycle and the load time spent in elation to the content of the page.* Table 2. shows the results for Load-Time Analysis Time per State:

Table 2.

Connect	45.52%
DNS	19.57%
Receive	18.19%
Wait	16.69%
Send	0.03%

From Table 2. it is evident that for the connection to the server the largest percentage of time is spent 45.52%, while sending data to server 0.03% of the time is spent. On the other hand, in Table 3. it can be seen that the loading of images and photos took the most, about 49% of the total time was spent. Table 3. shows the results for Time Spent per Content Type:

Table 3.

Image	49.01%
HTML	22.01%
Script	16.11%
CSS	12.88%

Application testing was made in a sample of 50 users via the MOS factor function where the system parameter is PLT. The results obtained are shown in Figure 5.

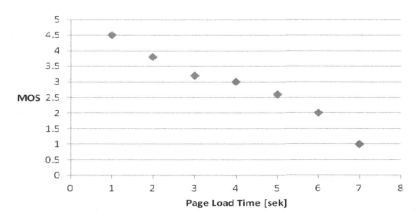

Fig. 5. MOS factor depending on the PLT

The coefficients k = -1.0101 and c = 4.5, are obtained from the graph and there are used in equation (1), and QoE is presented in the new form:

$$QoE = -1.01010 \cdot \ln(t) + 4.5 \qquad (1)$$

Form (2) shows the relationship between the page load time and the user QoE, which follows a logarithmic dependence of QoE on PLT.

4   Discussion

When a user accesses CCMS through a web application, he/she has the feeling that the system does not have an adequate response if the website load time is very large i.e. more than 10 seconds. In such situations, the MOS factor has a value between 0 and 0.5. For the page loading values less than 1sec the MOS factor is about 4.5.

The number of lost packets on the other hand is closely associated with the application designed for the packet speech. This paper is not focused on the voice applications, but the performance analysis of available networks and the Internet service provider for the Doboj area was tested, and it was determined that the existing telecommunications network packet loss was 75 %, while the MOS factor was 4.32 with a download speed of 3.09Mbps and upload speed of 0.82Mbps. QoS was around 98 % and the RTT (Round Trip Time) 280ms. By analyzing the page load times depending on the content it can be seen that the size of the images has a dominant impact on the load time. According to Table 3. for uploading images ther was used 49 % of the time. When designing Web applications that contain a lot of photographs, it is common to optimize the photographs. For the optimization of them JPG, GIF, PNG images formats are usually used.

5   Conclusion

The main research results of this paper is the multimedia web application with ITS, created as a software interface between the user and the CCMS. To assess the level of satisfaction of the customer specific systems or applications, it is suitable to analyze the QoS parameters such as the data speed, the speed of the connection, the communication security, the number of lost packets, the speed of loading web pages. By linking the measurable QoS parameters and the quality based on QoE which is a measurement of the user satisfaction with the offered application or system, it is possible to estimate the level of the user satisfaction. For the analysis QoE of multimedia application with ITS, the paper proposes a distributed multimedia information system (DMIS) used for the experiment, intended to actively monitor and control the public intercity bus transport. Using the MOS methods and WQL hypothesis, the evaluation of the degree of satisfaction with the implemented system or service was assessed. The analyzed performances were: the speed of loading web pages, and the number of lost packets in the streaming, which are important from the user's standpoint who accesses the web application with ITS and including the quality of services and experience from the interaction between the users and the applications it based on open source technologies such as HTML5, JavaScript and PHP language. In order to test the application, XAMPP package with the Apache web server was used and its associated modules on the platform (m: smart phone based on Android OS) with services (m: Smart phone Quad Allview P5), Google's web service "Distance Matrix API", and the tools available on the web pages http://tools.pindom.com; http://www.webpagetest.org; http://gtmetrix.com.

References

1. Kim, H.J., Lee, D.H., Lee, J.M., Lee, K.H., Lyu, W., Choi, S.G.: The QoE Evaluation Method through the QoS-QoE Correlation Model. Published in NCM 2008 Proceedings of the 2008 Fourth International Conference on Networked Computing and Advanced Information Management, vol. 02, pp. 719–725. IEEE Computer Society, Washington, DC (2008)
2. Reichl, P., Egger, S., Schatz, R., D'Alconzo, A.: The Logarithmic Nature of QoE and the Role of the Webwer-Fechner Law in QoE Assessment. In: IEEE ICC 2010 Proceedings (2010)
3. Schatz, R., Hossfeld, T.: Web QoE Lecture1: Quality of Experience. Phd School Krakov, Cost TMA, February 13-17 (2012)
4. Hoßfeld, T., Hock, D., Tran-Gia, P., Tutschku, K., Fiedler, M.: Testing the IQX Hypothesis for Exponential Interdependency between QoS and QoE of Voice Codecs iLBC and G.711. In: 18th ITC Specialist Seminar on Quality of Experience, Karlskrona, Sweden (May 2008)
5. https://play.google.com/store/apps/details?id=fr.herverenault.selfhostedgpstracker
6. Seo, J.W., Hwang, D.G., Lee, K.H., Kim, K.S., Jung, I.J.C., Kim, Y.: An Intelligent Bus Status Informing Scheme Exploiting Smartphone Application. International Journal of Smart Home 7(3) (May 2013)
7. http://www.raserabwehr.de/1/post/2011/11/speedcam-2011-ist-erschienen.html-22.02.2014
8. Egger, S., Hossfeld, T., Schatz, R., Fiedler, M.: Waiting times in Quality Experience for Web based Services. In: Forth International Workshop in Quality of Multimedia Experience (QoMEX), Melbourne, Australia (2012)
9. Prokkola, J.: QoS Measurements Methods and Tools. In: Easy Wireless Workshop, IST Summit Budapest, July 05 (2007)
10. Nguyen, L.T., Harris, R., Jusak, J.: Based on Varying Network Parameters and User Behavioar. In: Internatioanl Conference on Telecommunications Technology and Applications, IACSIT, Singapore, vol. 5 (2011)
11. http://www.000webhost.com (December 12, 2013)
12. http://www.apachefriends.org/download.html (November 10, 2013)
13. http://tools.pindom.com/ (January 12, 2014)
14. http://www.webpagetest.org (December 29, 2013)
15. http://gtmetrix.com (January 18, 2014)
16. Hossfeld, T., Hock, D., Tran-Gia, P., Tutschku, K., Fiedler, M.: Testing the IQX Hypothesis for Exponential Interdependency between QoS and QoE of Voice Codecs iLBC and G.711, University of Wuerzburg, Tech. Rep. 442 (March 2008)
17. Chen, K.-T., Tu, C.C., Xiao, W.-C.: OneClick: A Framework for Measuring Network Quality of Experience. In: Proceedings of IEEE INFOCOM 2009 (2009)

Auto-adaptation of Genetic Operators for Multi-objective Optimization in the Firefighter Problem

Krzysztof Michalak

Department of Information Technologies,
Institute of Business Informatics,
Wroclaw University of Economics, Wroclaw, Poland
krzysztof.michalak@ue.wroc.pl

Abstract. In the firefighter problem the spread of fire is modelled on an undirected graph. The goal is to find such an assignment of firefighters to the nodes of the graph that they save as large part of the graph as possible.

In this paper a multi-objective version of the firefighter problem is proposed and solved using an evolutionary algorithm. Two different auto-adaptation mechanisms are used for genetic operators selection and the effectiveness of various crossover and mutation operators is studied.

Keywords: operator auto-adaptation, multi-objective evolutionary optimization, graph-based optimization, firefighter problem.

1 Introduction

The firefighter problem was introduced by Hartnell in 1995 [14]. It can be used as a deterministic, discrete-time model for studying the spread and containment of fire, containment of floods and the dynamics of the spread of diseases.

The problem definition is as follows. Let $G = \langle V, E \rangle$ be an undirected graph, and $L = \{'B','D','U'\}$ a set of labels that can be assigned to the vertices of the graph G. The meaning of the labels is $'B' =$ burning, $'D' =$ defended and $'U' =$ untouched. Let $l : V \to L$ be a function that labels the vertices. Initially, all the vertices in V are marked $'U'$ (untouched). At the initial time step $t = 0$ a fire breaks out at a non-empty set of vertices $\emptyset \neq S \subset V$. The vertices from the set S are labelled $'B'$: $\forall v \in S : l(v) = 'B'$. At every following time step $t = 1, 2, \ldots$ a predefined number d of yet untouched nodes (labelled $'U'$) become defended by firefighters (these nodes are labelled $'D'$). A node, once marked $'D'$, remains protected until the end of the simulation. Each time step finishes with the fire spreading from the nodes labelled $'B'$ to all the neighbouring nodes labelled $'U'$. The simulation ends when either the fire is contained (i.e. there are no undefended nodes to which the fire can get) or when all the undefended nodes are burning. The goal is to find an assignment of firefighters to d nodes per each time step $t = 1, 2, \ldots$, such that, when the simulation stops, the number of saved vertices (labelled $'D'$ or $'U'$) is maximal.

E. Corchado et al. (Eds.): IDEAL 2014, LNCS 8669, pp. 484–491, 2014.

In this paper a multi-objective version of the firefighter problem is tackled in which there are m values $v_i(v)$, $i = 1, \ldots, m$ assigned to each node v in the graph. Each v_i value can be interpreted as a worth of a node with respect to a different criterion (e.g. the financial worth vs. the cultural value). The objectives f_i, $i = 1, \ldots, m$ attained by a given solution are calculated as follows:

$$f_i = \sum_{v \in V : l(v) \neq 'B'} v_i(v) \qquad (1)$$

where:
   $v_i(v)$ is the value of a given node according to the i-th criterion.

Many papers published to date on the firefighter problem deal with theoretical properties and concern specific types of graphs and specific problem cases. Obviously, such results are not applicable in the general case when specific assumptions may not be guaranteed to be true. In the paper [8] a linear integer programming model was proposed, however it was only a single-objective one. To date, few papers have been published on using metaheuristic methods for solving the firefighter problem. A recent paper [4] states even, that before its publication not a single metaheuristic approach has been applied to the firefighter problem. In the aforementioned paper an Ant Colony Optimization (ACO) approach was proposed for a single-objective case. In this paper a multi-objective evolutionary algorithm with operator auto-adaptation is used.

The rest of this paper is structured as follows. Section 2 presents the algorithm used for solving the multi-objective firefighter problem. In Section 3 the experimental setup is presented along with the obtained results. Section 4 concludes the paper.

2   Algorithm

Evolutionary algorithms are often used for multi-objective optimization. The advantage of this type of optimization methods is that they maintain an entire population of solutions which may represent various trade-offs between the objectives. Since the problem presented in this paper is multi-objective the NSGA-II algorithm [7] is used which is used in the literature in many areas including engineering applications [13] and operations research [18]. The optimization problem considered in this paper involves m objectives which represent the value of the graph nodes with respect to different criteria. In the algorithm an $m + 1$-th objective is added which represents the number of nodes saved (i.e. labelled either 'D' or 'U' at the end of the simulation). This objective is added in order to promote solutions that allow the fire to be contained early.

The genotype used in the algorithm is a permutation P of N_v elements. This permutation represents the order in which nodes of the graph are defended by firefighters. During the simulation of the spreading of fire firefighters are assigned to d nodes of the graph at the time in the order determined by the permutation P. If a given node is already labelled $'B'$ then a firefighter is assigned to the next untouched node from the permutation P. At a given time step d firefighters are always assigned, even if some nodes are skipped because they are already marked $'B'$.

Genetic Operators: A set of 10 crossover operators and 5 mutation operators is used for genetic operations. The crossover operators are: Cycle Crossover (CX) [17], Linear Order Crossover (LOX) [9], Merging Crossover (MOX) [1,16], Non-Wrapping Order Crossover (NWOX) [6], Order Based Crossover (OBX) [19], Order Crossover (OX) [11], Position Based Crossover (PBX) [19], Partially Mapped Crossover (PMX) [12], Precedence Preservative Crossover (PPX) [3,2] and Uniform Partially Mapped Crossover (UPMX) [5]. The mutation operators are: displacement mutation, insertion mutation, inversion mutation, scramble mutation and transpose mutation.

Auto-Adaptation Mechanism: The effectiveness of genetic operators may vary for different problems, different instances of a given problem and even may change at different phases of the optimization process. In order to choose the best performing crossover and mutation operators an auto-adaptation mechanism can be used [15]. In this paper two auto-adaptation mechanisms are compared. The first mechanism (RawScore) uses a raw score calculated as the number of times b_i when a given operator produced an improved specimen. Note, that in the case of the crossover operator each offspring is compared to each parent, so if one offspring is better than both parents then $b_i = 2$ and if both offspring are better than both parents then $b_i = 4$. Also, in the case of a multi-objective problem an improvement along any of the objectives is counted.

The second mechanism (SuccessRate) is based on the success rate of the operators. This mechanism counts the number of times each operator was used n_i and the number of improvements obtained b_i. Similarly as with the first mechanism each offspring is compared to each parent, so a maximum value of $b_i = 4$ can be obtained (per objective). The success rate is calculated as $s_i = b_i/n_i$ if $n_i \neq 0$ or $s_i = 0$ if $n_i = 0$.

Each of the N_{op} operators is given a minimum probability P_{min} and the remaining $1 - N_{op}P_{min}$ is divided proportionally to either the raw scores b_i (in the RawScore method) or the success rate values s_i (in the SuccessRate method) obtained by the operators. In both methods operators are selected randomly using a roulette-wheel selection principle. The selection of crossover and mutation operators is performed separately with separate probability assignment procedures.

3   Experiments and Results

In the experiments the optimization of firefighter assignment was performed for graphs with various density of edges. The graphs were built as follows. First, a number N_v of vertices were created. Then, edges were added with a probability P_{edge} of creating an edge between any given two vertices. The density of the graph heavily impacts the progress of the simulation. If the density is low the fire is easily contained. If the density is high the fire is very hard to contain and all nodes burn except those protected by firefighters (no nodes with the label 'U' at the end of simulation). Therefore, the value of P_{edge} was selected based on a preliminary round of tests in order to ensure that on one hand some of the nodes are left in the untouched state 'U' and on the other hand the fire is not stopped immediately. The number of vertices N_s at which the fire started was set to 1, $0.04 \cdot N_v$ and $0.1 \cdot N_v$. The number of firefighters N_f assigned at each time step was set to $N_f = 2 \cdot N_s$. The parameters of the test instances are presented in Table 1.

Table 1. Parameters of the test instances

N_v	P_{edge}	N_s
50	0.05	1, 2 and 5
100	0.02	1, 4 and 10
500	0.0055	1, 20 and 50
1000	0.0025	1, 40 and 100

The parameters of the evolutionary algorithm were set as follows in the experiments. The number of generations was set to $N_{gen} = 250$. In order to allow larger populations for larger problem instances the population size was set to be equal to the number of the nodes in the graph $N_{pop} = N_v$. Specimens for a new generation were generated by applying a crossover operator and then mutation operator. The number of new specimens generated by the crossover operator was equal to the population size N_{pop} and the probability of mutation was set to $P_{mut} = 0.05$. The minimum probability of selecting a particular operator in the auto-adaptation mechanism was set to $P_{min} = 0.02$ for crossover auto-adaptation and to $P_{min} = 0.05$ for mutation auto-adaptation.

The auto-adaptation mechanism changes the probabilities with which individual operators are selected. At various stages of the optimization process different operators may be used more often than others. An example of this effect is visible in Figure 1 which presents the success rates of crossover operators plotted against the generation number for the instance with $N_v = 1000$ vertices and $N_s = 100$ fire starting points. Clearly, around generation 40 the MOX crossover works best followed closely by the PPX. Near the end of the search the OBX and PBX crossovers work best.

The results of multi-objective optimization performed using different algorithms are often hard to compare because individual solutions may dominate one another with respect to different objectives. In order to obtain numeric values that represent the quality of the generated results various measures of the Pareto front quality are used. Hypervolume indicator introduced in [20] is very often used for that purpose. It has been proven in [10] that maximizing the hypervolume is equivalent to achieving Pareto optimality. Table 2 presents the hypervolume values obtained in the experiments for both auto-adaptation methods. The presented values are averages calculated from 10 runs of the test performed for each set of parameters.

In the experiments using the auto-adaptation based on the raw score the scores b_i were recorded for each operator $i = 1, \ldots, N_{op}$ separately for crossover

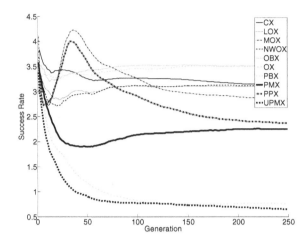

Fig. 1. The success rates of crossover operators plotted against the generation number for the instance with $N_v = 1000$ vertices and $N_s = 100$ fire starting points

Table 2. Hypervolume values attained by both auto-adaptation methods

N_v	P_{edge}	N_s	RawScore auto-adaptation	SuccessRate auto-adaptation
50	0.05	1	$4.6143 \cdot 10^6$	$\underline{5.2143 \cdot 10^6}$
		2	$5.5699 \cdot 10^7$	$\underline{5.6312 \cdot 10^7}$
		5	$5.2817 \cdot 10^7$	$\underline{5.4063 \cdot 10^7}$
100	0.02	1	$4.6310 \cdot 10^8$	$\underline{7.2057 \cdot 10^8}$
		4	$4.5708 \cdot 10^8$	$\underline{5.2746 \cdot 10^8}$
		10	$3.9548 \cdot 10^8$	$\underline{5.1009 \cdot 10^8}$
500	0.0055	1	$2.0798 \cdot 10^7$	$\underline{2.1406 \cdot 10^7}$
		20	$3.3457 \cdot 10^9$	$\underline{3.5775 \cdot 10^9}$
		50	$\underline{1.1501 \cdot 10^{10}}$	$0.9089 \cdot 10^{10}$
1000	0.0025	1	$\underline{1.0167 \cdot 10^8}$	$1.0016 \cdot 10^8$
		40	$\underline{2.7972 \cdot 10^{10}}$	$2.6013 \cdot 10^{10}$
		100	$\underline{1.1303 \cdot 10^{11}}$	$1.0949 \cdot 10^{11}$

and mutation operators. Recorded scores are presented in Tables 3 and 4. Among the crossover operators the CX crossover seems to work best as it has obtained the highest score most often. However, other crossover operators also produced good results on some instances. In the case of mutation operators the insertion mutation clearly performs best.

In the experiments using the auto-adaptation based on the success rate the success rates s_i were recorded for each operator $i = 1, \ldots, N_{op}$ separately for crossover and mutation operators. Recorded success rates are presented in Tables 5 and 6. Similarly as with the first auto-adaptation method the CX crossover achieves high success rates on some instances. Also, the OBX and PBX crossover operators seem to work well. Again, the insertion mutation achieves the best success rate outperforming, on average, all the other mutation operators.

Table 3. Scores obtained by crossover operators in the case of the auto-adaptation method based on raw scores

N_v	P_{edge}	N_s	CX	LOX	MOX	NWOX	OBX	OX	PBX	PMX	PPX	UPMX
50	0.05	1	_339.7_	172.9	312.5	210.7	272.7	148.0	182.6	167.9	140.4	231.3
		2	_703.9_	232.9	159.5	252.4	210.5	121.5	126.7	340.6	335.0	326.6
		5	197.9	120.6	_598.7_	227.2	215.6	270.1	205.8	112.2	149.5	93.6
100	0.02	1	_770.7_	314.1	269.7	532.3	248.8	106.4	609.1	475.4	447.0	258.8
		4	1112.6	949.7	1107.3	_1440.8_	1160.9	458.6	871.5	801.8	643.8	449.1
		10	917.8	_1361.1_	605.0	828.8	620.6	263.2	1084.6	1267.2	847.1	392.4
500	0.0055	1	4905.4	2907.4	4662.4	3410.5	6970.6	1293.1	8531.8	3733.0	_8706.2_	4891.3
		20	20910.5	12966.6	25450.5	15624.4	17796.1	1330.2	_29291.9_	7203.9	24431.8	2608.8
		50	_28899.5_	15642.5	14174.3	15757.6	26394.8	2181.3	22251.9	6649.2	11675.2	1408.3
1000	0.0025	1	12264.4	7023.0	9990.5	5643.3	17829.5	2150.1	17231.3	8418.1	_20872.6_	8802.2
		40	_60152.2_	38021.3	36766.7	47273.8	44378.3	3968.4	47085.1	15657.3	58720.6	2654.3
		100	_65961.7_	33573.6	37139.4	41497.6	59235.0	2671.5	61871.8	15297.6	33518.0	2224.6
AVERAGE			_16428.0_	9440.5	10936.4	11058.3	14611.1	1246.9	15778.7	5010.4	13373.9	2028.4

Table 4. Scores obtained by mutation operators in the case of the auto-adaptation method based on raw scores

N_v	P_{edge}	N_s	displacement	insertion	inversion	scramble	transpose
50	0.05	1	_16.6_	8.7	8.1	8.4	2.3
		2	6.3	_18.6_	3.8	10.8	5.8
		5	_10.1_	8.0	4.1	3.1	5.0
100	0.02	1	16.3	6.1	_22.7_	15.3	3.8
		4	17.3	25.3	13.8	_26.6_	9.9
		10	_26.5_	23.7	12.3	_26.5_	10.4
500	0.0055	1	_113.7_	5.8	111.0	86.8	3.4
		20	121.1	_401.7_	63.5	62.8	163.9
		50	80.3	_581.1_	25.8	63.8	124.9
1000	0.0025	1	132.1	6.6	113.8	_162.1_	3.5
		40	402.8	_1107.6_	112.5	176.3	353.0
		100	243.2	_1165.5_	53.0	138.7	339.6
AVERAGE			98.9	_279.9_	45.4	65.1	85.5

Table 5. Success rates of the crossover operators generated by the auto-adaptation method based on success rates

N_v	P_{edge}	N_s	CX	LOX	MOX	NWOX	OBX	OX	PBX	PMX	PPX	UPMX
50	0.05	1	0.356	0.329	0.346	0.341	0.411	0.251	_0.424_	0.331	0.347	0.338
		2	0.434	0.422	0.425	0.394	0.432	0.312	0.433	0.409	_0.456_	0.359
		5	_0.395_	0.330	0.336	0.343	0.356	0.280	0.357	0.318	0.331	0.278
100	0.02	1	0.380	0.325	0.366	0.353	_0.396_	0.190	0.386	0.292	0.390	0.323
		4	_0.797_	0.718	0.640	0.720	0.696	0.348	0.697	0.644	0.648	0.444
		10	_0.865_	0.802	0.730	0.787	0.743	0.342	0.756	0.706	0.753	0.461
500	0.0055	1	0.795	0.713	0.839	0.713	_1.076_	0.305	1.068	0.648	1.015	1.016
		20	2.601	2.532	2.575	2.540	2.670	0.612	_2.684_	1.983	2.495	0.875
		50	2.563	2.611	2.644	2.606	2.865	0.563	_2.874_	2.033	2.297	0.885
1000	0.0025	1	0.803	0.671	0.798	0.676	0.990	0.361	0.999	0.574	_1.009_	0.849
		40	_3.149_	2.924	2.957	2.929	3.142	0.665	3.146	2.101	3.130	0.693
		100	3.151	3.082	2.770	3.090	3.399	0.558	_3.407_	2.140	2.299	0.520
AVERAGE			1.357	1.288	1.286	1.291	1.431	0.399	_1.436_	1.015	1.264	0.587

Table 6. Success rates of the mutation operators generated by the auto-adaptation method based on success rates

N_v	P_{edge}	N_s	displacement	insertion	inversion	scramble	transpose
		1	0.054	0.050	0.053	0.033	0.048
50	0.05	2	0.057	0.086	0.031	0.039	0.058
		5	0.023	0.050	0.021	0.023	0.049
		1	0.056	0.036	0.035	0.041	0.022
100	0.02	4	0.051	0.102	0.058	0.055	0.076
		10	0.068	0.089	0.052	0.067	0.045
		1	0.057	0.019	0.044	0.049	0.010
500	0.0055	20	0.097	0.213	0.060	0.073	0.131
		50	0.074	0.231	0.048	0.051	0.154
		1	0.049	0.011	0.041	0.040	0.003
1000	0.0025	40	0.118	0.275	0.078	0.094	0.186
		100	0.080	0.263	0.043	0.055	0.168
AVERAGE			0.065	0.119	0.047	0.052	0.079

4   Conclusion

In this paper two auto-adaptation mechanisms for operator selection are tested on the firefighter problem. One of the auto-adaptation mechanisms uses raw scores calculated as the number of times each operator produces an improved specimen. The second mechanism uses scores normalized by the number of tries (i.e. the number of times each operator was applied). The first mechanism seems to work better on larger instances, while the second one produces better results on smaller instances.

Auto-adaptation based on raw scores seems to favor the CX crossover, while the auto-adaptation based on success rates gives a bit higher priority to the OBX and PBX crossovers. In the case of mutation both methods give the highest scores to the insertion mutation operator. The average score and success rate is clearly the highest in the case of insertion mutation. The inversion, scramble and transpose mutation operators obtain very poor results and are rarely used.

The fact that success rates of crossover operators vary over the duration of the optimization process motivates using auto-adaptation mechanisms because the performance of any individual operator may deteriorate over certain periods of time.

References

1. Anderson, P.G., Ashlock, D.: Advances in ordered greed. In: Dagli, C.H. (ed.) Proceedings of ANNIE 2004 International Conference on Intelligent Engineering Systems through Artificial Neural Networks, pp. 223–228. ASME Press, New York (2004)
2. Bierwirth, C., Mattfeld, D.C., Kopfer, H.: On permutation representations for scheduling problems. In: Ebeling, W., Rechenberg, I., Voigt, H.-M., Schwefel, H.-P. (eds.) PPSN 1996. LNCS, vol. 1141, pp. 310–318. Springer, Heidelberg (1996)
3. Blanton Jr., J.L., Wainwright, R.L.: Multiple vehicle routing with time and capacity constraints using genetic algorithms. In: Proceedings of the 5th International Conference on Genetic Algorithms, pp. 452–459. Morgan Kaufmann Publishers Inc., San Francisco (1993)

4. Blum, C., Blesa, M.J., García-Martínez, C., Rodríguez, F.J., Lozano, M.: The firefighter problem: Application of hybrid ant colony optimization algorithms. In: Blum, C., Ochoa, G. (eds.) EvoCOP 2014. LNCS, vol. 8600, pp. 218–229. Springer, Heidelberg (2014)

5. Cicirello, V.A., Smith, S.F.: Modeling GA performance for control parameter optimization. Morgan Kaufmann Publishers (2000)

6. Cicirello, V.A.: Non-wrapping order crossover: An order preserving crossover operator that respects absolute position. In: Proceedings of the 8th Annual Conference on Genetic and Evolutionary Computation, pp. 1125–1132. ACM, New York (2006)

7. Deb, K., Pratap, A., Agarwal, S., Meyarivan, T.: A fast and elitist multiobjective genetic algorithm: NSGA-II. IEEE Transactions on Evolutionary Computation 6, 182–197 (2002)

8. Develin, M., Hartke, S.G.: Fire containment in grids of dimension three and higher. Discrete Appl. Math. 155(17), 2257–2268 (2007)

9. Falkenauer, E., Bouffouix, S.: A genetic algorithm for job shop. In: Proceedings of the 1991 IEEE International Conference on Robotics and Automation, pp. 824–829 (1991)

10. Fleischer, M.: The measure of pareto optima. applications to multi-objective meta-heuristics. In: Fonseca, C.M., Fleming, P.J., Zitzler, E., Deb, K., Thiele, L. (eds.) EMO 2003. LNCS, vol. 2632, pp. 519–533. Springer, Heidelberg (2003)

11. Goldberg, D.: Genetic Algorithms in Search, Optimization, and Machine Learning. Addison Wesley (1989)

12. Goldberg, D.E., Lingle Jr., R.: Alleles, loci, and the traveling salesman problem. In: Grefenstette, J.J. (ed.) Proceedings of the First International Conference on Genetic Algorithms and Their Applications, pp. 154–159. Lawrence Erlbaum Associates Publishers (1985)

13. Haghighi, A., Asl, A.Z.: Uncertainty analysis of water supply networks using the fuzzy set theory and NSGA-II. Engineering Applications of Artificial Intelligence 32, 270–282 (2014)

14. Hartnell, B.: Firefighter! an application of domination. In: 20th Conference on Numerical Mathematics and Computing (1995)

15. Michalewicz, Z.: Genetic Algorithms + Data Structures = Evolution Programs. Springer, Heidelberg (1994)

16. Mumford, C.L.: New order-based crossovers for the graph coloring problem. In: Runarsson, T.P., Beyer, H.-G., Burke, E.K., Merelo-Guervós, J.J., Whitley, L.D., Yao, X. (eds.) PPSN 2006. LNCS, vol. 4193, pp. 880–889. Springer, Heidelberg (2006)

17. Oliver, I.M., Smith, D.J., Holland, J.R.C.: A study of permutation crossover operators on the traveling salesman problem. In: Proceedings of the Second International Conference on Genetic Algorithms on Genetic Algorithms and Their Applications, pp. 224–230. Lawrence Erlbaum Associates Inc., Hillsdale (1987)

18. Sadeghi, J., et al.: A hybrid vendor managed inventory and redundancy allocation optimization problem in supply chain management: An NSGA-II with tuned parameters. Computers & Operations Research 41, 53–64 (2014)

19. Syswerda, G.: Schedule optimization using genetic algorithms. In: Davis, L. (ed.) Handbook of Genetic Algorithms. Van Nostrand Reinhold, New York (1991)

20. Zitzler, E., Thiele, L., Laumanns, M., Fonseca, C.M., da Fonseca, V.G.: Performance assessment of multiobjective optimizers: An analysis and review. IEEE Transactions on Evolutionary Computation 7, 117–132 (2002)

Business and Government Organizations' Adoption of Cloud Computing

Bilal Charif[1] and Ali Ismail Awad[1,2]

[1] Department of Computer Science, Electrical and Space Engineering
Luleå University of Technology, Luleå, Sweden
[2] Faculty of Engineering, Al Azhar University, Qena, Egypt
`{bilal.charif,ali.awad}@ltu.se`

Abstract. Cloud computing is very much accepted and acknowledged worldwide as a promising computing paradigm. On-demand provisioning based on a pay-per-use business model helps reduce costs through sharing computing and storage resources. Although cloud computing has become popular, some business and government organizations are still lagging behind in adopting cloud computing. This study reports the status of cloud utilization among business and government organizations, and the concerns of organizations regarding the adoption of cloud computing. The study shows that some government agencies are lagging behind in cloud computing use, while others are leading the way. Security is identified as the major reason for delay in adopting cloud computing. The outcomes of the data analysis process prove that some security measures such as encryption, access authentication, antivirus protection, firewall, and service availability are required by clients for adoption of cloud computing in the future.

1 Introduction

Cloud computing is a promising computing paradigm where virtualized and dynamically scalable resources are provided to clients as services over the Internet. It has emerged as a popular computing model for big data processing. The cloud in a broad sense refers to the Internet, and in practice it refers to the data center hardware and software that support the clients' needs, often in the form of data storage and remotely hosted applications. The idea behind cloud computing started in the 1960s when John McCarthy thought of computation as a public utility [1], [2], [3], [4] [5]. Google, with its several services such as Google Docs is an example that gave a great push and public visibility to cloud computing. Moreover, Eucalyptus, OpenNebula, and Nimbus were introduced as the first open source platforms for deploying private and hybrid clouds [6], [7], [8].

Cloud computing provides services to the customers via its three major layers that form the technology stack. These layers are: Infrastructure (Infrastructure as a Service or IaaS), which provides managed and scalable resources as services to the client; Application platform (Platform as a Service or PaaS), which provides computational resources as a platform where applications and services

E. Corchado et al. (Eds.): IDEAL 2014, LNCS 8669, pp. 492–501, 2014.

can be developed and hosted; and Application software (Software as a Service or SaaS), where–in this model–applications are hosted as a service to the customers who access them via the Internet. In addition, the cloud itself has four deployment models: public, private, community, and hybrid [9], [10], [11], [12].

Although cloud computing is becoming more and more acceptable every day, some business and government organizations are lagging behind in adopting cloud computing, while others are leading the way. The term 'adoption' here refers to the use of cloud resources. The problem of cloud adoption has been observed in many business and government organizations. Addressing the reasons behind the lag in cloud adoption will benefit both clients and cloud providers.

The goal of this research is to discover the reasons why some business and government organizations are far behind others in adopting cloud services and resources, and what factors might induce them to move ahead. The study examines the current status of the cloud and the movement of organizations toward using cloud services. With businesses the major factor holding them back at first was cost and security. However, the cost has gone down so much that using the cloud actually saves most clients' money, and security is improved. Nevertheless, government organizations are still not progressing much toward cloud adoption.

A survey was created after examining the literature in order to get a definitive view of the current state of cloud adoption and some ideas of why there are differences in adoption between business organizations and government organizations. The questions were created in order to support what was found in the literature and to identify some factors involved in cloud computing. This survey was taken only by individuals in selected positions within organizations, since most other professionals have little or nothing to do with Information Technology (IT) or cloud adoption decisions and no responsibility for security. This paper focuses on the lag in cloud adoption as an issue in some organizations, and the gap between business and government organizations. Furthermore, the paper investigates how security responsibilities are perceived in the view of employees, and the expected security measures from the cloud service providers.

The remainder part of this paper is organized as follows: Section 2 is dedicated to explaining the definitions, the benefits, and the issues of cloud computing technology. In addition, it sheds light on organizations' adoption of cloud computing. Investigative work concerning data collection in survey and interviews, and data analysis toward understanding the lag in adoption by organizations, are reported in Section 3. Conclusions are drawn and future work is outlined in Section 4.

2   Cloud Computing

There are several definitions for cloud computing. However, the core of these definitions relies on the scalable and elastic IT resources provided as a service to the Internet customers [13], [14], [15]. This definition can be expanded to show the essential characteristics that cloud computing provides. IT capabilities such as networks, servers, storage, applications, and services can be individually obtained by a cloud user without the need for human interaction with the service provider. Resources thus provided can be dynamically allocated and released. Regardless of the end-user platform, users can benefit from the cloud. Cloud providers support multiple clients using a multi-tenant model. Resource usage is monitored, controlled, and reported for the services used [2], [15], [16].

Cloud computing facilities are provided at three different levels depending on the required resources. The customer can simply request software as a service (SaaS), or go deeper to platform (PaaS), or even the whole infrastructure (IaaS) which gives more control to the allocated hardware resources. Basically it is a matter of who controls the network, storage, server, services, applications and data. If the customer intends to use infrastructure as a service, then IT support and security responsibilities rely on the client, and almost no intervention from the cloud provider is required [9], [10], [12]. The cloud deployment model depends on the usability and allocation of the IT resources. This can be a public cloud, provided by a third party provider, or private (internal) cloud hosted in an internal network. In addition, community and hybrid clouds emerged from both public and private clouds [10], [13], [17], [18]. Aside from cloud types, a cloud developer can support multiple roles like cloud auditor, cloud service provider, cloud service carrier, cloud service broker, and cloud service consumer [19].

In terms of security, the cloud has both advantages and disadvantages. Comparing public cloud to privately owned resources, Wood et al. [20] showed cost reductions of up to 85% for taking advantage of disaster recovery services through cloud computing platforms. In addition, the cloud offers 24/7/365 security monitoring capabilities, which small or even medium-sized organizations might not be able to provide. Looking at the other side of the coin, Subashini and Kavitha [21] urged the use of strong encryption techniques and fine-grained authorization to control data access and for secure cloud usage.

2.1   Adoption of Cloud Computing

Cloud adoption refers to embracing the cloud services by following one or more deployment and delivery models based on the requirements. On-demand service based on the pay-per-use business model helps reduce costs through sharing computing and storage resources [22]. It is expected that cloud computing will become an essential step in the growth of information technology [23]. Cloud benefits may be perceived differently from the view of IT and non-IT personnel. Non-IT personnel sees mobility and access on all kinds of devices from anywhere in the world as a big advantage of cloud. Additionally, running applications without having to install them and eliminating the need for storage and constant upgrades are also appreciated. For IT personnel, cloud offers flexibility, elasticity of storage and access, as well as easy expansions and improvements for their networks. Denial-of-Service (DoS) attacks [24] would be greatly diminished in cloud, since security responsibilities are shared with or dedicated to the cloud provider. Server requirements, equipment to be upgraded and network capacity require less effort from an IT department when using cloud. In fact, separating public access from internal operations gives a sense of security and stability for a network from the perspective of IT personnel.

It is now apparent that the lag in adopting cloud computing in business and government organizations is an interesting problem. The adoption lag serves as a barrier between organization and cloud computing providers, and it leads to the benefits provided by cloud computing going to waste. Therefore, the focus of this research is on finding the hidden reasons behind the lag occurring in both business and governmental organizations. It is assumed that loss of direct control over data is a concern for many organizations, as is data integrity and safety. Moreover, data security in the cloud is considered a major issue, which acts as a barrier for organizations towards cloud computing adoption.

3   Investigation Work

The global variety of political systems and bureaucratic structures is so large that even if we only consider developed countries, there are far too many variables concerning socio-cultural groups and political systems among the various countries, and presenting them all is beyond the scope of this study. This research fits into the quantitative approach in which questionnaires were used; furthermore, the numerical data was collected and analyzed using mathematically-based methods. In addition, the short anonymous follow-up interviews used to obtain in-depth understanding of the problems framed for this study added a qualitative layer to the research. Considering the usage of two or more independent sources of data or data collection methods is useful for increasing the research findings reliability [25], [26], [27], [28].

3.1   Data Collection

The primary data collection methods used for this study are interviews and questionnaire for key people. The secondary data for this study was obtained from peer reviewed databases, science sites, online and physical libraries, and the Internet. The information from the literature review serves as basic information supporting both the reason for this research and the questions created for primary investigation. In order to focus upon the class of organizations of interest and the problems being investigated, the questionnaires/interviews are targeted at only individuals involved in IT security or management and stakeholder managers with input into decisions concerning the adoption of cloud. The study was not limited to government organizations, but the demographic questions allowed for separation in order to be able to find similarities and differences between business and government organizations.

This research relies on close-ended questionnaires and semi-structured interviews. The questionnaire mainly looks for information regarding the organization's background and respondent's position, how the organization's data is maintained, and cloud plans and related security measures. Surveys were distributed over several weeks and were administered by the researcher after making contact with prospective participants via email. Those contacts were asked for referrals with positions and qualifications matching the requirements for this

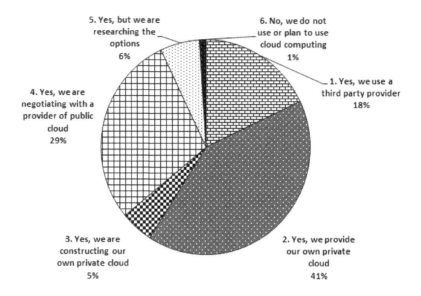

Fig. 1. Business organizations' adoption of cloud computing

study. Over several weeks, 110 qualified responses were received; out of these, 28 agreed to an interview. The questionnaires were followed up with short anonymous follow-up interviews via Skype with those individuals who agreed to participate. Fourteen participants with higher responsibility for decisions out of those agreeing to interviews were asked several follow-up questions concerning the possibilities of cloud adoption. Questions were created to identify the various problems and reasons underlying the slow rate of cloud adoption by government agencies around the world. The variables needed took the positions and responsibility of participants into consideration to ensure that they were knowledgeable concerning company policy and actions.

The follow-up interviews were done to enrich the data collection; in the view of Louise Barriball and While [29], using personal interviews as a method for data collection has a number of advantages. It can compensate for questionnaire survey in terms of the poor response rates. It is very useful when trying to explore attitudes, beliefs, values and motives [30]. It can, when discussing sensitive issues, provide the researcher with the chance to evaluate the validity of the respondent's answers by monitoring non-verbal indicators [31]. Moreover, comparability might be easier since the researcher can ensure that all questions are answered by each respondent. In fact, it ensures that responses are formulated by the respondent without receiving support from others.

In light of the research question, the survey questions requesting a measure of attitude or opinion were created using and arbitrary scale developed on ad hoc basis through the researchers' own subjective selection of items. In order to focus the respondents upon primary factors that affected decisions concerning cloud adoption, multiple choice answers were created that included all the possibly

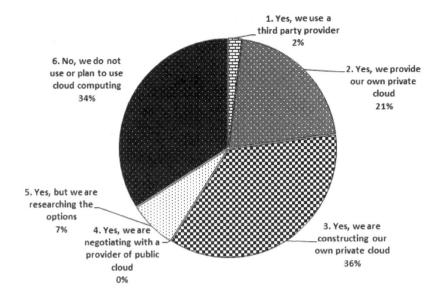

Fig. 2. Government organizations' adoption of cloud computing

relevant general responses. Before creating any question the expected answers were anticipated and the range of possibilities was carefully considered. Basically, the survey was aimed at getting a good picture of the company's current position regarding the use of cloud and the stance concerning security, data backup and cloud use from each respondent. A survey like this, however, is limited to just those answers allowed. As agreed, the recordings were only kept long enough to transcribe them and then destroyed in order to insure anonymity. Furthermore, the recorded responses were compared in order to ascertain similarities and differences.

3.2   Data Analysis Strategy

According to Price and Chamberlayne [32], statistics are classified into three types: descriptive, multivariate, and inferential. Descriptive statistics provide a summary of large amounts of information in an efficient and easily understood format. Multivariate statistics allow comparisons among factors by isolating the effect of one factor or variable from others that may distort conclusions. Finally, inferential statistics use a drawn sample from a population to suggest statements about that population. In short, the relevant statistics type for this survey was the descriptive one, since the research goal was to summarize the data on hand in order to provide valuable information as an outcome of this study.

Interviews in this study were transcribed and documented at the beginning. Keywords and important information that served the goal of this research were extracted and organized into concepts. The extracted information was reformulated in order to understand the relationship between one another. The data

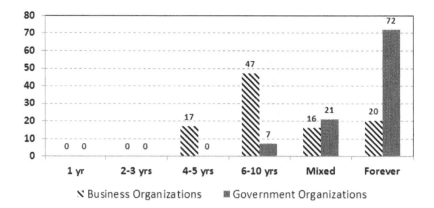

Fig. 3. Longevity of data storage in cloud

was evaluated against several available sources to measure the relevance and further legitimize the data. The findings from the collected data were reported and conclusions were drawn. Burnard [33], however, elaborated the above steps to define the process of analyzing qualitative interview transcripts into fourteen stages, and how it can be performed using a personal computer.

3.3   Data Analysis Outcomes

Data collection and analysis include a number of factors addressed in the survey and interviews. However, we report here only the most significant factors in cloud computing adoption we found. These factors encompass business organizations' adoption status, governmental organizations' adoption status, longevity of data storage, reasons for not adopting cloud, and the required security measures.

The status of business organization adoption is shown in Fig. 1. The figure demonstrates different levels of adoption. However, the most significant levels are as follows: (41%) of business organizations use their own cloud (private), (5%) are in process of extending their own cloud, and (29%) are already negotiating with a public cloud provider, and in the process of cloud adoption. On the other side, different adoption ratios are apparent for government organizations shown in Fig. 2. The figure illustrates the usage of private cloud in government organizations (21%), and (36%) with plans to provide private cloud. On the other hand, (34%) has no plans to use cloud, which is considered a lag in cloud adoption for government organizations.

Longevity of data storage can be considered one of the reasons behind the lag in cloud computing adoption. A comparison between both business and government organizations is shown in Fig. 3. It is obvious that government organizations look for long term backups which, of course, imposes a higher security risk. Security concerns in cloud computing appear clearly in Fig.4 as the most significant reason for not adopting to cloud computing. The business participants indicated

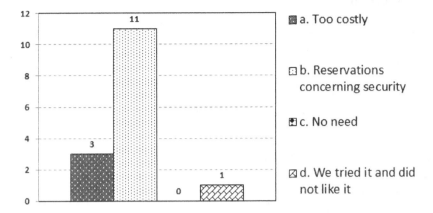

Fig. 4. Reasons for not adopting of cloud computing

that they tried but did not like the cloud. The rest of these responses were from participants who work for government organizations. Not only security and cost, but also service outage, performance, compliance, and private clouds compatibility requirements can be considered as adoption issues.

As a supplement to this research, we also investigated the security requirements of using cloud computing by both types of organizations. The findings are reported in Fig. 5. A deeper look at this figure, it becomes evident that all participants indicated that their organizations require at least encryption, access authentication, antivirus protection, firewall protection, and 24/7 access monitoring using artificial intelligence with human backup. Businesses were not as interested in triple redundancy data backup or triple layer access controls, which were popular among government agencies. Multiple analyses are possible using this chart, and the implications suggest that government organizations are far more security-conscious than private industry.

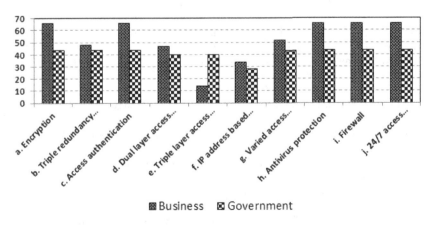

Fig. 5. Required security measures from cloud providers

4   Conclusions

Cloud computing is a promising computing paradigm that provides infinite resources to clients over the Internet. Although cloud computing offers cost reductions and ease of use, business and government organizations are not adopting cloud computing equally. This research has presented the outcomes of addressing the problem of lag in cloud adoption. The supporting research contained both survey data and interviews with personnel from business and government organizations. A descriptive data analysis approach was conducted for processing the collected data. Our findings confirm that government organizations lag behind in cloud adoption due to data storage and data security concerns. As a future solution, negotiating the agreement on level of service with the cloud providers will become a crucial step towards guaranteeing the security of data, and hence, increasing the possibly of cloud adoption.

References

1. Anthes, G.: Security in the cloud. Communications of the ACM 53(11), 16–18 (2010)
2. Krogstie, J.: Model-Based Development and Evolution of Information Systems: A Quality Approach. Springer (2012)
3. Hsu, D.F., Marinucci, D.: Advances in Cyber Security: Technology, Operations, and Experiences. Fordham University Press (2013)
4. Rimal, B.P., Choi, E., Lumb, I.: A taxonomy and survey of cloud computing systems. In: Proceedings of the Fifth International Joint Conference on INC, IMS and IDC, NCM 2009, pp. 44–51. IEEE Computer Society, Washington, DC (2009)
5. Zhang, Q., Cheng, L., Boutaba, R.: Cloud computing: State-of-the-art and research challenges. Journal of Internet Services and Applications 1(1), 7–18 (2010)
6. Peng, J., Zhang, X., Lei, Z., Zhang, B., Zhang, W., Li, Q.: Comparison of several cloud computing platforms. In: Second International Symposium on Information Science and Engineering (ISISE), pp. 23–27 (2009)
7. Sempolinski, P., Thain, D.: A comparison and critique of eucalyptus, opennebula and nimbus. In: IEEE Second International Conference on Cloud Computing Technology and Science (CloudCom), pp. 417–426 (2010)
8. Ogrizovic, D., Svilic, B., Tijan, E.: Open source science clouds. In: Proceedings of the 33rd International Convention (MIPRO), pp. 1189–1192 (2010)
9. Brunette, G., Mogull, R.: Security Guidance for critical areas of focus in Cloud Computing V2.1. CSA (Cloud Security Alliance), USA, vol. 1 (2009), http://www.cloudsecurityalliance.org/guidance/csaguide.pdf (last visit on June 9, 2014)
10. Krutz, R., Vines, R.: Cloud Security: A Comprehensive Guide to Secure Cloud Computing. Wiley (2010)
11. Hill, R., Hirsch, L., Lake, P., Moshiri, S.: Guide to Cloud Computing: Principles and Practice. Computer Communications and Networks. Springer (2012)
12. Juels, A., Oprea, A.: New approaches to security and availability for cloud data. Communications of the ACM 56(2), 64–73 (2013)
13. Mell, P., Grance, T.: The NIST Definition of Cloud Computing (Draft), p. 7. National Institute of Standards and Technology (2010)

14. Takabi, H., Joshi, J.B.D., Ahn, G.-J.: Security and privacy challenges in cloud computing environments. IEEE Security & Privacy 8(6), 24–31 (2010)

15. Dillon, T., Wu, C., Chang, E.: Cloud computing: Issues and challenges. In: 24th IEEE International Conference on Advanced Information Networking and Applications (AINA), pp. 27–33 (2010)

16. Grobauer, B., Walloschek, T., Stocker, E.: Understanding cloud computing vulnerabilities. IEEE Security & Privacy 9(2), 50–57 (2011)

17. Winkler, V.: Securing the Cloud: Cloud Computer Security Techniques and Tactics. Elsevier Science (2011)

18. Zissis, D., Lekkas, D.: Addressing cloud computing security issues. Future Generation Computer Systems 28(3), 583–592 (2012)

19. Aceto, G., Botta, A., de Donato, W., Pescapé, A.: Cloud monitoring: A survey. Computer Networks 57(9), 2093–2115 (2013)

20. Wood, T., Cecchet, E., Ramakrishnan, K.K., Shenoy, P., van der Merwe, J., Venkataramani, A.: Disaster recovery as a cloud service: Economic benefits & deployment challenges. In: Proceedings of the 2nd USENIX Conference on Hot Topics in Cloud Computing, HotCloud 2010, p. 8. USENIX Association, Berkeley (2010)

21. Subashini, S., Kavitha, V.: A survey on security issues in service delivery models of cloud computing. Journal of Network and Computer Applications 34(1), 1–11 (2011)

22. Armbrust, M., Fox, A., Griffith, R., Joseph, A.D., Katz, R., Konwinski, A., Lee, G., Patterson, D., Rabkin, A., Stoica, I., Zaharia, M.: A view of cloud computing. Communications of the ACM 53(4), 50–58 (2010)

23. Kim, W., Kim, S.D., Lee, E., Lee, S.: Adoption issues for cloud computing. In: Proceedings of the 7th International Conference on Advances in Mobile Computing and Multimedia, MoMM 2009, pp. 2–5. ACM, New York (2009)

24. Hoque, N., Bhuyan, M.H., Baishya, R., Bhattacharyya, D., Kalita, J.: Network attacks: Taxonomy, tools and systems. Journal of Network and Computer Applications 40, 307–324 (2014)

25. Saunders, M., Lewis, P., Thornhill, A.: Research methods for business students. Prentice Hall (2009)

26. Muijs, D.: Doing Quantitative Research in Education with SPSS. SAGE Publications (2010)

27. Dawson, C.: Practical Research Methods: A User-friendly Guide to Mastering Research Techniques and Projects. How to books. How To Books (2002)

28. Creswell, J.: Qualitative Inquiry and Research Design: Choosing Among Five Approaches. SAGE Publications (2012)

29. Louise Barriball, K., While, A.: Collecting data using a semi-structured interview: A discussion paper. Journal of Advanced Nursing 19(2), 328–335 (1994)

30. Richardson, S., Dohrenwend, B., Klein, D.: Interviewing: Its forms and functions. Basic Books (1965)

31. Gorden, R.: Interviewing: Strategy, Techniques, and Tactics. Dorsey Press (1987)

32. Price, J., Chamberlayne, D.W.: Descriptive and multivariate statistics. In: Gwinn, S.L., Bruce, C., Cooper, J.P., Hick, S. (eds.) Exploring Crime Analysis: Readings On Essential Skills, 2nd edn., pp. 179–183. BookSurge (2008)

33. Burnard, P.: A method of analysing interview transcripts in qualitative research. Nurse Education Today 11(6), 461–466 (1991)

Erratum: Ensemble-Distributed Approach in Classification Problem Solution for Intrusion Detection Systems

Vladimir Bukhtoyarov[1] and Vadim Zhukov[2]

[1] Siberian State Aerospace University, Department of Information Technologies Security,
Krasnoyarsky Rabochy Av. 31, 660014 Krasnoyarsk, Russia
vladber@list.ru
[2] Siberian State Aerospace University, Department of Information Technologies Security,
Krasnoyarsky Rabochy Av. 31, 660014 Krasnoyarsk, Russia
vadimzhukov@mail.ru

E. Corchado et al. (Eds.): IDEAL 2014, LNCS 8669, pp. 255–265, 2014.
© Springer International Publishing Switzerland 2014

DOI 10.1007/978-3-319-10840-7_60

The following acknowledgement was inadvertently omitted from the paper starting on page 255 of this volume:

This work was supported in part by the Russian Government under President's grant for young PhDs # 4.124.13.473-MK (04.02.2013).

The original online version for this chapter can be found at
http://dx.doi.org/10.1007/978-3-319-10840-7_32

Author Index